Table Olives

Table Olives

Production and processing

A. Garrido Fernández
Department of Food and Biotechnology,
Instituto de la Grasa (CSIC), Seville, Spain

M. J. Fernández Díez
Formerly Head of the Department of Food and Biotechnology,
Instituto de la Grasa (CSIC), Seville, Spain

and

M. R. Adams
School of Biological Sciences, University of Surrey, UK

Springer-Science+Business Media, B.V.

First edition 1997

Originally published by Chapman & Hall in 1997.
Softcover reprint of the hardcover 1st edition 1997

Typeset in 10/12pt Palatino by Columns Design Ltd, Reading, Berkshire

ISBN 978-1-4899-4685-0 ISBN 978-1-4899-4683-6 (eBook)
DOI 10.1007/978-1-4899-4683-6

A catalogue record for this book is available from the British Library

Library of Congress Catalog Card Number: 96–86320

♾ Printed on acid-free text paper, manufactured in accordance with ANSI/NISO Z39.48-1992 and ANSI/NISO 239 48–1984 (Permanence of Paper).

To Elena, who was patient enough to share with me complete devotion to table olives for four decades.

M. J. Fernández Díez

To my parents, for their early encouragement, and to my family for tolerating the loss of free time to table olive research.

A. Garrido Fernández

Contents

Preface

Table olive processing has long been, and still remains, a very important tradition and industry in Spain, Greece, Turkey and many other countries around the Mediterranean Basin. From the beginning of this century these countries were joined by the USA and others and table olives are now a well-known product all over the world.

With the exception of certain types and elaborations, table olives are fermented products, and so the study of the fermentation process and its physical, chemical and microbiological control is the fundamental basis for understanding and improving the preparation, storage, and safety of the final product. This has been the subject of serious and continuous research during the past 60 years, mainly at the University of California, Davis, by Professors Cruess, Vaughn and co-workers; the Instituto de la Grasa, in Sevilla, by Professor Rodríguez de la Borbolla and co-workers; and the University of Athens, in Greece, by Professor Balatsouras and co-workers.

The authors, M. J. Fernández Díez and Antonio Garrido Fernández, consider themselves very lucky to have been members of the Spanish group for 40 and 25 years respectively, to work on such an exciting subject and to share research results with American and Greek colleagues through correspondence, meetings and friendship. The third author, Martin Adams, has shared this interest in table olives since 1990 when he visited the Instituto de la Grasa.

Current knowledge on table olives is at an exciting stage. It covers practically all aspects of the product from removal from the tree until it reaches the consumer. In the following pages, the authors have tried to gather most of this information with the aim of presenting to readers the current status of the science and technology of table olives, though only limited aspects of the engineering aspects of the factory design are considered. The book covers the definitions of all commercial presentations, characteristics of the main cultivars devoted to table olive preparation, world production, consumption and international trade, a botanical description of the tree and fruit, composition of the raw material and end-product, harvesting and post-harvesting changes, the different processing methods, the physical, biochemical and microbiological changes that occur during the various processes, and the most common methods for controlling fermentation and final product quality. Special attention is paid to the environmental impact of the wastewaters from this industry. Lastly, the volume also includes detailed information on the Spanish and International legislation on table olives.

The book has two main objectives: (i) to supply a basic knowledge of the product, and a practical guide for agricultural scientists, food scientists, technologists, students, technicians and processors who are interested in or may be involved in the production of table olives; and (ii) to act as a reference book which will encourage food scientists to carry out more research in this area. In spite of the huge volume of information generated on table olives up to now – and the current, relatively good, understanding of the processes – there is still considerable scope for new researchers in the field to continue to enlarge our knowledge and improve further the quality of the different products. This will increase the incomes of producers and processors in the table olive regions; but, especially, will help to continue an ancient tradition and to supply consumers with healthier and better foods.

Special thanks are given to *Grasas y Aceites* (published by Instituto de la Grasa), which has published most of the Spanish research on table olives, and other journals for permission to reproduce various tables and figures. Finally, the authors wish to express their deep gratitude to all their co-workers from the Instituto de la Grasa, with whom Fernández Díez and Garrido Fernández have worked for so many years. This book is a tribute to their work, which has been an invaluable aid to the development of the industry, and an attempt to overcome the lack of understanding on the part of the scientific authorities which has been particularly apparent over the past few years.

The Authors

Introduction 1

1.1 ABOUT PICKLED PRODUCTS

Pickled vegetables, which include table olives, can be defined as those products in which preparation and preservation are carried out by a combination of salting, fermentation and/or acidification (Fernández Díez, 1984). This system of processing has a number of advantages. It permits preservation of an otherwise perishable raw material over extended periods of time. It also consumes relatively little energy, since the principal preservative factors are the combined effects of the salt, low pH and organic acidity and the product will often not require a heat treatment to ensure stability. This can also have the beneficial effect of retaining important heat-labile nutritional and physical properties such as colour and texture. However, frequently nowadays the preference for milder pickles means that a mild pasteurization treatment is necessary for shelf stability.

As far as today's consumer in the developed world is concerned however, the most important aspect of fermentation/pickling is its ability to generate products with very special and well-defined sensory characteristics which are widely popular.

In general, vegetable fermentation is carried out by homo- and heterofermentative lactic acid bacteria and/or yeasts. The initial processing of the raw material and the subsequent changes brought about by microbial activity give rise to a microbial sequence that leads ultimately to the dominance of these organisms and to the required product characteristics. Both microbial groups are preceded in the natural sequence of events by Gram-negative aerobic and facultatively anaerobic bacteria which generally constitute a far larger component of the raw material's natural microflora. Usually, these Gram-negative organisms are inhibited by the salt levels employed, but their early activity may help prepare the way for their successors by producing anaerobic conditions and a moderate decrease in the initial pH of the medium. In situations where the growth of Gram-negatives is not checked, they can produce some kinds of gaseous spoilage.

Any deviation from the required ambient conditions can alter the microbial balance and results in a defective or unacceptable product. For example, in some circumstances oxidative yeasts and moulds or members of the genera *Clostridium*, *Propionibacterium* or *Bacillus* can become significant, producing a variety of defects such as softening or off-odours.

Depending on the product, the microflora of lactic acid bacteria can also undergo sequential changes during fermentation. Faster-growing heterofermentative species can dominate the early stages, metabolizing carbohydrate to produce a mixture of carbon dioxide, lactic acid, acetic acid and ethanol. The carbon dioxide produced helps to establish anaerobic conditions which, combined with the developing acidity, prevents further growth of the Gram-negative microflora. It is essential that anaerobic conditions are maintained throughout the process to prevent later surface growth of moulds and oxidative yeasts which will decrease the product's protective acidity.

In the normal course of events, the growth of the heterofermenters slows with the increasing acidity and they are supplanted by homofermentative species. These produce predominantly lactic acid to give the high acidity levels necessary for satisfactory preservation. Failure to achieve this dominance of homofermenters can cause problems arising from the lower molar yield of lactic acid produced by heterofermenters and their depletion of carbohydrate which would otherwise be fermented by the homofermenters.

Fermentative yeasts often accompany the lactic acid bacteria in the natural fermentation, and in some cases are predominant. Sometimes they may be responsible for problems of softening, abnormal colours or undesirable gaseous fermentations. In others though, they can be the main agents responsible for the fermentation and, if well controlled, give rise to the highly appreciated sensory qualities of the final product.

This subtle and intricate balance between the various microorganisms present is controlled by a variety of factors, several of which are, in their turn, affected by microbial activity:

- medium pH
- free acidity
- combined acidity (the buffering capacity of the medium)
- acid composition (the balance between acetic and lactic acids)
- sodium chloride content
- temperature
- redox potential (the degree of anaerobiosis)
- the composition of the raw material
- the natural microflora of the raw material and the possible use of starter cultures

By improving our understanding of these complex relationships through scientific study of existing natural fermentations and their microbial successions, it is becoming increasingly possible to define the features which determine particular product characteristics. With this knowledge and the selection of appropriate product cultivars and the optimum harvest time, the range of pickled vegetable products is almost unlimited – a prospect which offers the food industry enormous scope for diversification and the consumer new and exciting food products.

1.2 ORIGINS OF RESEARCH INTO TABLE OLIVES

1.2.1 FROM HOME RECIPE TO COMMERCIAL PROCESSING

Like many other food products, the table olive industry started as a domestic, small-scale industry using home recipes which had been transmitted down through successive generations. Those products which proved popular with consumers began to extend further into the market, establishing small family businesses which later developed into cooperatives, small, medium and large processing industries. Although improvements were slowly introduced, the technology employed remained rudimentary and process failures, accompanied by important economic losses, remained a frequent occurrence. Despite this, these traditional industries persisted and grew in importance until the point when the techniques of modern food science and technology could be applied to their primitive formulae.

This route was followed more or less independently in the United States, Greece and Spain, the three countries which currently comprise the bulk of the table olive industry. A large diversity of practical recipes, coming from antiquity and local creativity, eventually gave rise to the three most important processes which currently dominate the international market for table olives.

1.2.2 AMERICAN SCHOOL

Around 1900, Mrs Freda Ehmann, a commercial packer, and Professor Bioletti, from California University, discovered independently that ripe olives, after a preliminary treatment to remove the bitter principle, could, like many other food products, be canned and preserved by a heat process (Cruess, 1958). A few years later, Mrs Ehmann had built a factory for this purpose in Oroville (California Olive Association, 1990), and a grower-owned cooperative called the Lindsay Ripe Olives Company had also been formed (Lindsay Olive Growers, 1990).

In the 1920s, another professor from California University, Professor Cruess, started a major research effort to establish the scientific and technological fundamentals of the process (Cruess, 1921, 1929; Cruess and Guthier, 1923; Cruess and Alsberg, 1934; Cruess *et al.*, 1939) – a programme that was successfully continued and developed by his co-workers.

1.2.3 GREEK SCHOOL

In Greece, the preparation of naturally black olives, either in brine or in dry salt, had developed in a similar fashion to elsewhere from small family businesses using traditional recipes, through to cooperatives and artisan factories.

The earliest recorded efforts to apply modern science and technology to improve production and introduce a consistent quality product in Greece were made by Kalogueria in the early 1930s (Kalogueria, 1932). However it was not until the beginning of the 1960s when a systematic research effort to put the industry on firm scientific foundations was initiated by Professor Balatsouras and co-workers from the Agricultural School of Athens University, and the technologists E. Aligizakis and G. Patroclos, who worked in the olive industry (Balatsouras, 1964; Balatsouras and Polymenakos, 1964).

1.2.4 SPANISH SCHOOL

The background of Spanish-style pickled green olives is similar to the Greek experience. The oldest written reference in Spain to the preparation of the fruit for eating dates back to the first century before the Christian era (Columela, 42 BC). The history of the international trade of these products began when olives of the Gordal variety were sent to America in 1540. These early exports were treated with water to debitter them before preparation with brine and aromatic herbs. It was not until the middle of the 19th century that more efficient debittering processes were described. First, green olives were treated with alkaline ashes to rid them of their bitter taste, and then in 1877, the use of sodium hydroxide was adopted. It was shortly after this development – from about 1890 – that the export of fermented green olives started to grow steadily (Barrera, 1985).

In 1923 Cruess made a study tour of Mediterranean countries with financial assistance provided by the California Ripe Olive Association (Cruess, 1924). As a result of this trip, he identified the pickled green olive as the product that, along with the established Californian ripe olives, showed greatest promise for the future development of the table olives industry and he initiated a research programme in this area (Cruess, 1930, 1937).

Cruess' work on green olives was subsequently taken up by Vaughn, also from University of California at Davis (Vaughn *et al.*, 1943). Research activity continued in the home of the Spanish green olive where in 1934 Moro Beato (1934) published the first technical report on the pickling of green olives, and where in 1947 Borbolla y Alcalá and co-workers started their continued and persistent research work on the process (Borbolla y Alcalá and Gómez Herrera, 1949; Borbolla y Alcalá *et al.*, 1949a,b; Gómez Herrera *et al.*, 1949).

1.2.5 OTHER TABLE OLIVE RESEARCH GROUPS

Although the activities of workers in the United States, Greece and Spain were central to the development of the different types of table olives, significant contributions were made by researchers elsewhere and there have been several research groups that have initiated work in this area fairly recently.

Argentinian scientists published their first papers on pickled green olives in the 1940s (Marsico, 1946; Merzari and Molina, 1949).

In Italy, the group of Panizzi *et al.* (1960, 1965) elucidated the structure of the oleuropein, the bitter principle and the main polyphenol of olive fruits, which is responsible for most of the specific characteristics of the olive fermentation. Another group dedicated to olive studies was led by Savastano and co-workers and the work of this group was continued under the direction of Dr Cucurachi (1971) when the Istituto Sperimentale per la Elaiotecnica was later established in Pescara. Since then the numerous contributions of them can be found in the Annali published by this Centre.

In the United States, Professors Fleming and Etchells (1967) were the first to demonstrate the occurrence of an inhibitor of lactic acid bacteria in olives and to test the effect of oleuropein and its derivatives on different species of microorganisms. Most of their work has, however, been centred on a different area of vegetable pickling, the cucumber fermentation.

In Israel, Juven *et al.* (1968) made notable contributions to the enhancement of the lactic acid fermentation of green olives, although this work did not continue. Later, Ben-Shalon *et al.* (1977) published important studies on catechol oxidase from olives and studied its role in some problems in table olives associated with the oxidation of polyphenols. However, their work was also isolated rather than part of an ongoing and concerted research programme on olives.

In the past 50 years many other Mediterranean countries have established institutions devoted to the development of their indigenous oil and table olive industries. Turkey has an institute in Bornova-Izmir with different departments involved in research on diverse aspects of the culture and processing of olives, including a pilot plant for table olive production. In North Africa, Tunisia has a research institute for olive studies,

Algeria has a section within its National Agronomic Research Institute dedicated to the tree and its fruits.

Portugal has recently initiated a programme in the Centro de Tecnologia Quimica e Biologica to improve its table olive production.

1.3 COMMON LINES OF RESEARCH

The three main schools – Cruess–Vaughn, Borbolla y Alcalá, and Balatsouras – used a common and useful approach in their efforts to apply science and technology to the development of the olive industry, an approach that is widely applicable to almost any form of natural product processing based on traditional recipes. It consisted of following step-by-step all the daily industrial operations of traditional practitioners and transferring them to the laboratory or the pilot plant where processes and improvements in technology could be investigated systematically.

Fortunately, the three schools did not act entirely independently of one another, wastefully duplicating effort. During the period 1969–1980 in particular, they developed a very active interchange of experience and knowledge which has permitted technologists to reach a quite wide, although still incomplete, understanding of the three principal methods of table olive manufacture.

The research undertaken to date has, among other things, permitted control of the most important spoilage problems, both microbial and enzymatic, which can result from a lack of control throughout the process.

Areas that have been, and remain, the principal foci of research effort are:

- the chemical composition of the fruit
- the natural sequences of microorganisms responsible for the fermentation process and the safe storage of the final product
- selection of those cultivars which, in different countries, are able to produce suitable olives for each type of processing
- the most representative characteristics of the fruits to fix an optimum range for the time of harvesting

Within these areas, the scientific and technological effort must not only achieve a consistent, high-quality product but must also aim to reduce manufacturing and trading costs. These are intrinsically high due to the very nature of the raw material, its mode of growth and its processing.

A large number of techniques currently employed by the modern olive industry around the world are a direct result of collaboration and cooperation between the various national research groups. This includes modern large-scale fermentation and storage techniques which have replaced old wooden barrels and tanks with vessels manufactured from various types of food-grade materials; the control of pH, salt content, free acidity, temperature, and microflora; the use of selected starter cultures for fermentation; the retail presentation of olives in small containers of different materials (glass, plastic, tin-plate, etc.); the mechanization and standardization of all the complementary operations for pitting and stuffing olives with pimento, anchovy, onion, etc., and pastes derived from those products; and, finally, the methods for re-using and reducing industrial wastes to minimise the environmental impact of olive processing.

The publication and wider dissemination of the results of research on table olives was initiated by Cruess in the 1920s (Cruess, 1924, 1930, 1937, 1941), and by Vaughn *et al.* (1943), Borbolla y Alcalá *et al.* (1949a, b) and Balatsouras (1966). Since then, a number of other texts have been published describing the work carried out in different countries (Fernández Díez, 1971, 1983, 1991; Fernández Díez *et al.*, 1972, 1985). The aim of this book is to update the information contained in these and provide the reader with the first comprehensive monograph on this topic in English. In what follows we hope to demonstrate the immense progress that has been made since those early beginnings and the fascinating insights obtained into the fundamental biology and chemistry of a process stretching back to antiquity. In addition, the authors hope that this book will also indicate the considerable distance that still has to be covered and the remaining problems that pose a challenge and incentive to young researchers for the future.

1.4 REFERENCES

Balatsouras, G.D. (1964) Composition chimique des olives noires de Grèce. Variation de quelques constituants en fonction de la region de production. *Inf. Oleic. Int.*, **28**, 131–56.

Balatsouras, G.D. (1966) *Contribution to the study of the chemical composition and the microflora for the stored in brine Greek black olives*, National Printing Office, Athens.

Balatsouras, G.D. and Polymenakos, N. (1964) Resultats preliminaires sur la fermentation des olives noires par acide lactique. *Inf. Olec. Int.*, **27**, 153–67.

Barrera, P. (1985) El comercio exterior de la aceituna de mesa, in *Proceedings of the 'Jornadas de Aceitunas de Mesa'*. Escuela Universitaria de Ingenieros Técnicos Agricolas, Sevilla (Spain). pp 129–48.

Ben-Shalon, N., Kann, V., Harel, E. and Mayor, A.M. (1977) Olive catechol-oxidase. Changes during fruit development. *J. Sci. Food Agric.*, **28**, 545–550.

Borbolla y Alcalá, J.M.R. de la and Gómez Herrera, C. (1949) La industria de aderezo de aceitunas verdes. *Rev. Cienc. Apl.*, **3**, 120–32.

Borbolla y Alcalá, J.M. de la, Gómez Herrera, C., Pérez Nuñez, F., and Castro Ramos, R. de (1949a) *Estudios sobre el aderezo de aceitunas verdes.II. La penetracion de sal en los frutos*. Instituto de la Grasa y sus Derrivados, Sevilla (Spain). Comunicación Técnica no. 7.

Borbolla y Alcalá, J.M. de la, Gómez Herrera, C., Pérez Nuñez, F., and Castro Ramos, R. de (1949b) *Estudios sobre el aderezo de aceitunas verdes.III. La fermentacion*. Instituto de la Grasa y sus Derrivados, Sevilla (Spain). Comunicación Técnica no. 8.

California Olive Association (1990) *California ripe olive. Nutritional facts and caloric information*. California Olive Association, Fresno (CA, USA).

Columela, L.J.M. (42 BC) *De Re Rustica* Vol II. Book XIII. Nestlé A.E.P.A. Eds., Santander (Spain). pp 209–13.

Cruess, W. V. (1921) *Factors affecting the sterilization of olives*. Univ. Calif., Agric. Expt. Sta. Bull. 333.

Cruess, W.V. (1924) *Pickling of olives in Mediterranean countries*. Circular 278, Agr. Expt. Sta., Univ. Calif., Davis (California).

Cruess, W.V. (1929) Effects of pH value on stability of color of canned olives. Proced. Oliv. Assoc. Techn. Conf. pp 35–45.

Cruess, W.V. (1930) *Pickling green olives*, Univ. Calif., Agr. Expt. Sta. Bull. 498.

Cruess, W.V. (1937) Use of starters for green olive fermentations. *Fruit Prod. J.*, **17**, 1–5.

Cruess, W.V (1941) Olive products. *Ind. Eng. Chem.*, **33**, 300–3.

Cruess, W.V. (1958) Pickling and canning of ripe olives, in *Commercial Fruits and Vegetables Products*, 4th edn., McGraw-Hill, New York.

Cruess, W.V. and Alsberg, G.L (1934) The bitter glucoside of the olive. *J. Am. Chem. Soc.*, **56**, 2115–17.

Cruess, W.V. and Guthier, E.H. (1923) *Bacterial decomposition of olives during pickling*. Univ. Calif., Agric. Expt. Sta. Bull. 368.

Cruess, W.V., el Saifi, A. and Develter, E. (1939) Changes in olive composition during processing. *Ind. Eng. Chem.*, **31**, 1012–14.

Cucurachi, A. (1971) 35 anni di attivita della Stazione Sperimentale di Olivicoltura e di olificio di Pescara. *Annali dell'Istituto Sperimentale per la Elaiotecnica*, **1**, 45–56.

Fernández Díez, M.J (1971) The Olive, in *The Biochemistry of Fruits and their Products*, Vol. 2, (ed. A.C. Hulme), Academic Press, London, pp. 255–79

Fernández Díez, M.J. (1983) Olives, in *Biotechnology*, Vol. 5 (eds H.J. Rehm and G. Reed), Verlag Chemie, Weinheim (Germany), pp. 379–97

Fernández Díez, M.J. (1984) Changes in the chemical components during the processing of table olives and their relation to the quality, in *Proceedings of the IUFoST international symposium*, Vol. I, Valencia, Spain, November 5–7, pp. 301–18.

Fernández Díez, M.J. (1991) Processing of table olives, in *Encyclopedia of Food Science and Technology* (ed Y.H. Hui), Wiley & Sons Publishers, New York.

Fernández Díez, M.J., Garrido Fernández, A., González Cancho, F., Durán Quintana, M.C. and Cordón Casanueva, J.L. (1972) *Elaboración de aceitunas negras de mesa*, Instituto de la Grasa, Sevilla (Spain).

Fernández Díez, M.J., Castro Ramos, R., Garrido Fernández, A., González Cancho, F., González Pellissó, F., Nosti Vega, M., Heredia Moreno, A., Mínguez Mosquera, M.I., Rejano Navarro, L., Durán Quintana, M.C., Sánchez Roldán, F., García García, P. and Castro Gómez Millán, A. (1985) *Biotecnología de la aceituna de mesa*, Consejo Superior de Investigaciones Científicas-Instituto de la Grasa, Sevilla (Spain).

Fleming, H.P. and Etchells, J.L. (1967) Occurrence of an inhibitor of lactic acid bacteria in green olive. *Appl. Microbiol.*, **15**, 1178–84.

Gómez Herrera, C., Castro Ramos, R. de, and Borbolla y Alcalá, J.R. de la (1949) Estudios sobre el aderezo de aceitunas verdes. I. Penetración de la sal en los frutos. *Anales Real Soc. Esp. Fís. Y Quím.*, **45B**, 795–806.

Juven, B., Zdenka, S., Henis, Y. and Jacoby, B. (1968) Enhancement of lactic acid fermentation of green olive by alkali and heat treatment. *J. Appl. Bacteriol.*, **31**, 200–7.

Kalogueria, S.A. (1932) *Table Olives* (in Greek). Ernou ed., Athens.

Lindsay Olive Growers (1990) *The Romantic History of California Olives*, Lindsay Olive Growers, Lindsay, CA, USA.

Marsico, D.F. (1946) *Elaboración de aceitunas verdes en salmuera*. Corporación Nacional de Olivicultura, Buenos Aires, Argentina.

Merzari, A.H. and Molina, J.S. (1949) Elaboracion de aceitunas tipo español en la Argentina. *Ciencia e Investigación*, **5**, 400–500.

Moro Beato, M. (1934) *Sobre el endulzado de la aceituna*. Sevilla (Spain).

Panizzi, L., Scarpati, M.L. and Oriente, G. (1960) Constituzione della oleuropiena, glucoside amaro e ad azione ipotensiva dell'olivo. Nota II. *Gazz. Chem. Ital.*, **90**, 1449–85.

Panizzi, L., Scarpati, M.L. and Oriente, G. (1965) Sulla oleuropeina. Nota III. *Gazz. Chem. Ital.*, **95**, 1279–92.

Vaughn, R.H., Douglas, H.C. and Gililland, R. (1943) *Production of Spanish type green olives*, Univ. Calif., Agr. Expt. Sta. Bull. 678.

Olives and table olives

2

2.1 OLIVES

2.1.1 A BRIEF BOTANICAL DESCRIPTION

The olive belongs to the family Oleaceae, genus *Olea* and to the species *Olea europaea*, L. It is the only species of *Olea* found in the tropical and sub-tropical regions of the world that produces edible fruits and these are used for both table olive processing and for oil extraction. Frequently, the cultivated olive is included in the sub-species *Olea europaea sativa* (Hoffm. and Link) to differentiate it from the wild sub-species (*oleaster*). Although this distinction has been abandoned in the formal classification, it is still often used in the table olive definition.

The olive is a small, slow-growing, evergreen tree with a long life-span (Figure 2.1); specimens of more than ten centuries can be seen in the Holy Land and other areas of the Middle East and the Mediterranean. It is easily regenerated from its long tap root and new trees can grow from an existing root system when the rest of the overground plant has decayed, been cut down or burned (Balatsouras, 1984). It is propagated asexually, is cross- and air-pollinated and relatively easily acclimatized to microclimates and microenvironments. The plant's hereditary mechanism appears to be quite flexible and subject to mutations, selection pressures or other type of adjustments. As a result, more than 600 olive tree cultivars have been developed through the centuries and grow all over the olive-producing countries.

Above the ground, the plant develops many branches with a dense assembly of limbs and a compact foliage (Figure 2.1). Olive leaves grow in pairs along the stalks (Figure 2.2), a feature that helps make them readily recognizable when represented in art. They are lanceolate, about 0.5–1.0 cm wide and 4–5 cm in length, with multicellular hairs on their surface. The leaf surface which faces the light displays an intense green colour, while the lower surface which has some stomata has a whitish-

Fig. 2.1 Old tree of Hojiblanca cultivar in Cabra (province of Córdoba, Spain).

Fig. 2.2 Olive stalks showing disposition of leaves and olive fruits.

green colour. Fruits are attached to terminal twigs (Figure 2.2). Since they lose little water, olive leaves are especially adapted to drought, a facility that permits survival in hot climes and areas close to desert regions. Typically the life of leaves is between 2 and 3 years, after which they become yellow (a coloration that may also be due to other physiological troubles) and abscise in spring.

Bud inflorescences are borne in the axil of each leaf from the previous season's growth and have around 20 small greenish white flowers. Only a few of these become fruits. Large crops may be expected when only one or two of those are perfect flowers. A more detailed description of the botany of the olive can be found in Loussert and Brousse (1980) and Martin (1994)

2.1.2 OLIVE FRUITS: ANATOMY AND GENERAL CHARACTERISTICS

The fruit of the olive tree is a drupe, similar to the other drupes of stoned fruits such as apricots or cherries and with the same anatomy. In fact the botanical term 'drupe' derives from *druppa*, the Greek and Latin word for ripe olive. Its component parts are: the epicarp or epidermis, the mesocarp or flesh, and the endocarp or pit, which consists of a woody shell enclosing one or, rarely, two almonds (Figure 2.3). However, olives differ from all other drupes in their chemical composition by having a relatively low concentration of sugars, 2–5% versus around 12%, a high oil content, 20–30% versus 1–2%, and in their characteristic strong bitter taste. This last quality is due to the presence in the olive of the glucoside, oleuropein, which does not occur in any other fruit or tissue in the plant kingdom.

Both epicarp and mesocarp of olive fruits – like any other food of plant origin – consists of parenchymatous cells which are free of collenchyma or sclerenchyma. Parenchymatous cells are loosely packed to leave empty intercellular spaces. They are big and isodiametric (with diameters of up to 1 mm), surrounded by a rigid cell wall mainly composed of cellulose and lignin. In their centre they have a large vacuole which is full of

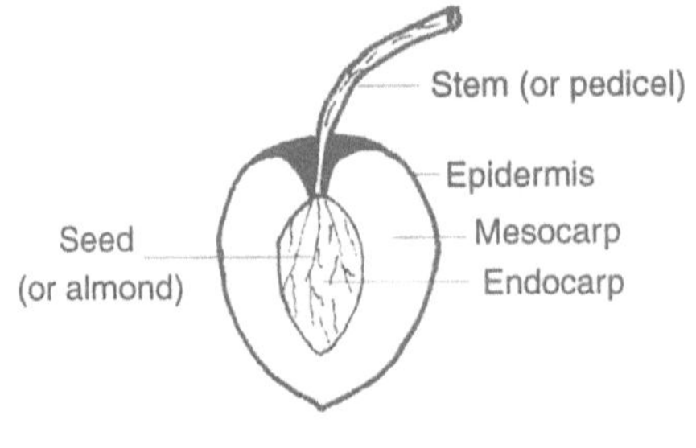

Fig. 2.3 Schematic diagram of an olive fruit, showing the component parts.

sap when the fruit is turgid, containing dissolved sugars, acids, water-soluble pigments and inorganic components. Between the vacuole and the cell wall lies the cytoplasm, a complex active mixture of proteins, fatty substances and other constituents in a jelly-like aqueous suspension. The cytoplasm is surrounded by the selectively permeable cytoplasmic membrane.

The epidermis consists of similar parenchymatous cells, which differ in being closely packed, without the empty spaces apparent between mesocarp cells. Their continuity is only interrupted by small openings, called lenticels, through which the fruit transpires and through which pathogenic agents such as fungal conidia can also sometimes gain access to cause infections (Balatsouras, 1980).

The natural bitterness of the fruit can be eliminated, or at least reduced, by processing to make it acceptable for human consumption. The oil content as well as the general composition of the pulp is highly variable, and is dependent on a multiplicity of factors such as climate, soil, culture systems, pruning, fertilizers, tree treatments and harvesting methods. Because of this, cultivated olive trees are classified into three categories, according to the type of fruit they produce and their principal use:

- cultivars producing fruit for table use
- cultivars producing fruit for oil extraction
- cultivars producing fruit for both purposes, also called double or dual use cultivars

This book will only be concerned with the use of the fruit as table olives.

2.1.3 THE ORIGIN AND SPREAD OF OLIVE CULTIVATION

Vavilov (1951) considered that the olive comes originally from an area corresponding to modern Syria and Iran. According to Almeida (1963), the olive originated in Afghanistan, where there are numerous other species belonging to the family of Oleaceae, and *Olea europaea* itself originated from *Olea cuspidata*. Simmonds (1976) on the other hand considered that the cultivated olive tree is related to three different *Olea* species: *Olea laperrini* (Sahara Mountains), *O. africana* (South Africa and Egypt), and *O. ferruginea (cuspidata)* (Iran). Authors agree on the hybrid origin of olives due to the diversity found in the species and the number and size of chromosomes of different species of *Olea* and genera of the family Oleaceae.

The origins of the cultivation of the olive tree lie rooted in legend and tradition. It probably started about 5000–6000 years ago within a wide strip of land by the eastern Mediterranean sea and in the adjacent zones comprising Asia Minor, part of India, Africa, and Europe (Fernández

Díez, 1971). Some authors consider Syria and Iran as the original site of cultivation (Condolle, 1982), while others (Zohary and Spiegel-Roy, 1975) believe it began in the Phoenician colonies of Lebanon and Palestine. The first reference to the olive tree in the Bible occurs in its very first book, Genesis, where the flight of the dove with the olive branch announcing the end of the flood is described.

Olive cultivation expanded from its origin to spread all around the Mediterranean sea as Oriental cultures diffused westwards. Numerous Mediterranean civilizations contributed to this process including the Phoenicians, Greeks and Hebrews. The Greek leader Solon in the sixth century BC banned the export of all agricultural products other than olives and their oil. This was intended as a measure to reduce the Greek dependence on imported corn, but had the opposite effect as it increased the already substantial areas devoted to olives. The Carthaginians were particularly active in developing their agriculture and promoted olive cultivation in their colonies. Subsequently, the Romans took advantage of this experience, diffusing olive growing throughout their empire. To assure their olive oil supplies, Roman Emperors gave incentives to olive farmers to establish the tree in North Africa and Spain. By that time olives were the most economically important agricultural product in these regions, reaching maximum popularity and extent during the second and third centuries (Loussert and Brousse, 1980).

During the 15th and 16th centuries, colonizers and Spanish Franciscan monks extended olive planting and the processing of table olives to various points in the New World. Olive cuttings were carried out to Peru by Spanish explorers in 1560. When Jesuits established missions in Mexico and Baja California in the early 1700s, olives were among the few fruits to be grown in those early settlements. Franciscan fathers founded their first mission at San Diego in 1769. A century later the commercial olive industry became established in California and later developed into what is now one of the world's most important after Spain, Turkey and Greece (Connell, 1994). In the same way, Italian and Spanish emigrants and missionaries spread the olive tree to Australia and South Africa.

Particularly interesting is the spread of olive cultivation in the Far East. According to Noro (1963), the first olives introduced to Japan were cultivated in Yokosuka, Kanagawa-Ken, around 1862. New plants from France arrived in 1872 and gave their first fruits in 1884. Around 1900 trees were planted in Shodoshima, which has since become the centre of olive culture in Japan, and new cultivars were later introduced in other areas such as Okayama, Hiroshima and Fukuoka.

As a result of this long period of expansion, the olive is now widely distributed in both hemispheres between latitudes 25 to 45° North and 15 to 35° South.

Readers interested in a more detailed history of the olive tree are referred to a number of authoritative accounts that have appeared over the years (Kalogueria, 1932; Molinari and Nicolea, 1947; Garoglio, 1950; Hartmann, 1953; Patac *et al.*, 1954; Foytik, 1960; Goor, 1966; Balatsouras, 1984).

All authors agree that use of the fruit to prepare table olives in the home can be traced back to antiquity (Vaughn, 1982). The oldest written reference to preparation of the fruit for eating is from the first century text by Lucius Columela (42 BC), *De Re Rustica* (On Country Matters) which appeared in about A.D. 65. It is one of the first known detailed reports on agricultural practices and food processing and describes various methods of preparing olives for eating. Systematic, scientific investigations of the processing of the olive started at the beginning of this century with the appearance of a number of reports considered classics in the field: Cruess (1924, 1937, 1941, 1958), Moro Beato (1934), Vaughn and co-workers (1943), Vaughn (1954), Borbolla y Alcalá and Gómez Herrera (1949), Borbolla y Alcalá and co-workers (1956), and Balatsouras (1966).

2.2 TABLE OLIVES

2.2.1 DEFINITIONS

Table olives are defined by the Unified Qualitative Standard Applying to Table Olives in International Trade as: 'the sound fruit of specific varieties of the cultivated olive tree (*Olea europaea sativa*, Hoffm., Link) harvested at the proper stage of ripeness and whose quality is such that, when they are suitably processed as specified in this standard, produce an edible product and ensure its good preservation as marketable goods. Such processing may include the addition of various products or spices of good table quality' (International Olive Oil Council, 1980).

The main purpose of processing is the removal, at least partially, of the natural bitterness of the fruit, to make it acceptable as a food or an appetizer.

Classification of the fruit, the traditional methods of processing and the modern technologies derived from them can be based on the ripeness of the fresh fruit and the colour of the final product. Four fundamental types have been established: green, turning colour, naturally black, and black olives. The first three types refer to the colour of the fruit used, which remains effectively unchanged throughout processing. Black olives, known in California as ripe or black-ripe olives, are made from fruit which is harvested during the stage of turning colour and which is later blackened by oxidation in an alkaline medium.

According to Spanish Standards (Spanish Government Presidency, 1983), the following definitions are given for the types of table olives mentioned above:

- **Green olives.** These are obtained from fruits harvested during the ripening period, prior to colouring but after they have reached normal size. Such olives shall be firm, sound, resistant to a slight pressure between the fingers and without marks other than the natural pigmentation, subject to the tolerances set out in those standards. The colour of the fruit may vary from green to straw yellow.
- **Turning colour olives.** Obtained from rose, wine-rose or brown-coloured fruits, harvested before the stage of complete ripeness is attained, which may or may not have been subjected to alkaline treatment and which are ready for consumption. Californian green ripe olives could be included in this type.
- **Naturally black olives.** Obtained from fruits harvested when fully ripe or slightly before full ripeness; they may, according to the production region and time of harvesting, be reddish black, violet black, deep violet, greenish black, or deep chestnut colour.
- **Black olives.** Obtained from fruits harvested when not fully ripe, which have been blackened by oxidation and have lost the bitterness by alkaline treatment. Those olives must be canned in brine and preserved by heat sterilization. Californian ripe olives could be included in this type.

International Olive Oil Council Standards (1980) consider only three types: green, turning colour, and black olives. However, under the heading 'Commercial Preparations', both standards clearly differentiate between alkali-treated (black) and untreated (naturally black) olives.

Bitterness in the fruit can be eliminated quickly and completely by alkaline hydrolysis; that is, by treatment of the olives with sodium hydroxide solutions before the fermentation, and by further storage in brine or in dry salt. Fruit treated in this way is generally called **pickled olives** in the literature and **treated olives** in the international standards cited above. Where there is no alkali treatment before fermentation and the olives are placed directly into brine or dry-salted, bitterness may also be removed slowly and partially during the acid fermentation. These products are generally known as **olives in brine** or **olives in dry salt**, and the word **untreated** appears in the description.

Numerous regional or national variations of the general processes described have resulted in a range of products of differing commercial importance and distribution. International and Spanish Standards try to include all of these under the denomination 'Commercial Preparations' or its equivalent 'Trade Preparations'. The complete name for each one of these includes information on:

1. the condition of the raw material (green, turning colour or black);
2. the procedure used to eliminate bitterness (treated or untreated); and
3. the preservation method (brine or dry salt).

Some of the commercial or trade preparations described in the standards are as follows:

Treated green olives in brine: these are treated in an alkaline lye and then packed in brine in which they undergo natural lactic fermentation, i.e. either complete fermentation (Spanish-style) or partial fermentation. Should the olives not undergo complete natural fermentation, their subsequent preservation at a pH value within the limit specified in the standards, may be ensured by sterilization or pasteurization, addition of preserving agents, refrigeration, or inert gas without brine.
Untreated green olives in brine: these are placed directly in brine and preserved by natural fermentation.
Treated, turning colour olives in brine: these are obtained after alkaline treatment and preserved by natural fermentation in brine or by heat treatment.
Untreated, turning colour olives in brine: these are placed directly in brine, preserved by natural fermentation and ready for consumption.
Olives darkened by oxidation: these are obtained from fruits which, when not fully ripe, have been darkened by oxidation and whose bitterness has been removed by treatment in alkaline lye. They have to be packed in brine and preserved by heat sterilization. Californian-style black-ripe olives correspond to this group.
Treated black olives: these are olives obtained after alkaline treatment and natural fermentation and preserved by one, or a combination of the following procedures: immersion in brine, sterilization or pasteurization, and addition of a preserving agent.
Untreated black olives: these are placed directly in brine. They have a fruity flavour which is more marked than that of the treated black olives, and they usually retain a slight bitter taste. They are subject to a natural fermentation and preserved through one or a combination of the following methods: immersion in brine, sterilization or pasteurization, and addition of a preserving agent. Greek-style naturally black olives in brine correspond to this group.
Shrivelled black olives: these are obtained from olives harvested just before they are fully ripe and which, after a brief immersion in a weak alkaline lye, are preserved by sprinkling them with dry salt and packing in wooden casks which are rotated daily until the olives are required.
Untreated, naturally shrivelled black olives: these are obtained from olives harvested when fully ripe, after they have become shrivelled on the tree. They are then packed directly in brine.
Treated black olives in dry salt: these are obtained from firm, practically ripe fruit which, following a slight alkaline treatment, are preserved in alternate layers of olives and dry salt, or by sprinkling dry salt over the olives.
Untreated black olives in dry salt: these are obtained from fruit harvested when fully ripe, placed immediately, or after partial drying, in alternate

layers with dry salt or by sprinkling dry salt over the olives. They retain a degree of bitterness and a fruity flavour more marked than those of treated olives in dry salt.
Untreated, naturally shrivelled black olives in dry salt: these are obtained from fruit harvested when fully ripe which, after they have become shrivelled on the tree, are preserved in alternate layers of olives and dry salt or by sprinkling dry salt over the olives.
Untreated, pierced black olives in dry salt: these are obtained from fruit harvested when fully ripe which, after the skin has been pierced, are preserved in alternate layers of olives and dry salt or by sprinkling dry salt over the olives.
Dehydrated black olives: these are obtained from ripe fruit, after having been blanched and partially dehydrated in salt by very gentle heating.
Bruised olives: these are obtained from whole fruit, fresh or previously placed in brine, subjected to a process whereby the flesh is opened without breaking the stone, which remains whole and intact within the fruit. They may be treated in weak lye and are preserved in brine, possibly flavoured, with or without the addition of vinegar. They are prepared from either green or turning colour-type fruit.
Split olives: these are green olives, turning colour olives or black olives split lengthways by cutting into the skin and part of the flesh, then placed in brine, with or without vinegar. Olive oil and aromatic substances may be added. They may have been treated with alkali or untreated.
Specialities: olives may also be prepared by means distinct from those specified in the standards. Such specialities retain the name 'olive' as long as the fruit used complies with the general definitions of those standards. The names used for specialities must be sufficiently explicit to prevent any confusion as to the origin and nature of the products and, in particular, with respect to the various designations in the standards.
According to the various 'types' and 'commercial preparations', olives may also be offered to the consumer in one of the following 'presentation forms':

- *Whole olives*: olives which have their natural shape and from which the stone or pit has not been removed (Figure 2.4).
- *Without stem*: whole olives from which the stem has been removed.
- *With stem*: whole olives retaining the stem.
- *Pitted olives*: olives from which the pit has been removed but which otherwise retain their basic natural shape (Figure 2.4).
- *Stuffed olives*: pitted olives stuffed with one or more suitable products (pimento, onion, almond, celery, anchovy, olive, orange or lemon peel, caper, etc.) or their prepared natural pastes.
- *Halved olives*: pitted or stuffed olives sliced into two approximately equal parts, perpendicularly to the major axis of the fruit.
- *Quartered olives*: pitted olives split into four approximately equal parts along the major axis of the fruit and perpendicularly to the same.

- *Divided olives*: pitted or stuffed olives cut lengthways into more than four approximately equal parts.
- *Sliced olives*: pitted or stuffed olives sliced into segments of fairly uniform thickness (Figure 2.4)
- *Chopped or minced olives*: small pieces of pitted olives of no definite shape and practically devoid of identifiable stem-insertion units as well as of slice fragments.
- *Olive paste*: exclusively olive flesh, finely crushed. Additives may be incorporated for its preservation.
- *Broken olives*: olives accidentally broken being pitted or stuffed. They normally contain pieces of the stuffing material.
- *Salad olives*: these are broken, unbroken and pitted olives with or without capers, plus stuffing material, but must constitute the major proportion of the entire product as marketed.
- *Olives with capers*: whole or pitted olives, usually small in size, with capers and with or without pimento, when they are the major proportion of the product marketed in this form.

Finally, according to the manner in which they are placed in the container, the olives may be presented as follows:

- *Place-packed olives*: when the fruits are placed, in the transparent rigid packs, either in orderly symmetrical fashion or forming geometrical shapes. Whole, pitted, stuffed and halved olives may be exported in this manner (Figure 2.5).
- *Random (thrown) packed olives*: when the fruits are not placed in orderly fashion in the packs. All styles may be exported in this manner.

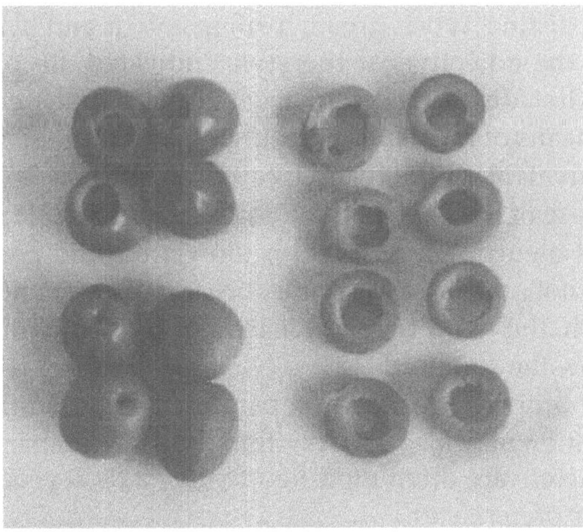

Fig. 2.4 Whole (or plain), pitted, and sliced olives.

Fig. 2.5 Place-packed olives from Manzanilla (left) and Gordal (right) cultivars stuffed with pimento.

2.2.2 PROCESSING METHODS

The three table olive preparation methods or styles of greatest importance in international trade are as follows (Fernández Díez, 1983) :

- Spanish (or Sevillian)-style (pickled) green olives in brine.
- Californian-style (pickled) black olives in brine.
- Greek-style naturally black olives in brine.

For each style, fruits are harvested at a different stage of ripeness, corresponding to distinct types, green, turning colour and black respectively. In each case, the adjective of the style indicates the place where the process was first developed and used. In the first two, bitterness is removed by means of an alkaline treatment, and for that reason the word 'pickled' (equivalent to the term 'treated' in standards) appears in the name. In the case of the Greek-style process, fruits are placed directly into the brine, and oleuropein removal is slow and only partial. The word 'pickled' does not appear in the name, because they are not 'treated' with lye. Henceforth, the word treated, instead of pickled, will be used if necessary, to follow the standards.

The sensory properties of each preparation are very different, and considerable variation may also occur within the same preparation since processing parameters are often modified to give a final product adapted to local consumer preferences.

For Spanish-style treated green olives in brine, a lactic fermentation is the principal feature, while for Greek-style naturally black olives in brine, the main organisms responsible for fermentation are yeasts, and lactic acid bacteria form only a small proportion of the total microflora.

Californian-style treated black olives in brine do not necessarily require a fermentation process at all. They may be directly treated with lye and oxidized, then washed, placed in brine, packed in cans and heat-sterilized. However, some of the crop may be placed directly into the brine for a variable period, in which case a fermentation similar to that of the naturally black olives in brine will occur.

Because of their major importance in international trade, the technology of these three methods has received most attention from scientists. For that reason, these will feature more in the following chapters, although the available information on other commercial preparations popular with consumers, will also be reviewed.

2.3 REFERENCES

Almeida, F.J. (1963) Acerca del melhoramiento da oliveira. *Bol. da Junta Nacional do zeite*,Lisbon.

Balatsouras, G.D. (1966) *Contribution to the study of the chemical composition and the microflora for the stored in brine Greek black olives*, National Printing Office, Athens.

Balatsouras, G. (1980) Nutritive and biological value of the Greek table olives, in *Proceedings of the III International Congress on the biological value of olive oil*. Chania (Creta, Grecia). pp. 485–521.

Balatsouras, G.D. (1984) *Contemporary Oleiculture, Vol I. The Olive Tree*, Mairomatis & Co, Athens (in Greek).

Borbolla y Alcalá, J.M.R. de la and Gómez Herrera, C. (1949) La industria de aderezo de aceitunas verdes. *Rev. Cienc. Apl.*, **3**, 120–32.

Borbolla y Alcalá, J.M.R. de la, Gómez Herrera, C., González Cancho, F., Fernández Díez, M.J., Gutiérrez González Quijano, R., Izquierdo Tamayo, A., González Pellissó, F., Vázquez Ladrón, R., and Guzmán García, R. (1956) *El aderezo de aceitunas verdes*, Consejo Superior de Investigaciones Científicas, Madrid.

Columela, L.J.M (42 BC) *De Re Rustica*. Vol II. Book XII, Nestlé A.E.P.A. eds., Santander, Spain, pp. 209–13.

Condolle, A. de (1982) *Origine des Plantes Cultivées*. Paris

Connell, J.H. (1994) Botany of the olive, in *Olive Production Manual* (eds L Ferguson, G.S. Sibbett and G.C. Martin), University of California, Division of Agriculture and Natural Resources, Oakland, CA, Publication 3353, pp. 1–9

Cruess, W.V (1924) *Pickling of olives in Mediterranean countries*. Univ. Calif. Agr. Expt. Sta. Circular 278.

Cruess, W.V. (1930) *Pickling green olives*. Calif. Agr. Expt. Sta. Bull. 498.

Cruess, W.V (1937) Use of starters for green olive fermentation. *Fruit Prod. J.*, **17**, 1.

Cruess, W.V (1941) Olive products. *Ind. Eng. Chem.*, **33**, 300–3.

Cruess, W.V. (1958) *Commercial Fruits and Vegetable Products*, 4th edn, McGraw-Hill, New York.

Fernández Díez, M.J. (1971) The olive, in *The Biochemistry of Fruits and Their Products*, vol. 2 (ed. A.C. Hulme), Academic Press, London, pp. 255–79.

Fernández Díez, M.J. (1983) Olives, in *Biotechnology*, vol. 5 (eds H.J. Rehm and G. Reed), Verlag Chemie, Weinheim, Germany, pp. 379–97.
Foytik, J. (1960) *California olive industry. Trends and outlook*, Calif. Agric. Expt. Sta. Circular 493.
Garoglio, P.J. (1950) El aceite de oliva y su industria, in *Tecnología de los aceites vegetales*. Vol II. Ministerio de Educación, Universidad de Cuyo, Mendoza, Argentina.
Goor, A. (1966) The place of the olive in the holy land and its history through the ages. *Econ. Bot.*, **20**, 223–43.
Hartmann, H.T. (1953) *Olive production in California*, Calif. Agric. Expt. Sta. Manual 7.
International Olive Oil Council, IOOC (1980) *Unified qualitative standard applying to table olives in international trade*, IOOC, Madrid.
Kalogueria, S.A. (1932) *Table Olives* (ed. Ermou), Athens (in Greek).
Loussert, R. and Brousse, G. (1980) Botanica, biología y fisiología, in *El olivo*, Ediciones Mundi-Prensa, Madrid, pp. 67–101.
Martin, G.C. (1994) Botany of the olive, in *Olive Production Manual* (eds L Ferguson, G.S. Sibbett and G.C. Martin), University of California, Division of Agriculture and Natural Resources, Oakland, CA, USA. Publication 3353, pp. 19–23.
Molinari, O.Ch. and Nicolea, H.G. (1947) *Tratado general de olivicultura*, El Ateneo, Buenos Aires, Argentina.
Moro Beato, M. (1934) *Sobre el endulzado de la aceituna*, Sevilla, Spain.
Noro, K. (1963) La situation actuelle de l'oliviculture au Japon. *Inf. Olei. Int.*, **22**, 39–60.
Patac de las Traviesas, L., Cadahía Cicuendez, P., and Campo Sánchez, E. del (1954) *Tratado de olivicultura*, Sindicato Nacional del Olivo, Madrid, Spain.
Simmonds, N.W. (1976) *Evolution of Crop Plants*, Longman, London, New York.
Spanish Government Presidency (1983) Reglamentación técnico sanitaria para la elaboración, circulación, y venta de las aceitunas de mesa. *Spanish State Official Bull.*, **108**, 12619–25.
Vaughn, R.H. (1954) Lactic acid fermentation of cucumbers, sauerkraut and olives, in *Industrial fermentations*, vol. 2 (eds L.A. Underkofler and R.J. Hickey), Chemical Publishing Co., New York, pp. 417–78.
Vaughn, R.H.(1982) The fermentation of olives, in *Industrial Microbiology*, 4th edn (ed. G. Reed), AVI Publishing Co., Westport, Connecticut, USA, pp. 207–36.
Vaughn, R.H., Douglas, H.C. and Gililland, R. (1943) *Production of Spanish type green olives*, Calif. Agr. Expt. Sta. Bull. 678.
Vavilov, N.I. (1951) Phytogeographic basis of plant breeding. The origin, variation, immunity and breeding of cultivated plants, *Chronica Bot.*, **13**, 1–366 (cited by Zohary, D. (1970) Centers of diversity and centers of origin, in *Genetic Resources in Plants* (eds Fraubel and Benet), I.B.P., Oxford, pp. 1–554).
Zohary, A. and Spiegel-Roy, P. (1975) Beginning of fruit growing in the old world. *Science*, **187**, 819–27.

Table olive cultivars 3

3.1 SELECTING A CULTIVAR

The chemical composition and physical properties of the olive are important factors determining the quality of the final product, and these are strongly influenced by both the cultivar and the time of harvest.

Cultivars for table olive production may be either autochthonous or imported. The first are generally better adapted to local conditions, although imported cultivars may be preferable to indigenous ones where changes in climate, soil and cultivation methods can make them more suited. However, good characteristics of any cultivar in the country of origin do not guarantee their maintenance in the new regions where they are imported.

It is necessary to select the most suitable cultivar for a particular olive type, style of processing or commercial preparation. Physical properties of fruits from different cultivars such as size, shape, flesh-to-pit ratio, ease of pitting, colour and texture are clearly of great importance, but a number of other factors must also be considered. These include:

- geographical situation of planting, plant age and distribution in the field;
- agricultural characteristics, such as productivity of the trees, ripening cycle and pest resistance;
- type of cultivation (irrigated or non-irrigated system);
- pruning methods; and
- harvesting procedure (manual or mechanical).

All these variables will have an important influence on composition, and hence the processing characteristics of the fruit.

Despite the obvious importance of cultivar, the different cultivated olive trees are, in most cases, not widely known and national or local names are generally used to identify them rather than their botanical names. Some of the principal cultivars in the main producing countries are described below.

3.2 CULTIVARS IN DIFFERENT COUNTRIES

Producing countries are considered here in decreasing order of their importance as table olive producers.

3.2.1 SPAIN

The main table olive producing regions of Spain are shown in Figure 3.1. Among the 22 dominant Spanish cultivars, the most important for the processing of table olives have been described by a number of authors (Patac de las Traviesas *et al.*, 1954; Ortega Nieto, 1963; Barranco Navero and Rallo Romero, 1984, 1985; Fernández Díez *et al.*, 1985).

Gordal or Sevillana

Also known as Gordal sevillano and Sevillano, its culture is mainly situated in the Low Andalucía, reaching its maximum extent in Sevilla province. It has big ellipsoidal to heart-shaped fruits (Figure 3.2), with an average size of 100–120 fruits/kg and a flesh-to-pit ratio of 7.5 : 1. They have a fine epicarp with characteristic white spots; a mesocarp with a good texture and a light green colour which changes to purple black when ripe; and right and regular endocarp suitable for mechanical depitting. The oil content is rather low, generally below 10% of the fruit weight. This cultivar ripens early and shows a preference for mild and warm climates.

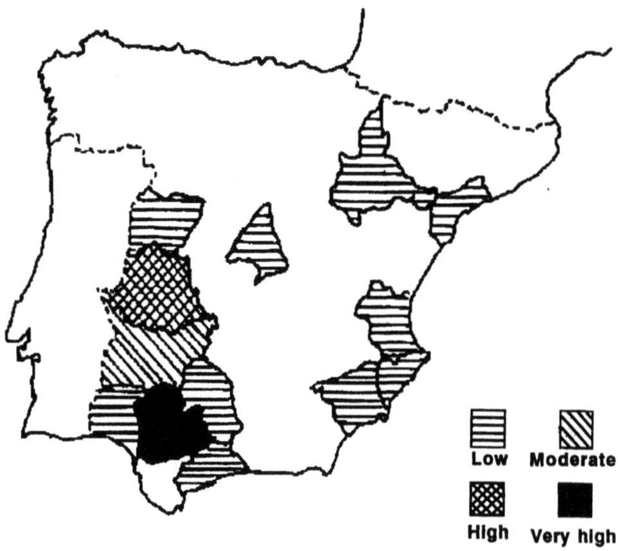

Fig. 3.1 The main Spanish table olive-producing areas.

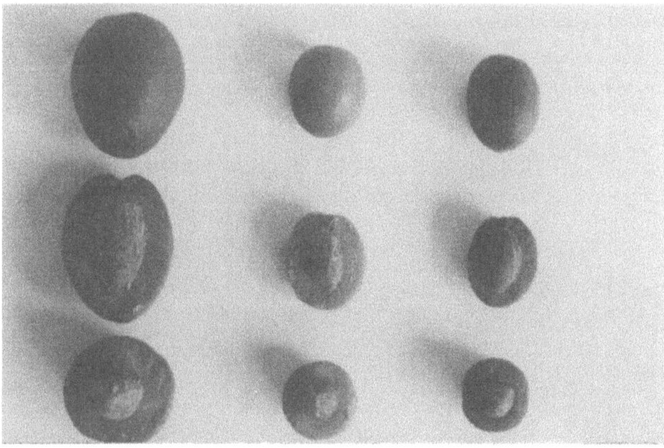

Fig. 3.2 Main Spanish table olive cultivars: Gordal (left), Manzanilla (centre), and Hojiblanca (right).

Gordal and the Manzanilla are the two best known cultivars for processing as treated green olives in brine, because of the quality of the flesh and the exceptional sensory properties of the final product.

Manzanilla

This is sometimes also known as Manzanillo, and is more widely cultivated in Spain than the Gordal cultivar. The fruits acquire their best quality with mild climatic conditions, close to the Guadalquivir river and the Sevillian zone. Figure 3.3 shows a typical Manzanilla cultivar orchard at Dos Hermanas (Sevilla). It has a number of different regional adaptations with local names such as Fina, Serrana and Carrasqueña. Manzanilla olives are most appreciated in the international market for their excellent organoleptic properties. The apple-shaped fruit (Figure 3.2) has a fine skin and an excellent texture. Their average size is between 200 and 280 fruits/kg with a flesh-to-pit ratio of 6 : 1. The pit is right and presents a very smooth surface. Manzanilla ripens later than Gordal and has a higher oil content, sometimes up to 15% by weight of the fruit. Its skin is resistant to white spots and the flesh is not prone to gaseous spoilage, but is liable to sloughing during the alkaline debittering treatment. This latter problem can be avoided by storing the fruits for 1–2 days in 20–24-kg boxes before immersion in alkali.

Morona

Very close to Manzanilla, it is perhaps the same cultivar. Its production is relatively small and located in the surroundings of Moron de la Frontera

Fig. 3.3 Typical Manzanilla cultivar orchard in Dos Hermanas (province of Sevilla, Spain).

(Sevilla province). It is used almost exclusively to prepare treated green olives in brine. It is very susceptible to gaseous spoilage.

Hojiblanca

This cultivar ripens later than the Gordal and Manzanilla cultivars and the oil content is higher, between 23 and 29%. Until the 1970s it had been grown solely for oil production, but it is currently the second most important Spanish cultivar in terms of volume, being the most appreciated for production of naturally black olives in brine or black-ripe olives in brine as well as being used in lesser quantities to produce treated green olives in brine. Hojiblanca is prone to gaseous spoilage, especially when mature and prepared as naturally black olives in brine, and to shrivelling during storage after picking as turning colour olives for the production of black-ripe olives. This cultivar was not used previously in table olive production due to its late ripening and the low temperatures in the growing zones when the fruit is green mature. These low temperatures can make fermentation difficult. However, new technologies to prepare treated green olives have overcome this difficulty and the introduction to Spain of naturally black olive and black-ripe olives, which do not need a lactic fermentation, have led to a massive extension of its use for table olives.

The main areas of cultivation are situated in Córdoba and Málaga provinces and, to a lesser extent, in Sevilla and Granada provinces. The shape of the fruits is regular (Figure 3.2), the pit is right, and the olives have a very variable size, ranging from 230 to 700 fruits/kg, the smaller sizes being used for oil extraction. Flesh-to-pit ratio varies between 4.9 and 6.6 : 1, depending upon the size of the fruit.

Cacereña

Quite similar to Manzanilla, although it generally has a lower average size and a slightly rougher texture. When mature it becomes soft and very susceptible to gaseous spoilage. It is cultivated mainly in the Western Cáceres province, from which it takes its name, and the south of Salamanca province. Together with Hojiblanca, it is the most suitable cultivar for producing black-ripe olives in brine. Adequately processed, it has a very characteristic taste which differentiates it from any other cultivar making it specially sought after in both national and international markets.

Verdial

Olives of this cultivar keep a very deep green colour until just before full maturity, changing later to red and reddish-black. It ripens later than the other cultivars described and the fruits are ellipsoidal, the pit is right and comparatively more elongated than the whole fruit. The ratio of flesh-to-pit averages approximately 6 : 1 and gives a high yield of oil, between 22 and 30%. Size is very variable, between 220 and 800 fruits/kg; only the larger sizes are used for treated and untreated green type olives and the rest of the crop is used for oil extraction.

The Verdial cultivar grows mainly in the provinces of the Western Andalucía, mostly in Sevilla province.

Picual

Also known as Picuda, this represents 90% of olive trees cultivated in Jaén province. Because of this, it is the premier cultivar in Spain, although it is used primarily for oil extraction. Fruit and pits both have a prominent tip at the button end. The oil yield is high, from 23 to 27% of the fruit weight, and the ratio flesh-to-pit varies between 3.8 and 6.1 : 1. Fruit size ranges approximately from 270 to 470 fruits/kg. Larger sizes are used for the production of green and naturally black olives types, generally for the national market.

There are a number of other cultivars that are grown to a relatively limited extent but are of sufficient commercial interest for table olive production to merit mention:

- **Lechín**: these are of high oil content, with an excellent black colour when fully ripe and very specific sensory properties if prepared as naturally black olives in brine; the small average size has limited their use in trade
- **Rapazalla**: this has a low flesh-to-pit ratio but is used for green type in local markets
- **Aloreña**: this is very similar to Hojiblanca cultivar.

3.2.2 TURKEY

The most important Turkish cultivars are as follows (Yenicesu, 1965; Fernández Díez, 1972a; Caran, 1989; Tunalioglu, 1994).

Domat

This is the most appreciated cultivar in Turkey for production of green-type olives. Its irregularly shaped fruits behave similarly to the Spanish cultivar called Aloreña. The texture is fibrous and the oil content is around 22% of the fruit weight. Average size varies between 180 and 190 fruits/kg.

Gemlik

This is the best Turkish cultivar for preparing naturally black olives in brine. Fruits have a pronounced tip at the end and their average size is 270–280 fruits/kg. The flesh-to-pit ratio is 6 : 1 and the oil content is high, around 24%. This cultivar accounts for about 50% of the Turkish market as a whole and represents 80% of the crop in the Marmara region.

Memeçik

This is located in Aegean region. Like Gemlik it has a pronounced tip but the average size is larger: 205–215 fruits/kg. It resembles the Spanish cultivar, Picual. The yield of oil is high, around 22%, and the flesh-to-pit ratio similar to that of Gemlik. It is the most extensively cultivated olive tree grown in Turkey, but is mainly used for oil extraction. It does however show some promise as a source of either green or naturally black olives.

Memeli

This is the fourth most important cultivar in Turkey. Fruits have a small tip at the end and have an average size of 200–210 fruits/kg. The flesh-to-pit ratio is 7 : 1 and the oil content around 25%. It is mainly used for green-type olives.

Edremit or Ayvalik

This is cultivated close to the Aegean coast, in a region situated between the two cities which give the cultivar its alternative names. Fruits are rounded, with an average size of 290–300 fruits/kg and a flesh-to-pit ratio of 5 : 1. It has a high oil content (25%) and good possibilities either as green or naturally black olives.

Among the other Turkish cultivars which are used to a lesser extent but have good potential for greater utilization in table olive production, the following are important:

- **Izmir Sofralik**: this has a good texture, pronounced tip, lower oil content (18%), average size of 140–150 fruits/kg, and a flesh-to-pit ratio of 6.5 : 1, and is suitable for green-type olives
- **Çilli or Bugarres**: this has rounded fruits, which resemble the Manzanilla cultivar, relatively low oil content (16%), a flesh-to-pit ratio of 6.5 : 1, and excellent texture when prepared as treated green olives
- **Celebi**: this has a very good average size, 90–100 fruits/kg, and a low oil content, around 16%, although flesh-to-pit ratio is not so favourable, 5 : 1
- **Uslu**: this only grows around Izmir, with an average size of 290–300 fruits/kg, and is used to produce naturally black olives.

3.2.3 GREECE

The three main cultivars of olives used for table olives in Greece are Conservolea, Kalamata and Chalkidiki (Balatsouras, 1967, 1984a; Fernández Díez, 1969a; Aligizakis, 1982).

Conservolea

This cultivar is by far the most important economically, being responsible for at least 80–85% of olive production in Greece. It grows throughout the country from sea level up to altitudes of 600 metres. Fruits are round- to oval-shaped and have properties in between those of the Manzanilla and Hojiblanca cultivars, being perhaps closer to the latter. Average size goes from 180 to 200 fruits/kg with a flesh-to-pit ratio of up to 8 : 1, with an oil content of 22–25% of the total fruit weight. Fruit maturation is slow. Some are picked when green mature (mid-September to mid-November) to produce Spanish-style green olives. The rest of the fruits are kept on the trees while the surface colour changes progressively to pink, violet and finally black (still retaining some white spots). This is the optimum colour for the preparation of naturally black olives in brine and is reached some time between November and January–February. This product has a good reputation all over the world. The fruit skin is thin, flexible and resistant to shrivelling, being able to withstand immersion in 12–19% NaCl brines, and the pulp has a fine, consistent texture.

Nychati Kalamon

This excellent Greek cultivar is grown less extensively, principally around the city of Kalamata in the South of the Peloponnese. It is the second most important cultivar used in the production of Greek table olives, although production is increasing as a result of a solid domestic market and increasing demand from other countries in the European Union, the USA and Canada for the product known as Kalamata olives.

It is mainly used for specialities within the naturally black olives in brine style. Fruits are cylindro-conical, curved, showing a prominent tip at the end. Sizes range between 180 and 360 fruits/kg, averaging 220–240 units/kg. The flesh-to-pit ratio is similar to that of Conservolea. The oil content is relatively high, around 25% of the fruit weight, although its sugar content is low, 3–5%. The skin is thin, flexible and has an intense black colour when mature, but still retains a good texture. The fruits are normally harvested at late ripeness in November–December.

Chalkidiki

This is the third most important cultivar used in the production of Greek table olives, accounting for about 3% of total production. It is cultivated in the Chalkidiki region, close to Salonika. It is not a well defined cultivar and is considered by some authors to be closely related to Conservolea. However, it differs from Conservolea in a number of respects: the fruits are larger, averaging 120–140 fruits/kg; they are more elongated, with a prominent tip at the bottom; the pit is slightly curved and the flesh does not have as good a texture. The better-sized fruits are used mainly for Spanish-style green olives in brine while the rest are used for oil extraction. Their oil content is about 19%. It is not suitable for the production of naturally black olives because it never reaches an adequate black colour. When used to produce green olives it is very liable to malodorous fermentations, due to the low sugar content of the flesh. Sometimes, the olives also develop a pink ring around the pit, thought to be due to a reaction involving phenols.

The remainder of the Greek production of table olives is distributed in small percentages among the following cultivars:

- **Megaritici**: this grows in the Attica region. It is a dual-use cultivar (oil extraction and table olive) with fruits of medium to small size (200–500 olives/kg). It is resistant to drought and is used for untreated green olives and in the production of naturally (wrinkled) black olives in salt. Also known as 'Greek-style olives', these were probably the first type of commercially prepared table olives.
- **Kothreici**: this is also a dual use variety, produced in the region of Focida, with fruits similar to, but smaller than, those of Conservolea. Experts consider it to be closely related to Conservolea, perhaps a clone of it. It is used for preparing naturally black olives in brine which have excellent colour and taste combined with a good texture. It is mainly processed for the Greek domestic market.
- **Karydolea**: this is also closely related to Conservolea. It is cultivated on the island of Eubea.
- **Thrubolea**: this is a cultivar from Crete. When completely mature, fruits are bronze to greenish and become sweet on the tree. It is used to prepare a special type of naturally black olives called 'naturally black olives Thruba style'. A clone of it cultivated on the island of Thasos, however, does not become naturally sweet, although it is used for the same purpose.

3.2.4 UNITED STATES OF AMERICA

California is the only State of the USA that produces table olives. Figure 3.4 shows the main producing areas. Five cultivars are cultivated in California (Hartmann, 1953; Foytik, 1960; Hartmann *et al.*, 1966, 1986; Fernández Díez, 1969b; Sutter, 1994): Manzanillo, Mission, Gordal or Sevillana, Ascolana and Barouni.

Manzanillo

This cultivar was introduced from Spain. In California, the trees are not tall and the fruit can be easily picked by hand. Highest yields are obtained in the San Joaquin Valley. Manzanillo is sensitive to cold and is currently harvested in September–October to prevent damage by early frosts. Fruits reach an average size of 200–220 fruits/kg, with a flesh-to-pit ratio of 8.2 : 1. These olives are mostly processed as black-ripe olives, although some fruits are also used to prepare Spanish-style green olives. The oil content is about 20% of the fruit weight, which is sufficiently high for the small and damaged fruits to be used for oil extraction. This possibility helps maintain the quality of those used in the production of table olives.

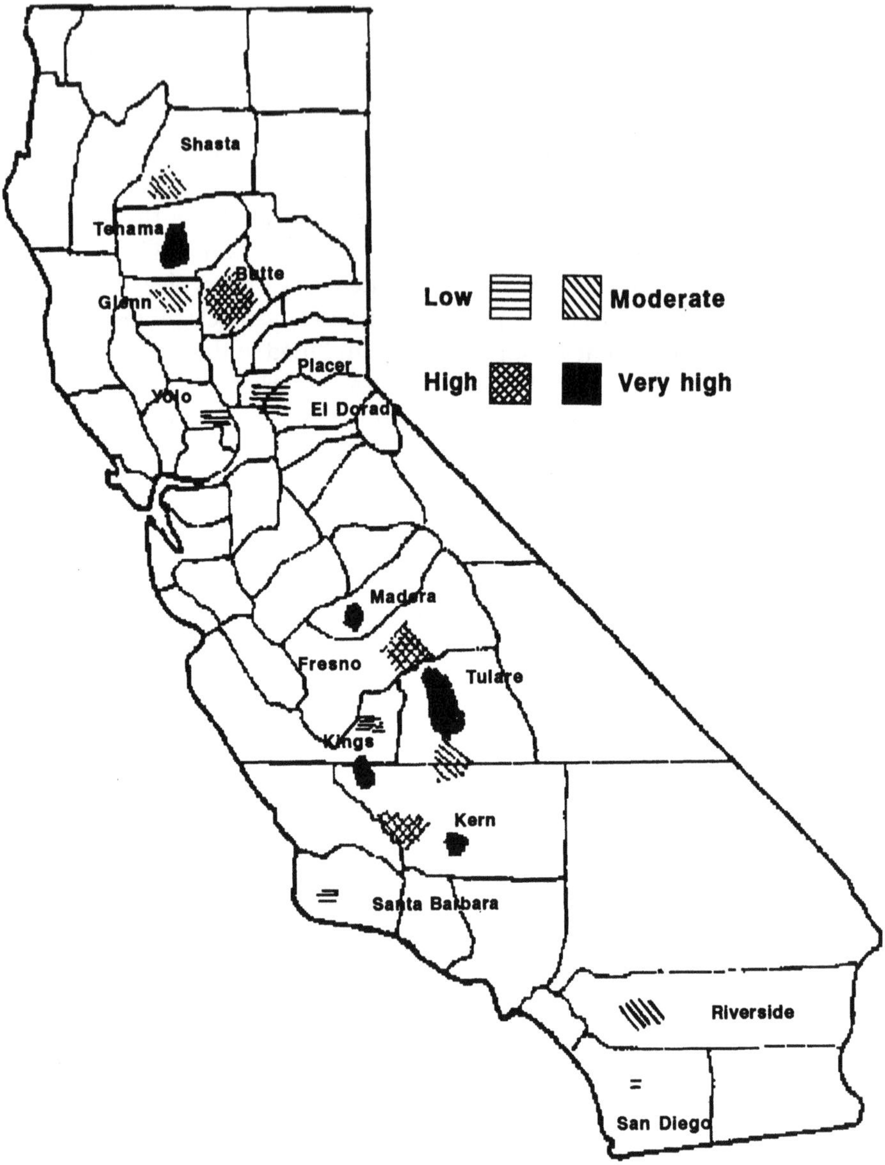

Fig. 3.4 Main table olive-producing areas of California (USA).

Mission

This is the longest-established olive cultivar in California, and was originally introduced from Mexico. Its trees are the tallest of the Californian cultivars and must be topped to facilitate harvesting. It is cold-resistant and fruits mature late in the season, October–November. The average size is 240–260 fruits/kg with a flesh-to-pit ratio of 6.5 : 1 in processed fruits and an oil content of around 22%. They produce very tasty olives with an appealing uniform black colour when used as black-ripe olives. When green, they are also used for producing Spanish-style green olives. It may also be used for oil extraction after harvesting in December–February.

Sevillano or Gordal

Another cultivar of Spanish origin, its trees are also taller than Manzanillo, although appropriate training can give smaller trees and easy hand harvesting. It is cold-resistant and planted mainly in northern California. Fruits reach an average size of 70–80 fruits/kg with a flesh-to-pit ratio for processed fruits of 7.3 : 1. The oil content is low, around 14%, which prevents its use for oil extraction. It is harvested in October and processed either as Spanish-style green or black-ripe olives, although the latter are more difficult to produce as the oxidized fruit has a propensity to bruise.

Ascolano

This is a cultivar imported from Italy. Its trees are rounded with a high centre. Fruits mature earlier than any other Californian cultivar, normally around September–October. They are large, with an average size of 110–120 fruits/kg, but with a soft texture which makes it difficult to pick them without bruising. The ratio of flesh-to-pit is 8.2 : 1 and the oil yield is rather low (19%) in comparison with Manzanilla and Mission cultivars. It is processed mainly as black-ripe olives, since the fruits become shrivelled when prepared as Spanish-style green olives.

Barouni

This cultivar comes from Tunisia and is the most recently introduced Californian cultivar (at the beginning of the 1900s). The trees are small, with spreading tops which makes them well suited to hand picking. The olive has an average size of 130–140 fruits/kg and a ratio of flesh-to-pit of 6.8 : 1. Oil content is low, 17%, which limits its use for oil extraction. Some of the fruits are sold fresh for home processing and the rest are processed as green-ripe olives, since they process poorly as canned black-ripe olives. This last factor has limited its planting in California making it one of the less important cultivars at present.

Other cultivars also present in USA are: **Oblonga, Allegra** and **Swan Hill**, although their commercial utilization is not significant.

3.2.5 ITALY

Italy's table olive production is of increasing interest. The main cultivars used are as follows (Baldini *et al.*, 1963; Cucurachi, 1977).

Ascolana

This includes three different cultivars: tenera, dura and semidura, tenera being the most widely cultivated. It is the best Italian cultivar for table olive production and, as a consequence, has been exported to many other countries including the USA, Mexico, Israel and Argentina. In Italy, it is mainly cultivated in the province of D'Ascoli Piceno. On the international market it is often confused with the Spanish cultivar Gordal because of its large size, although in other characteristics they are really quite different. It has oval, almost spherical-shaped fruits which may reach average sizes of 100–180 olives/kg presenting a very favourable flesh-to-pit ratio. Its oil content is about 17%, too low for oil extraction. Skin and pulp are very delicate, so these olives have to be processed very carefully. It is resistant to cold and has a green to yellow surface colour when it reaches the optimum degree of maturation for its main use, the preparation of Spanish-style green olives.

Grossa di Spagna

This is also known as **Oliva di Spagna** and **Oliva di Cerignola**. The cultivar produces fruits with a very good average size, around 100 fruits/kg. They are ellipsoidal, slightly asymmetric with a pronounced tip at the bottom. The fruit mature early and are not cold-resistant. When completely mature, the olives develop a black surface colour with white spots. Together with the Ascolana cultivar, it is used mostly for production of Spanish-style green olives and directly brined olives (without lye treatment). Production is limited and is dedicated to the domestic market.

Nocellara etnea

This cultivar is cultivated in eastern Sicily, particularly in the zone close to Mount Etna. Its oil content is about 15–20%, which is relatively low and prevents its use for oil extraction. Fruit size is around 200 fruits/kg with characteristic white striations and big lenticels on the surface when the fruit is mature. This occurs quite late around October time and the fruit is used principally for producing Spanish-style green olives.

Sant'Agostino

This is cultivated in Puglia and particularly in the region of Andria, giving rise to its alternative designations as 'oliva di Andria' and 'oliva grossa andriesana'. It is not resistant to cold and matures early (mid-September). The fruit are ellipsoidal and asymmetric with a medium size of about 135 fruits/kg. Its oil content is too low (about 14%) for it to be used for oil extraction and it is mainly used for preparing Spanish-style green olives.

Nocellara de bellice

This cultivar has attracted the interest of the industry recently. It averages about 150 fruits/kg and its flesh-to-pit ratio is around 7 : 1. It is cultivated together with the Guiarraffa cultivar to assure cross-pollination. Their fruits have good characteristics for use as table olives, mainly without lactic fermentation (Castelvetrano system)

Among the 296 Italian cultivars mentioned by Garoglio (1950), the following must also be considered as relatively important as sources of table olives:

- **Bella di Spagna**, quite similar to **Grossa di Spagna** and frequently indistinguishable from it;
- **Cucco**, very widely cultivated, of smaller size and used for green type olives;
- **Itrana** and **Uovo di Piccione**, with small and large fruits respectively and both mainly used for black-type olives. Uovo di Piccione is included within the group of cultivars known as Barouni, which are grown widely in Algeria and Tunisia with the names of **Grosse du Hamma** or **Bidh el Hamman**.

A complete description of cultivars cultivated in Sardinia can be found in Aggabio and Mulas (1987) and a complete enumeration of Italian cultivars and their uses in Marsilio (1994).

Finally, there are some other cultivars which mainly are used for oil extraction and only a minor part of the crop is destined to table use. Among them, **Leccino**, **Razzola** and **Frantoio** are the most important.

3.2.6 MOROCCO

The **Picholine marocaine** or **Zitoun** cultivar comes from Algeria, although it may have originated in France. It is a relatively resistant cultivar which adapts easily to soils of varied characteristics, but is particularly well adapted to the Moroccan soils. Its fruits are dual-use, although it is the main cultivar used in Morocco for table olives. These are mainly prepared as naturally black olives and canned black-ripe olives but Spanish-style green olives are also produced. Average fruit size is around 280 fruits/kg, with a range of 300–500 fruits/kg. They have an elongated ovoid shape, with a slightly asymmetric and curved pit. The oil content is around 17% and the ratio of flesh-to-pit is 5 : 1. The pulp is tasty and delicate. The fruit matures relatively late, at the end of November for green olives and December–January for naturally black olives.

Other Moroccan cultivars include **Gordal** and **Hojiblanca**, both of them Spanish origin, and two double-use cultivars, **Soussia** and **Meslala**.

For details of Moroccan cultivars, see Loussert and Brousse (1978) and Loussert (1989).

3.2.7 SYRIA

- **Dan** is a dual-use, resistant cultivar grown in the region of Damascus. Trees are big and produce medium-sized fruits (around 3.5 g/fruit). Its oil content is about 18%.
- **Djlt** is cultivated in the same areas as Dan. It produces big trees which yield big olives (7–10 g/fruit). Its oil content (7–10%) is too low for economic extraction and the fruit are entirely dedicated to the production of Spanish-style green olives and naturally black olives.
- **Abou-Satl** is cultivated in the Palmira oasis. A few vigorous trees produce big fruits (7–10 g/fruit) used for preparing Spanish-style green olives and naturally black olives.
- **Al-Kaissy** is cultivated in the north of the country and in Aleppo. It is a drought-resistant cultivar that produces medium-sized fruits (4–6 g/fruit). It is used mainly for Spanish-style green olives, although it has an oil content of about 18% and can be used for oil extraction.
- **Massabi** is cultivated around Damascus as small vigorous trees which yield big fruits (5–8 g/fruit). Its oil content is only about 10–12% so it is used almost exclusively to produce table olives.
- **Al-Sorani**, **Al-Doeibly** and **Al-Khodeiry** are three dual-use cultivars.

For details of Syrian cultivars, see Patac de las Traviesas (1954), Balatsouras (1984a) and Karabij (1992).

3.2.8 ARGENTINA

Arauco is the main cultivar in Argentina and is named after the region to which it adapted most easily after importation from Spain. Also called Criolla, it has a high oil content of 22–24%. The fruit is elongated with a pronounced tip and slightly curved pit and a flesh-to-pit ratio of 7 : 1. It generally reaches good commercial sizes of 125–300 fruits/kg. The physical and chemical characteristics of the fruits of this cultivar are quite variable depending upon the region in which they are cultivated. According to the degree of maturation, it may be used for Spanish-style green olives, canned black-ripe olives and naturally black olives, as well as for oil extraction.

Among the other cultivars of Spanish origin, **Empeltre**, a cultivar very resistant to cold climates; **Gordal** or **Sevillana**; and **Manzanilla**, are significant. Several adapted Italian cultivars grown include **Ascolano**, with big fruits and delicate pulp, **Cerignola** and **Santa Catalina**, but all of them have a very restricted area of cultivation.

For details of Argentinian cultivars, see Molinari and Nicolea (1947), Garoglio (1950), and Salvarredi (1987).

3.2.9 PORTUGAL

Among the Portuguese cultivars used for green-type olives, two are preeminent: **Carrasquenha** and **Redondil**. These are related to the Spanish Carrasqueña and Manzanilla cultivars respectively and share very similar characteristics with them.

The most extensively grown cultivar is **Galega Vulgar** which, when prepared as naturally black-type olives, gives a final product of excellent quality, although the average size of around 430 fruits/kg is relatively small. The ratio of flesh-to-pit is about 4 : 1 and the oil content around 19%.

Other interesting cultivars for table use are **Cordovil**, **Bical**, **Negrina**, **Verdial**, **Gordal** and **Manzanilla dos Hermanos**. All of them are of relatively low exploitation at present.

For details of Portuguese cultivars, see Montero Marín (1972) and Abela and Fernades Serrano (1987).

3.2.10 MEXICO

The olive cultivars planted in Mexico have followed the same pattern as those of Southern California in the United States. Approximately 95% of olive area of Mexico is located in Baja California and Sonora, and only minor production zones may be found in the States of Durango, Guanajuato and Aguascalientes (Emerson, 1980).

Until the 1920s, the **Mission** cultivar was predominant. Since then, new plantings have been made of **Manzanilla**, which is considered the best cultivar for processing, either as Spanish-style green olives or Californian-style green-ripe olives, a product that is more popular than the black-ripe olives. At the present time these two cultivars count for the vast majority of the crop. A small proportion of **Sevillana** or **Gordal** is also grown and smaller quantities of the **Ascolana** and **Barouni** cultivars.

3.2.11 ISRAEL

The main cultivars grown for table olives in the interior valleys are **Merhavia** and **Manzanilla** and, to a lesser extent **Nouvo di Sicrone** and **Sdeh Elianhu**. The local cultivar, **Souri**, used primarily for oil extraction is also used for table olives. The main characteristics of the most important cultivars are as follows.

Trees of the **Souri** cultivar could be the oldest in the world. Some of them in the region of Galilee and Judea are claimed to have been there since Biblical times. The biggest and oldest are known as **Rumi**, meaning that they were planted during the Roman colonization. Fruits of Souri cultivar are of late maturation (October–November), high oil content (greater than 30%) and small size (300–400 units/kg).

Nuovo di Sicrone is a cultivar probably imported from Italy, with big fruits (up to 12 g/fruit) and an oil content between 13 and 22%. Due to its good shape, it is used mainly for processing as black olives. When prepared as Spanish-style green olives, the flesh may become brown if the NaOH solution does not penetrate to the pit.

Merhavia takes its name from the region where it is mostly grown. It was probably introduced from abroad and its excellent characteristics for processing as naturally black olives and canned black-ripe olives favoured its further propagation. Fruits mature very early at the end of August, are of good size (100–300 units/kg) and have a very low oil content (less than 10% when green mature and about 16% at full maturation). The fruit shape is irregular. Some authors believe that this is a particular characteristic of the cultivar while others think that it is due to a pest. The flesh is easily detached from the pit, a characteristic that favours its utilization for Spanish-style green olives only.

Sdesh Eliauhu also bears the name of the region where it is cultivated. These trees are very sensitive to wind and must be protected by interplanting with other trees or by other protective measures. It matures late, October–November, and produces medium size fruits (150–250 units/kg). The fruit flesh is of excellent organoleptic quality and is used mainly to prepare Spanish-style green olives. Its oil content is high (up to 30%) which permits the selection of the biggest fruits for table olive processing, with a subsequent higher-quality product, and use of the smaller fruit for oil extraction.

According to Lavee (personal communication, 1982), other cultivars used in Israel for table olives are **Santa Catalina** and **Uovo di piccione**, for green type; and **Kalamata** for naturally black type.

Special interest has been paid to **Kadesh** which was developed for Spanish-style green olives (Lavee, 1978). Its fruits have a rather uniform oval shape, an average size of about 150 units/kg and a flesh-to-pit ratio of 5 : 1. Its growth is uniform from mid-June to black-ripening in mid-October. Its oil content is always less than 3% at the time of green maturation and about 8% at full black maturation. It gives a product with a reasonable texture and taste, comparable with that from Merhavia.

Nabali is a dual-use cultivar. Some other cultivars are being grown for experimental purposes including **Giarraffa**, **Cucco**, **Amigdalolea**, **Hojiblanca**, **San Francisco** and **Sant Agostino**.

For details of Israeli cultivars, see Glickson (1960).

3.2.12 TUNISIA

Meski, a local cultivar grown in the north, is the main cultivar used for table olives in Tunisia. Its trees should be interplanted with other cultivars to permit cross-pollination. Its fruits present, in general, a good average size and a favourable flesh-to-pit ratio, although published data are not available. The colour of the fruit is not intense when mature and the texture is fairly good with a low natural bitterness. It is used either for green or black types which are highly appreciated in local markets.

Among the other indigenous cultivars **Besbessi** and **Marsaline**, in the north of the country, and the dual-use southern cultivar **Chemchali**, all make significant contributions. There are also a number of important imported cultivars such as **Picholine**, introduced from France and excellently adapted to the region, **Lucques**, also from France, **Ascolana**, from Italy, **Manzanilla**, from Spain, and **Barouni**, which is notable for producing the biggest fruits.

For details of Tunisian cultivars, see Long and Bonet (1951), Trigui (1993) and Kachouri (1994).

3.2.13 ALGERIA

Sigoise is normally known as 'the olive of the country', and is cultivated mainly in the region of Oran and, to a lesser extent, in Constantine. The shape of the fruit is quite regular and well proportioned with a generally pronounced tip and good flesh-to-pit ratio. Sizes vary from 160–170 to 350–400 fruits/kg. The oil content, which is of good quality, ranges from 14 to 17%. The fruit ripens quite late, a feature that undoubtedly favours its use as table olives, and it has sensory similarities with the smaller fruit of the Spanish **Lechin** cultivar.

Sevillana was imported from Seville, Spain and is grown mainly in the area around Oran where it ripens early. Its size varies between 70 and 160 fruits/kg. The pulp is consistent, though a little hard and slightly granular, and adheres strongly to the pit. In these respects it is inferior to the original Spanish cultivar.

Azeradj produces fruits with a pronounced tip, reaching an average size of 200 units/kg. Only about 10% of the crop is used to produce table olives. It is mainly cultivated in the areas of Constantine and Algiers.

Other cultivars which account for only a small percentage of the production of table olives are the local cultivars: **Bouchouk, Aghenfour, Grosse du Hamma, Rougette** and **Verdial**, with average fruit sizes between 250 and 350 fruits/kg, **Picholine, Lucques, Cornicabra** and **Blanquette**, among those of Spanish, French and Italian origin.

For details of Algerian cultivars, see Fernández Díez (1972b), Balatsouras (1984b), Giametta (1985) and Benkeblia (1992).

3.2.14 CHILE

Azapa is the most popular Chilean cultivar for table olive processing and is named after the valley in which it grows, close to the Peruvian frontier. It represents about 90% of the olives grown in that area of the province of Tarapaca. The pulp and skin are fine with a high flesh-to-pit ratio which may reach 11 : 1. The average size is about 120 fruits/kg, with an overall range of 60–300/kg. It is also known as Azapeña and Sevillana from Azapa. These cultivars may come from the Spanish cultivar Gordal, although the Chilean fruits are more elongated and present a prominent tip at the bottom which does not appear on the original cultivar. General characteristics of the fruits are relatively close to those of the Argentine Arauco cultivar.

Huasco, also known as Sevillana from Huasco, is mainly cultivated in Atacama province. It reaches an average size of 220–300 fruits/kg, with an overall range of between 180 and 500. Its pit content is higher than that of Azapa, with a flesh-to-pit ratio of 5 : 1. The excellent microclimate in the Azapa valley must account for the differences between these two cultivars, which both originate from the same stock.

Among the other Chilean cultivars, **Manzanilla** and **Empeltre**, from Spain; and **Ascolano, Liguria** and **Santa Emiliana**, from Italy, account for a small percentage of processed table olives produced for local consumption.

For details of Chilean cultivars, see Fernández Díez (1973) and Tapia and Bastias (1987).

3.2.15 CYPRUS

The only Cyprian cultivar which is an important commercial source of table olives is **Ladoelia**, used mainly to produce naturally black olives in brine. Probably of Greek origin, the average size is 330 units/kg, and the flesh-to-pit ratio about 4.6 : 1. The oil content varies between 18 and 21%.

The Spanish **Gordal** and **Manzanilla**, the Italian **Cucco**, and the Greek **Kalamata** cultivars may also be found, but these have been grown mainly for experimental purposes.

3.2.16 FRANCE

In terms of market share, **Picholine** is the most important French table olive cultivar. Mainly located in Gard Department, it is a very adaptable cultivar which reaches full ripeness rather late. The fruits are ovoid, elongated, slightly curved and with a pronounced tip at the bottom. Average oil yield is close to 17%, and the flesh-to-pit ratio is around 4.8 : 1. Average size is typically 275 fruits/kg.

Tanche is another important French cultivar which, because of its more widespread cultivation, is also known as the olive from Nyons, and sometimes as the olive from Carpentras. Tauche is very rich in oil, containing on average about 25%. The fruit has a good average size, 250 units/kg, a rounded shape and a pit with very marked grooves. The flesh-to-pit ratio is less favourable, but the pulp is fine and the taste of the final processed product very agreeable and characteristic.

Lucques is probably of Italian origin. Mainly cultivated in the Herault Department, it is of secondary and decreasing importance. It has a similar oil content to Picholine, although its average size is smaller, its shape is more irregular and curved and the flesh-to-pit ratio a little lower, 4.5 : 1.

Amellan is also cultivated in the Herault Department. Almond-shaped, light green in colour with white spots on the skin, it has a more straight and elongated pit than the other French cultivars. It is also generally bigger, 230 units/kg, and has a flesh-to-pit ratio of 4 : 1. It ripens late and has a rather low oil content, around 14%.

Other French cultivars grown to a lesser extent include **Verdale**, **Aglaudaou**, **Germaine**, **Bouteillau**, **Pruneau de Cotignoc** and **Belgientieroise** (in the Departments of Var and Salonenque) which are used only to supply local markets.

For details of French cultivars, see Long and Bonet (1951) and Fady (1960).

3.2.17 AUSTRALIA

The most extensively grown cultivar in Australia is currently **Verdial**. It is a relatively tall tree with a height intermediate between those of the Spanish cultivars Gordal and Manzanilla. The ratio of flesh-to-pit is less favourable than these but the final product, as green-type olives, has a good flavour and an excellent texture.

Second most important is the **Manzanilla** cultivar, which during the past two decades has been considered the most suitable for new plantations under irrigation. **Gordal** or **Sevillana** also offer good possibilities in this respect. The general characteristics and properties of these cultivars, which grow under irrigation are completely different from those seen in Spain.

For details of Australian cultivars, see Fernández Díez (1972c).

3.2.18 (THE FORMER) YUGOSLAVIA

The most important indigenous cultivar is called **Oblitza** or **Oblica**. It grows across all the central part of the country, the main orchards being situated close to the Adriatic coast. It is resistant to drought and cold. The fruits vary between apple- and heart-shaped fruits, depending upon the different areas, with white spots on the skin. They have an average size of around 200 units/kg with a good flesh-to-pit ratio (6.5 : 1). Oil yield reaches up to 22% of the whole fruit weight and they are processed either as Spanish-style green olives (September–October) or canned black-ripe olives (end of October).

Other local cultivars, used mainly for oil extraction, are: **Istarka bjelica** with two different types of fruits **Bianchera-Tar** and **Plemenita belica**, located in the North (Istria); **Crnica** cultivated in Istria: **Slivnjaca** cultivated in Croatia and on the islands of the Krarner archipelago; and **Juditza** grown in Montenegro. The main imported cultivars now developing in the South include **Ascolana tenera**, **Cucco** and **Itrana**, from Italy, and **Picholine**, from France. For details of Yugoslavian cultivars, see Modun (1960) and Skarika (1989).

3.2.19 SOUTH AFRICA

The most commonly grown cultivar in South Africa is **Mission**, from California. It has fruits of a very good average size, asymmetric in form and with a pronounced tip. Small quantities of Spanish **Manzanilla** and **Gordal** cultivars, **Kalamata** from Greece, and **Azapa** from Chile, are also cultivated and used for table olive processing. For details of cultivars, see Impey (1974).

3.2.20 IRAQ

Ashasi cultivar is considered to be the best in the country for the preparation of table olives. It is distributed throughout northern Iraq. Fruits have an average size of around 180 fruits/kg, with an oil content of between 15 and 16%. The texture is quite good and the bitterness of the fresh fruit is rather low.

3.2.21 OTHER COUNTRIES

Specific data on the cultivars from the other countries, which share a significant fraction (6%) of world production of table olives are not available in the literature. Those countries are: Peru (2.1%); Egypt (1.7%); Jordan (1.2%); Lebanon (0.7%); Libya (0.2%) and Brazil (0.1%).

It is known that one of the most important Peruvian cultivars which grows in Tacna Valley, close to the frontier with Chile, is quite similar to the Azapa cultivar in the Arica region of Chile.

3.3 THE MAIN CULTIVARS FOR THE DIFFERENT PROCESSES

After this survey of the most important olive cultivars used for processing in the principal producing countries, some bold generalization seems appropriate. In the opinion of the authors, given the best conditions of cultivation, harvesting and processing, the Manzanilla cultivar is probably the most suitable for the preparation of Spanish-style treated green olives in brine; Kalamata for Greek-style naturally black olives in brine; and Mission for Californian-style black-ripe olives in brine. The same applies to the procedures related to and derived from these three principal industrial methods.

This general rule is not invariably true, other than in the countries which are traditionally seen as the origin of the cultivar and the corresponding process. Because of the different factors indicated at the beginning of this chapter, the evolution of a cultivar after it is exported and adapted may well produce a marked change in the general characteristics of the fruit in a way that is quite unpredictable. Two contrasting examples of this are particularly illustrative.

Plants of the Spanish Manzanilla cultivar carefully planted and grown in Turkey 20 years ago, developed some unfavourable characteristics, mainly textural, for green-type processing. On the other hand, the Spanish Verdial cultivar, first exported to California and, later on, from there to Australia where they were cultivated under irrigation, acquired in Australia some positive new characteristics, such as a larger size, better flesh-to-pit ratio and a more delicate texture that have never been seen in the original cultivar. These examples emphasize the importance for the scientist and technologist to study and select the most appropriate cultivar *in situ*.

3.4 REFERENCES

Abela V. and Fernades Serrano, J.M. (1987) Adaptación y comportamiento de siete variedades de olivo en la finca estatal de Abobada. *Olivae*, **18**, 11–19.

Aggabio, M. and Mulas, M. (1987) Le olive da mensa delle Sardegna: recupero e valoratione di cultivar locali. *Rivista di Frutticultura*, **XLIX**, 37–44.

Aligizakis, E.M. (1982) *Processing of Table Olives* Athens (in Greek).

Balatsouras, G.D. (1967) Processing the naturally ripe (black) olives, in *Proceedings of the International Olive Oil Seminar*, Perugia-Spoleto, Italy, pp 491–550.

Balatsouras, G.D. (1984a) The olive tree, in *Contemporary Oleiculture*, Vol I, Mairomatic & Co, Athens (in Greek).

Balatsouras, G.D. (1984b) *L'Oleiculture en Algerie. Problems et perspectives*, FAO, Athens, Greece.

Baldini, E., Carritano, D., Ciccarene, A., Crescimanno, F.G., Jacobeni, N., Martelli, G., Morettini, A., Parisi, A., Pisani, P.L., Savastano, G., Scaramuzzi, F. and Zito, F. (1963) *Olive da tavola*. Edagricole, Bologna, Italy.

Barranco Navero, D. and Rallo Romero, L. (1984) *Variedades de olivo cultivadas en Andalucía*. Junta de Andalucía, Consejería de Agricultura, Pesca y Alimentación, Sevilla, Spain.

Barranco Navero, D. and Rallo Romero, L. (1985) Las variedades de olivo cultivadas en España. *Olivae*, **9**, 16–20.

Benkeblia, N. (1992) Physico-chemical and biochemical characteristics of the Segoise table olive variety in different preparations. *Olivae*, **44**, 32–3.

Caran, D. (1989) Olive harvesting in Turkey. *Olivae*, **27**, 19–22.

Cucurachi, A. (1977) La oliva di tavola in Italia: produzione, transformazione e normativa. *Riv. Ital. Sost. Grasse*, **LIV**, 259–70.

Emerson, L.P.B. Jr (1980) *Mexico's Expanding Olive Industry*. Foreign Agricultural Service, USDA, U-249, pp. 1–14.

Fady, C. (1960) L'Olive de table en France. *Inf. Oleic. Int.*, **9**, 18–36.

Fernández Díez, M.J. (1969a) Elaboración de aceitunas de mesa en Grecia. *Grasas y Aceites*, **20**, 12–24.

Fernández Díez, M.J. (1969b) Table olive production in California. *Grasas y Aceites*, **20**, 194–205.

Fernández Díez, M.J. (1972a) Elaboración de aceitunas de mesa en Turquía. *Grasas y Aceites*, **23**, 138–45.

Fernández Díez, M.J. (1972b) Table olive processing in Algeria. *Grasas y Aceites*, **23**, 248–53

Fernández Díez, M.J. (1972c) Table olive processing in Australia. *Grasas y Aceites*, **23**, 40–4.

Fernández Díez, M.J. (1973) Table olive processing in Chile. *Grasas y Aceites*, **24**, 101–5.

Fernández Díez, M.J., Castro Ramos, R., Garrido Fernández, A., González Cancho, F., González Pellissó, F., Nosti Vega, M., Heredia Moreno, A., Mínguez Mosquera, M.I., Rejano Navarro, L., Durán Quintana, M.C., Sánchez Roldán, F., García García, P. and Castro Gómez Millán, A. (1985) *Biotecnología de la aceituna de mesa*. Consejo Superior de Investigaciones Científicas-Instituto de la Grasa y sus Derivados, Madrid-Sevilla, Spain.

Foytik, J. (1960) *California olive industry. Trends and outlook*. Calif. Agric. Expt. Stn. Circular 493.

Garoglio, P.J. (1950) El aceite de oliva y su industria, in *Tecnolgia de los aceites vegetales*. Vol II. Ministerio de Educación, Universidad de Cuyo, Mendoza, Argentina.

Giametta, G. (1985) Mechanical harvesting of table olives in Algeria. *Olivae*, **9**, 23–35.

Glickson, D. (1960) Les olives de table en Israel. *Inf. Oleic. Int.*, **10**, 24–31.

Hartmann, H.T. (1953) *Olive production in California*. Calif. Agric. Expt. Stn. Manual 7.

Hartmann, H.T. and Opitz, K.W. (1966) *Olive production in California*. Calif. Agric. Expt. Stn. Circular 540.

Hartmann, H.T, Opitz, K.W. and Bentel, J.A. (1986) Olive production in California. *Olivae*, **11**, 24–66.

Impey, L.H. (1974) *A handbook for South African olive growers* The S.A. Olive Growers Association, Paarl, S. Africa.

Kachouri, M. (1994) Table olives in Tunisia. *Olivae*, **53**, 7–10.
Karabij, M. (1992) The olive in Syria. *Olivae*, **43**, 19–20.
Lavee, S. (1978) Kadesh Table Olive. *HortScience*, **13**, 62.
Long, J. and Bonet, P. (1951) *L'Olivier a fruit de table*. Ministère de L'Agriculture, Montrouge, Seine, France.
Loussert, R. (1989) La oleicultura marroquí. Situación actual y perspectivas futuras. *Olivae*, **25**, 8–12.
Loussert, R. and Brousse, G. (1978) *L'Olivier*. G. P. Maisonneuve et Larose, Paris.
Marsilio, V. (1994) Production, processing and legislation of table olives in Italy. *Olivae*, **49**, 6–16.
Modun, E. (1960) L'Oliviculture en Yugoslavie. *Inf. Oleic. Int.*, **12**, 25–8.
Molinari, O. Ch. and Nicolea, H.G. (1947) *Tratado general de olivicultura*. El Ateneo, Buenos Aires, Argentina.
Montero Marín, M. (1972) Expériences sur divers types de preparation d'olives de table au Portugal. *Inf. Oleic. Int.*, **56–57**, 199–208.
Ortega Nieto, J.M. (1963) *Las variedades del olivo cultivadas en España*. Ministerio de Agricultura, Madrid.
Patac de las Traviesas, L., Cadahía Cicuendez, P. and Campo Sánchez, E. del. (1954) *Tratado de olivicultura*. Sindicato Nacional del Olivo, Madrid.
Salvarredi, E.M. (1987) La olivicultura en la República Argentina. *Olivae*, **18**, 20–7.
Skarika, B. (1989) La olivicultura in Yugoslavia: situación actual y perspectiva. *Olivae*, **29**, 14–25.
Sutter, E. (1994) Olive cultivars and propagation, in *Olive Production Manual* (eds L. Ferguson, G.S. Sibbett and G.C. Martin), University of California, Division of Agriculture and Natural Resources, Oakland, CA, USA, Publication no. 3353, pp. 23–31.
Tapia, I. and Bastias, M.F. (1987) Parámetros productivos del olivo (*Olea europaea*, L.) en el valle de Azapa, I Región (Chile), durante la temporada 1983–84. *Olivae*, **15**, 8–13.
Trigui, A. (1993) *Le Secteur oléicole: potential, production et evolution*. Institut de Olivier, Sfax, Tunisia.
Tunalioglu, R. (1994) General overview of the table olive economy in Turkey. *Olivae*, **50**, 18–20.
Yenicesu, I. (1965) L'Agronomie oleicole en Turquie. *Inf. Oleic. Int.*, **29**, 49–55.

World production, international trade and consumption of table olives

4

4.1 CHARACTERISTICS OF TABLE OLIVE PRODUCTION

There are two distinctive features which distinguish table olive production from that of most other pickled vegetables. The first of these is the long lead time before the effect of new cultural practices on production and quality can be assessed. It takes from 4 to 10 years before an olive tree becomes productive and this is a serious impediment to rapid innovation. Secondly, the pronounced variability of olive harvests gives rise to market instability. This was summed up by the International Olive Oil Council which stated: 'the essential feature of olive production lies in the irregularity and cyclic nature of olive harvests and market supply, which results in fluctuations in the value of production, instability of prices and export earnings and considerable ups and downs in producers' income'. These two problems have affected the table olive industry for centuries and have been serious obstacles to its development, both at the national and international level.

4.2 INTERNATIONAL AGREEMENT ON OLIVE OIL AND TABLE OLIVES

It was just after the Second World War that the first international measures to improve and develop the olive industry were introduced. This took the form of an International Agreement on Olive Oil and Table Olives within whose framework a number of actions were undertaken in the areas of international cooperation, modernization of cultivation practices, expansion and standardization of international trade, as well as a series of consumer activities. Like all the international commodity agreements, this was legally underpinned by principles recommended by the

United Nations Economic and Social Council. These were set out in Chapter VI of the Havana Charter negotiated in 1948 at the United Nations Conference on Trade and Employment which had been convened specifically to adopt principles encouraging the expansion of international trade.

The olive agreement had a number of objectives: to improve production of olives, olive oil and table olives; to develop international cooperation in this area; to regularize and establish rules for international trade; and to expand the consumption of table olives and olive oil.

This was the first of a series of long-term international commodity agreements concluded under the auspices of the United Nations with the aim of bringing together producers and consumers to solve problems within their respective sectors. This would lead to greater market transparency, better quality product and higher consumption levels. In the case of olives, the Agreement is administered by the International Olive Oil Council (IOOC).

The role played by the IOOC during the past 35 years has been central to the development of olive cultivation and processing. In particular, its work has led to designations and definitions of olive oils and table olives which have been the foundations for international rules and regulations that have contributed to preventing confusion or fraud in domestic and international trade. Now commerce is more transparent and buyers and consumers can be confident that they know exactly what they are purchasing or eating.

A further development that has facilitated international trade has been the development of standards for olive oil and table olives. In this, the IOOC cooperated with the Joint FAO/WHO Codex Alimentarius Commission to develop the Codex standard for table olives, as well as producing its own trade standard, the Unified Quality Standard Applying to Table Olives in International Trade.

From the point of view of trading practices, the IOOC offers Standard International Contracts for transactions, the International Conciliation and Arbitration Board and an international accreditation programme for testing laboratories.

Collectively these efforts have contributed substantially to achieving a secure framework for expansion of the national and international trade in table olives, avoiding disputes arising from people using differing standards, methods of analysis or quality control systems.

Finally, the IOOC has also developed a substantial database of statistics on national and world trade in olive products. It contains detailed data on aspects such as table olive production, exports and consumption over the period of the Institution's work and is the source of the figures cited below.

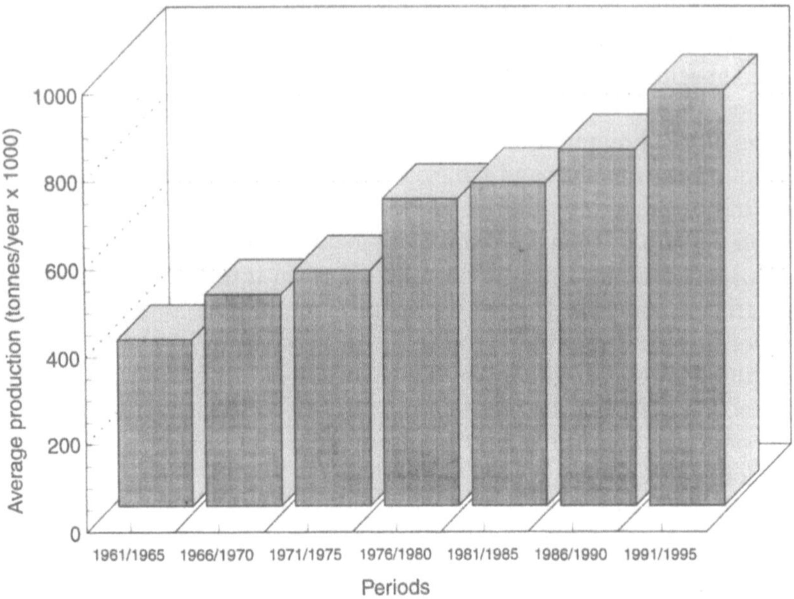

Fig. 4.1 Average production of table olives during the past 35 years using, with the exception of the last datum point, a 5-year average. (From Anonymous, 1994b; International Olive Oil Council, 1960–1990.)

4.3 TABLE OLIVE WORLD PRODUCTION

Table olives are currently the most important fermented vegetable product in the developed world (Buckenhüskes *et al.*, 1990). Estimated world production for the 1994/1995 season is in excess of 900 000 tonnes (Anonymous, 1994b).

The evolution of production is best studied by looking at the means of certain time periods to mitigate the effects of the cyclical nature of production. Figure 4.1 shows such data over the past 35 years using, with the exception of the last datum, a 5-year average (Anonymous, 1991a,b, 1992a,b, 1993a,b, 1994a,b). The data for the 1993/1994 and 1994/1995 seasons are based on a provisional figure and a forecast, respectively (Anonymous, 1994a,b). These figures indicate that production has increased almost linearly over the whole period to reach a current level that is more than double that of the early 1960s.

A number of factors have been cited as contributing to this observed expansion (Fernández Díez, 1985). For example, increased competition from seed oils has led to the progressive utilization of traditional oil-producing cultivars for table olive production in the main producing countries, particularly Spain. Producing countries are also increasingly diversifying into new styles so that Spain, in the last 20 years, has begun to prepare naturally black olives and canned black-ripe olives, and Greece is now producing green olives in addition to the traditional naturally black olives in brine. Developing countries have also expanded production to secure an increased share of an international trade traditionally dominated by others. This has been aided by the rapid transfer to developing countries of the benefits of recent scientific research leading to improvements in processing and preservation methods. In this, the IOOC has played an important role by supporting numerous training courses employing highly qualified specialists as instructors.

Fortunately, this increased production has been paralleled by a corresponding increase in consumption, although this is, in the main, a result of population growth since per capita intake has remained stable.

On this basis and projecting the trend shown over the past three decades, an average annual increment between 2 to 3% in production and consumption can be anticipated for the foreseeable future. This could however be increased by improving quality and diversifying into new value-added products and presentation forms. Primarily this represents a challenge to food technologists, but the IOOC is also developing the important task of promotion of table olives in existing markets and in regions with less tradition of olive consumption. Its activities through information leaflets and campaigns on television, in the press and other media focusing on the consumer appeal of olives as an attractive and nutritive food product, will further help to increase table olive consumption.

However, there are some problems arising in developed countries which may present a serious obstacle to further expansion of table olive production. The processing of table olives generally produces only a moderate volume of wastewater, but this does however have a high pollution potential. Environmental protection guidelines in the European Union (EU), the United States of America, and many other states are becoming progressively more strict and table olive industries are facing increasing problems with regulatory agencies. Since treatment of these wastes is difficult and costly, the way forward will depend on the development of new, less-polluting processes or cost-efficient wastewater treatment systems.

Producers in developed countries such as Spain and the USA also have the problem of high production costs which are making them less competitive in world markets than producers in North Africa and in Central and South America, where labour costs are lower.

Lastly, recent evidence of the health-promoting effects of olive oil has led to an increasing market demand for oil and a consequent price increase. Thus, olive fruits from dual-use cultivars may now be more profitably dedicated to oil extraction; reversing the tendency observed over recent years which led to increased table olive production.

At present these problems are affecting a limited number of producing countries, mainly from the EU and the USA, where production may be static in the near future. They are, however, considerably less important elsewhere so that global production may, as the IOOC forecasts, continue to increase.

4.4 DISTRIBUTION OF THE PRODUCTION OF DIFFERENT TYPES OF OLIVES

Available data on the distribution of table olive production, according to the three types described in the International Standards (1980), are summarized in Figure 4.2, expressed as percentages of total production (IOOC, 1960 to 1990). The figures represent three, 5-year averages and the 1992/1993 season (Anonymous, 1994a,b), since data from the last two years of the 1991/1995 period are not available at the time of writing. A full interpretation of these figures would have required a classification

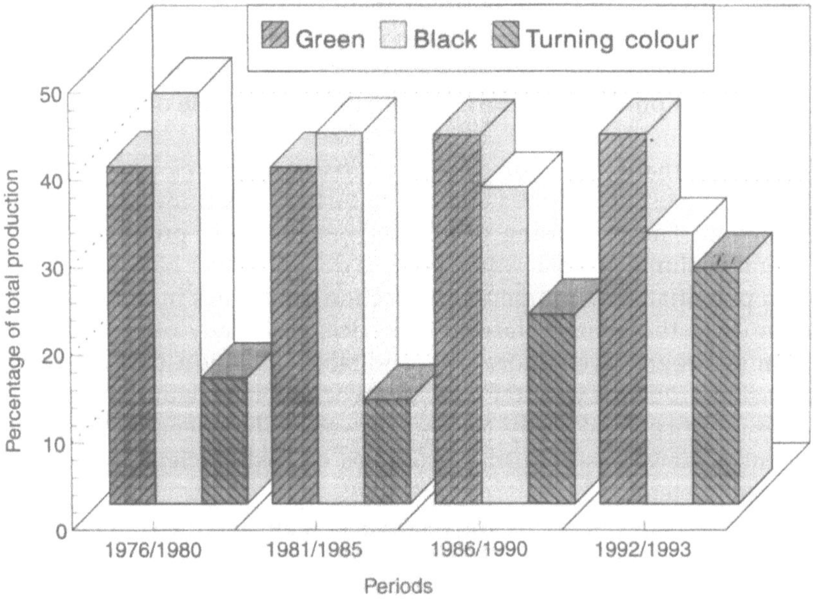

Fig. 4.2 Distribution of the total world olive production into the three main types. Black olives include mainly naturally black olives; turning colour olives represent almost exclusively canned Californian ripe olives. (From Anonymous, 1994b; International Olive Oil Council, 1960–1990.)

into four types; green, turning colour, naturally black and ripe-black (darkened by oxidation), but unfortunately such distinctions are not made in the IOOC data. However, a reasonable approximation to apply to these figures is that most of the green olives are Spanish-style pickled green olives in brine; that the majority of black olives in brine and in dry salt are 'untreated' or naturally black olives; and that the destination of most of the turning colour olives is for processing as 'treated' black-ripe olives in brine, prepared by alkaline oxidation.

As is evident from the figures, most table olive production was originally based on naturally black olives, followed by green olives. During the 1976/1990 period there was a decrease in the proportion of naturally black olives and a slight increase of green olives. However, the most outstanding changes have occurred with turning colour olives whose percentage has increased markedly from 1986. This tendency has continued into the 1990s, as figures from the 1992/1993 season demonstrate. As will be seen later, this position is largely due to the increasing production of this type in countries such as Spain and Morocco.

The decrease in production of naturally black olives arises from problems associated with the late picking required for this product. A late harvest increases the risk of shrivelling due to winter frosts and does not favour good following crops. Also, because of the higher content of soluble organic compounds in mature fruits, there are greater weight losses during fermentation and brining and an increased tendency to defects such as gas pocket formation within the fruit.

Thus, these most recently observed changes are forecast to continue.

4.5 PRODUCTION OF OLIVES IN DIFFERENT COUNTRIES

Table 4.1 shows the contribution of different countries to world production over the past 20 years (IOOC, 1960 to 1990; Anonymous, 1994a,b). Relative contributions are rather heterogeneous and show a concentration of production in a few countries. The first 10 countries listed account for about 90% of the world crop, Spain, Turkey, USA, Greece, and Italy being the most important.

The distribution of the different types of olives is shown in Tables 4.2–4.4. Spain is the main producer of green olives, although data in Table 4.2 clearly demonstrate a slight, but steady diminution of its contribution during the period. This is due to increased production in other countries in response to increased consumer demand rather than to a decrease in Spanish production.

In the case of naturally black olives in brine and in dry salt, Turkey, Greece and Italy account for most of the world production. Turkey, the largest individual producer, does however seem to have a decreasing dominance in this area.

Table 4.1 Average production of table olives in different countries, expressed as percentages of the total production

	Periods			Season
Country	1976/1980	1981/1985	1986/1990	1992/1993
Spain	23.4	23.6	25.1	22.2
Turkey	17.1	15.5	12.9	10.0
Greece	9.2	11.3	9.7	6.0
USA	9.6	8.8	9.5	14.7
Italy	10.3	9.0	8.1	7.0
Morocco	6.9	6.1	8.0	8.0
Syria	4.2	5.0	7.2	8.0
Argentina	4.1	4.5	4.0	3.3
Portugal	2.8	2.8	2.4	1.6
Peru	1.7	2.4	2.1	1.1
Egypt	1.1	0.9	1.7	6.0
Mexico	1.2	1.4	1.5	1.0
Jordan	1.2	1.2	1.2	1.2
Israel	1.1	1.8	1.2	1.2
Tunisia	1.1	1.4	1.1	1.3
Algeria	1.0	0.9	0.9	1.5

Other countries normally producing less than 1.0% are: Lebanon, Chile, Cyprus, France, Australia, Libya, Brazil, former Yugoslavia and South Africa, in decreasing order.

Table 4.2 Average production of green olives in different countries, expressed as percentages of the total

	Periods			Season
Country	1976/1980	1981/1985	1986/1990	1992/1993
Spain	49.2	43.9	40.9	34.4
Syria	6.0	7.8	11.6	11.3
Argentina	9.4	10.2	8.1	6.6
Morocco	5.5	4.5	7.1	7.1
Greece	2.9	5.7	6.2	5.0
Italy	6.9	5.7	4.3	3.3
Israel	2.3	3.2	4.0	3.0
Turkey	1.9	2.2	3.6	3.0
Mexico	2.7	3.5	3.5	2.2
Portugal	2.9	2.9	2.1	1.4
Egypt	1.2	0.9	1.6	11.7
USA	2.1	1.3	1.0	0.0
Tunisia	0.9	0.9	1.0	2.1
Jordan	1.0	1.0	0.9	1.2
Algeria	0.9	0.9	0.9	1.1

Other countries normally producing less than 1.0% are: Chile, Peru, Australia, France, Brazil, Cyprus, Lebanon, Libya, former Yugoslavia and South Africa, in decreasing order.

Table 4.3 Average production of naturally black olives in brine and in dry salt in different countries, expressed as percentages of the total

	Periods			Season
Country	1976/1980	1981/1985	1986/1990	1992/1993
Turkey	32.4	32.1	25.8	25.9
Greece	15.9	19.2	19.2	10.6
Italy	16.2	15.6	14.6	16.7
Morocco	7.5	4.6	7.7	12.7
Syria	4.1	4.4	6.7	7.7
Egypt	1.4	1.2	3.9	1.0
Portugal	3.4	3.7	3.5	3.2
Spain	4.9	2.2	3.3	3.5
Peru	3.3	5.1	3.2	2.2
Jordan	1.7	2.0	2.3	2.4
Lebanon	1.3	1.4	1.6	1.6
Tunisia	1.2	1.8	1.3	0.6
Argentina	1.1	1.3	1.2	1.6

Other countries normally producing less than 1.0% are: Israel, Cyprus, Chile, Libya, Algeria, France, Yugoslavia, Australia, USA and Mexico, in decreasing order.

Table 4.4 Average production of turning colour olives (including black-ripe olives) in different countries, expressed as percentages of the total

	Periods			Season
Country	1976/1980	1981/1985	1986/1990	1992/1993
USA	60.9	43.4	46.8	54.8
Spain	14.1	29.8	28.6	24.7
Morocco	7.5	12.8	7.4	3.7
Turkey	8.3	4.4	3.8	2.3
Syria	0.0	0.7	3.8	2.3
Greece	4.2	4.1	3.2	2.2
Peru	1.0	0.6	3.0	1.3
Italy	0.0	0.0	1.4	1.5
Tunisia	1.5	1.4	0.9	0.4

Other countries normally producing less than 1.0% are: Portugal, Chile, Algeria, Mexico, Israel, Lebanon and France, in order of decreasing importance.

The USA has maintained a predominance in the production of turning colour olives due to its canned black-ripe olives, oxidized in an alkaline medium. Nonetheless, its relative contribution has decreased over the years, mainly as a result of increased production in Spain. Other countries where production of black-ripe olives is increasing are Morocco, Turkey, and Syria, who are currently building modern factories for processing this style.

Table 4.5 Average annual exports of table olives (quantities in tonnes × 1000/year and %) and their distribution among the three main types

	Periods						Season	
Country	1976/1980		1981/1985		1986/1990		1992/1993	
Olive types	Quantity	%	Quantity	%	Quantity	%	Quantity	%
Green olives	110	57.6	116	52.7	133	52.6	109	53.4
Black olives in brine	54	28.3	56	25.5	72*	28.5	55*	26.6
Black olives in dry salt	13	6.8	6	2.7	–	–	–	–
Turning colour olives†	14	7.3	42	19.1	48	18.9	41	20.0

* Black olives in brine and in dry salt, in total.

† Including black-ripe olives.

4.6 EXPORTS

Table 4.5 shows average total exports of table olives, as well as their distribution according to the different types of olive (IOOC, 1960 to 1990; Anonymous, 1994a,b). Absolute values for green olives, mainly Spanish-style pickled green olives in brine, are generally increasing, except for the decrease in the 1992/1993 season caused by a crop shortfall in Spain. As a percentage of the whole export market though, its share remains relatively constant with perhaps even a slight decrease.

The same is true of black olives, which are mainly exported as Greek-style naturally black olives in brine. In contrast, there is a relatively pronounced increase in the export of turning colour olives during the past 15 years due to the general tendency already noted for increased production and consumption of canned black-ripe olives (Californian-style) in brine. Export figures represent around 27% of the total export market for the period 1976/1980, 29.9% for 1981/1985, 31% for 1986/1990, and about 23% for the 1992/1993 season.

The total amount of olive production dedicated to international trade is about 200 000 tonnes; most of this (c. 150 000 tonnes) is exported to other producing countries to supply their own domestic markets. Only about 25% is exported to non-producing countries. This distribution means that there are still considerable possibilities for expanding international trade by increasing consumption in countries where olives are not grown. However, export data for the last quinquennium (Figure 4.3) show the

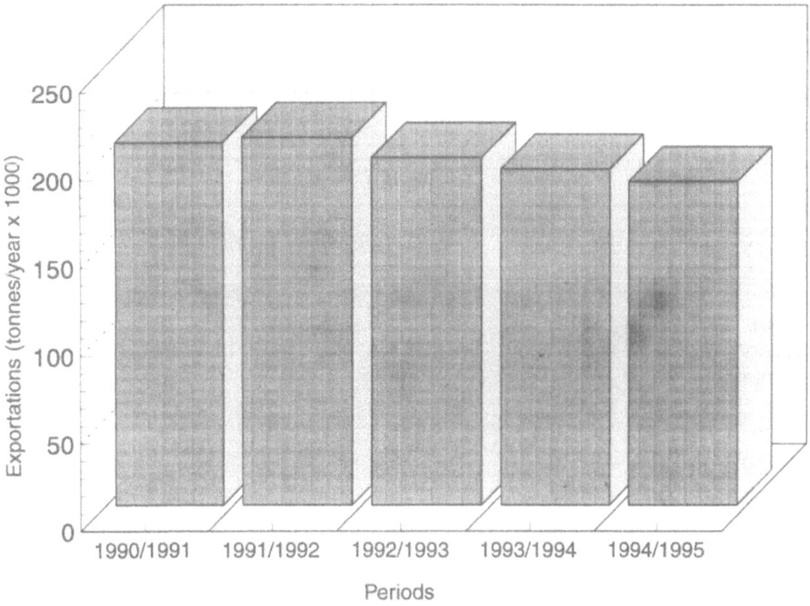

Fig. 4.3 Total table olive exportations during the past 5 years. Data concerning season 1994–1995 are based on forecasts. (From Anonymous, 1994b.)

disturbing trend of a steady decrease since the 1991/1992 season (Anonymous, 1991a,b to 1994a,b). This may be explained by increased consumption in the domestic markets of some producer countries, but nonetheless reflects a reduction in the international trade which should be reversed.

The main table olive exporting countries are shown in Table 4.6 (IOOC, 1960 to 1990). Spain, responsible for about half of the international trade, has become increasingly dominant over the years. Together with Greece, whose market share has decreased slightly, Morocco, which has heavily promoted the export of olives, and Argentina, it accounts for about 90% of the total export trade. Forecasts for the 1994/1995 season show a similar pattern, although the market share of Moroccan exports continues to increase (Figure 4.4). The number of exporting countries also appears to be decreasing, demonstrating perhaps that production is still mainly dedicated to domestic consumption in most olive growing areas.

The figures for different types of olives, presented in Tables 4.7–4.9 (IOOC, 1960 to 1990), underline the importance of Spain, Argentina, Greece and Morocco, for green olives; Greece, Morocco and Turkey, for naturally black olives; and Spain and Morocco for turning colour olives, which, in the case of Spain, are exclusively exported as oxidized olives. Figures for the EU as a whole illustrate its significance within the international trade of table olives, since EU exports for green, turning colour, and black olives account for 92.7, 56.0 and 22.0% respectively of the world total.

Table 4.6 Exporting countries for table olives, expressed as percentages of the total exports

	Period			Season
Country	1976/1980	1981/1985	1986/1990	1992/1993
Spain	43.0	46.8	50.6	68.7
Greece	21.5	23.0	19.1	17.1
Morocco	17.8	13.7	14.1	22.7
Argentina	9.4	8.9	7.3	8.8
Turkey	1.7	3.2	3.5	6.4
France	1.3	1.3	1.2	1.3
Portugal	1.0	1.3	1.2	1.3
Israel	0.4	1.3	1.0	1.0
USA	1.1	1.2	0.8	0.5
Italy	1.1	0.7	0.6	0.7
Algeria	1.3	0.5	0.1	0.0
Jordan	–	0.9	0.4	0.5
Tunisia	0.3	0.3	0.2	0.5

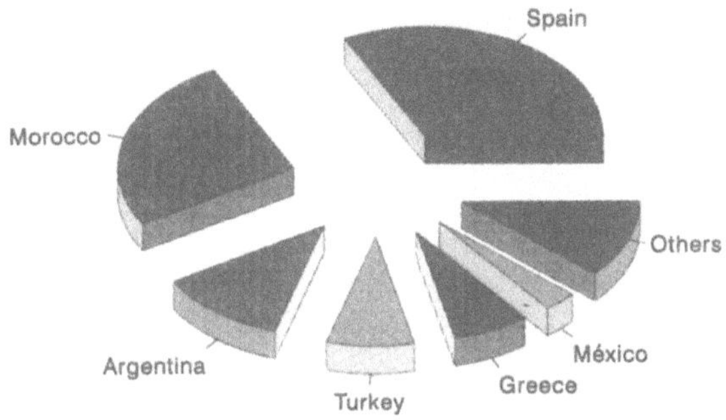

Fig. 4.4 Forecast of table olive exportations for the 1994–1995 season. (From Anonymous, 1994b.)

Table 4.7 Exporting countries for green table olives, expressed as percentages of the total

	Period			Season
Country	1976/1980	1981/1985	1986/1990	1992/1993
Spain	65.6	64.7	59.7	*
Argentina	13.2	14.1	12.0	14.7
Greece	7.3	10.6	13.3	14.7
Morocco	9.1	7.9	9.4	16.5
Israel	0.6	0.9	1.4	1.8
France	1.4	1.3	1.1	*
Italy	1.0	1.1	0.8	*
Portugal	0.7	0.6	0.6	*
Turkey	–	0.6	0.7	1.0
Jordan	–	0.4	0.5	0.5
Algeria	0.4	0.1	0.4	0.0
Tunisia	0.1	0.3	0.3	0.9
Germany	–	0.2	0.1	*
Netherlands	–	0.1	0.1	*

* Statistics of these countries are included in those of the European Union in these seasons. Green olive exports from the EU represent about 56%, mainly from Spain.

Table 4.8 Exporting countries for naturally black olives in brine and in dry salt, expressed as percentage of total exports

	Period			Season
Country	1976/1980	1981/1985	1986/1990	1992/1993
Greece	47.0	55.1	52.8	*
Morocco	27.4	14.1	26.7	48.7
Turkey	5.0	9.8	9.1	22.2
Argentina	4.8	4.9	3.1	3.7
Spain	9.1	6.1	2.2	*
Portugal	1.5	3.3	2.0	*
France	1.4	1.9	1.6	*
Jordan	0.4	2.1	1.1	0.9
Italy	0.5	0.5	0.8	*
Israel	–	0.5	0.4	0.0
Lebanon	–	0.1	0.1	1.8
Tunisia	0.2	0.4	0.1	0.0
Algeria	2.9	1.3	–	0.0
Cyprus	0.3	0.2	–	0.0

* Statistics for these countries were included in those for the EU. EU exports accounted for 22.2%, mainly from Greece.

Table 4.9 Exporting countries for turning colour table olives (including black-ripe olives), expressed as percentages of total exports

	Period			Season
Country	1976/1980	1981/1985	1986/1990	1992/1993
Spain	23.1	56.6	78.6	*
Morocco	35.1	27.3	13.6	4.5
USA	18.6	6.1	4.3	2.4
Greece	16.0	5.8	2.6	*
Turkey	–	0.7	0.4	0.0
France	0.1	0.2	0.2	*
Israel	1.1	2.9	0.2	0.0
Tunisia	4.2	0.4	0.1	0.0
Portugal	1.6	–	–	*
Algeria	0.2	–	–	0.0

* Statistics of these countries were included in those for the EU, which accounted for the 92.7% during this season, mainly from Spain.

4.7 IMPORTS

Table olive-importing countries are distributed according to the figures shown in Table 4.10 (IOOC, 1960 to 1990). The USA, Italy, France and Brazil are the most important, representing around 70% of the total. Canada, Romania, Germany, Bulgaria and Libya share another 20%. The rest is distributed in small proportions among all the other countries.

Available data differentiating between the types of olive are incomplete (Tables 4.11–4.13), especially for black and turning colour olives. However, the USA, Brazil, Italy and France are well-known importers of Spanish-style green olives in brine; the last three countries, with Bulgaria and Romania, are also major importers of Greek-style naturally black olives in brine. Italy and the USA are important markets for turning colour olives, in the case of the USA mainly processed as canned black-ripe olives (Californian-style).

Most United States imports come from Spain, as do those of Australia. Brazil imports mainly from Argentina and Portugal, France from Morocco and Bulgaria from Greece. Percentages of the different types imported indicate local preferences, reflecting the indigenous product which is produced at a level insufficient to meet demand, or in some cases the presence of a significant expatriate population from countries such as Greece.

Table 4.10 Importing countries for table olives, expressed as percentages of the total imports

	Period		
Country	1976/1980	1981/1985	1986/1990
USA	22.6	22.3	29.0
Italy	13.4	13.1	14.8
France	15.6	13.7	11.9
Brazil	10.4	19.5	9.0
Canada	5.8	4.8	4.8
Romania	4.4	4.8	5.1
Germany	3.8	4.6	4.0
Bulgaria	5.6	4.5	2.8
Libya	1.2	2.9	2.8
Former USSR	–	–	1.1
Egypt	0.6	1.5	0.9
United Kingdom	1.1	1.0	0.8
Australia	2.0	1.6	0.7
Former Yugoslavia	0.8	0.4	0.4

Other importer countries, in decreasing order of importance are: Belgium, Netherlands, Switzerland, Venezuela, Cyprus, Lebanon, Jordan, Israel, Denmark, Dominican Republic, Panama, and others (about 10%).

Table 4.11 Importing countries for green table olives, expressed as percentages of the total amount

	Period		
Country	1976/1980	1981/1985	1986/1990
Spain	34.6	36.3	46.0
Brazil	15.3	14.6	12.2
Italy	12.3	14.9	11.4
France	13.2	13.2	11.0
Germany	1.6	4.4	3.9
Libya	0.3	0.9	1.7
United Kingdom	0.5	1.1	0.5
Egypt	0.2	0.4	0.7
Lebanon	0.2	0.5	0.7
Australia	1.6	0.5	0.3

Other importing countries in decreasing order of importance are: Jordan, Netherlands, former Yugoslavia, Denmark, Israel, Belgium, Chile, Eire, Portugal, and others (10–20%).

Table 4.12 Importing countries for naturally black olives, expressed as percentages of the total imports

	Period		
Country	1976/1980	1981/1985	1986/1990
France	17.8	19.8	18.4
Italy	13.0	11.7	16.6
Brazil	4.6	7.0	8.1
Libya	2.9	7.7	6.7
Germany	1.9	6.5	6.3
USA	1.6	2.6	5.6
Egypt	1.6	4.2	2.1
Australia	2.7	4.2	2.0
Cyprus	1.0	1.1	1.2
Former Yugoslavia	1.5	1.1	1.1
United Kingdom	0.5	1.2	1.1
Israel	1.0	1.1	0.3
Jordan	1.6	0.4	0.1

Other importing countries in decreasing order of importance are: Belgium, Netherlands, Denmark, Lebanon, and others (30–50%).

The most popular presentation forms for exports are, in decreasing order of importance, Spanish-style green olives stuffed with pimento, anchovy, almond and onion, and whole, pitted, and sliced olives in the case of black-ripe olives. Natural olives are generally exported as whole olives (plain).

Table 4.13 Importing countries for turning colour olives (including black-ripe olives), expressed as percentages of the total amount

	Period		
Country	1976/1980	1981/1985	1986/1990
Italy	9.5	9.1	13.9
USA	10.7	6.9	12.1
France	21.0	0.9	3.1
Germany	0.4	0.8	2.3
Netherlands	–	–	0.5
United Kingdom	–	–	0.5
Belgium	–	–	0.5
Denmark	–	–	0.1
Other countries	58.4	82.3	67.0

4.8 CONSUMPTION

The relative importance of the various olive-consuming countries is shown in Table 4.14 in terms of their share of world consumption (IOOC, 1960 to 1990). This is broken down into the different types of table olives in Tables 4.15–4.17 (IOOC, 1960 to 1990). In general, producing countries use about 600 000 tonnes per annum. Other countries produce olives themselves but in insufficient quantities to meet home demand and so have to import around 200 000 tonnes. Consumption in countries which do not produce olives is approximately 50 000 tonnes.

With the exception of France, the main consumers are also the most important producers, although not in the same order of relative importance due to differences in population, customs and technology.

The USA is the fourth most important olive producer but is the largest consumer (Table 4.14), mainly because of its large population. The table olive is a king of fancy food, ubiquitous as an accompaniment to drinks and in salads, pizzas and Mediterranean-style meals. It is highly appreciated both for its sensory characteristics and its nutritive and biological value. Consumers show a general preference for Californian-style black-ripe olives, which are the main indigenous product (Table 4.17), followed by Spanish-style green olives, mainly imported from Spain (Table 4.15). Total annual olive consumption in the USA is 180 000 tonnes, which may account for about 20% of total world production.

Italy is the fifth largest producer. Its relatively large population and its important tourist industry probably account for it being the second largest consumer of olives as an appetizer and as a complement for Italian food (Table 4.14). Italian people generally prefer naturally black olives (Table 4.16) and green olives (Table 4.15).

Table 4.14 Consumer countries for table olives, expressed as percentages of total consumption

	Period		
Country	1976/1980	1981/1985	1986/1990
USA	14.7	15.5	18.3
Italy	13.5	12.8	12.9
Turkey	17.5	15.6	12.4
Spain	11.1	9.8	9.4
Syria	4.2	5.2	7.2
Greece	3.6	3.7	4.0
France	3.9	3.8	3.3
Morocco	2.3	2.3	2.9
Brazil	3.0	3.1	2.8
Peru	1.7	2.4	2.1
Portugal	2.6	2.4	2.0
Egypt	1.3	1.3	1.9
Argentina	2.1	1.9	1.8
Israel	1.1	1.4	1.6
Canada	1.4	1.4	1.4
Mexico	1.2	1.4	1.4
Romania	1.3	1.6	1.2
Germany	1.0	1.3	1.2
Jordan	1.4	1.1	1.1
Tunisia	1.0	1.3	1.1
Libya	0.6	1.1	1.0
Bulgaria	1.5	1.3	0.8

Other consumer countries in order of decreasing proportions are: Lebanon, Algeria, Chile, Cyprus, Australia, former USSR, former Yugoslavia, United Kingdom, Dominican Republic, Belgium, Netherlands, Switzerland, Denmark, etc.

Turkey, the second-largest producer of table olives, ranks third in terms of total consumption (Table 4.14), but first in terms of naturally black olives, either in brine or in dry salt (Table 4.16). The principal reasons for this are the relatively high population, but also the Turkish custom of using table olives as an important component in meals, especially breakfasts.

Spain, the foremost producing country, is the fourth largest consumer (Table 4.14). Not surprisingly, it shows a preference for Spanish-style green olives (Table 4.15), its own main product, followed by an increasing consumption of turning colour olives (Table 4.17), locally processed as black-ripe olives. This country also shares the Mediterranean tradition of using olives as an appetizer, a component in salads and in numerous other aspects of its cuisine.

Despite its relatively low population, Syria is currently the fifth most important consumer (Table 4.14), mainly of green (Table 4.15) and

Table 4.15 Consumer countries for green olives, expressed as percentages of total consumption

Country	Period		
	1976/1980	1981/1985	1986/1990
USA	16.3	16.4	18.8
Spain	21.2	18.0	14.5
Italy	11.2	12.1	12.7
Syria	6.0	8.0	10.1
Brazil	6.8	6.4	5.0
France	5.0	5.4	3.9
Argentina	4.6	4.6	3.8
Turkey	1.9	2.1	3.6
Mexico	2.8	3.5	3.4
Israel	2.1	2.7	3.0
Morocco	2.1	1.6	2.8
Portugal	2.7	2.7	2.0
Greece	0.5	1.3	1.6
Egypt	1.2	1.1	1.5
Germany	0.6	1.7	1.4

Other consumer countries in decreasing order of importance are: Algeria, Jordan, Tunisia, Chile, Libya, Peru, Lebanon, Australia, United Kingdom, Cyprus, Belgium, Netherlands, former Yugoslavia, Denmark and others (5–10%).

Table 4.16 Consumer countries for naturally black olives, expressed as percentages of the total consumption

Country	Period		
	1976/1980	1981/1985	1986/1990
Turkey	32.6	31.9	28.8
Italy	18.8	17.8	18.5
Greece	6.4	6.3	6.3
Syria	3.9	4.4	6.1
France	3.4	3.7	4.2
Peru	3.2	5.0	3.7
Egypt	1.7	2.1	3.4
Portugal	3.1	3.1	3.0
Morocco	2.2	1.8	2.1
Jordan	1.9	1.6	2.0
Libya	0.8	1.9	1.9
Brazil	0.9	1.4	1.8
Lebanon	1.4	1.4	1.5
Tunisia	1.2	1.7	1.4
Germany	0.4	1.3	1.4
Spain	2.9	1.2	1.5
USA	0.5	0.6	1.2

Other consumer countries in decreasing order of importance are: Cyprus, Israel, Chile, Australia, Argentina, Algeria, former Yugoslavia, United Kingdom, Belgium, Netherlands and others (5–10%).

Table 4.17 Consumer countries for turning colour olives (including black-ripe olives), expressed as percentages of the total consumption

	Period		
Country	1976/1980	1981/1985	1986/1990
USA	61.1	50.9	48.3
Spain	10.8	12.8	13.1
Turkey	8.8	4.8	4.1
Morocco	2.9	5.6	4.3
Italy	1.0	1.8	3.3
Greece	2.7	2.9	2.4
Syria	–	0.7	2.9
Peru	1.1	0.6	2.0
Tunisia	1.1	1.4	1.0
Algeria	0.3	0.5	1.0
France	2.6	0.3	0.5

Other consumer countries in proportion less than 1%, in decreasing order of importance, are: Chile, Portugal, Germany, Israel, Mexico, Lebanon and others (5–20%).

naturally black olives (Table 4.16). This reflects the central role of the olive in the Syrian diet, similar to that in Turkey.

Greece is both the third most important producer and consumer of naturally black olives (Tables 4.1 and 4.16), its main product. It has a relatively low population and table olives are used mainly as an appetizer and a complement to meals. Consumption of Spanish-style green olives has increased steadily in the past few years (Table 4.15). However, in terms of total olive consumption it ranks sixth, consuming only about 4% of world production.

France, with a substantial population and tourist industry as well as a Mediterranean-influenced diet, is an important consumer among the lesser-producing countries (Table 4.14), with a percentage similar to that of Greece. The presence of a significant population from former North African colonies may also be an important contributory factor to this.

Estimated *per capita* consumption during the past 5 years for which figures are available is extremely variable, being heavily influenced by differing customs and living standards in different countries (Table 4.18). Syria, probably the origin of the cultivated olive, has the highest rate of consumption. This is distributed among all the olive types, although green olives are predominant. This is rather surprising since the majority of countries in the region are more accustomed to naturally black olives either in brine or dry salt. Another outstanding *per capita* consumption figure is that of Cyprus, which, not surprisingly, is due mainly to naturally black olives.

Table 4.18 Average *per capita* consumption of the different types of table olives (kg/year) during the period 1986/1990

Country	Total	Green	Black	Turning*
Syria	6.3	3.8	1.9	0.6
Cyprus	6.5	1.6	4.2	0.7
Greece	3.3	0.6	2.3	0.4
Israel	3.3	2.6	0.7	–
Jordan	2.6	0.8	1.7	0.1
Libya	2.5	0.8	1.7	–
Turkey	2.0	0.2	1.6	0.2
Spain	2.0	1.3	0.1	0.6
Lebanon	1.9	0.5	1.3	0.1
Italy	1.8	0.7	0.9	0.2
Portugal	1.6	0.7	0.8	0.1
Tunisia	1.3	0.3	0.7	0.3
Morocco	1.1	0.4	0.3	0.4
Peru	0.9	0.1	0.6	0.2
USA	0.7	0.3	–	0.4
Bulgaria	0.7	–	0.7	–
Canada	0.5	0.3	–	0.2
France	0.5	0.2	0.2	0.1
Argentina	0.5	0.5	–	–
Romania	0.4	–	0.4	–
Chile	0.4	0.2	0.2	–
Egypt	0.4	0.1	0.3	–
Algeria	0.4	0.2	0.1	0.1
Switzerland	0.3	0.3	–	–
Brazil	0.2	0.1	0.1	–
Mexico	0.2	0.2	–	–
Australia	0.2	0.1	0.1	–
Belgium	0.2	0.1	0.1	–
Denmark	0.2	0.1	0.1	–
Former Yugoslavia	0.1	0.1	–	–
Germany	0.1	0.1	–	–

* Including black-ripe olives.

Greece (mainly naturally black olives) and Israel (treated green olives) have consumption rates about half those of Syria and Cyprus. Those of Jordan, Libya, Turkey and Spain are between 2 and 3 kg *per capita* and between 1 and 2 kg in the Lebanon, Italy, Portugal, Tunisia and Morocco.

Bearing in mind that the *per capita* consumption in most olive-producing countries is still low, there is still considerable scope for increasing consumption world-wide if the product is adequately promoted. The low consumption of turning colour olives in many countries is also notable, suggesting that considerable future expansion could also occur in this area.

4.9 REFERENCES

Anonymous (1991a) Situación y evolución del mercado internacional de las aceitunas de mesa. *Olivae*, **36**, 9–10.

Anonymous (1991b) Situación y evolución del mercado internacional de la aceituna de mesa. *Olivae*, **38**, 17–19.

Anonymous (1992a) Perspectivas a corto plazo del mercado internacional de las aceitunas de mesa. *Olivae*, **40**, 15–16.

Anonymous (1992b) El mercado internacional de las aceitunas de mesa. *Olivae*, **43**, 13–16.

Anonymous (1993a) Situación del mercado internacional de las aceitunas de mesa. *Olivae*, **45**, 11–13.

Anonymous (1993b) Situación y evolución del mercado internacional de la aceituna de mesa. *Olivae*, **48**, 18–20.

Anonymous (1994a) Situación y evolución del mercado internacional de la aceituna de mesa. *Olivae*, **50**, 18–20.

Anonymous (1994b) Situación y evolución del mercado internacional de la aceituna de mesa. *Olivae*, **54**, 18–21.

Buckenhüskes, H., Jensen, H.A., Adersson, R., Garrido Fernández, A. and Rodrigo, M. (1990) Fermented vegetables, in *Processing and Quality of Food, Vol. II, Avenues to Healthy and Nutritious Products* (eds P. Zeuten, J. C. Cheftel, C. Ericksson, T.R. Gormley, P. Linko and K. Paulus), Elsevier Applied Science, Barking, UK, pp. 2.162–2.187.

Fernández Díez, M.J. (1985) Table olives. Quality through research. *Olivae*, **9**, 11–15.

International Olive Oil Council (IOOC) (1960–1990) *Yearly reports on production of table olives*, IOOC, Madrid.

International Olive Oil Council (IOOC) (1980) *Unified Qualitative Standard Applying to Table Olives in International Trade*, IOOC, Madrid.

Physical and chemical characteristics of the olive fruit

5

5.1 OLIVE FRUIT GROWTH

Only a fraction of the flowers, even those originally pollinated and fertilized, ultimately produce an olive fruit. This is due mainly to competition among flowers in each inflorescence rather than to pistil abortions. Factors affecting fruit set are diverse, including absence of viable pollen, incompatible pollinating cultivars and nitrogen availability. Environmental conditions are also important. High winds diminish fruit set while certain combinations of temperature, rain, etc. may increase it.

Absence of viable pollen results in parthenocarpic fruits. The proportion of these is variable among cultivars. They mature earlier but their final size is 10–20% that of normal olives. Some table olive cultivars may produce appreciable quantities of these fruits, known in Spain as 'azofairones' and as 'shootberries' in the USA, which may remain on the tree up to picking time. Cross-pollination seems to reduce their incidence.

There can be problems related to pollination. For example, in some cases, the pollen of Manzanilla cultivars grows slowly and does not fertilize the ovary which is only short-lived after full bloom. Although some cultivars are self-compatible, cross-pollination generally favours better crops but problems can sometimes arise with incompatibilities between cultivars.

The proportion of flowers yielding a fruit may range from 1–2% to around 10%, depending on whether there is a high or low inflorescence population that season. Usually only one fruit per inflorescence is retained. This reduction is achieved by successive abscissions in the 4–6 weeks after full bloom. Additional fruit abscissions may be produced by pests or adverse environmental conditions (Martín *et al.*, 1994).

Like most drupes, olive growth follows a double sigmoidal curve with three steps. During the first there is rapid cell multiplication leading to rapid fruit growth, with the pit almost reaching its normal volume. In the second, the embryo grows quickly and reaches its normal size. This period, which lasts from mid-July to August, is characterized by the complete lignification of the endocarp. The third phase, initiated immediately after (September) and lasting to full maturation, is mainly characterized by an increase in the size of cells. Normally the fruits reach their maximum size when they change their surface colour from slightly pink to purple pink or black. The initiation of the first phase appears to be influenced mainly by temperature, although factors such as humidity, crop volume and nitrogen status may also play a significant role.

The ripening process begins when the first small pink spots appear on the surface and lasts until the skin and flesh become purple or black, depending on the cultivar (Tous Marti, 1990).

5.2 TABLE OLIVE CROP CONTROL

Olive crops tend to be cyclical with the number of fruits on a tree varying from one year to another. When there is a large fruit set, the fruits may fail to reach a profitable size. Their excessive carbohydrate absorption represses shoot growth, which reduces the potential of the following year's crop. This alternation of the olive tree's natural productivity can cause difficulties for table olive processing and marketing as a result of market uncertainty and price fluctuations.

To guarantee farmers a regular return, it is therefore desirable to prepare olive trees to produce a regular crop volume of an appropriate fruit size which also permits enough shoot growth for the following year's crop. This can be achieved by appropriate crop husbandry, although the actual load per tree will depend on the cultivar, its age and environment.

Normally, the best method of crop control is fruit thinning; the removal of excess fruits (Maranto and Krueger, 1994). This is most commonly done chemically by the application of a solution of naphthalene acetic acid (NAA) 12–18 days after full bloom or when fruits have the appropriate size (3–5 mm). NAA appears to act by absorption through the leaves and fruits from where it is transported to the abscission zone of the fruit stems, stimulating olive drop.

NAA solution is applied only once a year and trees should be completely covered for the best results. During this operation, tree stress should be minimized and only healthy trees treated.

Obviously there is always a risk in this procedure since, at the time it is applied, the ultimate crop size cannot be estimated accurately. On some occasions it may lead to an excessive loss of fruit. High temperatures (around 40°C) tend to increase the effect and the number of fruits released, causing greater thinning than expected. Alternatively, insufficient thinning can result in more fruits than expected and a poor fruit size. Responses to chemical thinning vary between cultivars and from season to season and any treatment should only be applied after careful study of the specific conditions. If applied in excess, the chemical may cause complete removal of the tree's leaves. Some of these uncertainties have prevented the rapid adoption of chemical abscission by the farmers, despite its unquestionable advantages when applied correctly.

The traditional alternative to chemical thinning, pruning, can reduce the fruit wood and enhance shoot growth for the following season. However, it cannot distinguish between fruits and leaves and so removes a large number of leaves. This has a negative effect on the leaf : fruit ratio and reduces the fruit size, since this ratio must be as high as possible to obtain the best results. The optimum time for pruning is late spring or early summer when the crop load can easily be seen (Maranto and Krueger, 1994). In general, pruning is an expensive cultural practice and is not as efficient as chemical thinning in eliminating the alternate bearing cycle.

To summarize, alternate bearing can be considerably reduced by improving cultivation techniques, by the use of the appropriate fertilizers, irrigation, chemical thinning or pruning and good control of pests.

5.3 PHYSICAL PROPERTIES

The olive is a round to ovoid, fleshy stone fruit 2–3 cm long with a transverse diameter of between 1 and 2 cm (Fernández Díez, 1983). It can weigh between 0.5 and 20 g, most commonly in the range of 3 to 10 g and is intensely bitter, although this varies with the cultivar and stage of maturity. During ripening the fruit colour changes from green to purple-red or purple-black, depending on the cultivar and its pigment composition.

The pulp accounts for 70–90% of the weight of the fruit and the endocarp accounts for the remainder. The seed weighs less than 10% of the weight of the stone.

5.3.1 SPECIFIC WEIGHT

The specific weight (weight/unit volume) is close to 1.0. Its changes during maturation of the Conservolea Greek cultivar were followed by Aligisakis and Patroclos (1971) who found a slight decrease until the end of October, when it reached the minimum, after which it increased until the end of maturation (Figure 5.1). This decrease may influence the farmer's decision to harvest. The period during which the specific weight decreases coincides with the appropriate degrees of maturation for green and turning colour olive types, so farmers can be reluctant to use fruits for these preparations, arguing that they could obtain more profit from their crop by delaying the picking moment. The higher prices obtained for table olives frequently compensate for this loss however.

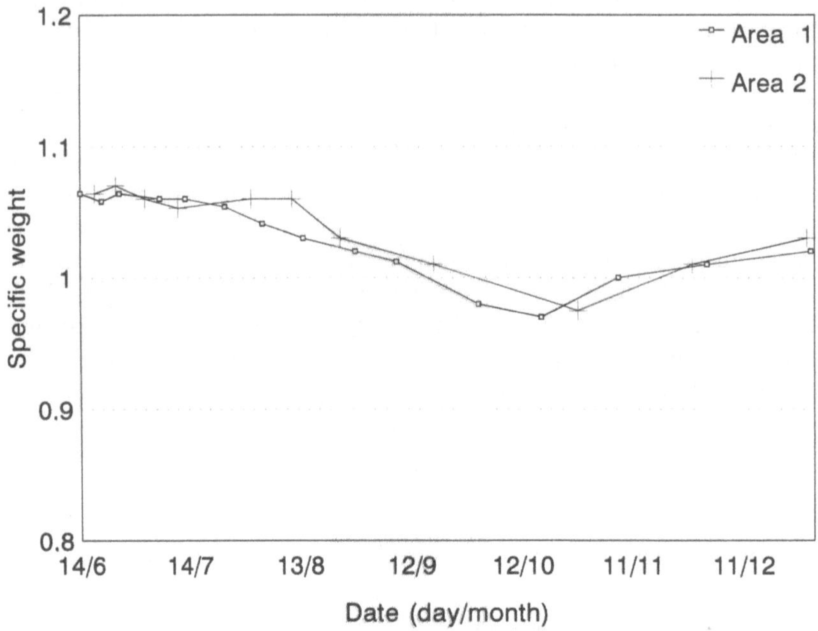

Fig. 5.1 Specific weight changes during maturation. Conservolea cultivar grown in two areas. (Redrawn from Aligisakis and Patroclos, 1971.)

Differences in specific weight before lye treatment have been proposed by Fernández Villasante and Fernández Díez (1968) as the basis of a method for classifying green olive raw material in both ungraded and size-graded fruits.

The specific weight of the pit is higher than that of the flesh and this difference is maintained throughout processing, brining and fermentation. It is used during the pitting and stuffing steps to separate pits and unpitted olives from correctly processed olives, which float in brines of adequate sodium chloride concentrations.

5.3.2 SIZE

Fruit size is another characteristic of importance in table olives. There is a rapid increase in fruit volume during July and August, which slows appreciably and progressively during the rest of the maturation period. Size changes have been followed by numerous authors (Hartman, 1949; Aligisakis and Patroclos, 1971; Garrido Fernández and Fernández Díez, 1972). Figure 5.2 shows one example of the average size changes for some Spanish table olive cultivars (Garrido Fernández and Fernández Díez, 1972). In general, during the period of interest for table olives, September

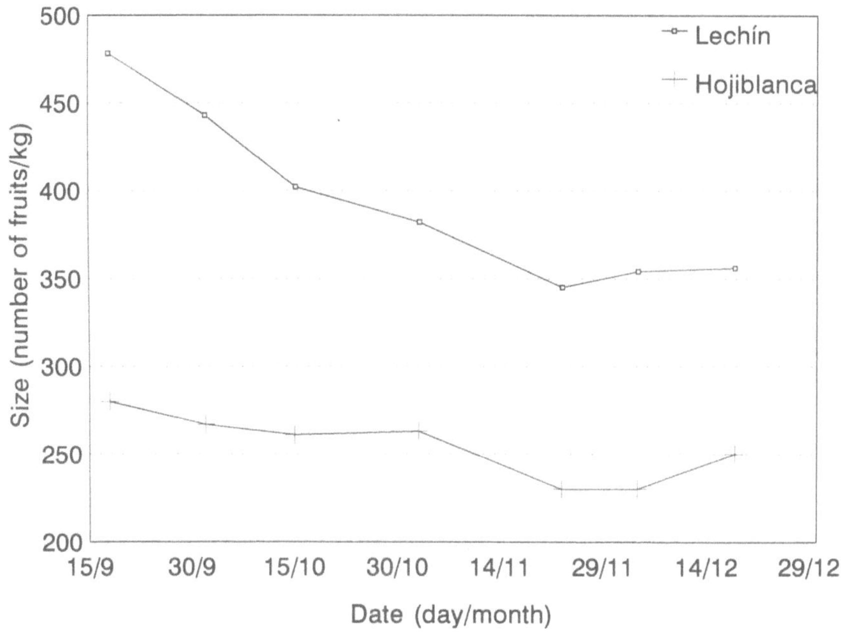

Fig. 5.2 Average fruit size changes during maturation of some Spanish table olive cultivars. (Redrawn from Garrido Fernández and Fernández Díez, 1972.)

to December, there is a slight increase in fruit size, as evidenced by the diminution of the number of fruits/kg. After this period, olive fruits may suffer the damage of winter frosts and the stress of low temperatures, which render fruits unsuited to processing, or can cause a slight decrease in fruit size (as in the case of Lechín, a frost-resistant cultivar). Excessive delay in harvesting may also adversely affect the tree's normal olive-bearing capacity and the following season's harvest. Unless special environmental conditions permit, table olives must, as a general rule, always be picked before December, even if they are going to be processed as naturally black olives.

Olive size is not uniform. Each tree will have fruits of different diameters, whose distribution approaches a slightly skewed, normal distribution with a mean value which increases slowly with time. Figure 5.3 shows these characteristics for Hojiblanca cultivar. Usually, fruits on the top of the olive tree and those most exposed to light are bigger, while those attached to low, north-facing or interior branches are less developed.

5.3.3 FLESH CHANGES

Changes in the proportion of flesh and its dry matter content are most marked at the beginning of growth, with only slight changes after

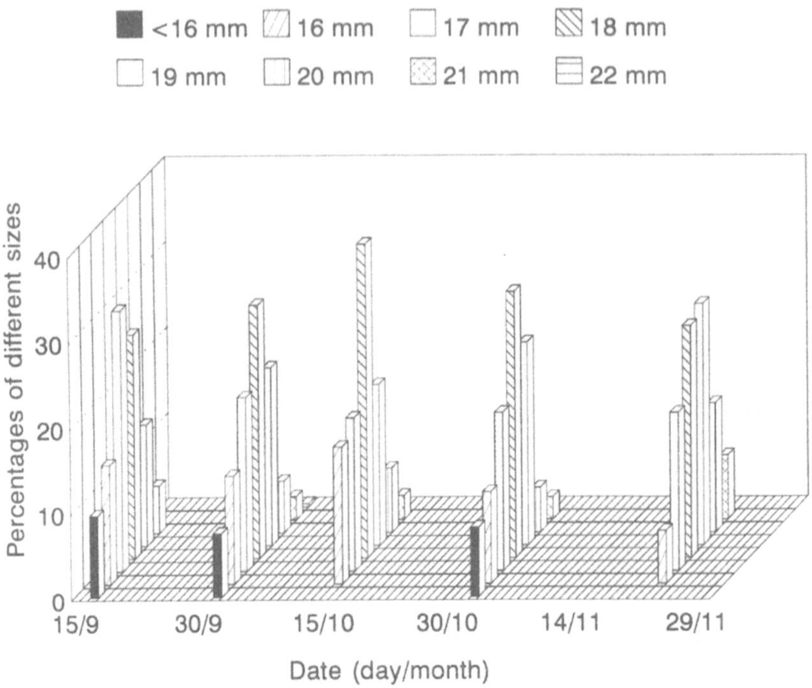

Fig. 5.3 Size distribution of Hojiblanca cultivar and its changes with time. (Redrawn from Garrido Fernández and Fernández Díez, 1972.)

September (Figure 5.4). The percentage of dry matter does continue to increase up to December as a result of rapid lipogenesis during this period. The flesh-to-pit ratio determined in the Conservolea cultivar (Balatsouras *et al.*, 1988) increased during maturation until the end of November and then decreased slightly (Figure 5.5). Other cultivars can be expected to behave similarly and this evolution could be a parameter of interest for fixing a picking date that would maximize a farmer's profits while retaining adequate quality in the final product.

5.3.4 FRUIT CONSISTENCY CHANGES

The texture of the edible flesh is very variable, depending upon the cultivar, its oil content, stage of maturity, type of culture, soil, climate, and a host of other factors which influence the fruit's chemical composition.

Objective methods for texture determination (Fernández Díez, 1979) which have been applied to fresh olives can be divided into two main groups:

1. Instrumental, or direct methods, based on a physical measurement (force or pressure) related to texture.

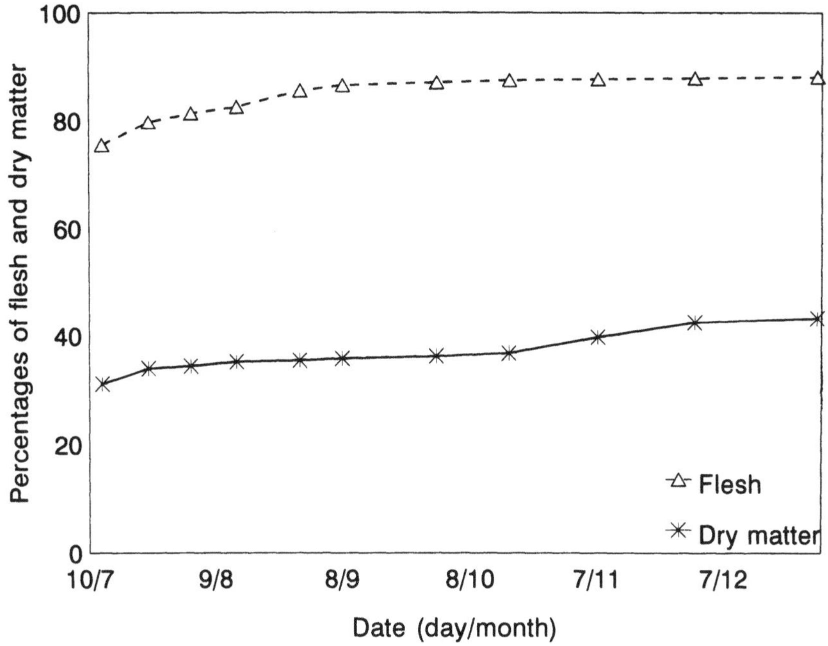

Fig. 5.4 Percentages of flesh and dry matter changes during olive growth. Conservolea cultivar. (Redrawn from Aligisakis and Patroclos, 1971.)

2. Chemical, or indirect methods, based on the analysis of certain components of the fruit related to cell structure and the mechanism of its degradation.

Within the first group, the Christel Texturometer, FFL Penetrometer, IC Texturometer and Kramer Shear Press (compression and shear/compression cells) have all been used with olives in the past. Data obtained with the same equipment are comparable, but usually differ when they come from different machines. Recent work, however, has used almost exclusively the Instron Universal Testing Machine, with the same compression and shear/compression cells. This is mainly due to its versatility, software availability and the fact that use of the same instrument permits data comparison across different research groups.

Among the chemical methods, those which involve quantification of pectic substances, their degradation products and pectinolytic enzyme are considered as generally the most reliable procedures.

A good understanding of the texture changes from September to December are central to obtaining good-quality final products. This is true of each table olive type, although appropriate values differ with processing method. In general, there is a progressive flesh softening during

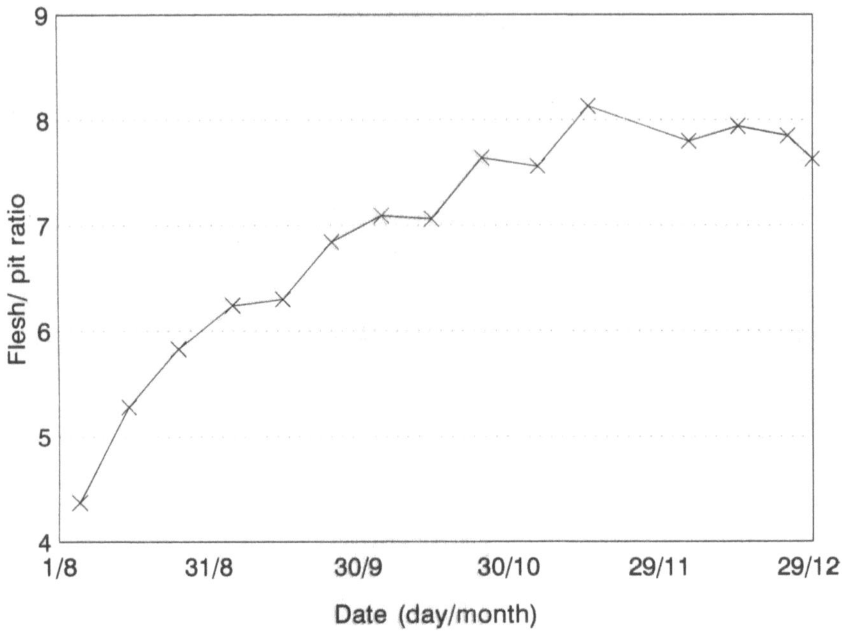

Fig. 5.5 Flesh-to-pit ratio changes in Conservolea cultivar. (Redrawn from Balatsouras *et al.*, 1988.)

this period, although its rate and extent are subject to the usual influences of cultivar, agronomic practices, etc. Figure 5.6 shows the textural changes in the Hojiblanca Spanish cultivar expressed as compression, shear-compression and texture index. Measurements of texture have been suggested as the most reliable procedure for fixing the picking period for the different table olive types (Garrido Fernández and Fernández Díez, 1972; Sánchez Raya *et al.*, 1981).

5.3.5 SURFACE COLOUR EVOLUTION

Fruit surface colour progressively changes through green, pale-green, straw-yellow, pink, purple-pink and black during maturation and ripening. This evolution may be followed by estimating the percentage of olives which have a specific colour. As with texture, the optimal value differs according to the type of table olive required. Usually bigger fruits, which are on the olive tree top, mature more rapidly and become pink quicker than smaller ones or those situated in the tree's interior.

Surface colour changes are usually relatively rapid and pale-green olives may become completely pink or black in a period of 6 weeks. Figure 5.7 shows these changes for Hojiblanca cultivar and different fruit

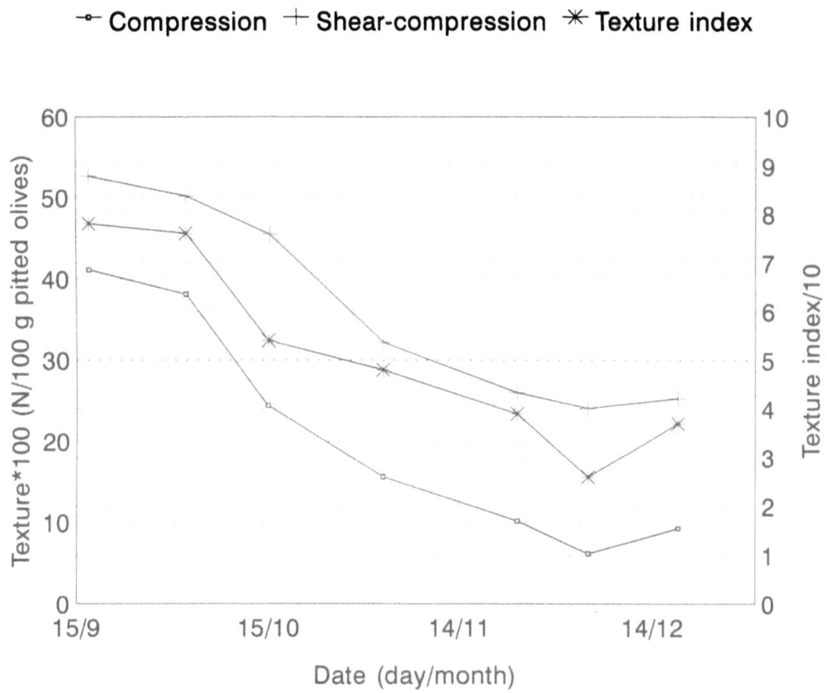

Fig. 5.6 Texture changes (compression, and shear-compression) from September to December; Hojiblanca cultivar. Data obtained with Allo Kramer Shear Press. (Redrawn from Garrido Fernández and Fernández Díez, 1972.)

sizes (Garrido Fernández and Fernández Díez, 1972). Normally, pink colouring is higher in the larger fruits and increases in parallel with texture loss. By the end of November, in the case of the example, olives of all sizes have reached 100% pink colour.

More complex visual methods to express these changes have also been developed, such as the Skin and Pulp Colour Index (CI) proposed by Sánchez Raya *et al.* (1981). In this case, each fruit of the sample (300 units) must be assessed according to the progressive colouring of skin and also of the mesocarp, giving the following CI scores:

0: live green skin
1: yellow green skin
2: red spotted yellowish skin
3: yellowish to reddish or light violet throughout the skin
4: black skin but colourless pulp
5: black skin with colourless inner half pulp
6: black skin with coloured pulp except for the proximity of the endocarp
7: black skin and entirely coloured pulp

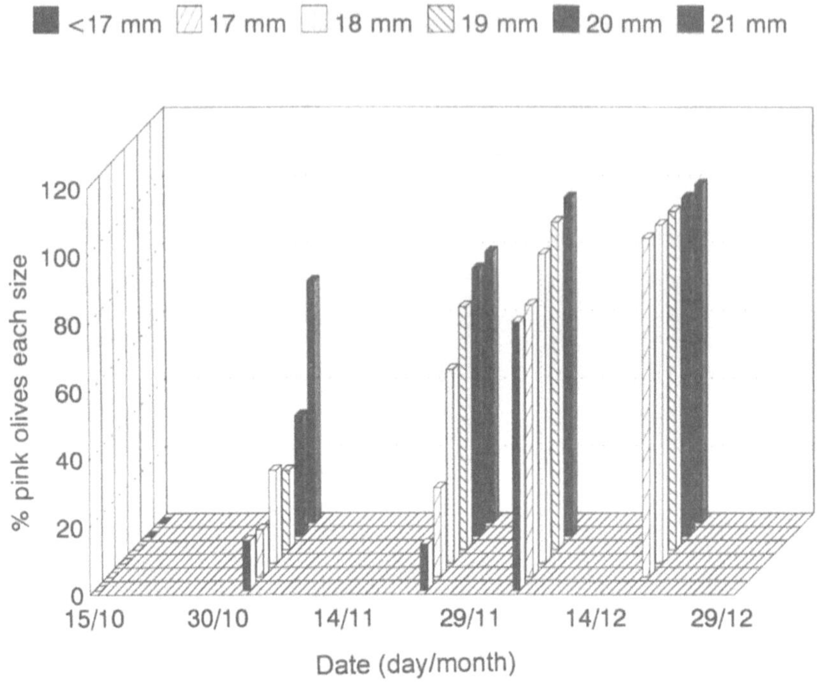

Fig. 5.7 Percentage of pink (and black) olive changes during ripening, according to olive size. (Redrawn from Garrido Fernández and Fernández Díez, 1972.)

The colour index (0 > CI > 7) is obtained by multiplying the fraction of olives assigned to each category by its value.

Fresh olive surface colour changes can also be measured objectively by the colorimetric method proposed by Fernández Díez and Garrido Fernández (1971). It is based on the percentages of reflectance of the fruit skin at 535 and 595 nm, and their rate of change. Measurements correlated well with subjective visual appreciations and were independent of cultivar. The proposed scale is shown in Table 5.1.

5.3.6 FRUIT REMOVAL FORCE

The force required to remove the fruit from the olive branch has also been studied as a method to follow maturation (Sánchez Raya *et al.*, 1981). It is, however, a less reliable guide than the other parameters discussed since it undergoes smaller changes during the period of interest for table olive processing.

Table 5.1 Relationships between visual and objective surface colour measurements of fresh olives, according to the maturation degree. (Reproduced from Garrido Fernández, A. and Fernández Díez, M.J., 1971)

Visual colour	R_{535}	R_{595}	R_{535}/R_{595}
Green	<28.5	<22.5	>1.22
Pale green	<29.0	>23.0	1.21–1.13
Slightly pink	28.5–17.0	26.0–19	1.12–0.90
Pink purple	9.5–8	10.5–8.0	
Purple-black	<7.5	<7.5	

5.4 CHEMICAL COMPOSITION AND CHANGES DURING GROWTH AND MATURATION

For table olives only the skin (epicarp or exocarp) and the flesh or pulp are of interest, since the pit (or stone) is not consumed and remains practically unchanged during processing. This is due to its possession of a lignocellulosic shell, which is resistant to the chemical and physical treatments applied. Thus, in what follows only the exocarp and mesocarp will be considered.

5.4.1 EPIDERMIS

The epidermis contributes 1.5–3.0% to the total weight of the fruit and its main function is to protect the fruit against pests and microbial infection. It is composed of parenchymatous cells which fit closely to each other, leaving no free spaces in between. This continuity is interrupted with small openings (stomata), through which the flesh releases the excess of CO_2 produced during the fruit respiration, absorbs the necessary O_2, and maintains the moisture equilibrium.

The stomata are formed by two occlusive cells and their ostioles belong to the amaviladaceas type. The ostiole opens on to a large and irregular substomatal chamber sunk within the mesocarp. When fruits are mature, the stomata are situated in the stomatal area, slightly protuberant and easily discerned by their whitish or yellowish colour.

A complete description of the morphological structure of olive fruit epicarp (and mesocarp) in Manzanilla and Zorzaleña cultivars can be found in Durán Grande and Izquierdo Tamayo (1965) and Izquierdo Tamayo and Durán Grande (1970). There is a morphological study of stomata and stomatal area in Durán Grande (1977).

The skin is abundant in cellulose and is impregnated with cutin, a wax-like substance impermeable to water and analogous to the suberine of cork cells. Frega and Lercker (1985) found that the epicarp was also coated with free acids, mostly triterpenic acids and the most mature samples showed the presence of 18- and 16-carbon fatty acids in addition to those with 28-, 26- and 24-carbon atoms. The triterpenic acids were identified as oleanolic, ursolic and maslinic acids with one, containing only one alcoholic group, remaining unidentified. Later, Bianchi *et al.* (1992) isolated the following compounds from olive skins: alkanes, C_{23}–C_{33} (3–7% of total extract); saturated and unsaturated alkyl esters (3–5%); aldehydes, C_{24}–C_{30} (3–5%); methyl phenyl esters (2–3%); triacylglycerols (17–25%); alcohols, C_{22}–C_{28} (10–18%); triterpenols (0–14%); fatty acids, C_{16}–C_{28} (9–10%); and pentacyclic triterpenols and triterpenoids acids (26–28%). All were present in both green and black olives, excepting the triterpenols which were detected only in the first stage of maturation. Other compounds previously identified from olive surface waxes cited by these authors are mainly hydrocarbons (n-alkanes, squalene, etc.) and fatty alcohols. These studies give some evidence of biosynthetic activity during maturation. Thus, a clear biosynthetic relationship was observed among aldehydes, alcohols and acids with only a partial correlation between these compounds and alkanes.

The epicarp must also be rich in phenols because it is a common observation that the surface of fruits from any of the table olive processing types becomes brown after a short exposure to air and darkens when alkali treated olives are aerated. Indeed, the oxidation step for producing ripe olives by alkaline oxidation (Californian-style) is based on this reaction.

In general these compounds are not all water soluble. Thus, the skin is an important barrier for most table olive processing treatments. The treatment of fruits with NaOH solutions must remove most of these protective substances, since skin permeability is considerably increased afterwards. However, in directly brined olives these substances remain present and consequently osmotic exchanges between the olive flesh and the surrounding brine are very slow. In these circumstances, fermentation can extend over a long period, as happens for example in the case of turning colour olives or naturally black olives in brine.

5.4.2 MESOCARP

General composition

The mesocarp consists of large parenchymatous cells, similar to those of the epicarp, although less closely packed, leaving free spaces occupied mainly by CO_2 and O_2 in proportions that are respectively higher and lower than those in the atmosphere. Inside each cell, both the vacuole and cytoplasm contain diverse and abundant compounds responsible for most of the biological and nutritive value of table olives, as well as for their characteristic behaviour during fermentation.

Olive pericarp (which is also referred to as pulp or flesh) accounts for 70–90% of the olive weight (wet basis); the endocarp (stone or pit) is the other 30–10%, which includes the seed (1–3% of the whole fruit). Table 5.2 shows the composition ranges of the main groups of compounds in the olive fruit and its fresh pulp. For table olives, the flesh-to-stone (or pit) ratio is important. Ideally it must be around 6.5 : 1. Higher values can produce soft final products; lower ones may be difficult to handle in the pitting and stuffing operations. In this respect, the Manzanilla cultivar is considered by most table olive processors as one of the best. In addition, its right pit facilitates considerably mechanical operations such as pitting and stuffing.

The main constituents of the flesh are water, lipids (oil) and sugars. Moisture content plays an important role in table olive processing because all the processing methods used require a series of exchanges between substances in the flesh and the surrounding solutions (NaOH solutions, brines, etc.). Experience has shown that calculations of equilibrium concentrations must be based on the hypothesis of a proportional distribution of the different solutes (NaCl, sugars, acids, etc.) between the brine and the so-called olive juice (effectively the fruit's moisture content). When the solutions in which the olive fruits are submerged have an excessive osmotic pressure (i.e. are too concentrated) in relation to the flesh, moisture is extracted from the fruits and a surface shrivelling affecting most of the skin occurs. The intensity of this increases with the concentrations used. Sensitivity among cultivars differs but, in general, olive fruits with a high flesh-to-pit ratio are more prone to this problem. Repair is difficult and skin may remain wrinkled throughout the whole process. The final product is thus depreciated and is sometimes even unmarketable. Gordal cultivar is very easily affected and the

Table 5.2 Composition of the olive fruit epicarp (Adapted from Fernández Díez *et al.*, 1985)

Component	Proportion (%)
Moisture	60–75
Lipids	10–25
Reducing sugars, soluble	3–6
Non-reducing sugars	≤0.3
Mannitol	0.5–1.0
Fibre	1–4
Crude protein (N × 6.25)	1–2
Ash	<1.0
Organic acids and their salts	0.5–1.0
Phenolic compounds	2–3
Pectic substances	≤0.6
Other compounds	3–7

concentration of NaOH and salt used with it are normally lower than those used with other cultivars. In contrast, with some olive types such as natural olives in dry salt, this effect is looked for and used as a means of enhancing the product's stability by reducing its water activity. In this case, the olive's shrivelled aspect is considered normal and the consumer is accustomed to it.

Olive respiration and crop distribution

The beginning of ripening coincides with a sharp increase of fruit respiratory activity and a change in the respiratory quotient. This is a quantitative rather than a qualitative change since the same substrates are used, albeit more quickly (Sánchez Raya *et al.*, 1981). Ripening rate has been modified by application of some compounds. Ethephon [(2-chloroethyl) phosphoric acid], when sprayed mid-way between the onset of pit hardening and the maximum respiratory rate, induces early pigmentation of drupes. If applied after the peak of respiratory activity, however, it only reduces the fruit detachment strength. These results are in agreement with the hypothesis that attributes a climacteric model to attached fruits and a non-climacteric model to the detached ones (Rugini *et al.*, 1982).

Full black maturation is dependent on crop load. Experience has shown that fruits with halved yield were bigger and reached this maturation stage 30 days earlier than those on more heavily loaded trees and that oil accumulated faster reaching higher final amounts (Barone *et al.*, 1994).

Oil changes during olive fruit growth

Accumulation of oil in the flesh of olives begins just after pit lignification and reaches a high rate during the second fruit growing stage (from early September). Oil drops inside the cells increase in size and progressively push the nucleus towards the cell's wall. The rate of oil biosynthesis may reach 30–50 mg oil per week during the period of maximal lipogenesis. The changes in oil content from June to full black maturation have been studied by several authors (Borbolla y Alcalá *et al.*, 1955; Uzzan *et al.*, 1956; Sandret, 1958; Donaire *et al.*, 1984; Aligisakis and Patroclos, 1971), who have always found an inverse relationship between oil content and moisture content, for the same degree of maturity. Although, as Table 5.3 shows, the decrease of water content is lower than the oil increase due to the increase in olive fruit size during this period.

Lipid composition also changes during the maturation period as shown in Table 5.4. At the beginning of lipogenesis, palmitic, linoleic and linolenic acids are abundant. However, as time progresses their proportions gradually diminish while the proportion of oleic acid markedly

Table 5.3 Moisture, oil, and reducing and total sugar changes during maturation. Gordal cultivar. All data are expressed as percentages of fresh flesh (Reproduced from Borbolla y Alcalá, J.M.R. de la *et al.*, 1955)

Date (Day/month)	Moisture	Oil	Reducing sugars	Total sugars
26/06	72.65	0.70	7.58	7.72
16/07	76.20	1.60	7.81	8.04
06/09	73.73	6.25	5.96	6.23
03/10			4.63	4.88
22/10	68.05	16.10	4.30	4.78
14/11	68.10	18.60	3.66	3.69
17/11	67.25	18.95	3.40	3.56
28/11	69.60	19.50	3.08	3.22

Table 5.4 Changes in the fatty acid distribution (% of fatty acids) of the olive lipids during growth (Reproduced with permission from Donaire, J.P. *et al.*, 1984)

Size (mg)	Palmitic 16 : 0	Stearic 18 : 0	Oleic 18 : 1	Linoleic 18 : 2	Linolenic 18 : 3
5	45.02	0.82	15.76	14.95	21.45
9	40.87	1.73	19.73	17.24	20.43
20	31.51	1.00	19.42	20.53	27.57
33	36.25	2.02	22.95	16.88	21.90
124	30.40	1.70	27.24	16.34	24.32
222	24.02	0.50	24.98	21.58	28.72
331	30.10	3.47	25.98	22.47	17.98
570	29.15	2.57	27.19	22.96	18.13
750	28.00	2.00	39.88	17.20	12.92
1350	25.12	2.72	51.16	12.33	8.78
2500	17.91	3.24	60.27	6.58	7.21
2900	15.90	2.52	80.05	0.71	0.80

increases. Stearic acid also increases slightly (Donaire *et al.*, 1984). The different types of table olive are collected over a relatively long period of time, from September to December, so differences in oil composition can be significant in some cases, especially between the extreme maturation degrees: green ripe and full black stages. Within the same type of table olive, however, such changes will be slight.

Olive also contains small amounts of palmitoleic, arachidic and eicosanoic acids as well as other lipid-soluble substances such as sterols and triterpenic acids. Most of the lipids in olive are triglycerides, but there are also diglycerides and free fatty acids (Vázquez Roncero and Mancha Perelló, 1965; Vázquez Roncero *et al.*, 1965). The percentage of triglycerides in the total fat in the olive fruits increases quickly up to 95%, while that of diglycerides changes from an initial 10% to less than 0.5% at the end of maturation. The amounts of di- and triglycerides per fruit increase during the whole period of growth (Vázquez Roncero and Mancha Perelló, 1970).

With respect to the distribution of the fatty acids during growth, the position 2 of the glycerol is initially mainly esterified by C_{18} of palmitic acid. This content decreases and reaches the normal value (1.5%) at normal full black maturation. Triglycerides containing only oleic (OOO) or oleic and a single palmitic residue (POO) always total more than 50% but they decrease continually. The L(linolenic)OO content, however, increases during fruit growth. Other triglycerides are present only in small quantities and there are no saturated triglycerides present at any stage (Vázquez Roncero and Mancha Perelló, 1970).

Oil content, as well as fruit size, is not homogeneous for all the olive crop. It is dependent on the fruit's position on the tree. Usually, olives in the upper part are bigger and have a higher oil content than olives lower down and in the interior of the tree. This fact prompted research to establish a relationship between light and fruit growth and lipogenesis (Sánchez, 1994).

Bearing in mind the green surface colour of olive fruits, some participation from the fruit's own photosynthetic activity in oil formation may be expected. Thus, although fruits are considered to be highly heterotrophic, acting as strong sink organs during the fruit growth, due to both the mitochondrial respiration of imported photosynthates and the impermeability of the skin, the fruit accumulates high levels of CO_2 in the interior of its free spaces which may be used for photosynthetic activity. This is also supported by the high activity of some enzymes, like phosphoenol pyruvate carboxylase, which can participate in the mechanism of refixing the CO_2 produced during respiration. The products formed may be stored in the vacuole as organic acids, mainly malate, and finally be transformed into lipids.

Lipid photosynthesis studies in olives have shown that the rate of ^{14}C-acetate incorporation into olive triglycerides is higher in light than in the dark. Also, $^{14}CO_2$ in a closed chamber was actively incorporated under illumination into lipids of the olive flesh but not into those of the seed. Thus, it seems that olive fruit photosynthesis plays an important role in the formation of olive flesh storage lipids. However, lipids in the seed are entirely dependent on the presence of the photosynthates imported from the leaves.

Due to the moderate degree of saturation of the fatty acids and its excellent quality, olive oil makes a significant contribution to the fact that table olives are considered a product of high biological and nutritive value (Gracian Tous, 1968; Mataix Verdú and Martinez de Victoria Muñoz, 1988).

Reducing sugar content and evolution

Soluble reducing and non-reducing sugars are the most important components with respect to the fermentation and preservation stages in all the types of table olive processing. Changes in total and reducing sugars during maturation are shown in Table 5.3. In general, reducing sugars are the most important of the olive flesh sugars (90–95%), with slight differences due to maturation degree. Also clear from Table 5.3 is a continuous decrease of these substances in the flesh during the period of interest for table olive picking. The inverse correlation between sugars and oil gave rise to a hypothesis on their biochemical relationship (Borbolla y Alcalá *et al.*, 1955; Uzzan *et al.*, 1956; Sandret, 1958)

It seems clear that soluble reducing and non-reducing sugars have an important role in oil biosynthesis. These compounds decrease in a continuous manner when oil is accumulated in the fruit, and statistical analysis of the results obtained for fat, reducing sugars and moisture contents in the fresh pulp of olives from two Spanish cultivars, Gordal and Zorzaleña, and one of Greek origin, Conservolea, during the growth and ripening period, gave rise to a highly significant multiple regression equation:

$$F = -0.82\,M - 2.72\,S + 82.01$$

in which F, M and S are oil, moisture and reducing sugars contents, respectively (Sánchez Gómez and Fernández Díez, 1991).

The level of mannitol, though relatively lower when compared with the total amount of soluble sugars, must also be important for the whole metabolic system (Wodner *et al.*, 1988). Some work has been done to elucidate the relationships between the diverse compounds related to lipid biosynthesis in olive fruits. Donaire *et al.* (1984) studied the changes during maturation of all the compounds related to the Krebs cycle, finding that the equilibrium between lipids, fixed acids and sugars could be shifted according to the maturation degree. In their later research, they showed that total Krebs cycle acids seem to be in an inverse relationship in leaves and fruits during growth, observing also the presence of high levels of uronic acids during full black fruit ripening (Donaire *et al.*, 1984). Wodner *et al.* (1988) found that mannitol was considerably lower in Kadesh, a low-oil-containing cultivar, than in any other. They deduced that mannitol and other polyols in the olive fruit may play an important role in the metabolic transformations and synthesis of olive oil, as happens with storage compounds in other higher plants. Thus, the relative proportion of mannitol in the total amount of olive sugars might indicate the potential of a cultivar for oil biosynthesis.

The principal sugars in the flesh are glucose, followed by fructose and, at a lower level, sucrose. Small concentrations of xylose and rhamnose are also present (Borbolla y Alcalá *et al.*, 1955; Fernández Díez, 1971; Fernández Díez *et al.*, 1981; Fernández Bolaños *et al.*, 1982).

The distribution of sugars in different Spanish cultivars and at different stages of the table olive picking period is shown in Table 5.5. The relationships appear similar for all cultivars at similar maturation degrees. As time progresses the loss in sugars affects mainly fructose and mannitol, which decrease markedly in relation to glucose. These changes are important for table olive processing; an excessive delay in the picking date may lead to low total sugars in the olive flesh and to deficient fermentations, especially in the case of Spanish-style green olives, whose preservation is strongly dependent on a strong lactic acid fermentation.

Evolution of all these components during the growth and ripening of the fruits has been studied for different Spanish cultivars by gas–liquid chromatography (Fernández Bolaños *et al.*, 1983), and later in Israel (Wodner *et al.*, 1988) by high-pressure liquid chromatography, using the 4-nitro-benzoate derivatives of sugars.

Cell wall polysaccharides and their changes; related enzymes

Fibre

Fibre represents the fundamental support for the structure of the fruit (Heredia Moreno, 1979, 1980). The fibre content in olives depends on the method used for its determination. In general, crude fibre determinations give values about half those obtained by acid detergent fibre (ADF) analysis (Heredia Moreno, 1980). Fibre fractions obtained from table olive cultivars at green ripe olive maturation degree are shown in Table 5.6. Chemical procedures, neutral detergent fibre (NDF) and ADF determination lead to lower values than those measured by enzymatic methods and as insoluble fibre (IF) (Guillén *et al.*, 1991, 1992).

Table 5.5 Main sugars* and polyols (expressed as percentages of fresh flesh) in different table olive cultivars (Reproduced from Fernández Bolaños, J.M. *et al.*, 1982)

Cultivar	Date	Glucose	Fructose	Mannitol
Manzanilla	August	2.97	1.33	1.41
Manzanilla	November	2.10	0.71	0.22
Gordal	September	2.40	1.10	–
Gordal	November	1.60	0.11	0.10
Hojiblanca	December	1.83	0.12	0.18
Verdial	December	1.48	0.14	0.20

* Sucrose concentration is usually around 0.2% during the table olive picking period.

Table 5.6 Fibre fractions (expressed as percentages of fresh flesh) in ripe green olive fruits (Adapted from Guillén *et al.*, 1991, 1992)

Fibre fraction*	Cultivar		
	Gordal	Manzanilla	Hojiblanca
ADF	2.1	2.4	2.6
NDF	2.8	3.2	3.7
IF	4.0	4.3	5.9
SF	1.0	0.7	1.1
TF	5.0	5.0	7.0

ADF, acid detergent fibre; NDF, neutral detergent fibre; IF, insoluble fibre; SF, soluble fibre; TF, total fibre (sum of SF and IF).

Table 5.7 Fibre fraction composition in green ripe olive fruits (expressed as percentages of fresh flesh) (Reprinted from *Food Chemistry*, **44**, Guillen, R. *et al.* Fibre fractions carbohydrates in *Olea europaea* (Gordal and Manzanilla varieties), pp. 173–178, copyright (1992), with kind permission from Elsevier Science Ltd., The Boulevard, Langford Lane, Kidlington, Oxford OX5 1GB, UK)

	ADF	NDF	IF
Gordal cultivar			
Cellulose	1.12	0.60	1.05
Hemicellulose	0.12	0.23	0.60
Lignin	0.83	0.83	0.83
Protein	0.07	0.48	0.73
Manzanilla cultivar			
Cellulose	0.56	0.54	0.81
Hemicellulose	0.07	0.24	0.36
Lignin	1.16	1.16	1.16
Protein	0.39	0.47	0.79

ADF, acid detergent fibre; NDF, neutral detergent fibre; IF, insoluble fibre.

With respect to the main components of the different fractions, Table 5.7 gives the proportion of cellulose, hemicellulose, lignin and proteins. In general, glucose was the main component of NDF, ADF and IF and arabinose was the major component in soluble fibre (SF). From Table 5.7, it is also evident that in spite of the fact that fibre extraction procedures always involve protein solubilization, some protein still remains in the fibre. This may be due to the existence of condensation products between proteins and polysaccharides and polyphenols or their formation during extraction. Currently such proteins tend to be considered as part of the fibre itself (Guillén *et al.*, 1991).

The major components for fresh green fruits in order of decreasing importance are: cellulose, lignin and hemicellulose, with cellulose accounting for more than 50% (Heredia Moreno and Fernández Díez, 1979). In some cases, differences in cultivar or fraction may be significant. Thus, the cellulose proportion in Manzanilla may range from 63 to 90% of the total fibre while in Gordal it may account for only 53–60%. Percentages of lignin are inversely proportional to those of cellulose, ranging for the same cultivars from 36 to 9% and 40 to 46% respectively. Although the crude fibre of the fresh olive flesh of these cultivars is composed mainly of cellulose, differences between them are enough to give diversity in consistency and palatability. Differences in texture are not always directly correlated with percentages of cell wall polysaccharides (Jiménez *et al.*, 1994). It seems that interactions between them must also play an important role in fresh fruit texture and changes during processing.

When the degree of maturity advances, enzymatic degradation may produce changes in the relationship between different components (Fernández Díez, 1991). In the Hojiblanca cultivar at different stages of ripeness, percentage of cellulose goes from 40–45% down to 23%; lignin remains almost steady, between 33% and 38%; and hemicellulose goes from 22–23% up to 41%, when fruits are completely ripe (Heredia Moreno and Fernández Bolaños, 1987). In general, the fibre content in fresh fruits is higher than in processed fruits, since some of its components are usually degraded during processing. This may be the result of enzymatic degradation or chemical solubilization by NaOH during the debittering process, or by the subsequent acidic conditions.

Cellulose and hemicellulose

The most studied fibre components have been cellulose and hemicellulose. The cellulose content seems to diminish with time during maturation. Changes in Hojiblanca were related to maturation degree and gave the following values expressed as percentages of ADF: 60.5% (green), 56.4% (turning colour), and 49.7% (full black maturation). The changes in cellulose content were strongly correlated to changes in cellulase activity (Heredia Moreno and Fernández Bolaños, 1985a).

Olive fruit hemicelluloses have recently been studied in Manzanilla and Gordal cultivars (Fernández Bolaños *et al.*, 1984) by means of their solubilization with aqueous alkaline solutions (10% NaOH). Polysaccharides obtained by neutralization of the extract with acetic acid were considered as hemicellulose A while those remaining in the solution which were precipitated with ethanol or acetone were classified as hemicellulose B. In the case of olive flesh, with its high lignin content, the polysaccharide extraction method must be carried out on samples previously delignified by acetic acid and $NaClO_2$.

Hemicellulose A from the Gordal cultivar contained 63% sugars, 57% of which was xylose. Its further purification with Cu(II) salts increased this percentage to 77.5%. The molecular weight, determined by gel filtration and by high-performance liquid chromatography, gave 37 000 and 40 700 Da, respectively. Partial acid hydrolysis showed a chain of β1–4-linked xylose units and some 4-*O*-methylglucuronic acid residues (Tejero Mateo *et al.*, 1985a). Later confirmation of the structure was carried out by methylation and gas chromatography–mass spectrometric analysis of the acetates of partially methylated alditols which were obtained by hydrolysis, reduction and acetylation. These studies showed that this xylan comprised a linear chain of xylose units with β1–4 links and 4-*O*-methylglucuronic acids linked to the main chain through the hydroxyl on the C-2 atom of the xylose (Tejero Mateo *et al.*, 1985b)

Another polysaccharide from hemicellulose A from the Hojiblanca cultivar with a molecular weight of 9500 Da and containing xylose and 4-*O*-methylglucuronic acid in the molar ratio 24 : 1 has also been isolated. Its structure is that of a polymer of β1–4-linked xylopyranosyl residues, having branches of 4-*O*-methylglucuronic acid groups on the O-2 atoms of the main chain (Gil Serrano *et al.*, 1986).

With respect to hemicellulose B, studies carried out without previous removal of the lignin permitted the separation of three fractions with the compositions shown in Table 5.8. Xylose is present in the first two, since Cu(II) precipitation affects selectively polysaccharides containing this compound and mannose by forming insoluble complexes (Fernández Bolaños *et al.*, 1984). Also shown in Table 5.8 are the respective molecular

Table 5.8 Composition of hemicellulose B obtained from Manzanilla and Gordal cultivars (percentages with respect to total sugars) (Reproduced from Gil Serrano, A.M. *et al.*, 1984b)

	Precipitate with Cu(II) salts		Solution after Cu(II) precipitation
Sugar	Manzanilla	Gordal	Gordal
Arabinose	7.7	5.6	1.9
Xylose	15.1	17.9	32.9
Mannose	17.8	23.4	–
Galactose	13.6	16.9	21.0
Glucose	45.6	35.4	15.3
Rhamnose	–	–	3.6
Yield*	2.0	0.8	1.2
Molecular weight (Da)	53 000	420 000	80 000

* With respect to the alkaline-extracted residue.

weights determined by filtration chromatography on Sepharyl S-300. These were 53 000, 420 000 and 80 000 Da for hemicellulose B from Manzanilla, Gordal and for non-Cu-precipitated from Gordal extracts respectively (Gil Serrano *et al.*, 1984a). Results of partial hydrolysis of hemicellulose B from the Manzanilla cultivar were studied by paper chromatography and gel filtration showing the presence of cellobiose and an oligosaccharide of glucose and mannose. It was identified as having the structure of a mannoglucan with residues of arabinose, xylose and galactose (Gil Serrano *et al.*, 1984b). Other polysaccharides have also been isolated from previously delignified Gordal extracts. One of these was a galactoglucomannan, with 85% sugars and a molar ratio of 3 : 3 : 1 of glucose, mannose and galactose. Its molecular weight was 39 400 Da. The structure is of a main chain of β1–4-linked glucose and mannose residues to which galactose is also α1–6-linked (Tejero Mateo *et al.*, 1986a). A xyloglucan has also been isolated. Structurally it consists of a main chain of glucopyranose units β1–4-linked, to which chains of β1–4-xylopyranose and some arabinofuranose and galactopyranose are linked to the C-6 atom hydroxyl of the main chain (Tejero Mateo *et al.*, 1986b).

Related enzymes

The presence of cellulase in fresh olives was investigated by Heredia Moreno and Mínguez Mosquera (1981). The enzyme was detected in the pulp of olives of the Hojiblanca cultivar. Enzymatic activity was negligible when fruits were green, developed when maturation reached the turning colour stage, and increased rapidly in full black mature fruits (Heredia Moreno and Fernández Bolaños, 1985a). The different factors influencing cellulase activity, such as incubation time, presence of sodium chloride, temperature, storage time, and degree of ripeness at fruit harvest, have been studied (Heredia Moreno and Fernández Bolaños 1985b; Heredia Moreno *et al.*, 1988). In general, greatest activity was shown at 30°C and the enzyme was completely inactivated when extracts were heated for 5–10 minutes at 70°C or over; the presence of sodium chloride increased the activity when fruits were green, but had a negligible effect with black olives. Increased duration of the storage period in air caused a rapid increase of activity with time which was more pronounced as maturation progressed.

Characterization and partial purification of the enzymatic complex 'cellulase' is also under research (Heredia Moreno *et al.*, 1989a). Purification of the enzyme requires partial or total previous elimination of polyphenols (Heredia Moreno *et al.*, 1990b). The existence of three enzymatic forms: endoglucanases, β-glucosidases and exoglucanases has been confirmed. Isolated endoglucanase showed the existence of eight subunits with molecular weights ranging from 18 000 to 72 000 Da. The presence of this enzyme, previously found only in microorganisms, leads to the conclusion that a similar cellulolytic complex may also be formed in olives or that, in full black mature olives, it may be – at least partially – microbial in origin (Heredia Moreno *et al.*, 1991). Cellobiose and, to a lesser extent, glucose competitively inhibit enzyme activity. Presence of ascorbic acid, tartaric acid and iron salts caused strong inhibition, but the effects of tartaric salts were only weak (Heredia Moreno *et al.*, 1989b). Polyphenols are also strongly inhibitory but this varied with different compounds (Heredia Moreno *et al.*, 1990a).

Other enzymes

The enzymes α-and β-galactosidase, α-mannosidase and α-arabinosidase have all been detected in olive fruits during growth and ripening. Since their extractions were performed at low ionic strength, they must not be cell wall-associated. High concentrations of NaCl strongly inhibited their activity, which had an optimum pH of between 4.3 and 5.3 and a relatively high thermal stability (Heredia Moreno *et al.*, 1992). Their activity was absent during the first phases of growth and appeared in the completely developed fruits (Heredia Moreno *et al.*, 1993).

Pectic substances and related enzymes

Composition and role of pectin substances in texture

The pectic substances are also thought to play an important role in the texture of plant tissue and fruits. Insoluble pectic compounds in the middle lamella bind the cells together and softening of the plant tissue is related to the degradation of pectic substances. The effect of maturity on pectin and texture in Californian olive cultivars was investigated by Chung *et al.* (1974). They found that as the maturity of raw olives advanced, the total pectin and protopectin decreased. Thus, Sevillano olives contained 0.451% (on a fresh weight basis) protopectin at the straw colour stage and only 0.306% at the black colour stage. As the protopectin content decreased, texture diminished markedly and there was a gradual increase in water-soluble pectin. The authors concluded that protopectin was thus the most important compound affecting firmness of these fruits.

The degree of esterification found in such pectins was 86% in unripe fruits and diminished as maturity of olives advanced. This was also accompanied by a gradual softening of the fruits. The acetyl content was 3.4% in the pectin of green firm olives but only 1.9% in that of black olives. The soft texture in overripe olives may thus be related to the gradual decrease in acetyl content.

Pectic substances in Spanish cultivars were studied by Fernández Díez *et al.*(1974) and Mínguez Mosquera *et al.*(1976) who found levels between 0.3 and 0.6% of the fresh pulp weight. They also found a parallel decrease in texture, the percentage of anhydrouronic acid content and the degree of esterification, which was completely absent in over-ripe fruits. However, these authors did not find any clear relationship with acetyl content. Marsilio and Solinas (1990) confirmed the close relationship between texture and the content of pectic substances by following changes in their level during maturation in Intosso, S. Agostino, Caroleo, Leccino and Grossa di Cassano Italian cultivars. By correlating texture with the different pectin fractions, they deduced that the main component responsible for consistency was the protopectin and that any treatment able to attack this compound could cause softening.

Pectinolytic enzymes

The presence of the pectin-degrading enzymes, pectin methyl esterase and polygalacturonase has been confirmed in Hojiblanca cultivar, but they were absent in Gordal, Manzanilla and Lechín (Castillo Gómez *et al.*, 1978a,b; Mínguez Mosquera *et al.*, 1978). Pectin methyl esterase appears in Hojiblanca at the turning colour stage and increases its activity until fruits reach the full black maturation. Maximum activity is around pH 7.5 and activity is easily inhibited by the salt (sodium chloride), being practically absent when concentrations exceed 6%.

Polygalacturonase activity has not been detected in green or turning colour olives and begins in black and over-ripe fruits. The existence of two isoenzymes was deduced from studies related to the effect of factors such as pH and NaCl on the activity. This opened the possibility of a double origin for the polygalacturonase; the fruit itself and the microbial population adhering to the olive surface. Their activity was optimal around pH 4.4 and was strongly inhibited by NaCl at concentrations of 4% or higher.

Due to the fact that both pectinolytic enzymes are susceptible to inhibition by NaCl, it is unlikely that they play an important role in fruit softening during processing. There is, however, a certain risk that their activity could cause problems during the period between brining and equilibration of the NaCl. A water-soluble and dialysable natural inhibitor has been reported in full black olives (Castillo Gómez *et al.*, 1979). It inhibits both the polygalacturonase isolated isoenzymes, although apparently by different mechanisms: competitive, for the isoenzyme present in the precipitate from the dialysed extracts; and non-competitive for that remaining in the soluble fraction.

The types of pectic enzymes and the changes they cause in pectic substances during ripening of the fruit are important and fix the time of harvest of certain cultivars and their processing (Mínguez Mosquera, 1982).

Phenolic substances

Phenolic compounds are responsible for several important olive fruit characteristics: bitterness, browning when injured and inhibition of microorganisms during fermentation. Their levels range from 1 to 3%, of the weight of the fresh pulp, expressed as caffeic acid (Fernández Díez *et al.*, 1972; Fernández Díez, 1983, 1991). The seed is also rich in phenols. Total polyphenol content has been recently determined routinely by spectroscopy of derivatives (Montaño Asquerino *et al.*, 1985).

The majority of those compounds are involved with colour changes in olives (González Cancho *et al.*, 1975; Garrido Fernández and Fernández Díez, 1976). Thus, some components are easily oxidized in an alkaline medium and are responsible for the blackening of the treated black olives (also known as ripe olives). Anthocyanins are responsible for the colour of the naturally black olives and the colour of green table olive brines is due to polyphenol polymerization.

Systematic study of the different polyphenols in olive fruit has been carried out only recently, although oleuropein was isolated at the beginning of this century by Bourquelot and Vintilesco (1908). Most of these compounds in olives are *ortho*-diphenols (Vázquez Roncero *et al.*, 1971). The first to be purified and its chemical structure elucidated was oleuropein. Using classical chemical techniques, Panizzi *et al.* (1960) established that it was a heterosidic ester of elenolic acid and (dihydroxyphenyl) ethanol (Figure 5.8). This was later confirmed by mass spectrometry (Inouye *et al.*, 1974).

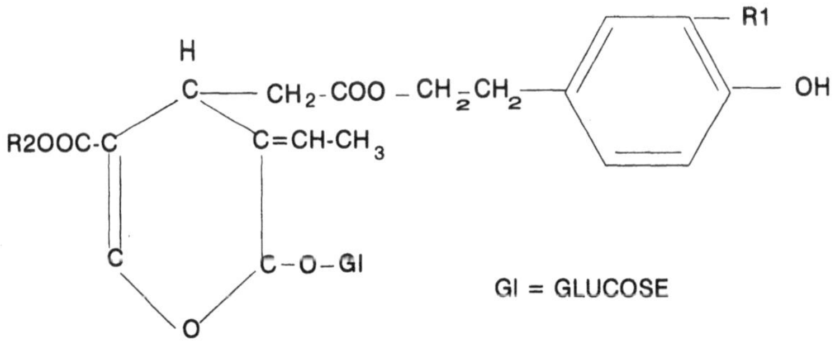

OLEUROPEIN R1 = OH R2 = CH_3

DEMETHYLOLEUROPEIN R1 = OH R2 = H

LIGSTROSIDE R1 = H R2 = CH_3

Fig. 5.8 Chemical structures of the different phenols related to oleuropein. (Redrawn from Bianco *et al.*, 1993.)

A new, very similar compound was isolated later by Ragazzi *et al.* (1973) and identified as demethyloleuropein. This differs from oleuropein in having a free carboxylic group instead of a methyl ester group (Figure 5.8), and was detected only in mature fruits and at much lower concentrations than oleuropein.

Recently, another phenolic compound known as ligstroside has been isolated from olive fruits (Kubo and Matsumoto, 1984). It too, is closely related to oleuropein with the difference that it has only one hydroxyl group and has thus lost the properties of an *ortho*-diphenol. Apparently, this compound disappears when olives reach normal size.

Other phenolic compounds are derivatives of hydroxycinnamic acid, verbascoside (caffeic acid and dihydroxyphenyl heterosidic ester) being the most important (Figure 5.9) (Fleuriet *et al.*, 1984). Bianchi *et al.* (1993) have recently reported the presence of another compound, cornoside (Figure 5.10), which seems to be related to the oxyphenylethyl alcohol moiety present in oleuropein and demethyloleuropein and can be easily transformed into halloridone.

Finally, Bianchi and Pozzi (1994) have reported the presence of 3,4-dihydroxyphenyl-glycol as the major C_6–C_2 phenolic compound in some varieties of *Olea europaea* fruits. In an Italian cultivar, Dritta, it was

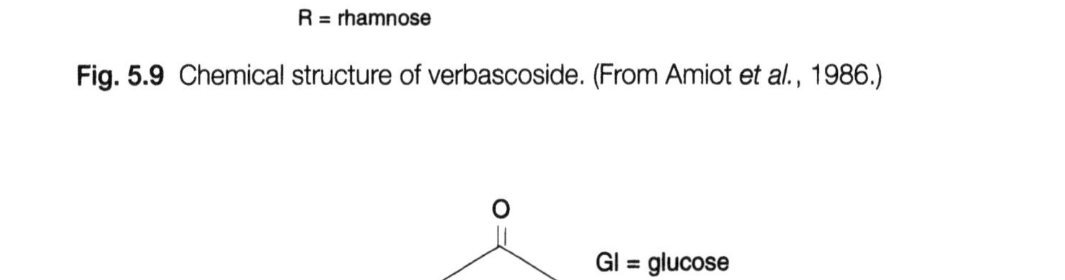

Fig. 5.9 Chemical structure of verbascoside. (From Amiot *et al.*, 1986.)

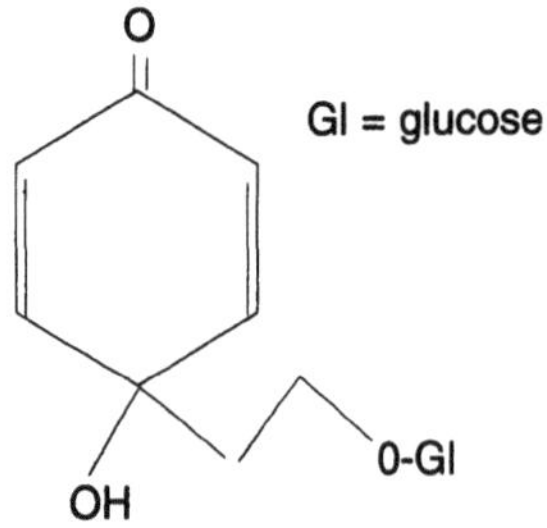

Fig. 5.10 Chemical structure of cornoside. (From Bianco *et al.*, 1993.)

detected as the major component of the water-soluble simple phenolic fraction. The finding of this substance in olive fruits is noteworthy because it occurs in animals including man in a ratio correlated to the sympathomimetic hormone, noradrenaline.

In addition to these very olive-characteristic phenols, this fruit also has other more common phenolics and products obtained from the hydrolysis of the precursors mentioned above. To study the different phenolics in olives, the development of thin-layer chromatography and later high-performance liquid chromatography methods has been essential (Ragazzi and Veronese, 1973a). A detailed study of the then-known phenolic compounds in olive flesh was performed by Vázquez Roncero *et al.* (1974) in both green and full black mature olives of most of the Spanish cultivars. He reported that the main phenolic was oleuropein; there were two glucosides of 3,4-dihydroxyphenylethanol (4-monoglucoside and 4-diglucoside) and an ester of caffeic acid (1-caffeil glucose), *p*-coumaric acid, a flavonol (quercitin 3-rutinoside) and three flavones (luteolin 7-glucoside, luteolin 7-glucoside and apeginin 7-glucoside). There were also the aglucone from oleuropein, β(3–4-dihydroxyphenyl)-ethanol (hydroxytyrosol), tyrosol and vanillic acid. When olives were mature, there were in addition anthocyanic pigments (these will be discussed later).

Polyphenol changes during maturation have been studied by several authors. Vázquez Ronquero *et al.* (1971) found in the Spanish cultivars Zorzaleña and Verdial a continuous decrease in phenolic (and *ortho*-diphenols) content from July up to full black maturation, from about the 7% of the dry matter (expressed as caffeic acid) in green fruits to less than 0.5% in mature fruits. Gordal cultivar was an exception, because it was always low in phenols (never exceeding 3%), and the content remained almost constant during growth. Similar behaviour was found by Solinas *et al.* (1975) with respect to total phenols, oleuropein and demethyloleuropein. However, marked changes in hydroxytyrosol and tyrosol were not observed.

Demethyloleuropein is very common in some Italian cultivars (Ragazzi and Veronese, 1973b) and its ratio to oleuropein was apparently characteristic for each cultivar. This led the authors to propose it as a parameter for their differentiation.

In the French cultivars Picholine, Lucques, and Salonenque, oleuropein reached up to 14% of dry matter in young fruits, diminished during growth as in other cultivars, but was still an important percentage at the harvest time for green picked cultivars. Small-fruit cultivars were characterized by low oleuropein content and high verbascoside (Amiot *et al.*, 1986). Normally, oleuropein loss is accompanied by the accumulation of two compounds, demethyloleuropein and elenolic acid glucoside (which is not phenolic and corresponds to the secoiridoid part of the oleuropein). Although more oleuropein disappears than can be accounted for by the derivatives, it is reasonable to assume a biochemical relationship between them due to their obvious chemical relatedness. There is a simultaneous increase in esterase activity which might be responsible for their formation. These derivatives could be useful markers for the start of green maturation, as well as for differentiation between cultivars since demethyloleuropein is not present in all cultivars (Amiot *et al.*, 1989).

During storage of harvested fruits in air and in modified atmospheres the amount of total polyphenols decreases. Raising the atmospheric CO_2 to 3% increases polyphenol retention, while a combination of 3% CO_2 and reduced oxygen (5%) resulted in the smallest polyphenol losses. This suggests that CO_2 may act as an antagonist of O_2 in polyphenol oxidation (Maestro Durán *et al.*, 1993). Thus, storage conditions can have an important influence on the quality of oil and table olives obtained.

The phenolic compounds in olive fruits are essential for the darkening that occurs during the processing of ripe olives (Brenes Balbuena *et al.*, 1992). They can also inhibit cellulase (Heredia Moreno *et al.*, 1990a) and other enzymes and have a marked antioxidant effect in olive oil (Tutour and Guedon, 1992). Oleuropein inhibited the growth and enterotoxin production of *Staphylococcus aureus* (Tassou and Nychas, 1994) or the germination and outgrowth of *Bacillus cereus* spores (Tassou *et al.*, 1991). Oleuropein also delayed onset of growth and sporulation of *Aspergillus parasiticus*, although it stimulated mycelial growth. Total production of aflatoxins was reduced by 83–93% in the presence of 6 mg/ml oleuropein (Gourama and Bullerman, 1987). The same behaviour was observed with another aflatoxin-producing mould, *Aspergillus flavus* NRRL 6555 (Mahjoub and Bullerman, 1987). Thus, olives are a poor substrate for mould development.

The inhibitory effect of phenolic compounds and some of their derivatives from olives on lactic acid bacteria has been well known for some time (Fleming and Etchells, 1967; Juvens *et al.*, 1968; Juvens and Henis, 1970; Fleming *et al.*, 1973; Walter *et al.*, 1973). Based on this wide antimicrobial spectrum some authors have seen oleuropein as an important part of the multi-chemical defence system of olive fruits together with other phenolic compounds like ligstroside and the physical barrier of oleanolic acid crystals (Kubo *et al.*, 1985).

Phenol-oxidase enzymes

In parallel with the presence of high levels of polyphenols, olive fruits are also rich in phenolic enzymes. Thus, catechol oxidase has been extensively described in olives and is the enzyme responsible for the change in colour of the fresh fruit from green to brown when olives are injured. It has been studied in detail during the ripening of the fruit by Ben-Shalon *et al.* (1977, 1978) and Angel and Ben-Shalon (1979). They have hypothesized that in green olives the enzyme was strongly bound to the chloroplasts whereas in the black fruits it is essentially soluble. During fruit development, the enzyme activity per fruit declines while the *ortho*-diphenols steadily increase. During the final stage of fruit development when the olives blacken the enzyme appeared predominantly in the soluble fraction. This change coincides with the appearance of anthocyanins and it is probably that at this stage, chloroplast membranes are disrupted and the lamellar structure disintegrates, facilitating the liberation of the enzyme. The browning potential of olives during fruit growth is determined by enzyme activity, substrate concentration and enzyme location. The enzyme has a molecular weight estimated as 42 000 Da and by electrophoresis gave seven to nine bands with catechol-oxidase activity. It did not oxidize mono-phenols and was strongly inhibited by diethyldithiocarbamate, while inhibition by chloride was pH-dependent, decreasing at pH values over 4.5.

Ben-Shalon *et al.* (1978) showed that browning of green olive fruits after mechanical injury was due to an enzymatic reaction catalysed by catechol-oxidase. This could not be prevented by inhibitors of the enzyme or by reducing agents due to the impermeability of the whole fruit to these compounds. However, dipping the fruits in weak sodium hydroxide solutions prevented the formation of the brown spots following mechanical injuries.

A detailed study of the olive catechol-oxidase was also carried out by Sciancalepore and Longone (1983a,b,c) with respect to its purification, properties, isoenzymes and electrophoretic behaviour. Comparative studies on susceptibility to browning have shown important differences among cultivars as well as the strong relationship between *ortho*-diphenol content and polyphenol oxidase activity. In contrast, there was no evidence linking peroxidase activity with browning (Sciancalepore, 1985). Among the Italian cultivars with lower rates of enzymatic browning were San Francisco and Passola, which can be considered as browning-proof since they have the lowest levels of both polyphenol oxidase and *ortho*-diphenols (Stefano and Sciancalepore, 1993a). These authors have found total inhibition of phenolases by $NaHSO_3$ at a concentration of 20 mg/l and also by 1.8 g/l of $CaCl_2$ (92% inhibition), probably due to the joint inhibition by Cl^- and Ca^{++}. Thus, this compound can be used to prevent olive browning, although there is still the problem of their slow penetration into the olive flesh from the surrounding solution. NaCl also showed an inhibitory effect, although concentrations needed were substantially higher (Stefano and Sciancalepore, 1993b).

Enzymatic browning studies in French cultivars (Picholin, Salonenque, Cailletier, Bouteillan and Lucques) agreed with previous results. The greatest browning potential occurred in green fruits and it was still considerable when olives were picked. The browning potential of olives resulted in this case from complex interaction between diphenol-oxidase activity and the oleuropein content. Inter-varietal comparison of olives picked at a similar stage of maturity showed that there was a positive correlation between browning and oleuropein content but no correlation with diphenol-oxidase activity (Goupy *et al.*, 1991). This is in contrast with most of the previous research on this enzyme.

Anthocyanins

The change in colour of olives from yellow to pink and black is due to the formation of anthocyanins. Vázquez Roncero and Maestro Durán (1970) studied these violet pigments in eight Spanish cultivars, finding that the colour was due to a mixture of five anthocyanins. The two found in greatest quantity were a cyanidin monoside and a cyanidin bioside which were identified as cyanidin 3-glucoside and cyanidin 3-rutinoside. The other three pigments were triosides, probably cyanidin 3-rutinoside and cyanidin 3-sophoroside derivatives. The third hexose, apparently glucose, may be linked to any of the free hydroxyl groups of the already attached sugars, forming linear or branched trisaccharides. The same authors, studying these pigments in Manzanilla olives, identified, in addition, another bioside, cyanidin 3-rutinoside acylated with caffeic acid and a further three triosides: cyanidin 3-glucosyl rutinoside (linear), cyanidin 3-glucosyl rutinoside (branched) and its caffeic acid acylated derivative (Maestro Durán and Vázquez Roncero, 1976).

The balance between the three types of glucosides is variable, depending on the cultivar and maturation stage. In general, anthocyanin formation begins when fruits reach the turning colour stage and increases rapidly to reach a maximum at about the end of November. It then decreases slightly during the rest of the period that the olive is attached to the tree. At first, the most abundant pigments are the triglucosides which decline up to mid-November and remain constant thereafter. Diglucosides increase in parallel with the loss of triglucosides and reach a steady state at the same time. Monoglucosides remain almost constant during the whole period. From December to full black-ripening, the ratio between mono-, di- and triglucosides is constant at around 15 : 60 : 25. The diglucoside : monoglucoside ratio is about 4.4 in Manzanilla, while in other cultivars such as Zorzaleña it can even approach unity. The relationship between the different anthocyanins appears to be characteristic of the maturation degree within each cultivar or group of similar cultivars.

The final content of anthocyanins may reach 0.5 g/100 g of fresh flesh, expressed as cyanidin, in practically all cultivars.

Similar results were obtained by Vlahov (1990a) who found that the proportion of polymeric anthocyanins increases with time during ripening and may reach 25% in advanced mature olives. This percentage may be strongly influenced by processing and, depending on cultivar, may increase to 70% when prepared as naturally black Greek-style olives, although the total content of anthocyanins decreases to below 50% of its original value (Vlahov, 1990b).

As in other products, anthocyanin biosynthesis in olives is strongly related to light. Olive fruits maintained in the dark during the growth period produced only a 10% of the amount found in similar fruits exposed to light. There is also a parallel delay in the general development of the fruit, which weigh less and are smaller. This means that, as with lipid biosynthesis, light plays an important role in olive maturation, not only when the fruit surface is green and there is endogenous photosynthetic activity, but also in more advanced stages of maturation. At least it is decisive for initiating the biochemical changes that lead to natural blackening.

Some agrochemicals seem to influence anthocyanin formation. Ethephon, when applied to the trees in September and in lower concentrations than those required to produce abscission of fruits, causes anthocyanins to appear earlier (Rugini *et al.*, 1982). Shulman and Lavee (1973) tested the effect of three cytokinins on green olives: kinetin, benzyl adenine and zeatin. All of them caused a marked increase in the anthocyanin content without inducing CO_2 production or softening. This suggests that cytokinins induce coloration directly and not via a ripening process since this is normally characterized by CO_2 and ethylene evolution as well as softening. NAA and 2,4,5-TP at concentrations of up to 200 and 300 p.p.m. respectively also caused anthocyanin accumulation but apparently as part of a general ripening induction. The same authors (1976) found considerable cytokinin activity in mature olive tissues. Its concentration suggests the presence of cytokinin-active nucleotides, probably due to an increase in ribonuclease and phosphatase activity in the maturing fruits and the decomposition of tRNA to compounds with cytokinin activity. However, there is no clear evidence as to whether the cytokinins in the mature olive have a biological function or are non-functional degradation products.

Acid hydrolysis of cytokinins from olive fruits gives products which retain their activity. This might indicate that they are complexes of simpler units, although they require a more severe hydrolysis to disrupt the complex than cytokinins from other fruits. As endogenous cytokinins induce anthocyanin accumulation in olive fruits and the level of endogenous cytokinin activity is considerably higher in black mature fruits than in green ones, it supports the speculation that cytokinins are causally involved in the process of olive fruit maturation.

Flavonoid compounds and their changes

Flavonoids are also present in the olive fruits, although their effect on table olive processing is still unknown. They have been studied by

Vázquez Roncero *et al.* (1970), Amiot *et al.* (1986) and Vlahov (1992). The main components are luteolin 7-glucoside, quercitin 4-rutinoside, apeginin 7-glucoside and quercitin 3-rhamnoside. In general, flavonoids are present from the first stages of fruit formation and levels progressively increase as maturation progresses.

Vlahov (1992) reported that the contents of quercitin 3-rutinoside and luteolin 7-glucoside at the different stages of maturation were higher in Cassano and Leccino cultivars than in Giarraffa. These varietal differences in flavonol and flavone composition were also paralleled by differences in anthocyanin formation. Thus, flavonoid composition might represent a useful contribution to biochemical characterization of olive fruit cultivars.

Chloroplastic pigments and their changes

Mínguez Mosquera and Garrido Fernández (1986, 1987, 1989) have studied the presence of chlorophylls and carotenoids in fruits of the Hojiblanca and Manzanilla olive cultivars. The qualitative composition is the same for both and does not change with maturation time. Chlorophylls are dominant, mainly chlorophyll a followed by chlorophyll b. Carotenoids are present as minor components in the following decreasing order of importance: lutein, β-carotene, violaxanthin and neoxanthin.

During growth and development of the fruits, a gradual and uniform decrease is observed in the individual concentrations of both chlorophylls and carotenoids. The Hojiblanca cultivar always shows a greater amount of pigments than Manzanilla cultivar, a fact which contributes to their differentiation. The authors consider that chlorophyll degradation during the ripening of the fruits could give rise to the biosynthesis of other components, such as anthocyanins. Table 5.9 shows distribution of chloroplastic pigment in both cultivars at different maturation stages and Table 5.10 shows the total pigment concentration (mg/kg) for the same stages.

The detection of de-esterified chlorophyll derivatives during table olive processing, as well as the drastic loss of chlorophyll pigments during olive oil extraction, have suggested the presence of chlorophyllase in olive fruits. Mínguez Mosquera *et al.* (1993, 1994) have found evidence for the activity of this enzyme. Olive chlorophyllase was detected in a crude enzymatic extract and showed maximum activity at 50°C and pH 8.5 for both chlorophyll a and b and pheophytins (a and b). Olive fruit chlorophyllase had a greater affinity for chlorophyll b, while the highest maximum rate of reaction occurred with pheophytin a. Substrate inhibition in the case of pheophytin b was also observed.

Table 5.9 Per cent distribution* of chloroplastic pigments in fruits of Hojiblanca and Manzanilla cultivars, for different maturation stages (Reproduced from Mínguez Mosquera, M.I. and Garrido Fernández, I. 1986)

Maturation stage	Chlorophyll a	Chlorophyll b	Lutein	β-Carotene	Violaxanthin	Neoxanthin
Manzanilla cultivar						
Intense green	68.2	16.0	7.2	4.0	2.4	2.4
Green	67.2	17.8	6.5	3.6	2.5	2.4
Yellowish green	66.8	18.0	6.6	3.6	2.5	2.5
Small reddish spots	66.9	18.1	7.4	4.0	1.9	2.0
Turning colour	63.0	20.1	8.5	4.6	2.0	1.9
Purple	59.3	13.4	14.4	7.7	3.1	2.1
Hojiblanca cultivar						
Intense green	66.2	17.3	6.7	4.0	2.6	2.9
Green	67.7	16.3	7.0	3.8	2.9	2.4
Yellowish green	68.0	16.8	6.7	3.9	2.5	2.2
Small reddish spots	66.9	16.8	7.6	4.1	2.4	2.5
Turning colour	62.5	18.2	9.9	5.2	2.1	2.6
Purple	49.5	14.9	19.7	9.5	2.7	2.7

* Percentages of total pigments.

Table 5.10 Total pigment concentration (mg/kg) in the fresh pulp of olives of Hojiblanca and Manzanilla cultivars, for different maturation stages (Reproduced from Mínguez Mosquera, M.I. and Garrido Fernández, J. 1986)

	Total pigments (mg/kg)	
Maturation stage	Hojiblanca	Manzanilla
Intense green	208.3	185.6
Green	165.8	137.5
Yellowish green	132.2	103.5
Small reddish spots	101.9	69.7
Turning colour	70.9	45.5

Protein content and its changes

Protein content of the fresh pulp is relatively low, generally between 1 and 3%, and remains almost constant during growth and ripening of the fruits (Fernández Díez, 1991). Non-protein or free nitrogen is only about 0.2–1.0% of total nitrogen. In the case of Koroneiki cultivar, its content was 1.1 mmol/g of fresh weight and its composition is described in Table 5.11 (Manoukas *et al.*, 1973). Arginine, alanine, aspartic acid, glutamic acid and glycine constituted approximately 60% of the free amino acids. Hydrolysis indicates that all the essential amino acids are present (Nosti Vega *et al.*, 1984).

The amino acid content of olive fruit protein may change with cultivar and degree of maturity. In Greek cultivars it ranged from 68.42 to 155.76 μmol/g of pulp fresh weight. Its composition is also shown in Table 5.11. Aspartic acid, glutamic acid and arginine were found predominant, comprising some 30% of the total amino acids. In general, it seems that the protein content is similar among cultivars and that its amino acid composition does not differ significantly, despite ecological and physiological differences.

Organic acids and their changes

Organic acids and their salts are present in the juice of fruits in concentrations between 0.5 and 1.0% based on the weight of the pulp. They are of great importance because of their buffering capacity during the fermentation stage and further storage of table olives. Fernández Díez and González Pellissó (1956) found oxalic acid (which ranged from 0.02% to 0.41%, fresh weight basis), malic acid (0.01–0.17%) and citric acid (0.01–0.24%) in Gordal, Manzanilla and Rapazalla cultivars. The total organic acid content tends to increase during growth from an average of 0.182% to 0.266% from June to September in the above cultivars. During

Table 5.11 Average amino acid composition of olive fruits. Koroneiki cultivar (Reprinted with permission from Manoukas, A.G. *et al.*, 1973)

Amino acid	Percentage of total free amino acids	Composition (g/100 g protein or g/16 g nitrogen)
Arginine	9.6	9.75
Histidine	0.7	2.69
Lysine	1.4	4.66
Methionine	1.1	1.32
Cystine	Traces	Traces
Phenylalanine	3.0	2.97
Tyrosine	2.0	2.39
Leucine	5.7	5.17
Isoleucine	5.0	3.09
Threonine	1.5	2.80
Valine	7.9	3.96
Alanine	13.1	5.18
Aspartic acid	15.6	11.46
Glutamic acid	9.9	11.70
Glycine	11.1	6.28
Proline	4.1	4.82
Serine	8.4	3.73

the same period, however, pH also increased slightly due to the fact that there was a greater parallel increase of the combined acidity, which changed from an average of 98.5 mM to 157.3 mM. Later, Vlahov (1976) studied the changes in oxalic (0.19 to 1.33% dry flesh weight basis), succinic (0.04–0.18%), malic (3.34–16.95%), tartaric (0.26–0.96%) and citric acid (1.16–7.65%) from August to December (full black maturation) in Ascolana and Leccino cultivars. Throughout this period there was a continuous increase of citric acid, while malic acid decreased. The malate : citrate ratio was inversely correlated to the oil formation. The maximum oil content was reached when the ratio was about 1; values lower than this would indicate over-mature fruits while higher values can indicate fruits that had still not reached the optimum growth. When this index is applied at a certain date, it can also indicate the relative maturation rate of each cultivar with respect to the others (low values were found for early maturation cultivars).

Mineral and vitamin content

Ash percentage varies from 0.6 to 1% of the fresh pulp. The most relevant elements include (in decreasing order of importance) K, Ca, P, Na, Mg and S and, to a lesser extent, Fe, Zn, Cu and Mn (Manoukas *et al.*, 1978; Nosti Vega *et al.*, 1984) (Table 5.12).

Table 5.12 Inorganic nutrients (p.p.m.) in the fresh pulp of some table olive cultivars in different stages of maturity (Adapted from Nosti Vega *et al.*, 1984; Manoukas *et al.*, 1978)

Mineral	Gordal (green)	Manzanilla (green)	Megaritiki (green)	Hojiblanca (black)
K	2820	2830	3222	2840
Ca	190	510	400	530
P	240	290	390	460
Na	170	80	40	20
Mg	90	140	180	180
S	50	40	–	150
Fe	19	39	14	13
Zn	5	5	8	7
Cu	5	4	–	6
Mn	1	2	2	1

The presence of vitamins such as carotenes, thiamin and riboflavin, have also been reported (Nosti Vega *et al.*, 1984; Fernández Díez *et al.*, 1985; Nosti Vega and Castro Ramos, 1985).

5.5 REFERENCES

Aligisakis, E.M. and Patroclos, G.A. (1971) Contribution a l'étude des changements survenant dans la composition chimique de l'olive de la variété Conservolea au cours de sa croissance et de sa maturation. Compte Rendu de la III Confèrence Internationale des Techniques Oléicoles. Torremolinos, Espagne.

Amiot, M.J., Fleuriet, A. and Macheix, J.J. (1986) Importance and evolution of phenolic compounds in olive during growth and maturation. *J. Agric. Food Chem.*, **34**, 823–6.

Amiot, M.J., Fleuriet, A. and Macheix, J.J. (1989) Accumulation of oleuropein derivatives during olive maturation. *Phytochemistry*, **28**, 67–9.

Angel, S. and Ben-Shalon, N. (1979) Research on olive quality in Israel for processing. I. Prevention of brown spotting in green olives for pickling. *Confructa*, **24**, 79–84.

Balatsouras, G., Papoutsis, G. and Papamichael-Balatsouras, V. (1988) Changes in olive fruits of 'Conservolea' during development viewed from the standpoint of green and black pickling. *Olea*, **19**, 43–5.

Barone, E., Gullo, G., Zappia, R. and Inglese, P. (1994) Effect of crop load on fruit ripening and olive oil (*Olea europaea*, L.) quality. *J. Horti. Sci.*, **69**, 67–73.

Ben-Shalon, N., Kahn, V., Harel, E. and Mayer, A.M. (1977) Olive catechol oxidase. Changes during fruit development. *J. Sci. Fd. Agric.*, **28**, 545–50.

Ben-Shalon, N., Harel, E. and Mayer, A.M. (1978) Enzymic browning in green olives and its prevention. *J. Sci. Fd. Agric.*, **29**, 398–402.

Bianchi, G. and Pozzi, N. (1994) 3-4-Dihydroxyphenylglycol, as a major C_6–C_2 phenolic in *Olea europaea* fruits. *Phytochemistry*, **35**, 1335–7.

Bianchi, G., Murelli, C. and Vlahov, G. (1992) Surface waxes from olive fruits. *Phytochemistry*, **31**, 3503–6.

Bianco, A., Scalzo, R.L. and Scarpatti, M.L. (1993) Isolation of cornoside from *Olea europaea* and its transformation into halleridone. *Phytochemistry*, **33**, 455–7.

Borbolla y Alcalá, J.M.R. de la, Fernández Díez, M.J. and González Pellissó, F. (1955) Cambios en la composición de la aceituna durante su desarrollo. I. *Grasas y Aceites*, **5**, 5–22.

Bourquelot, E. and Vintilesco, J. (1908) Sur l'oleuropein, nouveau principe de nature glucosidique retiré de l'olivier (*Olea europaea*, L.) *Compt Rend. Hebd. Acad. Sci.* (Paris), **147**, 533–5.

Brenes Balbuena, M., García García, P. and Garrido Fernández, A. (1992) Phenolic compounds related to the black colour formed during the processing of ripe olives. *J. Agric. Food Chem.*, **40**, 1192–6.

Castillo Gómez, J., Mínguez Mosquera, M.I. and Fernández Díez, M.J. (1978a) Presencia de poligalacturonasa (PG) y su relación con el ablandamiento en algunos productos empleados en la industria del aderezo (pimiento y aceitunas). *Grasas y Aceites*, **29**, 97–101.

Castillo Gómez, J., Mínguez Mosquera, M.I. and Fernández Díez, M.J. (1978b) Presencia de poligalacturonasa (PG) en la aceituna negra madura. Factores que influencian la actividad de dicha enzima. *Grasas y Aceites*, **29**, 333–8.

Castillo Gómez, J., Mínguez Mosquera, M.I. and Fernández Díez, M.J. (1979) Presencia de inhibidores de poligalacturonasa en la aceituna negra madura (variedad Hojiblanca). *Grasas y Aceites*, **30**, 11–18.

Chung, J.I., Sekamura, S. and Luh, B.S. (1974) Effect of harvest maturity on pectin and texture of canned black olives (*Olea europaea*, L.). *Confructa* **19**(5), 227–36.

Donaire, J.P., Belver, A., Rodríguez García, M.I. and Megías, L. (1984) Lipid biosynthesis, oxidative enzyme activities and cellular changes in growing olive fruit. *R. Esp. Fisiología*, **40**, 191–204.

Durán Grande, M. (1977) Estudio morfológico de los estomas y de las areas estomáticas del fruto de *Olea europaea*, L. *Bol. R. Soc. Española Hist. Nat.*, **75**, 23–41.

Durán Grande, M. and Izquierdo Tamayo, A. (1965) Estudio sobre la estructura histológica del fruto *Olea europaea*, L. I. Variedad Zorzaleña. *Bol. R. Soc. Española. Hist. Nat. (Biol)*, **63**, 271–91.

Fernández Bolaños, J., Fernández Díez, M.J., Rivas Moreno, M. and Gil Serrano, A. (1982) Azúcares y polioles en aceitunas verdes. II. Identificación y determinación cuantitativa por cromatografía sobre papel. *Grasas y Aceites*, **33**, 208–11.

Fernández Bolaños, J., Fernández Díez, M.J., Rivas Moreno, M., Gil Serrano, A. and Pérez Romero, T. (1983) Azúcares y polioles en aceitunas verdes. III. Determinación cuantitativa por cromatografía gas-líquido. *Grasas y Aceites*, **34**, 168–71.

Fernández Bolaños, J.M., Gil Serrano, A.M. and Tejero Mateo, M.P. (1984) Polisacáridos en aceitunas. I. Aislamiento y purificación de hemicelulosas en aceitunas de las variedades Manzanilla y Gordal. *Grasas y Aceites*, **35**, 1–3.

Fernández Díez, M.J. (1971) The Olive, in *The Biochemistry of Fruits and Their Products*. Vol 2 (ed. A.C. Hulme), Academic Press, London, pp. 255–79.

Fernández Díez, M.J. (1979) Texture of table olives and pimentos. *Journal of Texture Studies*, **10**, 103–16.

Fernández Díez, M.J. (1983) Olives, in *Biotechnology*, Vol 5 (eds H.J. Rehm and G. Reed). Verlag Chemie, Weinheim, Germany, pp. 379–97.

Fernández Díez, M.J. (1991) Olives, in *Encyclopedia of Food Science and Technology*, Vol. 3 (ed Y.H. Hui), Wiley & Sons, New York, pp. 1910–25.

Fernández Díez, M.J. and Garrido Fernández, A. (1971) Ripe black olives. I. Colour as a criterion for maturity in fresh fruits and in the final product. *Grasas y Aceites*, **22**, 193–9.

Fernández Díez, M.J. and González Pellissó, F. (1956) Cambios en la composición de la aceituna durante su desarrollo. II. Acidez y pH del jugo. Determinación de ácidos oxálico, cítrico y málico. *Grasas y Aceites*, **7**, 185–9.

Fernández Díez, M.J., Garrido Fernández, A., González Cancho, F., Durán Quintana, M.C. and Cordón Casanueva, J.L. (1972) Elaboración de aceitunas negras de mesa. Instituto de la Grasa, Sevilla, Spain.

Fernández Díez, M.J., Mínguez Mosquera, M.I. and Castillo Gómez, J. (1974) Composición péctica y presencia de pectinesterasa en algunos productos del aderezo. *Grasas y Aceites*, **25**, 331–40.

Fernández Díez, M.J.; Fernández Bolaños, J.; Heredia Moreno, A.; Rivas Moreno, M. and Iglesias Guerra, F. (1981) Azúcares en aceitunas verdes. I. Determinación enzimática de glucosa. *Grasas y Aceites*, **32**, 83–5.

Fernández Díez, M.J., Castro Ramos, R., Garrido Fernández, A., González Cancho, F., González Pellissó, F., Nosti Vega, M., Heredia Moreno, A., Mínguez Mosquera, M.I., Rejano Navarro, L., Durán Quintana, M.C., Sánchez Roldán, F., García García, P. and Castro Gómez Millán, A. (1985) *Biotecnología de la aceituna de mesa*. Consejo Superior de Investigaciones Científicas, Instituto de la Grasa, Madrid y Sevilla.

Fernández Villasante, J. and Fernández Díez, M.J. (1968) Classification des olives vertes d'après leur degré de maturité par flottage en saumure. *Inf. Oleic. Int.*, **42**, 59–63.

Fleming, H.P. and Etchells, J.L. (1967) Occurrence of an inhibitor of lactic acid bacteria in green olives. *Appl. Microbiol.*, **15**, 1178–84.

Fleming, H.P., Walter, W.M.J.R. and Etchells, J.L. (1973) Antimicrobial properties of oleuropein and products of its hydrolysis from green olives. *Appl. Microbiol.*, **26**, 777–82.

Fleuriet, A., Marcheix, J.J., Andary, C. and Villemur, P. (1984) Mise en evidence et dosage par chromatographie liquide à haute performance du verbascoside dans le fruit de six cultivars d' *Olea europaea*, L. *C. R. Acad. Sci. Paris*, t. 299, Série III, no. 7, pp. 253–6.

Frega, N. and Lercker, G. (1985) La composizione dei lipidi della drupa di olivo durante la maturazione. *Agrochimica*, **XXIX-N.2-3-4**, 300–8.

Garrido Fernández, A. and Fernández Díez, M.J. (1972) Black table olives. Changes in the main physico-chemical characteristics of olive fruits during growth and ripening. *Grasas y Aceites*, **23**, 309–18.

Garrido Fernández, A. and Fernández Díez, M.J. (1976) Estudio crítico de la determinación colorimétrica de polifenoles y su aplicación a las salmueras de aceitunas negras de mesa. *Grasas y Aceites*, **27**, 107–12.

Gil Serrano, A.M., Tejero Mateo, M.P. and Fernández Bolaños, J.M. (1984a) Polisacáridos en aceitunas. II. Determinación de los pesos moleculares medios de hemicelulosas aisladas en aceitunas de las variedades Manzanilla y Gordal. *Grasas y Aceites*, **35**, 89–92.

Gil Serrano, A.M., Tejero Mateo, M.P. and Fernández Bolaños, J.M. (1984b)

Polisacáridos en aceitunas. III. Hidrólisis parcial y acetolisis de una hemicelulosa B aislada en aceitunas de la variedad Manzanilla. *Grasas y Aceites*, **35**, 224–7.

Gil Serrano, A.M., Mateos Matos, M.I. and Tejero Mateo, M.P. (1986) Acidic xylan from olive pulp. *Phytochemistry*, **25**, 2653–4.

González Cancho, F., Nosti Vega, M., Durán Quintana, M.C., Garrido Fernández, A. and Fernández Díez, M.J. (1975) El proceso de fermentación en las aceitunas negras maduras en salmuera. *Grasas y Aceites*, **26**, 297–309.

Goupy, P., Fleuriet, A., Amiot, M.J. and Macheix, J.J. (1991) Enzymatic browning, oleuropein content and diphenol oxidase activity in olive cultivars (*Olea europaea*, L.). *J. Agric. Food Chem.*, **39**, 92–5.

Gourama, H. and Bullerman, L.L.B. (1987) Effects of oleuropein on growth and aflatoxin production by *Aspergillus parasiticus*. *Lebensm. Wiss. u. Technol.*, **20**, 226–8.

Gracian Tous, J. (1968) Chemistry and analysis of olive oil, in *Analysis and Characterization of Oils, Fats and Fat Products*, Vol. 2 (ed. H.A. Boekenoogen). Interscience, London, pp. 315–606.

Guillén, R., Heredia Moreno, A., Felizón, B., Jiménez, A. and Fernández Bolaños, J. (1991) Preparación y caracterización de fracciones de fibra en aceitunas (variedad Hojiblanca). *Grasas y Aceites*, **42**, 334–8.

Guillén, R., Heredia, A., Felizón, B., Jiménez, A., Montaño, A. and Fernández Bolaños, J. (1992) Fibre fractions carbohydrates in *Olea europaea* (Gordal and Manzanilla cultivars). *Food Chemistry*, **44**, 173–8.

Hartman, H.T. (1949) Growth of the olive fruit. *Proc. Am. Soc. Hort. Sci.*, **54**, 88–94.

Heredia Moreno, A. (1979) Composición de fibra, I. Métodos de determinación de lignina y celulosa. *Grasas y Aceites*, **30**, 105–11.

Heredia Moreno, A. (1980) Composición de fibra. IV. Una comparación de métodos de fibra en aceitunas. *Grasas y Aceites*, **31**, 251–4.

Heredia Moreno, A. and Fernández Díez, M.J. (1979) Composición de fibra, II. Lignina y celulosa como componentes de la fibra de aceitunas. *Grasas y Aceites*, **30**, 141–3.

Heredia Moreno, A. and Fernández Bolaños, J. (1985a) Celulasas en aceitunas y su posible influencia en los cambios de textura. II. Actividad celulolítica en la variedad Hojiblanca. *Grasas y Aceites*, **36**, 130–3.

Heredia Moreno, A. and Fernández Bolaños, J. (1985b) Celulasas en aceitunas y su posible influencia en los cambios de textura. III. Estudio de algunos factores que pueden modificar su actividad. *Grasas y Aceites*, **36**, 171–6.

Heredia Moreno, A. and Fernández Bolaños, J. (1987) Evolución de la fibra alimentaria durante el desarrollo y maduración del fruto del olivo (*Olea europaea arolensis*). *Alimentaria*, **181**, 19–22.

Heredia Moreno, A. and Mínguez Mosquera, M.I. (1981) Celulasas en aceitunas y su posible influencia en los cambios de textura. I. Estudio de estos enzimas. *Grasas y Aceites*, **32**, 319–25.

Heredia Moreno, A., Fernández Bolaños, J. and Guillén Bejarano, R. (1988) Actividad celulolítica en aceitunas durante su maduración y ablandamiento. *Alimentaria*, **198**, 49–52.

Heredia Moreno, A., Fernández Bolaños, J. and Guillén Bejarano, R. (1989a) Caracterización y purificación parcial de enzimas celulolíticas en aceitunas. *Grasas y Aceites*, **40**, 190–3.

Heredia Moreno, A., Fernández Bolaños, J. and Guillén Bejarano, R. (1989b)

Inhibitors of cellulolytic activity in olive fruits (*Olea europaea*, Hojiblanca var.). *Z. Lebens Forsch.*, **189**, 216–18.

Heredia Moreno, A., Fernández Bolaños, J. and Guillén, R. (1990a) Cellulase inhibition by polyphenols in olive fruits. *Food Chemistry*, **38**, 69–73.

Heredia Moreno, A., Fernández Bolaños, J., Guillén, R., Vioque, B. and Fernández, J.C. (1990b) Cellulases in cell-wall degrading extracts of *Olea europaea arolensis*. *Z. Lebensm. Unters Forsch.*, **190**, 22–4.

Heredia Moreno, A., Fernández Bolañós, J. and Guillén, R. (1991) Identification of endoglucanases in olives (*Olea europaea arolensis*). *Z. Lebensm. Unters Forsch.*, **192**, 1531–4.

Heredia Moreno, A., Guillén, R., Jiménez, A. and Fernández Bolaños, J. (1992) Olive fruits glycosidases: factors affecting their extraction. *Z. Lebensm. Unters Forsch.*, **194**, 561–5.

Heredia Moreno, A., Guillén, R., Jiménez, A. and Fernández Bolaños, J. (1993) Activity of glycosidases during development and ripening of olive fruit. *Z. Lebensm. Unters Forsch.*, **196**, 147–51.

Inouye, H., Yoshida, T., Tobita, S., Tanaka, K. and Nishioka, T. (1974) Uber die Monoterpenglucoside und verwandte Naturstoffe. XXII. Absolutstrukturen des Oleuropeins, Kingisids und Mononisids. *Tetrahedron*, **30**, 201–9.

Izquierdo Tamayo, A. and Durán Grande, A. (1970) Estudios sobre la estructura histológica del fruto *Olea europaea* L. II. Variedad Manzanilla. *Biol. R. Soc. Española Hist. Nat. (Biol.)*, **68**, 167–79.

Jiménez, A., Guillén, R., Fernández Bolaños, J. and Heredia, A. (1994) Cell wall composition of olives. *J. Food Sci.*, **59**, 1192–6.

Juvens, B. and Henis, Y. (1970) Studies on the antimicrobial activity of olive phenolic compounds. *J. Appl. Bact.*, **33**, 721–32.

Juvens, B., Samish, Z. and Henis, Y. (1968) Identification of oleuropein as a natural inhibitor of lactic fermentation of green olives. *Israel J. Agric. Res.*, **18**, 137–8.

Kubo, I. and Matsumoto, A. (1984) Molluscicides from olive *Olea europaea* and their efficient isolation by countercurrent chromatographies. *J. Agric. Food Chem.*, **32**, 687–8.

Kubo, I., Matsumoto, A. and Takase, I. (1985) A multichemical defense mechanism of bitter olive *Olea europaea* (*Oleaceae*). Is oleuropein a phytoalexin precursor. *J. Chem. Ecol.*, **11**, 251–63.

Maestro Durán, R. and Vázquez Roncero, A. (1976) Colorantes antocianicos de las aceitunas manzanillas maduras. *Grasas y Aceites*, **27**, 237–43.

Maestro Durán, R., García, J.M. and Castellano, J.M. (1993) Changes in polyphenol content of olives stored in modified atmospheres. *HortScience*, **28**, 979.

Mahjoub, A. and Bullerman, LL.B. (1987) Effects of nutrients and inhibitors in olives on aflatogenic moulds. *J. Food Protection*, **50**, 959–63.

Manoukas, A.G., Mazomenos, B. and Patrinou, M.A. (1973) Aminoacido composition of three cultivars of olive fruit. *J. Agr. Food Chem.*, **21**, 215–17.

Manoukas, A.G., Grimanis, A. and Mazomenos, B. (1978) Inorganic nutrients in natural and artificial food of *Dacus oleae* larvas (Diptera: Tephritidae). *Ann. Zool. Ecol. Anim.* **10**(1), 123–8.

Maranto, J. and Krueger, W. H. (1994) Olive fruit thinning, in *Olive production manual*. Publication 3353. University of California, Division of Agriculture and Natural Resources, Oakland, California, USA, pp. 87–9.

Marsilio, V. and Solinas, M. (1990) Presenza e comportamento delle sostanze pectiche nelle olive da tavola. Influenza della varieta e della tecnologia, in *Proceedings of the International Seminar on Olio d'Olivo e Olive da tavola: Tecnologia e qualitá*. Istituto Sperimentale per la Elaiotecnica. Pescara, Italy, pp. 331–56.

Martín, C.G., Ferguson, L. and Polito, V.S. (1994) Flowering, pollination, fruiting, alternate bearing, and abscission, in *Olive production manual*. Publication 3353. University of California, Division of Agriculture and Natural Resources, Oakland, California, USA, pp. 51–6.

Mataix Verdú, F.J. and Martínez de Victoria Muñoz (1988) El aceite de oliva (Bases para el futuro). Junta de Andalucía. Consejería de Agricultura y Pesca. Dirección General de Investigación y Extensión Agraria, Sevilla, Spain.

Mínguez Mosquera, M.I. (1982) Evolución de los constituyentes pécticos y de las enzimas pectinolíticas durante el proceso de maduración y almacenamiento de la aceituna Hojiblanca. *Grasas y Aceites*, **33**, 327–33.

Mínguez Mosquera, M.I. and Garrido Fernández, J. (1986) Composición y evolución de clorofilas y carotenoides durante el desarrollo y maduración de los frutos del olivo. *Grasas y Aceites*, **37**, 337–42.

Mínguez Mosquera, M.I. and Garrido Fernández, J. (1987) Diferenciación de las variedades de olivo Hojiblanca y Manzanilla según su contenido pigmentario. *Grasas y Aceites*, **38**, 4–8.

Mínguez Mosquera, M.I. and Garrido Fernández, J. (1989) Chlorophyll and carotenoid presence in olive fruit (*Olea europaea*). *J. Agr.Food Chem.*, **37**, 1–7.

Mínguez Mosquera, M.I., Castillo Gómez, J. and Fernández Díez, M.J. (1976) Variaciones en la composición péctica durante la elaboración, fermentación, y conservación de productos aderezados (pimiento y aceituna). *Grasas y Aceites*, **27**, 27–32.

Mínguez Mosquera, M.I., Castillo Gómez, J. and Fernández Díez, M.J. (1978) Presencia de pectinesterasa y su relación con el ablandamiento en algunos productos del aderezo. *Grasas y Aceites*, **29**, 29–36.

Mínguez Mosquera, M.I., Gandul Rojas, B. and Gallardo Guerrero, L. (1993) De-esterification of chlorophylls in olives by activation of chlorophyllase. *J. Agr. Food Chem.*, **41**, 2254–8.

Mínguez Mosquera, M.I., Gandul Rojas, B. and Gallardo Guerrero, L. (1994) Measurement of chlorophyllase activity in olive fruit (*Olea europaea*). *J. Biochem.*, **116**, 263–8.

Montaño Asquerino, A., Rejano Navarro, L. and Sánchez Gómez, A.H. (1985) Determinación de polifenoles totales en extractos de pulpa de aceitunas y en salmueras de fermentación por espectroscopia de derivadas. *Grasas y Aceites*, **36**, 274–9.

Nosti Vega, M. and Castro Ramos, R. de. (1985) Composición y valor nutritivo de algunas variedades españolas de aceitunas de mesa. VII. Aceitunas negras oxidadas. *Grasas y Aceites*, **36**, 203–6.

Nosti Vega, M., Castro Ramos, R. de and Vázquez Ladrón, R. (1984) Composición y valor nutritivo de algunas variedades españolas de aceitunas de mesa. VI. Cambios debidos a los procesos de elaboración. *Grasas y Aceites*, **35**, 11–14.

Panizzi, L., Scarpati, M.L. and Oriente, G. (1960) Costituzione della oleuropeina, glucoside amaro e ad azione ipotensiva dell'olivo. Nota II. *Gazz. Chim. Ital.*, **90**, 1449–85.

Ragazzi, E. and Veronese, G. (1973a) Quantitative analysis of phenolic compounds after thin-layer chromatographic separation. *J. Chromatogr.*, **77**, 369–75.

Ragazzi, E. and Veronese, G. (1973b) Indagini sulla presenza di demetiloleuropeina in olive di cultivar diverse. *Ann. Chim.*, **63**, 21–7.

Ragazzi, E., Veronese, G. and Guiotto, A. (1973) Demetil oleuropeina, nuovo glucoside isolato da olive mature. *Ann. Chim.*, **63**, 13–20.

Rugini, E., Bongi, G. and Fontanazza, G. (1982) Effects of Ethephon on olive ripening. *J. Am. Soc. Hort. Sci.*, **107**, 835–8.

Sánchez, J. (1994) Lipid photosynthesis. *Prog. Lipid Res.*, **33**(1/2), 97–104.

Sánchez Gómez, A.H. and Fernández Díez, M.J. (1991) Correlación entre materia grasa, azúcares reductores, y humedad en la pulpa de aceitunas. *Grasas y Aceites*, **42**, 414–19.

Sánchez Raya, A.J., Gómez, M. and Leal, A. (1981) Criterios indicativos de la maduración de la aceituna. *Anales de Edafología y Agrobiología*, **40**, 905–10.

Sandret, F.G. (1958) La maturation de la pulpe de la olive. *Oleagineux*, **13**, 459–64.

Sciancalepore, V. (1985) Enzymatic browning in five olive cultivars. *J. Food Sci.*, **50**, 1194–5.

Sciancalepore, V. and Longone, V. (1983a) Researches on catechol oxidase from green olives. Note I. Partial purification and some properties of the enzyme. *Industrie Alimentari*, **22**, 549–53.

Sciancalepore, V. and Longone, V. (1983b) Researches on catechol oxidase from green olives. Note II. Isoenzymes. *Industri Alimentari*, **22**, 640–4.

Sciancalepore, V. and Longone, V. (1983c) Researches on catechol oxidase from green olives. Note III. Polyacrylamide gel electrophoresis analysis. *Industri Alimentari*, **22**, 745–8.

Serrano, A., Mateos Matos, M.I. and Tejero Mateo, M.P. (1986) Acidic xylan from olive pulp. *Phytochemistry*, **25**, 2653–4.

Shulman, Y. and Lavee, S. (1973) The effect of cytokinins and auxins on anthocyanin accumulation in green Manzanilla olives. *J. Exp. Botany*, **24**, 655–61.

Shulman, Y. and Lavee, S. (1976) Endogenous cytokinins in maturing Manzanilla olive fruits. *Plant Physiol.*, **57**, 490–2.

Solinas, M., Giovacchino, L. di and Cucurachi, A. (1975) I polifenoli delle olive e dell'olio di oliva. Nota I: Variazione che subisceno alcuni polifenoli delle olive col procedere della maturatione. *Ann. Ist. Sper. Elaiot.*, **5**, 129–54.

Stefano, G. and Sciancalepore, V. (1993a) Polyphenol oxidase in five olive cultivars. *Industrie Alimentari*, **XXXII**, 731–3.

Stefano, G. and Sciancalepore, V. (1993b) Effect of some substances on green olive phenolase. *Industrie Conserve*, 68–138.

Tassou, C.C. and Nychas, J.E. (1994) Inhibition of *Staphylococcus aureus* by olive phenolic in broth and in a model food system. *J. Food Protection*, **57**, 120–4.

Tassou, C.C., Nychas, G.Y.E. and Board, R.G. (1991) Effect of phenolic compounds and oleuropein on the germination of *Bacillus cereus* T spores. *Biotech. Appl.Biochem.*, **13**, 231–7.

Tejero Mateo, M.P., Gil Serrano, A.M. and Fernández Bolaños, J.M. (1985a) Polisacáridos en aceitunas. IV. Estudio de una xilana aislada de la pulpa de aceitunas de la variedad Gordal. *Anales de Química de la Real Sociedad Española de Física y Química*, **81**, 214–16.

Tejero Mateo, M.P., Gil Serrano, A.M. and Fernández Bolaños, J.M. (1985b) Polisacáridos en aceitunas. V. Análisis estructural de una xilana aislada de la pulpa de aceitunas de la variedad Gordal. *Anales de Química de la Real Sociedad Española de Física y Química*, **81**, 217–19.

Tejero Mateo, M.P., Gil Serrano, A.M. and Fernández Bolaños, J.M. (1986a)

Polisacáridos en aceitunas. VI. Estudio de una galactoglucomanana aislada de la pulpa de aceitunas de la variedad Gordal. *Anales de Química de la Real Sociedad Española de Física y Química*, **82**, 155–7.

Tejero Mateo, M.P., Gil Serrano, A.M. and Fernández Bolaños, J.M. (1986b) Polisacáridos en aceitunas. VII. Estudio de una xiloglucana aislada de la pulpa de aceitunas de la variedad Gordal. *Anales de Química de la Real Sociedad Española de Física y Química*, **82**, 158–61.

Tous Marti, J. (1990) Características botánicas y fisiológicas, in *Olivo*, Exma Diputación, Tarragona, Spain, pp. 103–21.

Tutour, B. le and Guedon, D. (1992) Antioxidative activities of *Olea europaea* leaves and related phenolic compounds *Phytochemistry*, **31**, 1173–8.

Uzzan, A., Pelle, C. and Mahdi, M. (1956) Determination de la maturité des olives. *Oleagineux*, **11**, 405–13.

Vázquez Roncero, A. and Maestro Durán, R. (1970) Los colorantes antocianicos de la aceituna madura. I. Estudio cualitativo. *Grasas y Aceites*, **21**, 208–14.

Vázquez Roncero, A. and Mancha Perelló, M. (1965) Componentes químicos de la aceituna. II. La presencia de diglicéridos naturales en la aceituna y el aceite de oliva virgen. *Grasas y Aceites*, **16**, 13–16.

Vázquez Roncero, A. and Mancha Perelló, M. (1970) Transformaciones de los glicéridos durante la maduración de la aceituna. Variaciones de los di y triglycéridos. *Grasas y Aceites*, **21**, 124–32.

Vázquez Roncero, A., Vioque, E. and Mancha Perelló, M. (1965) Componentes químicos de la aceituna. III. Variaciones de los componentes liposolubles durante la maduración. *Grasas y Aceites*, **16**, 17–23.

Vázquez Roncero, A., Maestro Durán, R. and Janer del Valle, M.L. (1970) Colorantes antocianicos de las aceitunas II. Variaciones durante la maduración. *Grasas y Aceitas*, **21**, 337–341.

Vázquez Roncero, A., Maestro Durán, R. and Graciani Constante, E. (1971) Changes in polyphenols during olive fruit growth. *Grasas y Aceites*, **22**, 366–70.

Vázquez Roncero, A., Graciani Constante, E. and Maestro Durán, R. (1974) Componentes fenólicos de la aceituna. I. Polifenoles de la pulpa. *Grasas y Aceites*, **25**, 269–79.

Vlahov, G. (1976) Gli acidi organici delle olive: Il rapportto malico/citrico quelle indice de maturazione. *Ann. Ist. Sper. Elaiot. II.*, 93–112.

Vlahov, G. (1990a) La polimenizzazione degli antociani nelle olive nere da mensa. Ruolo delle tecnologie. *International Seminar on Olio d'Oliva e Olive da Tavola. Tecnologia e qualitá. Istituto Sperimentale per la elaiotecnica*, Pescara, 25–28 April, pp. 357–69.

Vlahov, G. (1990b) Gli antociani polimeri delle olive nere. Loro determinatione in funzione del processo di maturazione. Nota I. *Industrie Alimentari*, **XXIX**, 11–18.

Vlahov, G. (1992) Flavonoids in three olive (*Olea europaea*) fruit varieties during maturation. *J. Sci. Food Agric.*, **58**, 157–9.

Walter, W.M., Fleming, H.P. and Etchells, J.L. (1973) Preparation of antimicrobial compounds by hydrolysis of oleuropein from green olives. *Appl. Microbiol.*, **26**, 773–6.

Wodner, M., Lavee, S. and Epstein, E. (1988) Identification and seasonal changes of glucose, fructose, and mannitol in relation to oil accumulation during fruit development in *Olea europaea* (L). *Scientia Horticulturae*, **36**, 47–54.

Picking and post-harvest operations

6

6.1 OPTIMUM PICKING PERIOD

Harvest operations for table olives are mainly dependent on the required olive type but are also influenced by cultivar, geographical situation, climate, and cultivation system (Fernández Díez *et al.*, 1985).

Olives for the green type, mainly used for the preparation of green Spanish-style table olives in brine and, to a lesser extent, untreated green olives in brine, are harvested when fruits have reached a green–straw yellow colour and the pit can be freed easily from the flesh by finger torsion once the fruit has been cut around the transverse diameter. At this stage, a characteristic white juice is also obtained when the fruit is gently squeezed with the fingers. If the fruit surface presents a more than light rose colour, some texture deficiencies may develop in the pickled product. In Spain, harvesting the earliest cultivar, Gordal, starts in the first 15 days of September, and for other cultivars about 15–20 days after that, depending upon the cultivation area.

Fruits for turning colour type are harvested when the epicarp presents a purple colour and the mesocarp is still white. Generally in Spain, the crop is picked between the end of September and the first 15 days of December, depending upon cultivar and production area.

Olives for black-ripe olive types (treated olives or ripe olives) are harvested at almost the same time as turning colour olives. Since they are further oxidized after the lye treatment, the permissible range of colour at picking is wider than for turning colour type. However, the timing of picking has still to be studied and fixed very carefully for each cultivar because deep-coloured fruits give a poor texture in the final product and those which are not ripe enough do not develop a good black colour after oxidation. In the USA the colour range considered ideal for picking extends from an even pale green with a minimum of whitish spots through to a straw colour. A red blush is permissible on Manzanilla, Mission and Sevillano olives early in the season, but such colour is considered undesirable late in the season (Martin *et al.*, 1994).

Olives used for naturally black type must be harvested when completely ripe; superficial colour is not a good criterion in this case and it is necessary to observe the flesh colour in representative samples by cutting along the fruit, close to the pit. A good stage of maturity is when the purple colour reaches to within 2 mm of the pit. Harvesting time for this type in Spain generally ranges from the middle of November to the end of February and even longer for some cultivars and areas.

In general, the harvesting date for each table olive type is constrained by the characteristics of the product, mainly their colour and texture. In the case of green olives, fruits are frequently picked before they reach full development and farmers are usually concerned about getting slightly less than the maximum potential crop weight. Often though, the higher prices obtained for these olives more than compensate for such losses. With naturally black olives the maximum fruit development has always been reached. However, losses might occur if olives are permitted to remain too long on the trees, since over-ripe fruits progressively lose moisture and frost damage may occur. Also, sensory examination of processed olive fruits picked at different maturation stages showed a slight increase in off-flavours as more mature olives were used for processing (Karaoulanis and Markakis, 1987).

With turning and black-ripe olives, picking dates are not as strict, and a certain flexibility is possible. This has permitted Ferguson *et al.* (1991) to establish a simple method based on fruit size to predict the timing of harvest and maximize farmers' profits. It is based on the fact that individual olives, within their size classification, do not change weight significantly over the harvest season and that weight bears no relation to crop value. On the contrary, value is strongly related to the percentages in the standard canning categories of medium, large and extra-large. By following the evolution of these percentages, the harvest date of maximum profit can be obtained. Normally this is when about 50% of the olives fall in these categories; picking must then begin in 1 week and last no longer than 2 weeks. The period is rather short and can cause farmers problems. The precise date may change with local conditions, but in California it is usually about mid-October.

6.2 HARVESTING

6.2.1 MANUAL HARVESTING

The manual system of picking olives still gives the best quality raw material. The oldest procedure, used for delicate olives such as those of the Gordal and Manzanilla cultivars, consists of picking the fruits one by one without spoiling any branch or leaf on the tree. Because of the high cost of this method, it was replaced by a manual 'milking' of the branches, drawing the fruits out and leaving them to fall into small, 8–10-kg capacity, baskets which workers have hanging from their necks and suspended in front of them at waist level. This method is very good at avoiding injury to the fruit surface, although it still presents the difficulty of high labour costs.

When the baskets are full, the fruits are released into the bug boxes (Figure 6.1) or bins used for transportation. Young trees and the lower parts of older trees can be harvested from the ground but the upper canopy must be picked from ladders. This can be a hazardous operation, so limiting the height of these ladders and keeping them well maintained are strongly recommended safety measures.

In spite of the pickers' skills and the careful handling of fruits, some tree and fruit damage is unavoidable. At each leaf torn off with the fruit a scar is produced through which the olive knot-fungus can enter. Fruit damage occurs not only during removal, but also when it is transferred into the bigger containers and during transportation.

In recent years, many devices to improve hand-picking and labour management innovations have been tested. However, none has significantly improved harvest rates or raw material quality. In general, motivated crews work better without any such aids.

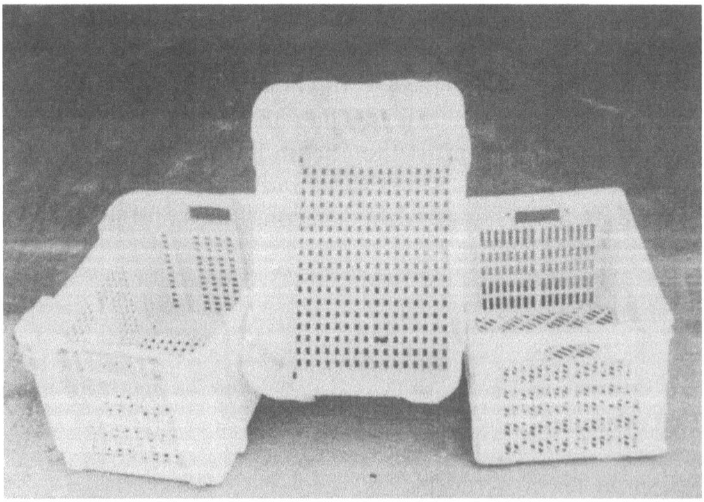

Fig. 6.1 Bug boxes used to transport the olive fruits from the orchards to the processing factories.

6.2.2 MECHANICAL HARVESTING

Mechanical systems have important economic advantages over traditional manual picking procedures, mainly due to a great reduction in labour costs and harvesting time. In essence, these consist of applying a vibration to the tree or scaffold limbs which is conducted to the points of fruit attachment (Figure 6.2). Fruits are thus shaken and eventually removed from the stem if the energy received exceeds the fruit removal force (FRF). Only a small fraction of the force applied reaches the fruit, since much energy is dissipated in the tree structure. The first study to apply shakers to the mechanical harvesting of olives was carried out by Lamouria and Brewer (1962) who established the upper and lower bounds on the effect of amplitude and frequency (shaking energy) on the detachment of olives. This is over four times that required for most other fruits and is different for each cultivar. It can reach values close to 10 N but decreases progressively during growth and ripening. The work of Friedly *et al.* (1973) is classical in this field.

Since then mechanical harvesting has been widely used mainly for fruits destined for oil extraction, when removal forces and the shaker energy required are relatively low. Badell Lapetra *et al.* (1969) described the different methods and machinery to be applied for this. The tree shaker or vibrator device gives the best performance. Its application to

Fig. 6.2 Mechanical harvesting. Shaker preparing for attachment to the trunk of an olive tree. (Courtesy D. Francisco Sánchez Roldán.)

black-ripe table olive harvesting in California was led by Hartmann *et al.* (1975) who developed a tree shaker specifically for olives which removed 80–85% of the fruits.

Humanes Guillén *et al.* (1981) studied in detail the action of mechanical harvesters and described the type of shaker that they developed at the Oleiculture Department of the Spanish Institute of Agricultural Research. Its application in green table olive harvesting gave similar yields to those reported for black-ripe. It has, however, been used almost entirely in harvesting fruits for oil extraction.

A number of factors facilitate maximum efficiency of mechanical harvesting. Among these are preparation of the soil to ease the movements of the shaker, the convenience of only one trunk per tree (Figure 6.3), a low trunk diameter, a low tree height, and a not too heavy crop per tree.

Successful application of mechanical harvesting requires that mature

Fig. 6.3 'One trunk per tree' olive tree prepared for mechanical harvesting.

orchards are adapted and unsuitable trees removed, or adequate new planting specifically designed for the purpose. In general, the tree structure especially influences fruit removal. Shakers work best with trees whose primary scaffold branches are thick (15–20 cm in diameter), grow upright and have the fruit-bearing surface on short twiggy growth around the top of the scaffold. In these conditions, branches transmit the energy of the shaker to considerable heights. Thus, mechanically harvested trees can be allowed to grow to heights that would be impractical for hand-harvesting. Trees with secondary scaffold branches that extend horizontally for a distance and then hang downward make fruit removal difficult because energy is poorly transmitted through these structures.

To facilitate the work of a shaker, it is necessary to remove all low lateral hanging branches that can interfere with the operator's view and the shaker clamp attachment. The number of primary scaffold branches should also be reduced, ideally to one, for a rapid harvest operation. Old orchards should be systematically pruned to reduce scaffold branches to no more than three, permitting at the same time an increase in the volume of the remaining primary branches to maintain productivity.

To avoid bark damage the angle of attachment between the shaker and the scaffold must be 90°. Clamp pressure has to be sufficient to prevent slippage on the trunk, but not so great as to crush the bark. On irrigated trees an interval of at least 2 weeks between the last irrigation and shaking is advisable. Finally, shaker pads should be well maintained (cleaned, rotated, etc.) and should be changed regularly since a shaker can go through four to six pads a day.

A complete description of the current use of this technology for black-ripe olives in California can be found in Martin (1986) and for olives destined for both green Spanish-style olives and oil extraction in Porras Piedras *et al.* (1994).

Although mechanical harvesting may appear rougher on the fruit than conventional manual picking, quality of the black-ripe processed product is equal to that of hand-picked fruits (Hartmann *et al.*, 1975). There are no reports in the literature about naturally black olives but, in principle, it should work well.

However, there are specific problems associated with green Spanish-style table olives, where brown spots can form on the fruit surface due to impacts the olives receive during vibration of the tree and in falling on to the collection cloth spread on the soil surface. This browning spoilage may become permanent; its effect progresses with time in the fallen fresh fruit and greatly decreases the quality of the final pickled product. A solution to this problem is urgently required before mechanical harvesting can be applied more generally.

The first work carried out on mechanical harvesting of green Spanish-style table olives (Herruzo *et al.*, 1975) showed unacceptable levels of

Table 6.1 Fruit removal force (FRF, in Newton) and efficiency of mechanical harvesting with green table olives. (Adapted from Herruzo *et al.*, 1975)

Shaking time	FRF (N)	Removed fruits			Unremoved fruits		
		Removed fruits (%)	Damaged fruits (%)	Size (fruits/kg)	Removed fruits (%)	Damaged fruits (%)	Size (fruits/kg)
Manzanilla							
Short (6 s)	6.61	39.3	16.3	348	60.7	9.7	361
Long (11 s)	6.51	44.0	17.0	331	56.0	17.0	336
Gordal							
Short (6 s)	9.92	47.0	50.9	–	53.0	22.6	–
Long (11 s)	9.86	55.6	59.3	–	44.4	31.0	–

damage both in green removed fruits and in those still attached to the tree (Table 6.1). Damage was always greater in removed fruits, supporting the dual origin of total damage. The relative contributions are approximately: 22% due to the shaking process, 31% from the fruit falling, and 8% due to canvas handling and dumping into the bug boxes. Longer shaking times increased harvest efficiency but also damage. Removal efficiency was relatively low in both cultivars, although it was lower with Manzanilla trees due to the small trunk diameters of this cultivar and the heavy crop, but mainly to the higher FR/fruit weight, 2.26 N in Manzanilla compared with only 1.28 N in Gordal.

Fruit damage increases with time as can be seen in Figure 6.4. Total damage is high (about 75%) immediately after shaking, and large intense green spots, which turn brown after fermentation, were already present in almost 50% of the fruits. As time progressed this increased. If fruits remain untreated for 24 hours or more the product is unacceptable for green Spanish-style olive processing. However, if treated with NaOH earlier, fruit damage could be maintained at acceptable levels. Experiments carried out with earlier NaOH treatments showed promising results as

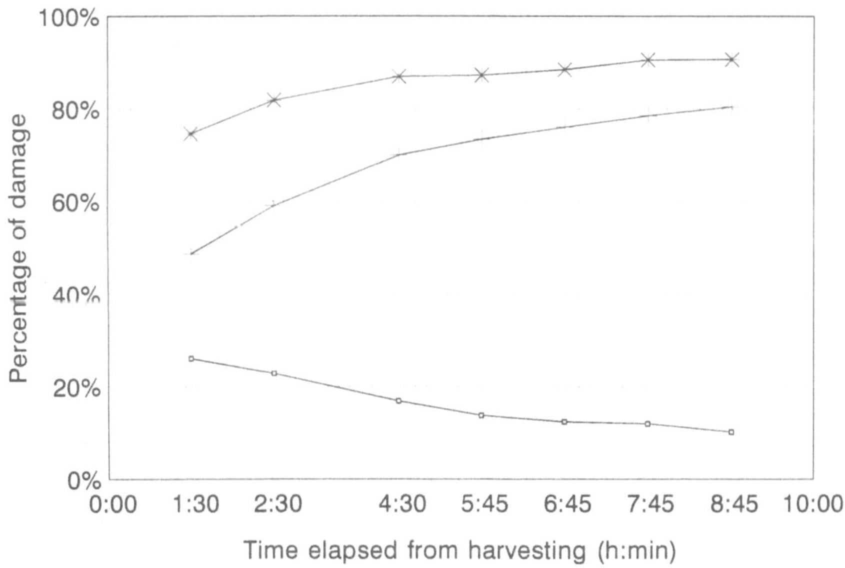

Fig. 6.4 Mechanical harvesting. Fruit damage changes with time after removal. □, minor damage; +, major damage; ×, total damage. (Redrawn from Pastor Muñoz Cobo, 1978.)

indicated in Table 6.2. Treated olives immersed in the lye solution 1 hour after harvesting had similar total percentages of damaged fruit as did manually picked fruits processed normally. The only appreciable difference was a slightly higher number of large spots in the mechanically harvested fruits. However, olive fruits which remained on the tree after shaking and were later hand-picked and processed conventionally were also seriously damaged because of the increased damage during the normal 24-hour waiting period before lye treatment (Pastor Muñoz-Cobo *et al.*, 1978). Furthermore, with the immediate lye treatment final quality was better than the original crop evaluation. Thus, the lye treatment may not only stop further deterioration but also has a positive effect in repairing existing damage.

Since a high percentage of fruits is damaged during falling, efforts have also been made to diminish the impact force of landing in receiving devices. Agabbio *et al.* (1986, 1989) used mechanical harvesting and nets, hand-suspended in the air, to reduce the falling distance with the Italian Nova cultivar (characterized by its strong consistency). Fruits were submerged immediately after removal in a 1.8% (w/v) lye solution. Results showed a nine-fold increase in productivity in relation to manual picking and less damaged fruits than with nets laid out on the ground. The procedure only improves results obtained with mechanical harvesting and data with other more sensitive cultivars were poorer (Table 6.3) (Lombardo, 1990). In all cases, the percentage of damaged fruits was significantly reduced compared with the method using nets spread on the ground. This opens the possibility of new developments such as: (i) the use of larger surface-area fruit reception devices, since with those in current use many fruits still fall directly to the ground; (ii) a more acute inclination in the base of the reception nets in order to facilitate the rapid removal of fruits to the conveyor belt after they fall; and (iii) the use of shock absorber bands, 20 mm in width and about 10 mm apart, placed just over the reception surface. To facilitate its use in the orchard, the whole device should be mounted in a 4 × 4 traction vehicle.

The fact that the slight brown spots disappear during lye treatment was the starting point of research undertaken by Ben-Shalon *et al.* (1977, 1978) and Angel and Ben-Shalon (1979), who demonstrated that the browning phenomenon produced by mechanical injuries of the fruit is an oxidation process catalysed by the enzyme catechol oxidase. It could be avoided by immersion of picked fruits into dilute aqueous solutions of sodium hydroxide, where the olives must remain until the lye treatment for debittering takes place. This procedure avoids applying the normal lye treatment immediately after fruit removal, which can be conducted in the factory where the facilities permit better control. It therefore helps avoid one of the main disadvantages of mechanical harvesting, although research is still required to solve a series of general and specific varietal problems which occur:

Table 6.2 Effect of time elapsed from fruit removal to lye treatment and the harvesting procedure on the fruit damage after fermentation. Verdial cultivar (Adapted from Pastor Muñoz-Cobo *et al.*, 1978)

Harvesting condition	Olive quality (%)						
	Good	Surface damage			Blisters*	Sloughing*	Gas-pocked*
		Minor	Major	Total			
Mechanical harvesting 1 h†	70.54	10.76	14.20	24.96	0.32	0.06	4.12
Mechanical harvesting 6 h	57.28	14.60	26.20	40.80	0.12	0.60	1.20
Mechanical harvesting 24 h	33.80	15.20	49.80	65.00	—	—	1.20
Fruits in tree after mechanical harvesting‡	46.00	10.00	43.20	53.20	—	—	0.80
Normal harvesting§	72.86	17.55	6.38	23.93	1.10	1.45	0.66

* Different olive spoilage which will be commented upon later on in the book

† Time elapsed from fruit removal to the sodium hydroxide solution treatment. Fruits then followed normal factory processing conditions.

‡ Manually picked and processed under normal factory conditions.

§ Processed according to normal factory conditions.

Table 6.3 Influence of the use of suspended canvas on fruit damage during mechanical harvesting (Adapted from Lombardo 1990)

Cultivar	Fruit removal force (N)	Harvesting procedure	Total fruit damage (%)	Weight (g)	FRF/ weight N/g)
Sant'Agostino	7.725	Manual	9.82	4.14	1.865
		On canvas	29.46		
		On nets (ground)	39.34		
Nocellara Etnea	8.160	Manual	13.98	4.42	1.846
		On canvas	44.20		
		On nets (ground)	62.61		
Nocellara Messinese	5.931	Manual	25.00	3.21	1.847
		On canvas	42.06		
		On nets (ground)	51.88		

- Establishing the maximum permitted time between harvesting and immersion of the olives into dilute sodium hydroxide solution (also called 'transport lye', since the treatment is made during transport from the orchard to the factory).
- Characteristics of the transport lyes, mainly sodium hydroxide concentration and temperature.
- Time range during which the olives can be maintained in the transport lye.

Several laboratory and field trials were made at the Instituto de la Grasa (Fernández Díez *et al.*, 1985) mainly with Manzanilla fruits, one of the most sensitive to mechanical harvesting. From the results the following conclusions were drawn:

1. The time interval between harvesting of the olives and their immersion in the transport lye must be as short as possible but never more than 1.5 hours.
2. The best NaOH concentration for a transport lye is between 0.6 and 0.8% (w/v) for non-irrigated fruits and between 0.2 and 0.3% for irrigated olives. It is quite important to avoid any lye penetration in the fruit flesh during the immersion time.
3. The most favourable immersion time to eliminate browning spots is between 3 and 6 hours. In the case of Manzanilla cultivar, it is better to use lower transport lye concentrations and increase the immersion time to 12–14 hours to avoid additional sloughing problems, characteristic of this cultivar (see Chapter 7).

6.2.3 FRUIT ABSCISSION

Manual picking is still mainly used for table olive fruit harvesting, in spite of the fact that it accounts for half or even more of the raw material price. Clearly, mechanical harvesting could reduce costs and increase competitiveness in more developed countries where labour costs are higher. A major reason for its low usage is the relatively poor removal efficiency as a result of the high strength of the union between fruits and stem at picking time. Thus, it was apparent from the earliest application of shakers to table olive harvesting that mechanical harvesting should be accompanied by chemical treatment to diminish the FRF and accelerate the normal process of natural fruit loosening, abscission. This could improve the proportion of detached fruits, currently about 50% with green Spanish-style table olives (Pastor Muñoz-Cobo *et al.*, 1978) or about 70% in that of black-ripe olives (Hartmann *et al.*, 1975), reduce fruit damage by decreasing the time and energy required, speed up harvesting, and, very importantly, reduce costs.

Some years ago, trials made in this area consisted only of empirical treatments of olive trees with different chemicals. More recently, using experience acquired with other fruits, the work has been directed to treat the olive tree with specific compounds which, once incorporated by the tree, give rise to ethylene production, an important influence on the ripeness and senescence of fruits.

Among these compounds, cycloheximide, 2-chloroethyl-phosphonic acid (Ethephon, Ethrel), 2-chloroethyl-tris-(2-methoxy-ethoxy)-silane (CGA-13586, Alsol), and 2-chloroethylsulphonic (Höl) require mention. Although they all produce fruit abscission, they also cause an excess of leaf fall and reduce flowering the following Spring (Hartmann *et al.*, 1972, 1976).

Only a detailed study of the abscission mechanism will provide efficient methods for practical control of the process. Several approaches have been followed over recent years. One is based on the known action of indoleacetic acid (IAA) or auxin, which delays fruit ripeness and senescence, and the presence in leaves and buds of the enzymatic system IAA-oxidase. Indirect measurement of such enzymatic activities in plant extracts could be made by following levels of certain micronutrients important in the enzyme composition. Vioque Pizarro and Albi Romero (1975) did this in the stem of the olive, from its formation up to full ripeness and the fall of the fruit. During two crops, they followed the variation in iron, copper, manganese and zinc levels in stems and fruits. No changes were observed in the olives and only the iron content in the stem varied, increasing in concentration to reach a maximum when the fruit had fully developed but still maintaining a green colour. After this point, the force required to pull the olives from the stem begins to decrease progressively.

The increase in iron in the stem was confirmed during three successive vegetative cycles and was associated with an increase in the concentration of the haem-containing IAA-oxidase system, which reduces the auxin concentration to a level to produce abscission. Further histochemical studies demonstrated the presence and accumulation of the enzyme system in the abscission zone of stem–fruit.

The same authors (Vioque Pizarro *et al.*, 1978a,b) studied the characterization of the enzymatic system and confirmed its participation in the last phase of ethylene biosynthesis (Vioque Pizarro *et al.*, 1980). In the presence of pyridoxal and IAA, enzyme extracts from olive tissues transformed 1-aminocyclopropane-1-carboxylic acid (ACC) into ethylene. They also attributed a dual function – oxidative and peroxidative – to the enzymatic system (Vioque Pizarro *et al.*, 1981).

The work of Albi *et al.* (1987a), who also studied the behaviour of ACC in the presence of different olive tissues (leaf, stem, and fruit), may have important implications for oleiculture. Their first results were promising, showing that, in the presence of exogenous ACC, the tissues liberated ethylene in greater quantities than controls using only endogenous ACC. Field trials (Albi *et al.*, 1987b,c) have confirmed the possibilities of ACC usage to ease fruit abscission and examined the influence of ACC concentrations in the range 500 to 2000 p.p.m. and the absorption, transport and metabolism of ACC in the tree.

The abscission process for olive fruits, leaves and inflorescences has also been studied in detail by researchers from the Pomology Department at UC Davis in the United States. A complete description of this work can be found in Martin *et al.* (1994).

Olive fruit abscission involves two possible zones: peduncle/pedicel and pedicel/fruit. The first is much like that of the leaf, with an easily identifiable separation zone of isodiametric cells which are smaller than the adjacent elongated cells. However, no abscission layer cells are apparent in the pedicel/fruit junction. The separation process produces cell plasmolysis in the pedicel pith and cortical regions first and then moves toward the phloem. The main biochemical changes observed are formation of starch grains, loss of pectin and cell wall polysaccharides and disintegration of the cell wall and middle lamellae as plasmolysis proceeds. After ethylene application, fruit separation occurs more frequently at the pedicel/fruit zone, where these transformations are less evident.

Similar changes have been observed in laboratory- and field-induced abscission. However, in the field, fruit respond more rapidly than do leaves to the application of ethylene-releasing compounds (ERCs). Among the factors influencing abscission effectiveness are environmental conditions at treatment time, the ethylene dosage and length of exposure, and the type of ERCs.

With respect to the environmental conditions, it is evident that CO_2 and O_2 levels have little influence in field applications, although it is known that high concentrations of both diminish abscission. In some cases the relative humidity (RH) can be important, as Ethephon decomposition has been shown to be reduced at 93% RH or higher and that its transportation into the olive tree is lower when the humidity is scarce (Kleien *et al.*, 1978). It has also been observed that fruit abscission is higher when orchards are at the end of the irrigation cycles (Lavee and Haskal, 1976) and the consensus is that higher efficiency is obtained when tree tissues suffer a slight water stress (scarce water availability).

Temperature seems to play an important role. Kleien *et al.* (1978) found better results at temperatures below 30°C. This is not consistent with the decomposition rate of Ethephon which, 5 hours after application, is 10% at 20°C and 40% at 30°C. The optimum temperature range is in fact between 16°C and 30°C and is related to the exposure time to ethylene, which is critical to the process. A rapid release of ethylene at higher temperatures gives the fruit a lower exposure time and consequently leads to a low abscission rate (Lang and Martin, 1985). To obtain a high rate of fruit abscission in a short time it is necessary to increase markedly the concentration of the ERCs.

Leaves and fruits show a different behaviour. In general, leaves need longer exposures so short exposures to high concentrations could improve the rate fruit removal and, consequently, facilitate treatment efficiency.

The main problem with chemical abscission treatments is related to leaf loss. In general, ERCs affect both fruits and leaves and an excessive loss of leaves will lead to a lower crop the following year as a result of reduced flower production. Experience has shown that leaf losses over 15% can sensibly alter the tree metabolism, reduce shoot growth and limit tree vigour. In these cases, flowers and fruit can be absent for at least 2 years. In addition, each leaf lost produces a scar through which the knot bacterium may enter. Thus, it is necessary to develop treatment strategies which permit a selective abscission of fruits rather than leaves.

In general, all ERCs that are effective in fruit release have the same effect on leaves, which are even more sensitive to ethylene. However, some differences found in the course of research into abscission could be exploited to overcome this problem. For example, the leaf surface has a complex composition and physical structure which do not favour the absorption of chemicals. If surfactants are absent in the solutions used for treatments, lower absorption through leaves can be expected. Another difference of potential importance is related to the natural structure of leaves and fruits. Unlike leaves, fruits have a cavity in the pedicel/fruit junction which can retain a small amount of the applied solution over extended periods of time. This can enhance both the concentration and exposure time to ERCs and may provoke a selective abscission of fruits.

A detailed study of the time and concentration effect on fruit and leaf abscission has been carried out by Lang and Martin (1989). From the response surface obtained, they found evidence of greater synchronization of fruit abscission at critical ethylene treatment thresholds, relative to that of leaves. At 4 to 10 ml/l of ethylene (C_2H_4) the exposure time required to induce the critical level of fruit abscission was 27 to 29 hours, while the range required to induce the critical level of leaf abscission was quite broad (25 to > 42 hours). This suggests that regulation of applied ethylene concentration at about 4 to 7 μl/l would allow variation in exposure duration without the danger of inducing excessive leaf abscission. In general, the fruit-to-leaf ratio response surface indicated that the optimal removal of 90% or more fruits with less than 25% of leaf abscission could be achieved in the region defined by a concentration of 3 to 5 μl/l over a period of 28 to 34 hours. Thus, an ERC which could release ethylene over this sort of time-scale could be expected to achieve the greatest fruit abscission with minimal leaf loss.

Recent research to improve Ethephon penetration in olive tissues (Denney and Martin, 1994) has suggested the use of low pH, since higher pH results in higher petiole penetration and greater leaf drop. The most appropriate application efficiency is obtained under relatively cool and low moisture conditions. Good fruit abscission is currently obtained with the product CGA 15281 [(2-chloroethyl) methylbis (phenylmethoxy) silane], whose main advantage over Ethephon is the rapid ethylene release; this could be decisive in enhancing fruit versus leaf removal. Field experiments have shown that the tree can be mechanically harvested only 3–4 days after application (compared with 6–8 days with Ethephon) and the solution is easy to prepare (Ben-Tal, 1987). The author felt that further studies on the application of this product could lead to successful application of mechanical harvesting to table olives.

The problem has not yet been solved, however. New research is underway to design ERCs with release characteristics suitable for pulse ethylene treatments. Ideally they would release ethylene for 8 hours, stop for 16 hours before resuming release again for 24 to 32 hours. This could give about 98% fruit removal with less than 6% leaf drop.

Differences in fruit and leaf responses to ethylene are apparent. Ethylene production by olive fruit is low and there is no autoenhancement produced by additional exterior ethylene. However, this is observed in the leaf 120 hours after application and after leaf drop. This supports the hypothesis that autoenhancement of ethylene may affect leaf senescence, but not abscission. There is also evidence of greater sensitivity of fruit to ethylene gas or ERCs sprays. Finally, recent research has shown that phosphorus can play an important role in leaf abscission. The mechanism of phosphorus-induced abscission seems to be independent of evolved ethylene and is accompanied by an accumulation of phosphorus in the abscission zone (Burnik-Tiefengraber *et al.*, 1994). Studies on the physiology of olive leaf abscission induced by phosphorus have proved that its action is independent of the endogenous ethylene evolution and that oxygen is necessary for phosphorus-induced abscission (Yamada and Martin, 1994).

In conclusion, important basic and applied knowledge on olive fruit abscission are available and some of the current ERCs (Ethephon, CGT 15281, etc.) can achieve good results, diminishing the FRF to levels at which mechanical harvesting could be applied to table olives. Nevertheless, this remains an open field for research and fresh approaches could help to provide a definitive solution.

6.3 TRANSPORT OF THE FRUITS TO THE FACTORY

In general, a first cleaning and separation of the culls is carried out in the orchard. This is done by passing the olives over sorting belts which eliminate any undersized fruits as well as leaves, twigs, stones, etc. Culls are sent to the olive mills. Commercial sizes of picked olives are placed in perforated plastic bug boxes of about 20–22-kg capacity, in which they are transported to the factory. Since fruits remain in the boxes until they are processed, perforation is important to avoid heating of the fruits.

The parallel, pipe-shaped metal bins have lateral plastic network walls, hold about 500 kg of olives, and are also used to transport the fruits. The base of the bin is sloped to a hatch door which allows the fruit to be unloaded easily. Bulk transportation in trucks must be avoided as much as possible in order to avoid damage to the fruits.

The most suitable container to carry mechanically harvested olives in transportation lye from the orchard to the processing plant still needs careful study. In principle, different types of plastic vats may be used.

6.4 QUALITY EVALUATION OF THE FRESH FRUIT

Prices to growers are based on total weight, size distribution and the quality of olives within each size classification. The season's crop size also has an important influence on total income, since a shortfall leads to increased prices and vice versa.

When the fresh fruits reach the factory, it is advisable to collect all the information which allows the lot to be identified. The most important data to be recorded are: cultivar, supplier, orchard, cultivation system (irrigated or non-irrigated trees), harvesting date, date of arrival at the processing plant, number of days which will elapse between arrival of the fruits and their processing, maturity, water content, net weight, size distribution and total quality.

To fix the quality and price of the lot, it is necessary to register the weight of small-sized olives (if they have not already been eliminated), which are not useful for table olive preparation and must be sent for olive oil extraction. The separation of this fraction is made with a conventional cable size-grader. At the same time, it is necessary to take a suitable sample to calculate the average size and the different defect percentages (injured, unacceptable colour, wrinkled, and insect- or pest-damaged fruits).

No systematic and scientific sampling procedure is generally adopted due to the relatively low price of the product, the lack of time during the days of peak activity, the labour needed and the fruit damage produced on sampling. A practical sampling procedure consists of taking a handful of fruits from each box (or from each representative part of the bin), to give a sample of at least 8–10 kg which, once homogenized, is further reduced to 5 kg to obtain the average size, expressed as fruits/kg. The real size distribution figures are obtained when the olives are graded after processing; in fact, despite the sampling difficulties the final values obtained are generally in good agreement with the initial estimate.

In California (USA), evaluation of the harvested olive fruit is more strict and is regulated by the Federal Marketing Order 932 administered by the California Olive Committee. The inspection service is maintained by a cooperative agreement between the Federal Department of Agriculture (Agricultural Marketing Services), the California Department of Food and Agriculture (Division of Inspection Services) and Food and Vegetable Quality Control (Federal-State Inspection Service). All canning olives are graded by licensed state inspectors upon delivery to the processor to assure that established standards are uniformly enforced. The California Fresh Olive Circular (Department of Food and Agriculture) contains a description of the procedures to be used and the documents the inspector should provide. Inspection is performed personally by the inspector or, more often, by the processor's personnel under the inspector's supervision. However, the inspector is the only person responsible to the California Olive Committee for the proper implementation of procedures, accuracy of data, weight, counts and records.

It is necessary to inform the Administration at least 48 hours before the beginning of the picking season. Later, only 24 hours is required after any temporary interruption.

Any lot grading should begin with the proper identification, determination of the 'natural condition olives' or 'canned ripe olives of the tree-ripened type', verification of the olive cultivar and its group, removal of immature or over-ripe culls, culled for machine- or hail-damaged fruits infested with *Parlatoria*, and, finally, size-graded. The number of bug boxes or bins to be inspected depends on the lot size and is specified in the Circular. The whole sample is reduced by splitting it several times until a final sample of 6 kg (Manzanilla) or 13 kg (Sevillano) is reached. The olives are then run over a sample grader and the number of olives per kg counted. If the olives meet the ranges designated by the US Standards and Marketing Order sizes, the lot is passed. If they do not, this is reported on the Obligation Credit form to indicate that the olives cannot be used for canning. An appeal for re-inspection can be made to the state inspection service and this is then made as soon as possible on a sample twice as large as the original.

A complete description of the procedure can be found in Ferguson and Welch (1994) and on the Circular available from the California Olive Committee or the Division of Inspection Services (California Department of Food and Agriculture).

6.5 OTHER POST-HARVESTING OPERATIONS

As living material, olive fruits detached from the tree continue their respiration and metabolism. They are very sensitive to damage by handling machinery, grader devices, knocks and other rough manipulation. Thus, post-harvesting operations should be reduced to a minimum. This is especially important in the case of green Spanish-style table olives, in which any olive surface scar may become a brown spot, blister, or gas-pocket when fermented. This sensitivity has prevented the use of a generally adopted size grading before processing, although it can markedly enhance lye treatment and can give exact knowledge about the availability of final product within each commercial size.

The application of a surface colour classification of the raw material for green olives has led to some disasters after fermentation, since these grading machines run at high speed and handling of the fruits is far from friendly.

Similar poor results were obtained with the use of a network of pipelines and pumps for the transportation of fresh green olives suspended in water or brine. Here, there were additional problems to the blemishes produced by impacts within the pipes as the fruit surface acquired many greenish marks due to the introduction of moisture under the skin caused by the high pressures produced in the system by the pumps.

However, a grading system based on the different specific weights of olives at different ripening degree stages may be of interest, since there are few moving parts in the machinery required. This was developed by Fernández Villasante and Fernández Díez (1968) and uses brines of different strengths to obtain homogeneous groups with respect to their degree of maturation. It has been applied to classifying raw material for green Spanish-style table olives in both ungraded and graded fruits satisfactorily. Table 6.4 shows an example in both classes of fruits. As is evident from the Table, even in the case of graded fruits, the application of this procedure may lead to a new more homogeneous group in relation to texture, size (expressed as number of fruits/kg), and surface colour. This technique has the advantage over other grading techniques, including those of the divergent cable type, of more friendly fruit handling which does not cause any damage or bruising.

Bearing in mind the high sensitivity of olive fruits, the floating classification could be advisable in those cases where careful preparation is needed, since application of this raw material grading process may:

- improve the sodium hydroxide treatment in the debittering step of Spanish-style green olives, rendering it more homogeneous;
- enhance fermentation due to a more uniform NaOH penetration in the flesh and provide a better control of the hydrolysis of inhibitory compounds;
- facilitate final size grading;
- diminish the proportion of broken olives during the pitting and stuffing operations; and
- enhance selection of final products with appropriate colour and texture characteristics, according to the market demand.

Table 6.4 Olive classification by flotation in different specific weight solutions. (Adapted from Fernández Villasante and Fernández Díez, 1968)

	Classification*				
	1	2	3	4	5
Ungraded olives					
Average texture, N/100 g pitted olives†	3060	3150	3250	3320	3350
Fruit sizes, number of fruits/kg	310	340	345	380	410
Graded olives (280/300 commercial size)					
Average texture, N/100 g pitted olives	2810	2990	3110	3270	3210
Percentage of pink olives	6	4	7	12	26
Fruit sizes, number of fruits/kg	411	416	422	424	440

* 1, Olives that floated in tap water; 2, olives which floated in 1°Bé brine; 3, olives which floated in 2°Bé brine; 4, olives which floated in a 3°Bé brine; 5, olives which did not float in 3°Bé brine.

† Texture determined in an Allo Kramer Shear Press.

Where olives are for preparation of black-ripe olives or naturally black olives, the impact of blemish problems in the final product can be diminished due to the black surface colour. However, careful handling of the olive fruits is still recommended, as the mark of any blow may become a soft centre or give rise to a gas-pocket fissure.

The time elapsed from harvesting to lye or brine immersion is, in general, not too large. During this period, fruit respiration continues (Maxie *et al.*, 1960), the intensity of which increases as the maturation degree changes from green to black. This biological activity uses part of the fermentable substrates in olives and may cause softening. In addition, there are also continuous moisture losses which can cause appreciable weight losses and shrivelling and which will persist through processing, causing subsequent quality deterioration. Thus, detached olive fruits must be processed as soon as possible after harvest.

The incidence of all these changes increases with temperature. Consequently, fresh olives must be situated in the coldest places of the factory and transported during the lowest temperatures of the day or, preferably, at night. This is particularly important in some cases of Spanish-style green olive preparations where a certain delay in lye treatment is necessary to avoid sloughing.

To avoid such undesirable biochemical transformations Shulman *et al.* (1974) reported the use of ethylene and Ethephon. Both green and half-black Manzanilla cultivar were exposed for 1 to 2 weeks to 1 to 1000 μl/l of ethylene and to 100 to 500 μg/l of Ethephon. All treatments inhibited anthocyanin accumulation and retarded softening of the pericarp. Similar effects were observed when picked olives were exposed to a CO_2-rich atmosphere. This very special reaction of the olive fruit to ethylene – when the normal behaviour is the enhancement of colour development and acceleration of fruit softening – is not an isolated phenomenon since other growth regulators (cytokinins, auxins, etc.) tested by the same authors also had unique effects. However, the effect of CO_2 was similar to that found with other fruits. This seems to point to an effect of CO_2 which is independent of its suggested action as a competitive inhibitor of ethylene, as in this case there was no such antagonism between CO_2 and ethylene in fruit ripening. From the practical point of view, these effects could be exploited to enhance picked olive preservation for short periods of time.

6.6 COLD FRESH OLIVE STORAGE

Traditionally, table olives have always been preserved in brine for later use. Nevertheless, the current difficulties for discarding brine and the increasing market for home pickling have pushed researchers to study the possibility of cold storage of olives in a similar way to that followed with other fruits.

The first studies on cold storage of olive fruits were carried out by Maxie (1964) and Woskow and Maxie (1965). They found that olive fruits were prone to chilling injury, a physiological condition common to many tropical and subtropical fruits. The olive fruits were severely injured at 0°C in as short a time as 1 week, with some differences among cultivars. This injury begins as a light tan colour adjacent to the pit, which becomes progressively darker, encompassing more and more of the flesh until the symptoms extend to the skin.

The fact that the storage life of any fruit is closely related to the respiratory rate, suggested to Maxie the possibility of slowing down respiration and with it ripening of olive fruits by reducing the O_2 and increasing the CO_2 proportion in the atmosphere around them. Experiments held in controlled atmospheres at 0–10°C showed that black-ripe olives deteriorated with time, mainly due to a progressive decrease in flavour quality.

Later storage experiments were done with the aim of removing from the surrounding gases possible off-flavour components produced during storage. In this case, there was no significant decrease in the flavour of olives held in purified air for as long as 5 weeks at 5°C. There was also little evidence of ripening. However, longer storage led to progressive injury in fruits.

The more conclusive results for the storage potential of fresh olives were recently obtained by Kader *et al.* (1990), who showed that the ideal pre-processing storage for fresh, green 'Manzanilla' olives is at 5–7.5°C and 90–95% RH. Under these conditions the olives can be stored for 6–8 weeks in air, or 9–12 weeks in a 2% oxygen atmosphere with no significant loss in quality. Exposure to 5% or more CO_2 can cause physiological injury to olives.

These procedures can thus be used for olive transportation over long distances and for the storage of olives destined for retail markets, notably for use in home pickling. They can also be used at times of peak olive arrivals at factories and for those cultivars which require a delay before being lye treated.

Cold storage has also been applied to olive fruits destined for oil extraction. The best storage conditions were similar to those for table olives (García and Streif, 1991; García Martos, 1993). However, such storage can influence the oil quality; Pérez-Camino *et al.* (1992) found that storage of fruits at 5°C yielded lower proportions of polar compounds in the oil extracted. At this temperature they could not find any influence of increasing CO_2 to 3% or decreasing O_2 to 5%.

6.7 REFERENCES

Agabbio, M., Dettori, S., Paschino, F. and Schirra, M. (1986) Raccolta meccanica di olive da mensa verdi con impiego in campo di soluzione alcaline. *L'Informatore Agrario* (Verona), **XLII**, 187–9.

Agabbio, M., Dettori, S., Paschino, F. and Schirra, M. (1989) Recolección mecanizada de aceitunas de mesa verdes mediante el empleo en el olivar de soluciones alcalinas. *Olivae*, **26**, 33–5.

Albi, M.A., López, A., Castellano, J.M., Vioque, B., Ysern, A. and Vioque, A. (1987a) Abscisión de la aceituna. IV. Utilización del ácido 1-aminociclopropano-1-carboxílico (ACC) para facilitar su recogida. *Grasas y Aceites*, **38**, 110–15.

Albi, M.A., López, A., Castellano, J.M., Vioque, B., Ysern, A. and Vioque, A. (1987b) Abscisión de la aceituna. V. Tratamiento de olivos con ácido 1-aminociclopropano-1-carboxílico (ACC): Influencia de la concentración del tratamiento. *Grasas y Aceites*, **38**, 154–61.

Albi, M.A., Castellano, J.M., López, A., Vioque, B., Fernández Maculet, J.C. and Vioque, A. (1987c) Absorción, transporte, y metabolismo del ácido 1- aminociclopropano-1-carboxílico (ACC) rociado sobre el olivo. *Grasas y Aceites*, **38**, 394–9.

Angel, S. and Ben-Shalon, N. (1979) Research on olive quality in Israel for processing. I. Prevention of brown spotting in green olives for pickling. *Confructa*, **24**, 79–84.

Badell Lapetra, I.G., Fernández del Pozo y Gironza, J., Serna Ferrer, R. de la, Alonso Castellanos, N., Bufalá Pérez, F., Gías Fuentes, M., Rubio de Hornedo, G., Acedo Villalba, F., Navarro Marín, G. and Enebral Casares, M. (1969) *Recolección mecanizada de la aceituna*, Ministerio de Agricultura, Madrid.

Ben-Shalon, N., Kahn, V., Harel, E. and Mayer, A.M. (1977) Olive catechol oxidase. Changes during fruit development. *J. Sci. Food Agric.*, **28**, 545–50.

Ben-Shalon, N., Harel, E. and Mayer, A.M. (1978) Enzymatic browning in green olives and its prevention. *J. Sci. Food Agric.*, **29**, 398–402.

Ben-Tal, Y. (1987) CGA 15281. Nuevo compuesto químico para facilitar la recolección mecánica de la aceituna. *Olivae*, **18**, 9–10.

Burnik-Tiefengraber, T., Weigs, K.G., Webster, B.D., Martin, C.G. and Yamada, H. (1994) Phosphorus effects on olive leaf abscission. *J. Am. Soc. Hort. Sci.*, **119**, 765–9.

Denney, J.O. and Martin, G.C. (1994) Ethephon tissue penetration and harvest effectiveness in olive as a function of solution pH, application time, and BA or NAA addition. *J. Am. Soc. Hort. Sci.*, **119**, 1185–92.

Ferguson L. and Welch, G. (1994) Grading canning olives, in *Olive Production Manual*, University of California, Division of Agriculture and Natural Resources, Oakland, CA, USA, pp. 129–31.

Ferguson, L., Gawad, H., Sibbett, G.S., Freeman, M. and Hatukeda, J.J. (1991) Using fruit size to predict harvest timing of California Manzanilla olives for ripe-black processing. *HortScience*, **26**, 26–9.

Fernández Díez, M.J., Castro Ramos, R., Garrido Fernández, A., González Cancho, F., González Pellissó, F., Nosti Vega, M., Heredia Moreno, A., Mínguez Mosquera, M.I., Rejano Navarro, L., Durán Quintana, M.C., Sánchez Roldán, F., García García, P. and Castro Gómez Millán, A. (1985) *Biotecnología de la aceituna de mesa*. Consejo Superior de Investigaciones Científicas, Instituto de la Grasa, Madrid, Sevilla, Spain.

Fernández Villasante, J. and Fernández Díez, M.J. (1968) Classification des olives vertes d'après leur degré de maturité par flottage en saumure. *Inf. Oleic. Int.*, **42**, 59–63.

Friedly, R.B., Mehlschan, J.J., Hartmann, H.T. and Logan, S.M. (1973) Mechanical harvesting of olives. *Transactions of the ASAE*, 58–61.

García, J.M. and Streif, J. (1991) The effect of controlled atmosphere storage on fruit and oil quality of Gordal olives. *Gartenbauwissenschaft*, **56**, 233–8.

García Martos, J.M. (1993) Efecto del CO_2 en la atmosfera de almacenamiento sobre la fisiología de la aceituna de molino. *Grasas y Aceites*, **44**, 81–4.

Hartmann, H.T., El-Hamady, M. and Whisler, J.E. (1972) Abscission induction in the olive by cycloheximide. J. *Am. Soc. Hort. Sci.*, **97**, 781–5.

Hartmann, H.T., Reed, W., Whisler, J.E. and Opitz, K.W. (1975) Mechanical harvesting of olives. *Calif. Agric.*, **6**, 4–6.

Hartmann, H.T., Read, W. and Opitz, K.W. (1976) Promotion of olive fruit abscission with 2-chloroethyl-tris-(2-methoxy ethoxy)-silane. *J. Am. Soc. Hort. Sci.*, **101**, 278–81.

Herruzo, B., Pastor, M. and Holgado, G. (1975) Recolección mecanizada de la aceituna de mesa en verde. II Seminario Oleicola Internacional. Córdoba (Spain) 6–7 Octubre.

Humanes Guillén, J., Herruzo Sotomayor, B. and Porras Piedra, A. (1981) Estado actual de la recolección mecanizada de aceituna. XVII Reunión Plenaria de la asamblea de Miembros del Instituto de la Grasa, Sevilla, Spain.

Kader, A.A., Nanos, G.D. and Kerbel, E.L. (1990) Storage potential of fresh Manzanilla olives. *Calif. Agric.*, **44**, 23–4

Karaoulanis, G.D. and Markakis, P. (1987) Relation between maturity and processing of olives Kalamon variety, in *Frontiers of Flavours*. Proceedings of the 5th International Flavour Conference, Porto Karras, Chalkidiki, Greece, 1–3 July, pp. 307–19.

Kleien, I., Epstein, E., Lavee, S. and Hutchinson, J.R. (1978) Environmental factors affecting ethephon in olive. *Scientia Hortic.*, **9**, 21–30.

Lamouria, L,H. and Brewer, H.L. (1962) The determination of selected bioengineering properties of olives. Winter Meeting American Society of Agricultural Engineers, Chicago, Illinois, USA.

Lang, G. and Martin, G.C. (1985) Ethylene releasing compounds and the laboratory modeling fruit abscission versus ethylene release. *J. Am. Soc. Hort. Sci.*, **110**, 207–11.

Lang, G.A. and Martin G.C. (1989) Olive organ abscission: fruit and leaf response to applied ethylene. *J. Am Soc. Hort. Sci.*, **114**, 134–8.

Lavee, S. and Haskal, A. (1976) Further field studies of the mode of application and efficiency of various ethylene releasing compounds to facilitate olive fruit harvest. *Riv. Ortoflorofructtic. Ital.*, **60**, 166–75.

Lombardo, N. (1990) Pruebas de recolección mecánica de aceitunas de mesa verdes. *Olivae*, **32**, 34–7.

Martin, G.C. (1986) Recolección de aceituna en California, Estados Unidos de America. *Olivae*, **11**, 11–22.

Martin, G.C., Klonsky, K. and Ferguson, L. (1994) The olive harvest, in *Olive Production Manual*, University of California, Division of Agriculture and Natural Resources, Oakland, CA, USA, pp. 117–28.

Maxie, E.C. (1964) Experiments on cold storage and controlled atmosphere. *Calif. Olive Assoc. Techn. Report*, **43**, 12–15.

Maxie, E.C., Catlin, P.B. and Hartmann, H.T. (1960) Respiration and ripening of olive fruits. *Am. Soc. Hort. Sci.*, **75**, 275–91.

Pastor Muñoz Cobo, M., Márquez González, J., Herruzo Sotomayor, B. and Porras Piedra, A. (1978) Recolección mecanizada de aceituna de mesa para el aderezo al estilo sevillano. XIV Reunión Plenaria de la Asamblea de Miembros del Instituto de la Grasa, Sevilla, Spain.

Pérez Camino, M.C., García, J.M. and Castellano, J.M. (1992) Polar compound concentrations in virgin oils from stored cultivar Picual olive fruits. *J. Agric. Food Chem.*, **40**, 2260–2.

Porras Piedras, A., Albenza Corral, J.M., Marcilla Goldaracena, I. and Moreno Valencia, A. (1994) *Recolección mecanizada de aceitunas*. International Olive Oil Council, Madrid, Spain.

Shulman, Y., Eric, A. and Lavee, S. (1974) Delay in ripening of picked olives due to ethylene treatment. *Scientia Hort.*, **2**, 21–7.

Vioque Pizarro, A. and Albi Romero, M.A. (1975) Elementos trazas y abscision de la aceituna. *Grasas y Aceites*, **26**, 73–8.

Vioque Pizarro, A., Albi Romero, M.A. and Vioque Cubero, B. (1978a) Abscision de la aceituna. II. Purificación parcial del sistema enzimático ácido indolacético-oxidasa del olivo. *Grasas y Aceites*, **29**, 391–5.

Vioque Pizarro, A., Albi Romero, M.A. and Vioque Cubero, B. (1978b) Abscisión de la aceituna. III. Algunas propiedades del sistema enzimático ácido indolacético-oxidasa del olivo. *Grasas y Aceites*, **29**, 397–406.

Vioque Pizarro, A., Albi Romero, M.A. and Vioque Cubero, B. (1980) Formación de etileno a partir del ácido 1-aminociclopropano-1-carboxílico por diversos tejidos del olivo y sus extractos. *Grasas y Aceites*, **31**, 196–205.

Vioque Pizarro, A., Albi Romero, M.A. and Vioque Cubero, B. (1981) Role of IAA-oxidase in the formation of ethylene from 1-aminocyclopropane-1-carboxylic acid. *Phytochemistry*, **20**, 1473–5.

Woskow, M. and Maxie, E.C. (1965) Cold Storage studies with olives. *Calif. Olive Assoc. Techn. Report*, **44**, 6–8.

Yamada, H. and Martin, G.D. (1994) Physiology of olive leaf abscission induced by phosphorus. *J. Am. Soc. Hort. Sci.*, **119**, 956–63.

Green olive type 7

7.1 INTRODUCTION

According to the Unified Qualitative Standard Applying to Table Olives in International Trade (International Olive Oil Council, 1980) and the Spanish Standard (Spanish Government Presidency, 1983), **green olives** are the type obtained from fruits harvested during the ripening period, before colouring but having reached normal size. Such olives shall be firm, sound, resistant to a slight pressure between the fingers and without marks other than the natural pigmentation, subject to the tolerances set out in those standards. The colour of the fruit may vary from green to straw yellow.

Within this type, the most important commercial preparation or trade preparation – which covers a high percentage of the international market (see Chapter 4) – is denominated Spanish (or Sevillian)-style (pickled) green olives in brine. These are defined in the Standards as 'treated in an alkaline lye and then put in brine, in which they undergo a natural and complete lactic fermentation'.

If the fermentation is only partial, the denomination pickled (or treated) green olives in brine is kept, but the term Spanish-style cannot be applied and preservation must be ensured by sterilization or pasteurization, addition of preservatives, refrigeration, or packing in an inert gas atmosphere, without brine.

Other green olives may be prepared without a previous alkaline treatment, placing them directly into the brine in which they undergo a natural fermentation. In that case, they receive the name of **natural** (or untreated) green olives in brine. Their commercial importance is markedly lower than that of Spanish-style green olives, but this process frequently gives rise to final products of excellent quality.

7.2 GENERAL OVERVIEW OF SPANISH-STYLE GREEN OLIVES IN BRINE

In this procedure (Fernández Díez, 1971; Fernández Díez *et al.*, 1985) green to yellowish green fruits are treated with a lye (a dilute aqueous solution of sodium hydroxide) to remove the greater part of the bitter glucoside known as oleuropein. Lye concentration is variable depending upon temperature, olive cultivar and degree of ripeness of the fruits. The treatment lasts until the lye penetrates the pulp to a depth of two-thirds to three-quarters of the distance between the skin and stone. The fruits are then washed with water for variable times to eliminate most of the lye. They are then placed into a brine of sodium chloride of 9 to 11 °Be in which the lactic fermentation takes place. Fermentation time mainly depends upon the characteristics of the previous lye treatment, olive cultivar, temperature and microbial population.

After the fermentable substrates are exhausted, the fruits are kept in the same brine until they are packaged and sold. This can be in bulk in plastic containers holding 150 to 300 kg of fruit and the same brine in which they were fermented, a freshly prepared brine or a mixture of both. For hermetically sealed consumer-oriented plastic, glass or tin packages, a freshly prepared brine is mostly used and the product may be heat-pasteurized. Before bottling, the stones of the fruit may be removed mechanically, or the fruit pitted and stuffed with pimento, onion, anchovy, and other food materials.

The general scheme of the process is shown in Figure 7.1.

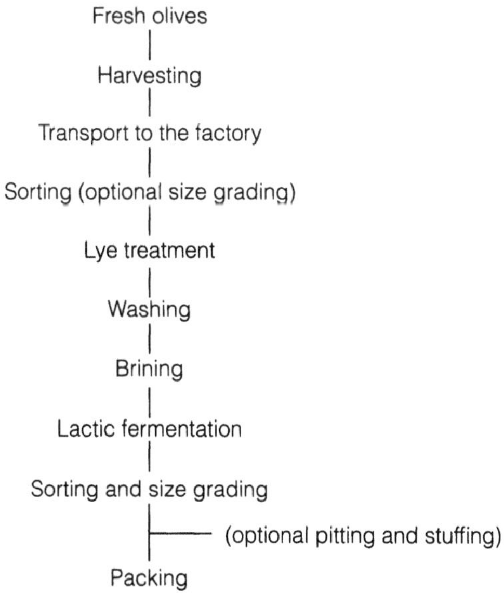

Fig. 7.1 Scheme for the processing of Spanish-style pickled green table olives in brine. (From Fernández Díez *et al.*, 1985.)

7.3 HARVESTING, TRANSPORTATION, SORTING AND SIZE-GRADING

The general rules for these early phases of the process were described in Chapter 6 and apply to all types of table olives. However, there are some features specific to this commercial preparation (Fernández Díez *et al.*, 1985):

1. Skin and pulp defects of the fresh fruit resulting from physical damage and other causes have a greater influence on the final product of this preparation, so special care must be taken during all the operations of harvesting and transport, sorting and grading stages to minimise the incidence of injured fruits.
2. Although the size-grading of the fresh fruits before lye treatment is generally useful in obtaining homogeneous lye penetration in the pulp, there are certain olive cultivars, mainly Gordal and Ascolano, which may be affected by the cables of the grader. In these cases, manual sorting is recommended. Alternatively, lye treatment must immediately follow any size or colour grading.
3. There are certain cultivars, mainly Manzanilla and Morona in Spain (and maybe others elsewhere), which have a tendency to lose the skin during lye treatment when they are processed soon after harvesting. A practical way to avoid this situation is to let the fresh fruits stay in the baskets for a period of 3 to 5 days before the lye treatment. An alternative procedure is the use of transport lyes described in Chapter 6. The scientific basis of the efficacy of these technical solutions to the problem are still under research although moisture content and pectic composition of the pulp must be implicated.

7.4 LYE TREATMENT

7.4.1 CHARACTERISTICS OF THE TREATMENT

Good control of this operation is essential for regulation of the speed, the effectiveness and the yield of the fermentation. However, good control is also essential for the organoleptic quality of the final product, principally for its colour and texture. During lye treatment, hydrolysis of oleuropein occurs and there is a loss of sugars and organic acids as well as the formation of acetic acid from sugars, which contributes to the volatile acidity that is always present in the fermented product. From a practical point of view the strength of the lye solution may be varied – generally between 1.3. and 2.6% (w/v) – and depends on a number of factors (Borbolla y Alcalá, 1981; Fernández Díez, 1983) among which the most important are cultivar, maturity of the fruit and temperature of the treatment.

The effect of concentration on penetration rate has been studied by several authors and found to be closely related to temperature. Borbolla y Alcalá *et al.* (1952a) found that the relative lye penetration was practically a linear function of concentration over the range of 1.08 to 2.64% (Figure 7.2). Barret and Bidan (1964), working with Picholine cultivar, found the results shown in Tables 7.1 and 7.2. This means that selection of appropriate lye concentration is essential for a suitable and practical process. There is general agreement on the establishment of the lye penetration as a fixed parameter of the debittering process and that it must reach about two-thirds to three-quarters of the distance from the olive surface to pit. This is especially important for Manzanilla olives and similar or more phenol-rich cultivars. Cruess (1930), in his first studies on green Spanish-style olives, reported that olives with lower lye penetration were bitter even after a year of brining, developed a brown surface colour, and frequently gave dormant or stuck fermentations. However, with Gordal or Sevillano, lye penetration was less decisive, probably due to its lower oleuropein content. Lye should not, however, penetrate to the pit since

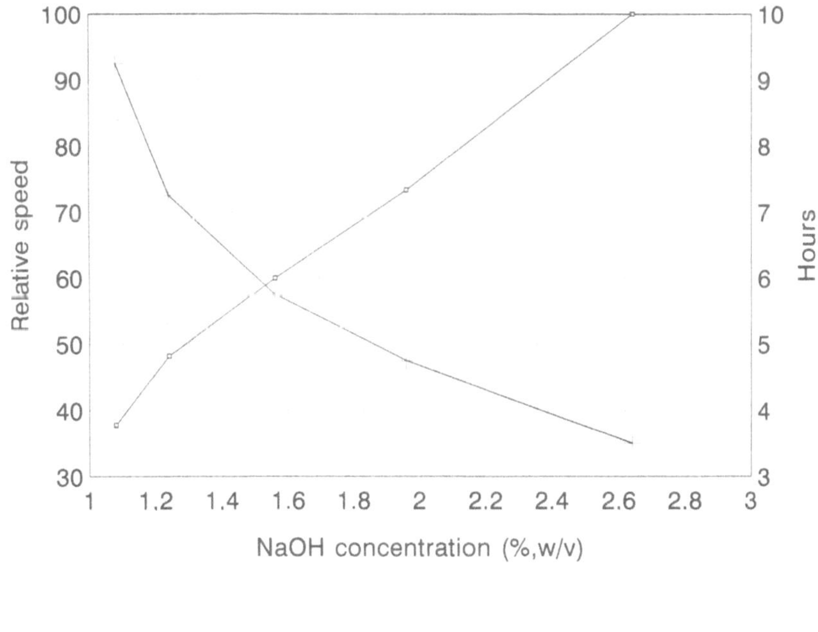

Fig. 7.2 Effect of NaOH concentration on the time elapsed (+) and relative speed (□) of the lye treatment (two-thirds of the flesh). (Redrawn from Borbolla y Alcalá *et al.*, 1951.)

Table 7.1 Influence of lye concentration on the penetration time to reach two-thirds of the flesh. Picholine cultivar of 330 fruits/kg size; temperature ca. 15°C (From Barret and Bidan, 1964)

Lye concentration (% NaOH, w/v)	Penetration time
0.90	1 month
1.20	1 month
1.50	1 month
1.80	45 hours
2.10	15 hours
2.40	13 hours 30 min.
2.70	11 hours
3.00	7 hours

Table 7.2 Influence of temperature on lye penetration (two-thirds of flesh). Picholin cultivar of 320 fruits/kg size; lye concentration, 1.8% (From Barret and Bidan, 1964)

Temperature (°C)	Time
7	112 hours
18	39 hours 30 min.
25	22 hours 30 min.
30	18 hours 30 min.

the small part of the flesh which is not treated seems to play an important role in the final organoleptic characteristics and the final mechanical operations.

Lye concentration is regulated so that the treatment takes a known number of hours to reach the suitable penetration for each cultivar. As a general rule, for the majority of cultivars, this is between 5 and 7 hours, except for Gordal and Ascolano which require around 9–10 hours with more dilute lyes, because of the texture and composition of the skin and flesh.

Bearing in mind the diverse factors that may influence the lye treatment, concentrations used in different countries vary according to temperature, cultivar and experience. In the USA, the concentration may range from 1.25 to 2%, although the higher concentration must be used carefully with cultivars like Sevillano where it can produce softening and blisters. In Spain, it ranges from 1.5 to 3.5%, depending on cultivars and production zones; in general, higher concentrations are used with Hojiblanca cultivar. Other concentrations recommended in different countries are: Israel 1.6–1.8%; Turkey 1.5–2.5%, Argentina 1.4–2.0%, Italy 1.0–2.0% and Greece 1.5–1.8% (Borbolla y Alcalá, 1981). Mondher and Chakib (1993) recommended recently the use of 2.0–2.5% for treating the Meski cultivar in Algeria. A complete study of the relationship between NaOH concentration, temperature and treatment time to penetrate two-thirds of the flesh can be found in Borbolla y Alcalá (1981).

When the fruits are treated for too long and with too low a lye concentration, colour development is only moderately acceptable and the fermentation will be poor because of residual bitterness and the inhibitory action of oleuropein and its hydrolysis products on microorganisms (Garrido Fernández and Vaughn, 1976; Fernández Díez *et al.*, 1985). On the other hand, a high lye concentration applied for a shorter period may produce texture deficiencies and a high loss of fermentable matter.

Amiot *et al.* (1990) have proposed following the kinetics of the debittering process by monitoring penetration of the lye in fruits and determination by high-performance liquid chromatography (HPLC) of the oleuropein : hydroxytyrosol ratio in the fruit and by estimating degradation products from oleuropein in the lye by spectrometry and its residual alkali level by titration. Results could allow the modelling and optimization of the debittering process according to the degree of maturity of the fruits before treatment and the optimal quality of correctly debittered olives. However, the difficulties of applying the procedure in the industry are evident.

Lye treatment is generally conducted at room temperature. The higher the temperature the lower the lye concentration required. It is recommended that the lye temperature does not exceed 20°C in normal processing in order to avoid fruits losing their skin during treatment.

7.4.2 LYE TREATMENT EFFECTS

In addition to debittering, lye treatment has a number of other effects that have not as yet been completely elucidated. Penetration resistance is higher in the outer layers of fruits which consume most of the treatment time, while the inner parts of the pulp need weaker concentrations. Borbolla y Alcalá (1981) also found that the different layers of the flesh neutralized progressively lower amounts of NaOH (Table 7.3). This may explain the relatively slow penetration at the beginning and the fast penetration once the outer layer has been passed. This high rate during the last stage of lye treatment must be carefully monitored because the lye could easily reach the pit if tests for penetration are not frequent enough.

Another effect found during lye treatment is the formation of organic acids. Borbolla y Alcalá *et al.* (1952a) detailed the formation of 122–185 mEq/kg after lye treatment of Gordal fruits and 168–231 mEq/kg in the Manzanilla cultivar. About 25% of the acids are volatile acids. Obviously,

Table 7.3 Quantity of NaOH required to neutralize the different layers of the olive flesh (Reproduced with permission from Borbolla y Alcalá, J.M.R. de la 1981)

Flesh layer	NaOH (m Eq/100 g tissue)
Outer layer (epicarp and adjacent mesocarp)	170
Mid-mesocarp	139
Inner mesocarp	113

Table 7.4 Effect of temperature on reducing sugar content, combined acidity (CA,mg NaOH/100 g) and texture (tenderometer) of the flesh. Lye concentration, 1.5% NaOH (w/v) (Reproduced from Sciancalapore, V. 1984)

Temperature (°C)	Reducing sugar content (%, w/w)	Combined acidity (mg NaOH/100 g)	Texture
Nocellara etnea			
3	0.11	110	68
10	0.18	125	62
15	0.25	153	54
20	0.35	175	45
25	0.40	191	38
30	0.46	220	31
Sant' Agostino			
3	0.25	145	80
10	0.33	164	77
15	0.45	187	64
20	0.49	207	55
25	0.53	232	48
30	0.68	250	34

the organic acids form salts in the alkaline medium and the final result is an increase in the combined acidity. The effect appears to increase with lye concentration, since it is well known that higher percentages of NaOH always yield higher combined acidity values. The formation of acids also depends on temperature and the washing process applied. Acid formation has an important role in the later fermentation process since it permits an initial pH decrease to values appropriate for the growth of lactic acid bacteria.

Sciancalapore (1984) found higher amounts of reducing sugars as temperatures rose in both Sant' Agostino and Nocellara Etnea Italian cultivars after 1.5% NaOH lye treatment (Table 7.4). This phenomenon can explain the easy fermentability of those olives treated with strong NaOH solutions. The same author observed that combined acidity (related to the formation of organic acids) also increased with temperature. Effect on texture was also evident, since consistency decreased with temperature.

However, the most important role of the lye treatment is apparently related to skin permeabilization, which increases with the NaOH concentration used. This can be observed by the different rate of NaCl absorption in olive fruits treated with different NaOH solutions (Figure 7.3). Mondher and Chakib (1993) found a similar behaviour in Meski cultivar (Table 7.5). This must be considered when giving the lye treatment since NaCl concentration should have reached levels tolerable for lactobacilli when this microorganism begins its growth.

From the structural point of view, Cucurachi *et al.* (1974) observed that the lye treatment caused a diminution of the skin thickness, softer flesh and alterations in the nucleus and cytoplasm that were more intense as the lye concentration increased. Such modifications were intensified during fermentation.

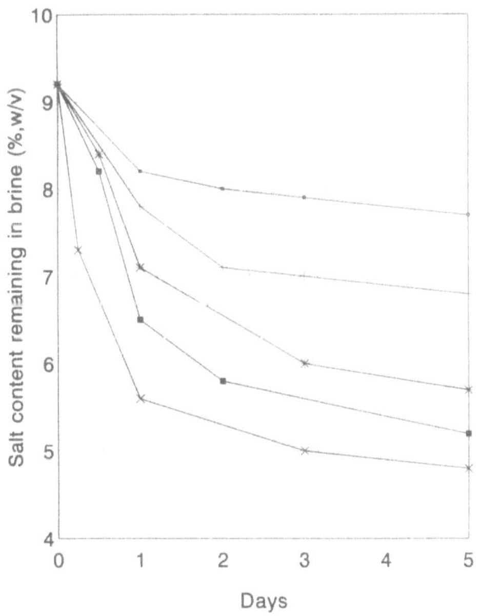

Fig. 7.3 Effect of lye concentration on NaCl absorption rate after brining. (Redrawn from Borbolla y Alcalá, 1981)

Table 7.5 Changes in NaCl concentration after 3 days of brining of olives treated with different lye concentrations. Meski cultivar. Initial brine concentration 10 °Bé (From Mondher and Chakib, 1993)

Lye concentration (%, w/v NaOH)	Brine concentration (°Bé)
0.7	8.4
1.42	7.4
2.11	5.95
2.77	4.75
3.45	4.45

7.4.3 MODIFICATIONS TO TRADITIONAL LYE TREATMENT PROCESS

Some modifications to the traditional lye treatment have been proposed by several authors. Samish *et al.* (1968) tried hot lye treatment to remove the skin and facilitate fermentability. In their experiments, fermentation changes in these olives, those peeled mechanically as well as pricked olives were compared with Spanish-style and untreated olives. Effects on remaining reducing sugars were evident. In all cases, hot peeled olives reached a high acidity, but remained bitter. This taste was removed by rinsing for 3–6 hours. However, this technique of peel removal led to olives with a soft and slippery surface. Consequently, the product was not attractive to consumers and both lye or mechanically peeled fruits were never commercialized.

Some changes in lye treatment have also been made with the aim of reducing sloughing, blistering and gas-pocket formation (Rejano Navarro *et al.*, 1978a). In general a decrease in spoilage was observed as NaOH concentration was reduced, although there was a lower limit below which fermentation was deficient. Working slightly over this limit might therefore be a good option. As temperature decreased, sloughing seems also to decrease. The use of different salts ($MgCl_2$, $CaCl_2$, NaCl) did not give conclusive results, although the presence of salt had a favourable effect on blistering and sloughing. This effect was also noticed by Samish (1954) who found that moderate NaOH concentrations and temperature reduced blister formation, but good results were difficult to achieve due to the hot weather during the Spanish-style green olive processing season in Israel. She solved the problem by adding 3–6% of salt to the solution used for the lye treatment. This significantly reduced blister formation and the practice has been used in many processing plants, although it is not so popular now.

Sánchez Gómez *et al.* (1990) have proposed the use of NaOH concentrations of 1.5–1.8% for Manzanilla olives and 1.8–2.0% for Hojiblanca at a controlled temperature of 30°C. This decrease of NaOH levels could help reduce the number of washings necessary while obtaining a reasonable combined acidity in the fermenting and final product. Results showed that a diminution of 56 and 26 mEq/l of combined acidity could be obtained, respectively, with the utilization of the lower percentages. However, when lye concentration was 2% or over no reduction was observed. Later fermentation was not affected by the modification.

The procedure requires a strict control of temperature and treatment conditions, since at 30°C a significant increase of sloughing in the more sensitive cultivars (e.g. Manzanilla) might be expected.

Olive fruits have been treated traditionally with fresh, recently prepared lye, which were poured into municipal sewage systems or public streams after each treatment. This causes a heavy pollution problem which, in some cases, has moved the authorities in charge of the wastewaters management to constrain the release of this effluent. An internal control measure to partially abate this problem has been the re-use of lyes. It was shown that their successive use, when the NaOH concentration was properly restored, did not have any negative effect on the later fermentation or final product quality (Garrido Fernández *et al.*, 1977). Furthermore, lactic acid yields in fermentations following this modification were increased, as was the proportion of lactic/acetic acids. Thus, this procedure actually enhances fermentation and its yield. Pollution can be diminished in proportion to the number of times a lye is re-used. In an experiment with 15 re-uses, the pollution charge only rose three-fold compared with the pollution charge of a lye from only one treatment. It is thus an interesting internal pollution control that is being progressively introduced in olive processing factories.

7.5 WASHING

7.5.1 OBJECTIVE AND TRADITIONAL SYSTEMS

After alkali treatment, fruits are washed with water to eliminate the major portion of the lye remaining in the flesh. The duration and number of washings necessary are also important factors to be considered (Borbolla y Alcalá and Rejano Navarro, 1978). Their main objective is the elimination of excess NaOH. However, many other soluble substances are also removed with these solutions. Reducing sugars are always present. Table 7.6. shows the effect of both the lye and washing waters. High number and long washing times lead to low reducing sugar content in the treated fruit which, in turn, will provoke low acid formation and deficient fermentations.

However, with moderate NaOH concentrations and washings, sugar concentrations can be sufficient to give appropriate final acidities. During

Table 7.6 Effect of lye concentration, number and total duration of washing waters on the remaining reducing sugar content in the flesh of treated Spanish-style green olives. (Reproduced from Borbolla y Alcalá, J.M.R. de la and Rejano Navarro , L. (1978)

Lye concentration (% NaOH)	Number of washes	Total duration of washes	Reducing sugar content (%, w/w)
Gordal			
3.67	2	4	1.59
3.67	5	25	0.68
1.92	1	1	2.25
1.92	4	24	1.18
Manzanilla			
3.67	2	5	1.15
3.67	5	26	0.55
1.92	1	2	2.75
1.92	4	25	1.47

part of the washing time the olive flesh remains in a alkaline medium and similar effects involving the formation of organic and volatile acids to those found with lyes have been reported (Borbolla y Alcalá *et al.*, 1952a). This is necessarily lower since the alkali concentration begins to diminish just after the lye removal. However, with high NaOH concentrations and long washing periods effects can be appreciable.

The other factor of interest for the washing operation is its effect on the removal of combined acidity, since excessively high values can prevent a proper final pH value being achieved, and a subsequent spoilage risk. Several combinations for obtaining a good result exist. Table 7.7 gives data obtained from Borbolla y Alcalá and Rejano Navarro (1978). Almost the same results are obtained with three and four washings. Furthermore, only long washings can achieve a reasonably low final combined acidity. If this were applied for a little longer than 16 hours a low combined acidity could still be reached. However, the washing periods should not be excessively extended since this can favour the growth of a high population of Gram-negative bacteria (Borbolla y Alcalá and Rejano Navarro, 1978), part of which will remain as an inoculum when the olives are brined. The elimination of such microorganisms during fermentation is difficult and the risk of gas-pockets may arise. This author recommends that the final combined acidity should be below 0.13 N. Otherwise, the acidity needed to reach a pH of 4.0 or lower would not be reached in normal conditions, given the relationship found by him relating both characteristics: $y = 0.0081 + 6.569x$, where x = combined acidity (expressed as N or Eq/l) and y = acidity (% of lactic acid, w/v). Similar equations can be obtained for other specific situations.

Table 7.7 Effect of different washing systems on the final combined acidity (Reproduced with permission from Borbolla y Alcalá, J.M.R. and Rejano Navarro, L. 1978)

Washing system	No. of washings 1	2	3	4	Final combined activity (Eq/1)
1	0.5*	–	–	–	0.180
2	16.0	–	–	–	0.121
3	0.5	16.0	2.0	–	0.110
4	3.0	16.0	1.0	–	0.098
5	0.5	2.0	16.0	8.0	0.080

* Washing duration in hours.

Increasing the number of washings can deplete the fermentable matter and nutrients such that it will be necessary to add these compounds to complete the fermentation process.

On the other hand, excessively short washing leads to the retention of high concentrations of bound organic acids (called 'residual lye' or combined acidity as previously mentioned) in the pickled product. As will be seen later, this prevents the attainment of suitable pH values necessary for the adequate conservation of the fermented product.

Relatively high temperature of the washing water must also be avoided to prevent the growth of undesirable Gram-negative bacteria during the final stage of washing. It is recommended not to exceed a temperature of 20°C for the washing water.

In summary, for each cultivar, stage of maturity, and temperature, a compromise must be found between lye treatment and washing system which permits fruits, when placed into the brine, to retain enough fermentable compounds to reach a suitable degree of free acidity, around 1% (w/v) of lactic acid, at the end of the fermentation, and a concentration of organic salts, normally around 0.10–0.13 M, able to maintain the right pH value for safe storage of the fruits in brine.

If the lye treatment is correct, a quick rinse immediately after treatment, followed by a first washing of 2–3 hours, and a second one of 10–12 hours may be a suitable schedule.

Incorrect washing can, however, be corrected later. When washes have been excessive and the combined acidity is normal or even low, but the fermentation will be deficient because of the lack of fermentable substrates, reducing sugars may be added after brining. When the washing periods are short, with the subsequent high residual bitterness, phenols, and mainly, high combined acidity, the brine will need an additional dilution to support a correct fermentation. Corrections will be further discussed under fermentation.

7.5.2 MODIFICATIONS TO THE TRADITIONAL WASHING SYSTEM

Etchells *et al.* (1966) showed that Manzanilla olives that had been lye-treated, washed and heated at 74°C for 3 minutes fermented quickly and perfectly when placed into a cold pasteurized brine and inoculated. The effect was attributed to the heat destruction of a naturally occurring antibiotic compound. However, Borbolla y Alcalá *et al.* (1969c) demonstrated that conventionally treated olives fermented as rapidly and reached the same acidity as those pasteurized. No further research has been conducted along this line and both washing and brining are carried out currently at room temperature.

Washing waters are also heavily polluted and their disposal causes considerable problems. Efforts for limiting their production during Spanish-style green processing have been attempted by the industry. Because of this constraint, the washing waters must be reduced to the minimum compatible with a good final quality. Reduction of the number of washings to only one was tested successfully by Garrido Fernández *et al.* (1977). Further research to eliminate both washings and their substitution by NaOH neutralization with HCl was partially successful (Garrido Fernández *et al.*, 1979). In some cases there were difficulties obtaining an adequate fermentation and, in some cases, the total substitution of washing waters by HCl, directly added to the brining solution, led to a very low initial pH and stuck fermentations (González Cancho *et al.*, 1983). These problems were solved by adding the HCl in two steps: the first in the brining solution and the second when the fermentation had finished (González Cancho *et al.*, 1984). With this modification it is possible to avoid using washing waters at all during the debittering process. However, their total elimination leads to high concentrations of substrate, polyphenols, etc. in fermentation brines and this gives rise to an excessively strong flavour which is rejected by some processors. Thus, although processing without washing waters is technically feasible, in practice it is reduced to the application of one-long duration wash (18–20 hours) and the subsequent correction of the eventual excess of residual combined acidity. In the absence of waste water disposal difficulties the traditional two–three washing waters is still recommended.

7.6 BRINING

When the olives have been washed they are placed into a brine which consists of an aqueous solution of sodium chloride, and in which the fermentation takes place. In general, microorganisms need a more or less complex medium for growth. They require minerals, fermentable substrates, amino acids, vitamins, etc. as well as appropriate pH, temperature, a_w, etc. In the course of the spontaneous fermentation, the sequence of microorganisms will depend on the respective nutritive requirements and physiology of the different species (Borbolla y Alcalá and Rejano Navarro, 1979). Initially, brines contain only sodium chloride and some impurities such as Ca, Mg, Fe, etc. Thus, at the beginning, this solution is not an adequate medium for the growth of the microorganisms currently found in olive fermentation, both desirable or undesirable. Its transformation into a suitable culture medium is achieved by the extraction of the different nutrients from the olive fruits. This process occurs at different rates depending on olive cultivar, previous lye and washing treatments, olive fruits : brine ratio, and the concentration and quality of the brine. Sufficient nutrients are present for growth of Gram-negative bacteria to start almost immediately. More fastidious organisms such as lactobacilli grow after about 48 hours if the pH is appropriate, and even earlier if the pH is artificially adjusted.

In general, osmotic exchange between fruits and brines is rapid because of the previous lye treatment and washing. The higher the NaCl concentration the faster the nutrient extraction from fruits. During the first period of contact there is an evident increase in the volume of the added brine (due to the fruit juice extracted) which is proportional to the brine strength, as shown in Table 7.8. The extraction is, however, apparently selective and the extraction rate of reducing sugars is the inverse to that of other organic or inorganic components, as shown in Table 7.9. Thus, a higher carbohydrates content in brine is obtained with weaker brines. High NaCl brines are not appropriate because they can provoke surface shrivelling and slower sugar diffusion. Finally, the presence of NaCl increases the ionic strength of the solution and thus helps also to achieve the necessary pH decrease. Excessively weak brines will not inhibit butyric bacteria which can cause malodorous spoilage at the beginning of the process. Some of them can tolerate 6% NaCl or even higher when the acidity is low. On the other hand, many Gram-negative bacteria will grow at 9.5% salt or even higher if the acidity is low.

Table 7.8 Volume increment of the brine added to previously lye-treated and washed olives (From Borbolla y Alcalá, J.M.R. de la 1979)

Initial NaCl concentration (%, w/v)	Initial brine increment (%)
4	0
6	1
8	3
10	6
12	7
14	10

Table 7.9 Effect of the initial brine concentration on sugar content in brine (Adapted from Borbolla y Alcalá et al., 1956)

Initial brine concentration (%, w/v, NaCl)	Sugar content in brines (%,w/v)	
	1 Day	3 Days
4	0.200	0.478
6	0.169	0.428
8	0.132	0.350
10	0.098	0.276
12	0.087	0.242
14	0.084	0.218

Lactic acid bacteria are sensitive to NaCl and initial salt content can influence the further fermentation as shown in Figure 7.4. It is apparent that a high NaCl level leads to both low acidification rate and final titratable acidity. Balatsouras (1985) studied in detail the salt tolerance of many lactic acid bacteria isolated from Spanish-style and natural black fermentations, finding that the 6% NaCl inhibited growth of one-third of isolates from green olives. Above this salt level growth rate diminished dramatically and no growth at all was observed over 9%, which seems to be the upper limit for the growth of lactic acid bacteria in the brine of Spanish-style green olives.

The penetration of salt into the fruits is very fast and, with the concentrations normally used, equilibrium is practically reached within 24 hours. However, the extraction of organic salts (combined acidity) is slower and reaches equilibrium after 15–20 days of brining.

Selection of the initial salt concentration is a compromise. From a practical point of view, the concentration of the initial brine is generally between 9 and 11 °Bé (Borbolla y Alcalá, 1979; Fernández Díez, 1983). Depending on the cultivar and the maturity of the fruit, the concentration of salt must be regulated within these limits.

After the initial decrease in salt concentration, it must be increased gradually during fermentation to maintain satisfactory fermentation, good fruit texture and storage stability, but not so high as to interfere with the growth of lactobacilli. As an approximation, the concentration of salt should be between 5 and 6% (w/v) during the greater part of the fermentation, rising to 7% at the end. Thereafter, it may be increased to 8% or even higher during storage to prevent growth of propionibacteria and a corresponding depletion of lactic acid. After each salt addition the brine must be homogenized to prevent stratification with layers of different salt concentration, especially in big fermenters.

Finally, in the case of the green Spanish-style olive production it is particularly important that the brine is free from iron salts, since these can

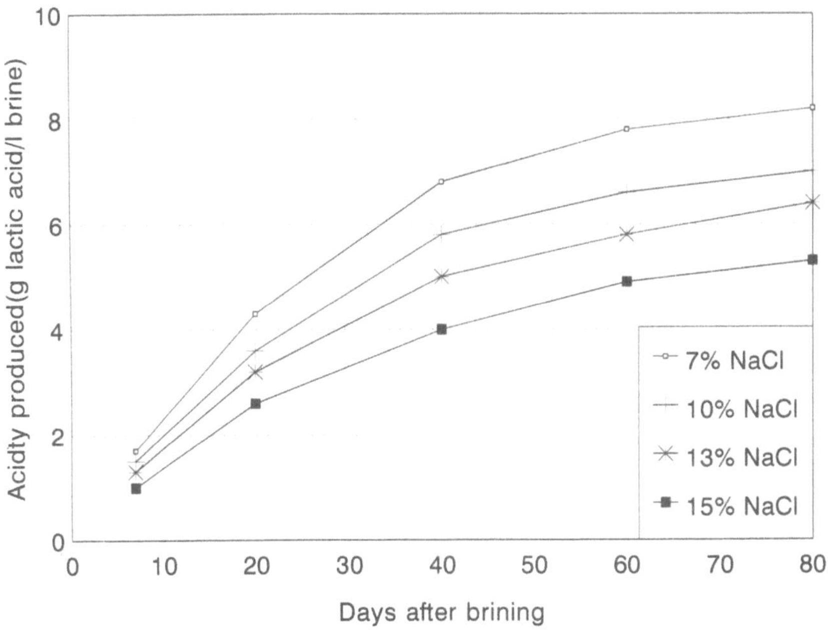

Fig. 7.4 Effect of the initial NaCl concentration on the acidity formed during the further fermentation. All olive fruits underwent the same lye and washing treatments. (Redrawn from Borbolla y Alcalá *et al.*, 1952)

form complexes with phenols and give rise to brown colours which lower the acceptability of the final product. All devices in contact with brine must therefore be of stainless steel or plastic to prevent any iron dissolution.

7.6.1 INITIAL BRINE TREATMENTS

Usually, the initial brine is composed simply of tap water and salt. The solution is added to the washed olive fruits at room temperature and the whole suspension is allowed to ferment spontaneously. However, some pre-treatments have been suggested in certain circumstances.

Etchells *et al.* (1966) studied an olive heat-shock of 74°C for 3 minutes before brining in a pasteurized brine. Inoculated olives fermented more rapidly than controls which followed the traditional process. The same procedure was later studied by Balatsouras *et al.* (1983) for enhancing the fermentability of the Greek Conservolea cultivar. He found that heating olives at 78°C for 3 minutes had an appreciable effect, although less marked than that reported by Etchells *et al.* (1966). Thus, Balatsouras concluded that heat-shock was too costly a procedure to be introduced in the industrial practice.

As already mentioned, the initial brine may sometimes contain acid to help neutralize the remaining NaOH in the flesh of the fruits. Traditionally, the recommended acids were weak concentrations of lactic or acetic acids (0.1–0.4%). Their presence in the final product was insignificant, since lactic acid is produced during the fermentation and the flavour of acetic disappeared during fermentation. Recently, the use of HCl has also been suggested when short washings are applied and a parallel diminution of the combined acidity is required (González Cancho *et al.*, 1983). The amount of this inorganic acid should not exceed reasonable levels because an excess may produce stuck fermentations which need costly further treatments. Moderate levels have not shown any negative effect on the fermentation process and, in fact, may help to inhibit Gram-negative and other spoilage microorganisms originally present in the brine (González Cancho *et al.*, 1984). Initial acidification is not, however, widely used unless an excess of NaOH is expected.

Acidification had little or no positive effect on difficult fermentations such as the Greek Conservolea (Balatsouras *et al.*, 1983), but the use of acetic acid was very effective at preventing gas-pocket spoilage in sensitive olives such as Morona and Sevillano (Borbolla y Alcalá *et al.*, 1959).

Another procedure for initial pH control is injection of the fermentation brine with CO_2. This has the advantage over strong acids of a limited pH decrease determined by the buffering capacity of the different anions (CO_3^{2-},HCO_3^-, organic salts, etc.) in solution. Like acetic acid addition, its utilization has been strongly recommended to prevent gas-pocket formation during the first days of brining through inhibition of Gram-negative bacteria. Apparently, maintaining a saturated atmosphere of CO_2 prevented the formation of gas-pockets, even in the presence of gas-forming Gram-negative bacteria known to produce this effect.

7.7 FERMENTATION

7.7.1 THE USE OF STARTER TO INITIATE THE FERMENTATION

The correct fermentation of Spanish-style green olives relies on the presence of lactic acid bacteria and their rapid initiation of growth. The natural microflora of olives is normally destroyed during the lye treatment, so unless appreciable numbers of lactic bacteria are introduced during subsequent washing and brining operations, successful lactic fermentation will probably not occur and undesirable bacteria from the washing water or brine may cause the olives to spoil. Thus, the use of a lactic starter culture seems indicated.

Tests comparing spontaneous and inoculated fermentations have been carried out since the initiation of scientific studies on table olives. Vaughn *et al.* (1943) recommended the Californian industry to use starter culture for the Manzanilla olive since this cultivar was difficult to ferment in the conditions of lye treatment and strong washing system used there. Later, Borbolla y Alcalá *et al.* (1952b) – after conducting a series of trials with starter cultures – stated that addition of a pure culture of lactobacilli was important to ensure the normal evolution of the traditional process and to prevent an anomalous fermentation. They envisaged a transformation of the Spanish olive industry based on the use of starter cultures. They also found that starter cultures could be useful for enhancing stuck fermentations and even for improving fermented olives stored for long periods of time, provided fermentable substrates were added at the same time (Borbolla y Alcalá *et al.*, 1953). Etchells *et al.* (1966) combined the use of a starter with a heat-shock for obtaining good fermentability of Manzanilla olives, although Spanish researchers later demonstrated that the heat-shock was unnecessary if proper lye treatment and washings were applied. The use of starter culture did however help (Borbolla *et al.*, 1969a,b). Balatsouras *et al.*(1983) found with the Greek Conservolea cultivar that starter cultures improved fermentation and that there were marked differences between starters of different origin, probably reflecting culture adaptation.

Brighigna and Pelagatti (1980), working with Italian cultivars, developed a procedure for preparing pure, spray-dried cultures of *Lactobacillus plantarum*. This preservation system had the advantage of an easy handling (at 2°C) with a high survival rate when revitalized. Its use in Nocellara Etnea Spanish-style fermentations led to improved process economics and better final quality. The best medium for reconstituting the culture was: NaCl, 8.5 g; glucose 10 g; yeast extract, 3 g in 1 litre of distilled water (pH 6.4) (Pelagatti and Brighigna, 1981). The preparation of improved commercial starter cultures specifically for table olives and their application to Spanish-style fermentations has recently been reported (Reig and Hernández, 1991). However, in spite of such good results and prospects, the use of starter cultures with Spanish-style green olive is still limited.

For a starter culture to succeed, it must itself be adequate and the medium must meet certain requirements:

- availability of fermentable substrates
- appropriate salt concentration in brine
- presence of necessary growth-promoting substances (amino acids and vitamins)
- suitable pH in the brine
- appropriate temperature in the fermenting medium
- absence of inhibitory substances
- suitable olive fruit cultivar

Failures with starter cultures in the past have been mainly due to lack of appropriate conditions for its growth. As always stressed, the availability of fermentable compounds, influenced by the lye treatment and washings, determines in part the fermentability of olives. Ordinarily, Gordal (in Spain) or Sevillano (in USA and other countries) olives respond rapidly to inoculation because of their low inhibitor content and more fermentable material remains at the time of brining. Manzanilla or Hojiblanca, on the other hand, may show little response to inoculation, even in the presence of enough sugars if the lye treatment was deficient due to the presence of an excess of phenols in the brine. The same can occur if the salt level is over 7–8% or the temperature is low. Thus, inoculation alone is frequently not sufficient to accelerate acid production, unless the appropriate conditions are met.

Bearing in mind the specific characteristics of the olive fruit, some physiological properties of a starter culture for olive fermentation are convenient. These include:

- Good resistance to the inhibitory effect of polyphenols.
- Low requirement for growth factors.
- Good survival against the wild strains of related species.
- Possible inhibitory effect against undesirable microorganisms.
- Good tolerance to high levels of salt and extreme pH values (both alkaline and acidic).
- Rapid acid production and complete utilization of fermentable substrates.
- Good revitalization characteristics from dried or frozen state.

Until now, strain selection has been based mainly on the capacity for lactic acid production without paying too much attention to the other requirements. Failures can thus often be attributed to this.

Currently, some research work is been carried out with some bacteriocin-producing strains of *Lact. plantarum* which, in addition to other favourable physiological characteristics, can compete with and dominate the indigenous microflora. Jiménez Díaz *et al.* (1993) checked 26 strains of lactobacilli isolated from green olive fermentations for antagonistic activities. Some were found with inhibitory effect towards some Gram-positive bacteria, including clostridia and propionibacteria as well as natural competitors of *Lact. plantarum* in olive fermentation brines. The inhibitory substances were isolated and classified as bacteriocins, which were designated as plantaricin S and T. Trials using this starter have shown that this strain can survive in fermenting brines throughout the process, while other wild strains grew and disappeared successively, being present for only limited periods of time (Ruiz-Barba *et al.*, 1994).

Where olives are properly processed (in both lye treatment and washing), the environmental conditions, particularly temperature, are good and there is an abundant flora of suitable wild lactic acid bacteria in the factory, there is no need for a starter culture. Alternatively, this may come from an actively fermenting brine of the same cultivar or from Gordal, which is well known to have good fermentation characteristics. In other cases, the use of a starter culture is recommended. Inoculation consists of the introduction of the culture into the fermenter brine sufficient to reach about 10^6 cells/ml after a few days of brining (when the pH of the medium is about 6.0). As the medium already contains similar microorganisms as part of the spontaneous fermentation, there ensues a competition between the inoculum and the wild species. In general the indigenous, wild lactic acid bacteria predominate with time, although there is usually a better acidification rate.

However, starter use does have some advantages; it:

- induces the correct development of the fermentation;
- favours a better control of organoleptic and physicochemical characteristics;
- may diminish the time elapsed from brining, to the lactic microflora growth and reduce the fermentation period; and
- limits the incidence of stuck fermentations.

With respect to the type of microorganism used, practically all starters available commercially are made from strains of *Lact. plantarum*, which is the predominant species isolated from Spanish-style green fermentations. Fermentations carried out with only this microorganism have, however, undesirable flavours. So, sterilization of the medium and further controls to permit only the growth of the inoculated species is not appropriate.

Although the presence of fermentative yeasts during the normal process of Spanish-style green olives is considered favourable, trials carried out with inocula of different species of *Saccharomyces* (*S. oleaginosus*, *S. cerevisiae*, *S. fructuum* and *S. italicus*) singly and as a mixture showed a diverse growth and influence on the final organoleptic characteristics (Castro Gómez-Millán *et al.*, 1989). Fermentations were inoculated 7 days after brining to permit the growth of the normal lactic acid bacteria (except in the case of the mixture). By that time, the yeast's population is still very low, so the yeast strain inoculated was in a good position to predominate. A further salt correction to 8.5% was made 20 days after brining to slow down the lactobacilli growth and facilitate yeast development. In general, *S. fructuum* showed very limited growth, *S. oleaginosus* and *S. cerevisiae* grew abundantly for a short period after which their population diminished, and *S. italicus* had an abundant growth for 2 months. It was obvious that the *Lact. plantarum* and the relatively low pH of the brine could, at least partially, inhibit the growth of the inoculum, since the fermenter that was maintained at 8.5% NaCl from the very beginning of the process and did not support lactic bacteria growth, showed an abundant growth of the mixed yeast culture. Organoleptic tests showed that all the olives from the yeast-inoculated fermenters scored lower than those from containers that had followed the spontaneous lactic process.

To our knowledge, this has been the only experience of inoculation with yeasts isolated from brines of the same olive type and results were rather disappointing. It is possible that better results could be obtained with other species or inoculation conditions, because the presence of a certain population of these organisms is necessary for achieving the typical final flavour of the product. It must also be emphasized that olives prepared with a pure culture of *Lact. plantarum* were also judged unacceptable by specialized panellists.

The spontaneous (or initially aided) process will continue in the near future until we have a better understanding of the ecology of the process. The relationships between the different microbial species involved in the green Spanish-type fermentation is clearly quite complex and microbiological control is still rather difficult. Consequently, further studies on the design of new starter are necessary. Addition of properties to those currently available in starters will undoubtedly favour their utilization in the future. Examples would be bacteriocin-producing strains and mixture cultures of lactic acid bacteria and selected yeasts.

7.7.2 MICROBIOLOGICAL ASPECTS OF THE FERMENTATION

Four phases have been identified during the spontaneous fermentation of Spanish-style green (pickled) olives in brine (Fernández Díez, 1983, 1991; Fernández Díez *et al.*, 1985). Each one is characterized by the development of a different microbial population as well as by the changes in the physicochemical characteristics of the medium.

The duration of each phase, as well as the relative importance of the growth of the different microorganisms, depends on a series of factors:

- Prior treatments of the olives (lye treatment, washing operations, and initial concentration of the brine).
- Capacity of the fermenters and, as a consequence, the ratio fruit weight : brine volume.
- Climatic and environmental conditions of the factory.
- Cultivar of the olives.
- Sanitary conditions of the plant and equipment.

7.7.3 FERMENTATION PHASES

First phase

This lasts from brining until the growth of lactobacilli, which in normal processes occurs after about 48–72 hours. As previously emphasized, it is of great importance for the further normal development of the fermentation. Thus, changes during this phase must be strictly controlled to provide the conditions necessary to enhance development of the desired organisms and, in parallel, inhibit the growth of those that may produce spoilage.

Many of the microorganisms present during the first hours are contaminants from fermenters, pipelines, pumps and other devices in contact with the brine and olives. Species identified belonged to the genera *Bacillus*, *Micrococcus* and many Gram-negative bacteria, mostly belonging to the Enterobacteriaceae family (González Cancho, 1960b). The most representative species of Gram-negative bacteria isolated from Spanish-style green olives are shown in Table 7.10. Gram-positive bacteria are also encountered, especially at the end of the phase.

It is advisable to avoid contamination with undesirable microorganisms during this period by applying the best sanitary conditions during the different operations. However, a certain population of Gram-negative bacteria must be considered normal and helps decrease the pH to values appropriate for lactobacilli growth.

Second phase

This phase begins when the pH of the medium has dropped to about 6.0 and usually lasts about 2 weeks. During the phase, the lactic cocci population increases, though it starts to diminish towards the very end of the period. Gram-negative bacteria counts decrease progressively throughout and their total disappearance is an indication of the end of this phase.

Table 7.10 Relevant species of Gram-negative bacteria identified from Spanish-style green olive brines (From Fernández Díez *et al.*, 1985)

Enterobacter cloacae
Citrobacter freundii
Enterobacter aerogenes
Escherichia coli
Aeromonas hydrophila
Flavobacterium diffusum
Flavobacterium balustinum

In general, Gram-negative organisms should not persist after 20–25 days (first and second phases). Delay in this may lead to the production of gas-pocket spoilage.

Characteristic microorganisms of this phase are lactic acid-producing cocci (González Cancho and Durán Quintana, 1981), mainly from the genera *Pediococcus* (homofermentative) and *Leuconostoc* (heterofermentative). The genus *Streptococcus* (homofermentative), also present, is not considered important. The growth of these organisms, together with other factors and the action of Gram-negative bacteria, influences the proportion of volatile acidity in the final product. The more relevant species identified in this phase and the duration of their presence in brine are shown in Table 7.11.

Third phase

This is characterized by an abundant growth of lactobacilli, mainly from the species *Lact. plantarum* which under normal conditions becomes predominant and is responsible for the typical fermentation of the Spanish-style olives (González Cancho, 1963). The same author has reported the presence of *L. delbrueckii*, although its significance is rather limited. Recently, Rodríguez Quiñones *et al.* (1983) have identified other species in Gordal cultivar brines. In decreasing order of importance, they were: *L. casei*, *L. fermentum*, *L. cellobiosus*, *L. coryniformis*, *L. casei* and *L. brevis*. The last three occurred at a very low frequency. This diversity of species may be related to the excellent fermentation characteristics of the Gordal cultivar and the physicochemical conditions used in the process. In California, Vaughn *et al.* (1943) isolated *L. brevis* from fermentation brines of the same cultivar. This reflects the low salt concentrations used with this olive since it was never isolated from Manzanilla, which is usually fermented in higher salt concentrations.

Thus, it is generally accepted that *Lact. plantarum* is the most common, abundant and characteristic species of *Lactobacillus* associated with the Spanish-style green olive fermentation. As a consequence of its growth, the acidity increases steadily throughout this period and the pH drops to

Table 7.11 Relevant species of lactic cocci isolated from fermenting green Spanish-style olive brines (From Fernández Díez *et al.*, 1985)

Species	Growth interval (days)
Streptococcus (Lactococcus) lactis	6–7
Pediococcus urinae-equi	4–23
Leuconostoc paramesenteroides	4–55

values around 4.0 or below. Numbers of lactobacilli decline when fermentable substrates are exhausted and the pH has reached values that may partially inhibit even *Lact. plantarum*, the most acid-resistant of the diverse lactic bacteria present and the one found to complete most vegetable fermentations.

Fourth phase

Propionibacterium is the only genus which grows during the fourth phase, if the salt concentration in the brine is not suitably controlled during storage (Borbolla y Alcalá *et al.*, 1975; Rejano Navarro *et al.*, 1978a).

This phase begins when the main lactic fermentation has finished and is characterized by an increase in volatile acidity and pH, with a decrease in lactic acid. The species identified has been *Propionibacterium acnes* (the most characteristic and abundant) and *P. acidi-propioni* (Castro and Rejano Navarro, 1990) which, due to their ability to metabolize free or combined lactic acid, are natural successors of the lactic bacteria. Their principal metabolites are acetic and propionic acids.

The extent of this phase is variable, producing in most cases a slight increase in the pH values (0.1 to 0.2 units). Only when the fermentation has been defective is the growth of propionibacteria abundant and this may cause pH increases of 0.4 or over, leading to eventual spoilage. To avoid this possibility, the propionibacteria activity must be controlled. This is achieved by raising the salt level to over 8% (w/v) in April or May (according to the environmental temperature) while the pH is maintained below 4.0.

7.7.4 PRESENCE OF YEAST DURING FERMENTATION

A variable and small population of fermentative yeasts co-exists with lactobacilli throughout the fermentation of Spanish-style green olives (González Cancho, 1960a, 1963, 1965) and probably contributes to the sensory properties of the final product. Thus, they are not considered harmful if their counts are not excessive since none of them has been proven to use lactic, acetic or citric acids in the anaerobic conditions prevailing (Ruiz Cruz and González Cancho, 1969). Table 7.12 shows the most common species found. In stuck fermentations, however, the yeast population increases markedly and lactobacilli diminish to the point of disappearance. The flavour of such fruits is then unsuitable for marketing as Spanish-style green. To enhance a lactic process corrections should be made to decrease the yeast population and increase the lactobacilli levels (see Fermentation control, p. 165).

Oxidative yeasts have also been found in fermenting green olive brines due to deficient cleanliness of surfaces in contact with the air. As their

Table 7.12 Relevant species of fermentative yeasts isolated from Spanish-style green table olive fermentation brines (From Fernández Díez *et al.*, 1985)

Candida krusei
Hansenula anomala
Saccharomyces chevalieri
Candida parapsilosis
Hansenula subpelliculosa
Candida pseudotropicalis
Saccharomyces cerevisiae
Candida tropicalis

Table 7.13 Relevant species of oxidative yeats isolated from Spanish-style green table olive fermenting brines (From Fernández Díez *et al.*, 1985)

Candida valida
Pichia membranaefaciens
Candida zeylanoides
Torulopsis inconspicua
Candida rugosa

metabolism uses the lactic acid previously formed they can cause an appreciable increase in pH which, in turn, may lead to malodorous spoilage. Relevant species identified are shown in Table 7.13. Film-forming oxidative yeasts must be avoided by a suitable sealed closing of the fermenters which maintains anaerobic conditions.

Moulds are usually found in the same films with aerobic yeasts. Balatsouras and Vaughn (1958) identified from these the genera *Fusarium*, *Penicillium*, *Aspergillus*, *Verticillium* and *Geotrichum*. However, as with the oxidative yeasts, they should not be involved in the normal fermentation process.

7.7.5 CHARACTERISTICS OF BULK FERMENTATIONS

The introduction of bulk fermentation has altered somewhat the classical evolution of the microorganisms. With this system, the growth and persistence of Gram-negative bacteria have been reduced and in some cases they are completely absent. This is due to the use of better sanitary conditions during processing, mainly in the glass-fibre fermenters (which are easily cleaned), less contamination during fruit handling and the use of chlorinated water.

Also, the cocci population has decreased and a perfect sequence of

Pediococcus, *Leuconostoc* and *Lactobacillus* genera occurs in the new containers, reaching their respective maxima at about 5–13, 12–24, and 19–30 days. Each of them has effectively disappeared when the following one has its maximum count.

Fermentative yeasts are also markedly lower and oxidative yeasts have practically disappeared because of the low brine surface area-to-volume ratio. In addition, new fermenters can be provided with almost hermetic closing devices which avoid completely the problem of film-forming yeasts and moulds.

Thus, the bulk fermentation produces a natural cycling even better than the traditional techniques. Each genus and species gradually adapts the characteristics of the medium to the requirements of the microorganisms which succeed it within the natural sequence, until the last one becomes dominant. The correct control of a reduced number of factors such as pH, free and combined acidity, salt concentration, soluble sugars content, temperature, degree of aerobiosis or anaerobiosis, and use of a pure culture as starter, permit the food technologist to vary the duration and relative importance of each of the phases and to avoid some of the microorganisms that can have negative effects on the quality of the final product (Rejano Navarro *et al.*, 1977; Borbolla y Alcalá and Rejano Navarro, 1979, 1981).

Acidification or injection of CO_2 into the brine during the first stage, to prevent the development of Gram-negative bacteria, and inoculation with a pure culture of *Lact. plantarum* or a well-fermented brine with a good population of active lactobacilli is a very efficient way of minimizing the first and second phases, accelerating the start of the fundamental third phase, and achieving a complete lactic fermentation.

7.7.6 INHIBITORY EFFECTS OF OLIVE PHENOLS

It has long been known that unsuitably lye-treated olives hardly support lactic acid bacteria growth and frequently lead to stuck fermentations. Manzanilla, Hojiblanca and similar cultivars were considered sensitive to this problem, although others, like Gordal and Sevillano, always ferment fairly well. Obvious differences in the fermentation rates of cucumbers and olives (Etchells *et al.*, 1966) moved Fleming and Etchells (1967) to explore possible explanations for this observation. They found that aqueous extracts of frozen Manzanilla olives were not fermented by lactic acid bacteria and that the fermentation rate of cucumber extracts was decreased when such olive extracts were added to cucumbers. The presence of inhibitory compounds in olives was demonstrated by adding olive tissues and extracts to seeded Cucumber Juice Agar (CJA) plates. A good correlation between the inhibitory effect and bitterness (probable oleuropein content) in the Californian cultivars was found, as well as the parallel disappearance of bitterness and inhibitory activity by adjustment of the pH to 10. These facts led them to assign responsibility for the inhibitory effect to oleuropein.

At the same time, Juven *et al.* (1968a) found a marked enhancement of the fermentation rates of intact Merhavia and Manzanilla green olives in brine following pretreatment with hot alkali. They attributed this to the destruction of an antibiotic factor by the alkali and developed a rapid method to detect it in the press juice of olives. Later, Juven *et al.* (1968b) extracted the antilactic factor which was identified as oleuropein. The compound had an inhibitory effect against *Lact. plantarum* [with a minimal inhibitory concentration (MIC) ranging from 300 to 1000 µg/ml] and *Leuconostoc mesenteroides* (MIC 330–670 µg/ml), among others. However, it was not inhibitory for any of the Gram-negative bacteria and yeasts tested. A synergistic effect with sodium chloride was also observed. Simultaneously, Fleming *et al.* (1969) reported the isolation of a inhibitory phenolic compound from Manzanilla olives whose effect was increased by freezing the olives before extraction. Juven and Henis (1970) found that there were two compounds in the ethyl acetate extract of green olive with antilactic activity which were isolated and identified as the glucoside oleuropein and its phenolic aglycone. However, it was Walter *et al.* (1973) who prepared a series of derivative compounds from oleuropein and had them tested individually (Fleming *et al.*, 1973). From their results they concluded that oleuropein itself was not inhibitory against lactic acid bacteria, although it certainly was for three other types of bacteria. However, the aglycone of oleuropein and the elenolic acid inhibited the growth of the four species of lactic acid bacteria tested, the effect being increased in the presence of 5% NaCl. In these conditions, 150 µg/ml of either compound prevented growth of *Lact. plantarum*. Another product of the oleuropein hydrolysis, β-3–4-dihydroxyphenylethyl alcohol (hydroxytyrosol) was found not to be inhibitory. These results disagreed with those obtained by Ruiz-Barba *et al.* (1991), who found oleuropein inhibitory to *Lact. plantarum* at concentrations of 0.4% (7.7 mM) as well as hydroxytyrosol and, to a lesser extent, β-4-hydroxyphenylethyl alcohol (tyrosol) (Ruiz-Barba *et al.*, 1990). The inhibitory effect increased as the concentration rose. The most recent studies on the inhibitory effects of phenolic compounds from green olives (Ruiz-Barba *et al.*, 1993) have shown that neither oleuropein, tyrosol, verbascoside, luteolin 7-glucoside nor vallinic acid had effective inhibitory effect against lactic acid bacteria when tested at concentrations similar to those found in fermenting brines. However, hydroxytyrosol and combinations of this with oleuropein or verbascoside, as well as the combination of verbascoside and oleuropein, had marked effects.

Apparently, although inhibition depends strongly on the concentration of each individual compound, their real effect on brines is due to a combined action. This hypothesis is also supported by the non-specific mechanism of action of these compounds which is discussed later. Discrepancies showed also that the inhibitory effect must be strongly related to specific composition and conditions of the testing medium. Differences found by Fleming *et al.* (1973) and Ruiz-Barba *et al.* (1993) could be explained by the presence or absence of nitrogenous compounds in the respective media.

In addition to the inhibitory effect of olive fruit phenols on microorganisms normally found in olive fermentations, these compounds also affect other pathogenic and non-pathogenic organisms. Thus, 0.1% oleuropein delayed the growth of *Staphylococcus aureus*, 0.6% was completely inhibitory, and concentrations over 0.2% prevented the formation of enterotoxin B (Nychas *et al.*, 1990; Tranter *et al.*, 1993). In model systems, inhibition was stronger in samples with low pH, low inoculum size and in the presence of milk extracts (Tassou and Nychas, 1994). Phenolic extracts from olives and oleuropein inhibited germination and outgrowth of *Bacillus cereus* T spores or significantly delayed outgrowth when added immediately after germination (Tassou *et al.*, 1991). MICs (μg/ml) of several olive phenols in a laboratory medium against *Staphylococcus aureus, Bacillus cereus, Escherichia coli* and *Salmonella typhimurium* ranged from 350 (caffeic acid) to >600 (*p*-coumaric and 4-hydroxyphenylacetic acids) according to Tunçel and Nergiz (1993).

The inhibitory effects of olive phenols on moulds has also been investigated. Coumarin and caffeic acid at 10 mM inhibited *Aspergillus flavus* growth while only 0.5% to 1.0% of olive extracts inhibited aflatoxin B_1 production by 90% (Paster *et al.*, 1988). Oleuropein at 6 mg/ml reduced the production of aflatoxin by *Aspergillus parasiticus* by 83–93% and delayed onset of growth and sporulation; however total mycelial mass was stimulated by the presence of oleuropein (Gourama and Bullerman, 1987). Tyrosol at 0.2–0.6% inhibited the growth of *Penicillium* and *Aspergillus* isolated from Moroccan olives, caffeic acid being less inhibitory than tyrosol (Gourama *et al.*, 1989). Fresh and frozen olive pastes supported poor growth of *Aspergillus parasiticus* and *Aspergillus flavus*, causing no aflatoxin production, while heat- and lye-treated pastes supported extensive growth but little aflatoxin production (Mahjoub and Bullerman, 1987). Thus, in general, olives are a poor substrate for mould development, or are inhibitory to growth and aflatoxin production.

7.7.7 INHIBITORY MECHANISM OF OLIVE PHENOLS ON LACTIC ACID BACTERIA

It has long been known that the action of phenols on microorganisms was due to their ability to denature proteins and act as surface-active agents. They cause leakage of cytoplasmic constituents from the bacterial cell, as a result of damage to the cell membrane affecting its permeability.

Lact. plantarum incubated in presence of oleuropein changes in size and shape. Cells become longer and wider and often deformed. The typical Gram-positive appearance is lost and a Gram-negative reaction is observed. These changes occur 30–60 minutes after incubation. This indicates that the disruption of the *Lactobacillus* peptidoglycan is promoted by oleuropein and that such effects can lead to cell death by destruction of the cell envelope. Nevertheless, based on these observations cell death by damage of other structures (e.g. cytoplasmic membranes) cannot be excluded (Ruiz-Barba *et al.*, 1991).

Detailed studies of oleuropein's effect on cells were carried out by Juven *et al.* (1972), who demonstrated that it induced changes in the membrane of red blood cells, leading finally to haemolysis. Since the concentration of the active agent was only 2 mg/ml, this effect could not be due to osmotic pressure but resulted from changes in the permeability of the cell membrane. Oleuropein decreased the surface tension of water by about 11%. In addition, it increased the leakage of inorganic phosphate, glutamic acid and potassium from cells of *Lact. plantarum*, an effect that might be due to mechanical injury as well as to damage of the regulatory mechanisms controlling cell membrane permeability. Utilization of nutrients leaking from damaged cells by undamaged cells may be responsible for the diminution of the inhibitory action of oleuropein when the size of the inoculum increases. Oleuropein had no effect on the rate of glycolysis in resting cells of *Lact. plantarum*, suggesting that it does not affect the activity of the glycolytic enzymes, although smaller amounts of ATP were present in the bacterium. The reduction in the ATP in the treated cells could be the result of the effect of oleuropein on the plasma membrane and leakage of ATP or phosphate from the cell. This would lead to less energy being available for metabolism and growth.

Ruiz-Barba *et al.* (1990) observed that the ultra-violet absorption at 278 nm of brines or solutions containing olive phenolic extracts decreased by 6, 11 and 19% after 30 minutes, 60 minutes and 24 hours respectively, of incubation with cells of *Lact. plantarum*. The cells separated from the medium showed an increased fluorescence intensity with incubation times. Isolated cell walls of *Lact. plantarum* also showed increased fluorescence with time after extended incubation, while this could not be detected in untreated cell walls. In electron micrographs, the cell surface of the untreated control cells appeared smooth, without any visible structure; however, after 24 hours of incubation in the presence of total phenolic compounds extracted from brines, the cells exhibit an irregular surface covered by projections or granules of different sizes giving them a rough appearance.

Ultrathin sections (Ruiz-Barba *et al.*, 1990) revealed that incubation for 30 or 60 minutes in the presence of phenolic compounds caused remarkable alterations in *Lact. plantarum*. Some projections appeared on the bacterial surface, becoming more numerous and longer with increasing incubation times. These projections seem to be responsible for the rough surface observed on scanning electron micrographs. After 3–11 days of incubation, the cell envelope disappeared completely. Another noticeable change in cellular structure was the presence of mesosomes in most treated cells which were not visible in untreated cells. After 11 days of treatment the majority of the cells had lost their rod shape, appearing spherical like protoplasts; some of them appeared broken and only a few remained intact.

Thus, results from both Juven *et al.* (1972) and Ruiz-Barba *et al.* (1990) agree that the bactericidal effect of phenolic compounds from olives occurs by a mechanism similar to that found for other phenols, although it seems that oleuropein and its derivatives are more selective than simple phenols, as seen from the variable sensitivity of different bacteria. Some discrepancies found among the results from different researchers can be explained by the diverse experimental conditions used. One of the difficulties arises from the presence of organic nitrogenous compounds in most of the culture media employed, since they can combine with phenols by means of hydrogen or covalent bonds. This scavenging effect obviously diminishes the free (and available) phenol content and can lead to erroneous results. This interference was observed and avoided by Ruiz-Barba *et al.* (1990) and possibly facilitated detection of the strong inhibitory effect of hydroxytyrosol, which had not been detected previously by other researchers. In general, the bactericidal effect seems to be due to alterations at two different levels of the cellular ultrastructure, cell wall and cytoplasmic membrane. These increase with contact time and inhibitor concentration, promote the leakage of glutamate, potassium, inorganic phosphorus and possibly other essential compounds, and lead finally to cell lysis and death.

7.7.8 EFFECT OF OLEUROPEIN AND RELATED COMPOUNDS ON FERMENTATION

A suitable lye treatment of olives permits a subsequent lactic fermentation. The alkali hydrolyses oleuropein and facilitates fermentability. Ruiz-Barba *et al.* (1991) demonstrated that alkali-treated oleuropein allowed *Lact. plantarum* survival for at least 24 hours with only a minor decrease in cell viability. However, other factors must influence the olive lactic fermentation because brines which have previously shown a strong inhibitory effect, even after heat treatment, allowed *Lact. plantarum* survival and growth when supplemented with MRS broth, a nitrogen-rich medium used to culture lactic acid bacteria. This effect might be due to the higher nutrient availability which permits faster growth than death (due to polyphenol presence) or to inactivation of the polyphenols by the added organic nitrogen (Ruiz-Barba and Jiménez Díaz, 1989). Apparently, the same effect was also responsible for only delaying growth of certain lactic acid bacteria in the presence of up to 0.4% oleuropein, which in the absence of glucose was utilized as carbon source for their growth (Garrido Fernández and Vaughn, 1976).

Juven and Henis (1970) also found that the inhibitory effect of oleuropein was augmented by reducing the nitrogenous compounds and the inoculum size. They also noted that survival was increased by treating the olives with hot lye. It seems that the brief alkaline pre-brining treatment caused a faster release into the brine of olive cell constituents (sugars, B_1 complex vitamins and amino acids). The concentration of phenolic compounds in the brine also increased rapidly, but their inhibitory activity towards lactic acid bacteria was counteracted by the high concentration of nutrients and their growth-enhancing effect. As result, a rapid predominance of *Lact. plantarum* in the brined alkali treated olives was observed (Juven and Henis, 1970).

From the above, differences between the ability of alkali-treated and untreated olives to support lactic acid fermentation are apparent, in spite of the fact that in both brines the main phenol present is hydroxytyrosol (Ruiz-Barba *et al.*, 1993) and its concentration (6–8 mM) is similar in both cases. However, in the case of Spanish-style green olives this level is reached at a high rate, reaching equilibrium after only 1 month, while in untreated olives the diffusion of phenols is markedly lower, reaching equilibrium after 6–8 months. The explanation must be that with alkali-treated olives there is simultaneously a rapid availability of all nutrients which permit a heavy growth of lactobacilli which the remaining active polyphenols (e.g. hydroxytyrosol) cannot inhibit at the same rate. Eventually, the active phenols may decrease due to reaction with dead cells and nitrogenous compounds released from them or the olives. In the case of untreated fruits, nutritional and phenolic neutralizing compounds take longer to emerge from the fruits and the inhibitory effect predominates.

7.7.9 NUTRITIONAL REQUIREMENTS OF *LACT. PLANTARUM* FROM SPANISH-STYLE GREEN FERMENTATIONS

Ruiz-Barba and Jiménez Díaz (1994) have studied the nutritional requirements of 24 wild strains of *Lact. plantarum* isolated from green olive fermentations. Most were able to grow in the complete defined medium containing all the amino acids considered essential for this bacterium, but nine failed. These must have had additional requirements which are present in the brines where they grow. Some strains were, however, able to grow in the absence of pyridoxal, *p*-aminobenzoic acid and nicotinic acid.

The nutritional profile of these microorganisms must be determined by the characteristics of Spanish-style green olive processing. Lye treatment of olive fruits promotes alkaline hydrolysis of proteins in the flesh, as well as making the skin more permeable. Consequently, the rapid leaching of nutrients into the brine provides all the necessary peptides and amino acids for *Lact. plantarum* growth. However, the alkali treatment may destroy some of the vitamins contained in the fruits. Other water-soluble B-complex vitamins like biotin, nicotinic acid, pyridoxal or *p*-aminobenzoic acid can be washed out of the starting brine during the removal of the lye and the subsequent washes with water that are part of the traditional brining procedure. So, the ability of organisms to synthesize these vitamins would clearly confer some advantage in olive brines. This can explain why some of these strains were prototrophic for more vitamins than is usually associated with *Lact. plantarum*. So, in summary, Spanish-style green olive brines must have all the nutritional requirements that are needed for lactobacilli growth. However, in spontaneous fermentations wild strains specifically adapted to possible vitamin deficiency are found. This may, at least in part, be responsible for the often-noted predominance of wild strains over inoculated ones which usually disappear a few days after their addition. Later on in the fermentation these vitamins may be excreted by other species in the microflora such as Gram-negative bacteria or yeasts.

7.7.10 FERMENTATION CONTROL

Analyses used to follow the fermentation changes are: pH, salt content (expressed as %, w/v), titratable acidity (% lactic acid, w/v), combined acidity (mEq/l) and reducing sugars (% glucose, w/v). Microbiological analyses are also generally useful if the factory has the facilities, but will be necessary with specific problems related to the microbial population. Free acidity and pH changes are the most important factors to control the fermentation from the beginning and they must be measured frequently during the first months of the process. Combined acidity reaches its equilibrium (flesh–brine) around 20 days after brining; a determination is required at that time to establish the remaining buffering capacity of the fermentation solution. It must also be checked after any brine correction. Reducing sugars give an idea of the potential free acidity after fermentation. Their determination is also important before any addition of starter culture, correction to enhance lactic acid production or packing. Microbial analyses are especially recommended in cases of anomalous fermentations or stuck (dormant) processes.

7.8 PHYSICOCHEMICAL CHANGES IN BRINES AND FRUITS DURING FERMENTATION

7.8.1 BRINES

Acidity and pH

During the normal process there is a progressive pH decrease (Table 7.14). The final value depends on the acidity produced and the combined acidity of each specific brine, as described earlier. For a combined acidity of about 140 mN , the titratable acidity needed to obtain a pH of 4.0 is 1.0%. If combined acidity figures are higher, obtaining enough acidity to reach pH 4.0 is difficult and brines may be exposed to higher pH values for long periods with the associated risk of spoilage. To get low combined acidities strong washings are necessary which, in turn, mean a greater loss of fermentable substrates. Where such washes are unavoidable, the potential acidity must be increased by addition of supplementary fermentable sugars or directly with lactic acid.

Fermentable substrates

Reducing sugars pass to the brine, where they are used by the microorganisms. The most desired transformation is the formation of lactic acid. The yield, however, depends on the microbial population and the type of lactic acid bacteria. An excessive number of Gram-negative bacteria or yeasts may reduce the substrate availability for lactate production, hence the advantage of reducing this population (by initial pH adjustment with acetic acid, CO_2 bubbling, etc.) to the minimum count compatible with the normal process. Lactate production also depends on the type of lactobacilli. Homofermentative strains (e.g. *Lact. plantarum*) are preferable since they produce twice the amount of lactic acid produced by heterofermenters.

Table 7.14 Changes in the main chemical characteristics of olive fruits during fermentation. Gordal cultivar (Reproduced with permission from Bravo Abad, F. and Iñigo, R.M. 1987)

Days after brining	pH in brine	Flesh-to-pit ratio	Fat content (%,w/w)	Nitrogen content (mg/100g flesh)	Moisture (%, w/w)
11	4.70	5.00	12.40	291	67.30
34	4.02	4.84	14.69	260	72.68
95	3.96	4.98	13.46	195	73.98
141	3.90	4.39	14.30	199	73.50
175	4.01	4.93	15.90	223	71.90
215	3.89	5.17	15.40	200	72.90

A complete analysis of the fermentable substrates and their transformation products by a selected strain of *Lact. plantarum* was conducted by Montaño Asquerino *et al.*(1993). After 30 days, complete consumption of mannitol, malic and citric acids had occurred. By that time, 90% of glucose and 93% of fructose had been used. Utilization of sucrose was slower (only 27%), although the initial level present was very low.

Glucose and fructose are metabolized via the glycolytic pathway. Malic acid was presumably degraded to lactate and CO_2 via the malolactic pathway and citric acid, the consumption of which did not begin until the end of the exponential phase of growth, could have been degraded via the anaerobic pathway of citrate metabolism to form CO_2, acetate, 2,3-butylene glycol, acetoin and diacetyl. However, neither of these two last compounds were found. Citric acid could also act as an electron acceptor for anaerobic mannitol metabolism, due to the formation of succinic acid and methanol. The end-products formed were mainly lactic acid, followed by acetic acid, succinic acid and ethanol. However, the final product obtained by pure fermentation was described as strange by taste panellists when compared with a spontaneously fermented control; a further demonstration that end-products from other microorganisms are also necessary for normal olive quality.

Phenolic compounds

The presence of polyphenols in olive fermentation brines is characteristic. Total polyphenols are easily determined by derivative spectroscopy (Montaño Asquerino *et al.*, 1985). The method utilizes the absorption maximum at 278 nm attributed to these compounds and uses the 2nd derivative of the spectrum for sample measurements, which must include a standard. Their studies showed that the most appropriate standard was pyrocatechin due to its spectrophotometric properties. The method is suitable for following the polyphenol changes of any type of table olives in both fruit and brines during fermentation. It is practically free from interference by other substances present and independent of pH within the normal ranges encountered.

In general, phenols in olives are hydrolysed during lye treatment and are partially lost during this operation and in subsequent washing. Thus, the concentration and types of phenols in treated and fermented fruits are different from those in the raw material. After brining, there is also a substantial diminution of phenol content in fruits and a parallel increase in the brines. Vázquez Roncero and Janer del Valle (1977) followed the qualitative changes in phenol composition during Spanish-style green processing. After the lye treatment, they found only hydroxytyrosol monoglucoside and diglucoside, apeginin-glucoside, luteolin 7-glucoside and caffeic derivatives among the hydrolysable compounds. The non-hydrolysable phenols were composed of hydroxytyrosol, tyrosol, caffeic, vanillic and *p*-coumaric acids as well as aglucones of flavonoids. The same compounds were also found in the lyes. Washing waters had a lower number of compounds containing only hydroxytyrosol monoglucoside and diglucoside, luteolin 7-glucoside and apeginin 7-glucoside, among the hydrolysable phenols and practically only hydroxytyrosol in the non-hydrolysable fraction. After fermentation, the same compounds were identified as well as tyrosol, vanillic acid and derivatives of hydroxybenzoic acid. The most important change was the virtual disappearance of oleuropein. The hydroxytyrosol glucosides are more stable to the alkaline treatment and do not suffer hydrolysis under these conditions. Due to their solubility they are found in both the flesh and the brine. In the time after brining they pass completely into the brines and are not detected in the fermented fruits. The rest of the esters behave similarly. Flavonoids are hydrolysed to a greater or lesser extent according to their structures; quercitin 3-rhamnoglucoside is completely hydrolysed while apeginin 7-glucoside and luteolin 7-glucoside are only partially attacked. Oleuropein hydrolysis yields hydroxytyrosol, which is abundant in all the successive solutions with which lye-treated fruits are in contact. It also remains partially in fruits. The glucoside of elenolic acid must also be present, although these researches did not detect it due to its lack of phenolic character. This hydrolysis of oleuropein accounts for the debittering effect of lye treatment because none of the compounds produced has this character.

Caffeic and *p*-coumaric acids are present in the lye and in the washing waters as sodium salts. However, with the rapid brine acidification these compounds are not easily dissolved and remain in the flesh. Finally, the aglucones of flavonoids are scarcely soluble in water and also remain in the flesh of the final product.

In later studies, Brenes *et al.* (1990) also found catechin, 3,4-dihydroxyphenyl acetic acid and 4-hydroxybenzoic acids in the brines of Spanish-style green olives during the olive storage phase.

Volatile compounds

Montaño Asquerino *et al.* (1990) have identified by gas chromatographic analysis the following compounds (in mg/l) in the volatile aroma components of olives: acetaldehyde (12.3); methanol (578); ethanol (523); 2-butanol (16.4); *n*-propanol (15.4); acetone and ethyl acetate.

Colour of brines

A method to follow the colour changes of brines is of interest since it could also be applied to packed olives, and brine colour is one of the quality attributes first observed by any consumer when glass or any other transparent container is used. Montaño Asquerino *et al.* (1988b) studied a series of possible parameters (L^*, a^*, b^*, c^*) to express this characteristic as well as absorbances at selected wavelengths. From their correlations with subjective assessments they selected the difference $A_{440}-A_{700}$ as the most appropriate index. In the fermentation brines, colour increases rapidly during the first days and then remains almost stable during the rest of the period in brine. There is a good correlation between colour and polyphenol content in brines, especially in packed olives.

7.8.2 FRUITS

Surface colour

Studies on the surface colour of Spanish-style green olives were initiated by Castro Ramos and Vázquez Ladrón (1976a,b) and Vázquez Ladrón and Castro Ramos (1976), who measured the reflectance values of Verdial, Manzanilla and Gordal cultivars. They found maxima at 560 nm (green zone) only in the case of the raw material and at 530 (green), 560 (green), 590 (yellow) and 635 nm (red) in fermented olives. Later, Sánchez Gómez *et al.* (1985) studied reflectances at different maxima and selected combinations of these. The best correlation with a subjective taste panel assessment was found to be the so-called 'colour index' (CI), defined as:

$$CI = (4R_{635} + R_{590} - 2R_{560})/3$$

where R values are reflectances at the mentioned wavelengths.

In general, the lye treatment, washing and fermentation produce a rapid increase in reflectances at 560 and 635 nm during the 50–60 days following brining. Then the increase slows down and the reflectance practically stabilizes up to the end of the fermentation and storage period. The reflectance at 590 nm increases more rapidly than that at other wavelengths during the first phase of brining, but then the rate of increase diminishes, although it continues to increase slightly while the olives are in brine.

Of the different treatments, the lye and water immersions always diminish the reflectance values, although the olive surface recovers reflectance during fermentation. Slight differences in NaOH concentrations or HCl corrections do not influence the colour significantly (Sánchez Gómez *et al.*, 1986).

Chloroplastic pigments

Other important chemical changes are related to the fruit-colouring pigments (Minguez Mosquera *et al.*, 1989). As stated in Chapter 5, the surface green colour of olives is due to the presence of chlorophylls (a and b) and carotenoids. Their complete analysis and quantification has recently been successfully achieved by HPLC (Mínguez Mosquera *et al.*, 1991). Spanish-style processing causes total transformations of the chlorophylls initially present in the fruits by two mechanisms, one, via an enzymatic reaction (due to chlorophyllase), which produces chlorophyllides during the first days of brining (when the pH is still rather alkaline); as the pH decreases this process stops. The other chemical transformation is produced by the subsequent progressive acidification of the medium which facilitates the exchange of magnesium ions for hydrogen ions in all the porphyrin compounds present. This leads to the formation of pheophorbides (from chlorophyllides) and pheophytins (directly from the remaining untransformed chlorophylls) (Mínguez Mosquera *et al.*, 1990, 1994).

Only those components of the carotenoid fraction sensitive to the acidic medium are affected (Mínguez Mosquera *et al.*, 1990). Thus, the only transformation detected in the structure of the carotenoids initially present in the fruits was the oxidation of the 5–6-epoxide groups to 5–6-furanoid groups, this being catalysed by the presence of acids. Consequently, violaxanthin and neoxanthin are transformed into auroxanthin and neochrome while the rest of the carotenoid components remain unchanged (β-carotene, lutein, and auroxanthin). In general, the total carotenoid content of fruits remains constant, indicating no other transformations (Mínguez Mosquera and Gandul Rojas, 1994).

As result of these changes, the olives lose their original green colour and become pale yellow due to the carotenoids. There are differences in the chloroplast pigment content between cultivars, which are maintained during fermentation. This facilitates varietal differentiation, even after storage and packing (Mínguez Mosquera *et al.*, 1990).

Texture

Olive texture is a very important quality attribute and also facilitates post-fermentation operations such as pitting and pitting and stuffing. The optimum final texture is only obtained when the correct maturity raw material is used, but subsequent treatment can also have an effect, particularly lye treatment. High NaOH concentrations and temperatures can degrade this attribute markedly. Also, excessive NaCl levels in the initial brine may cause shrivelling that can persist into the final product.

In general, suitable treatments improve fruit texture as measured by compression tests. The elasticity of the skin increases and it becomes more resistant to breaking by finger pressure. However, shear-compression texture is always lower than in the original raw material. Lye treatments always produce a marked loss of texture which persists during washing but recovers partially during fermentation due to the effect of salt and the absorption of divalent cations (Mg^{++} or Ca^{++}) from the brine.

Currently, most green Spanish-type olives are sold pitted or pitted and stuffed with products like pimento, anchovy, onion, cucumber, etc. High levels of broken olives in these operations may produce big losses. Minimizing breakage seems to be related more to maturity and lye treatment than to specific texture values (Montero Marín and Fernández Díez, 1976). However, Fernández Díez and Vidal Sigler (1968) found a good correlation between the texture index (compression/shear-compression) and the level of broken olives (Table 7.15).

Texture may be seriously damaged if fruits are left in contact with pectinolytic surface yeasts and moulds. Consequently, they should be kept well submerged in the brines and any surface growth of microorganisms avoided.

Removal of pesticide residues

Green Spanish-type olive processing also has a favourable effect on pesticide residues in the raw material. Table olives, like any other fruit, must be protected from insect attack since their activity produces surface punctures which can lead to blemishes in the final product, reducing its market value. Chemical control of pests usually consists of preventive treatments at fixed times. Normally, picking is delayed for a period sufficient to allow degradation and disappearance of any pesticide residue. In these cases no problems arise. However, this safety period might not be observed and olives with pesticide residues processed. To assess the risks associated with this possibility, trials with maximum doses of pesticides and reduced waiting time have been carried out. Albi (1981) treated olives with 6 l/tree of a 1.5 g/l solution of carbaryl 85 and picked the

Table 7.15 Relationship between texture index (compression/shear-compression) and broken olives during pitting (Reproduced with permission from Fernández Díez, M.J. and Vidal Sigler, A. 1968)

Texture index interval	Average broken olives (%)
0.61–0.80	9
0.81–1.00	19
1.01–1.20	30
1.21–1.40	42

olives only 24 hours after application. The results when such fruits were processed are shown in Table 7.16. Lye treatment and washing were enough to eliminate all the pesticide residue. Albi and Rejano Navarro (1982) studied the elimination or degradation of dimethoate, diazinon and imidan phosphorylated insecticides during the same process. Dimethoate and imidan were destroyed by the sodium hydroxide treatment and subsequent washing processes. Diazinon was partially eliminated by these treatments, although a low level remained in the olives after 6 months of fermentation (Table 7.17). Russo *et al.* (1987) followed the changes in residues of the pesticides chlorpyriphos, methidation, parathion-me, dimethoate and parathion, finding that sweetening the olives in a 1% NaOH solution fully eliminated all insecticides examined (Table 7.18).

Table 7.16 Persistence of carbaryl residue* in olive flesh during Spanish-style green table olive processing (Adapted from Albi and Rijano, 1982)

	Average carbaryl content (p.p.m.)	
Fruit processing stage	1st season	2nd season
After 24 hours' tree treatment	11.13	14.74
After lye treatment	0.32	0.57
After two washings	0.00	0.00
After 1 month fermentation	0.00	0.00
After 6 months' fermentation	0.00	0.00
After 6 months' fermentation (brines)	0.00	0.00

* Trees were treated with 6 l/tree of 1.5 g/l solution of carbaryl 85. Waiting period: 24 hours.

Table 7.17 Persistence of dimethoate, diazinon and imidan phosphate insecticides in olive flesh during Spanish-style green table olive processing (Adapted from Albi and Rejano, 1982)

	Insecticide content. (Average of two seasons, p.p.m.)		
Fruit processing stage	Dimethoate	Diazinon	Imidan
After 24th tree treatment	0.94	1.34	1.27
After lye treatment	0.17	0.63	0.00
After two washes	0.07	0.40	0.00
After 1 month fermentation	0.02	0.39	0.00
After 6 months' fermentation	0.00	0.22	0.00
After 6 months' fermentation (brines)	0.00	0.00	0.00

* Tree treatment consisted of application of 5 l/tree of a solution containing: 4 g/l (of Roger L 396 with 20% of pure product), 1.5 m l/l (Diazinon 60E with 60% of pure product), and 3 g/l (Fosmet with 50% of pure product), respectively. Olives were picked and processed 24 hours after treatment.

Table 7.18 Persistence of chlorophyriphos, methidation, parathion-me dimethoate and parathion insecticides* in the olive flesh during Spanish-style green table olive processing (Adapted from Russo *et al.*, 1987)

Insecticide	Insecticide content (p.p.m.)	
	Fruit before processing	Fruit after 1 month fermentation
Chlorpyriphos	1.8363	0.0043
Methidathion	0.2987	0.1040
Parathion-me	0.0203	0.0083
Dimethoate	1.8400	0.0000
Parathion	0.0010	0.0000

* Olive trees (Nocellara Messinese) were treated with 0.06% of active product. Fruits were debittered with 1% NaOH solution and brined in a 7% NaCl solution.

Other results of Garrido and Cert (unpublished data) have corroborated these findings, as well as the elimination during processing of endosulphan and phenytoin. However, fosalone had a higher degree of persistence due probably to its solubility in the oil. Treatment with this pesticide must therefore be avoided. Obviously, pesticide residues in fruits picked after the recommended waiting periods must have lower levels of pesticides than those reported in these researches. Thus, it can be stated that as a result of their processing, green Spanish-type olives are usually free of pesticide residues. This was confirmed by the Food and Drug Administration Pesticide Programme in 1988, in which 124 samples of olives were analysed and only two were found to violate the legislation because there were no tolerances in USA for the low residues found (FDA, 1989).

Other fruit characteristics

Changes in other fruit characteristics are shown in Table 7.14. Changes in flesh-to-pit ratio are minimal. Moisture and fat content increase slightly throughout fermentation, although nitrogen content diminishes through the fermentation. This was consistent for different cultivars and may be due to a continuous leaching from the fruits (Bravo Abad and Inigo, 1987).

Finally, Hassapidou *et al.* (1994) have followed the changes of minor components such as tocopherols and tocotrienols. They have found the former only in green olives. The different Spanish-style green processing stages did not modify consistently the amounts of α-tocopherol in Conservolea but levels were reduced in the case of Halkidiki. On the other hand, γ-tocopherol was present only in Halkidiki and did not

Table 7.19 α-, and γ-tocopherol content (mean ± SE) changes during Spanish-style green table olive processing. (Reprinted from *Food Chemistry*, **50**, 111–14, Hassapidou, M.N. *et al.* Effect of processing upon the tocopherol and tocotrienol composition of table olives. Copyright (1994), with kind permission from Elsevier Science Ltd., The Boulevard, Langford Lane, Kidlington, OX5 1GB, UK)

Compound	Processing stage	Conservolea (μg/g pericarp)	Halkidiki (μg/g pericarp)
α-Tocopherol	Raw	21.6 ± 1.7	29.2 ± 1.9
	Debittered	24.2 ± 1.6	21.6 ± 2.4
	Fermented	23.8 ± 2.6	6.4 ± 2.4
γ-Tocopherol	Raw	Traces	5.1 ± 1.9
	Debittered	Traces	7.2 ± 1.4
	Fermented	Traces	5.1 ± 0.6

change (Table 7.19). More studies on these compounds and their changes are necessary to be able to reach any definitive conclusion.

Table 7.20 summarizes the most relevant compositional changes which are produced during the Spanish-style process (Fernández Díez, 1983, 1991).

7.9 ACCELERATION OR CORRECTION OF FERMENTATION

The rate of acid production is an important aspect of producing an adequate final product and avoiding spoilage. Assuming all other requirements are met, the activity of lactic acid bacteria depends strongly on temperature. The climatic conditions of Seville during the green olive picking season were the reason for the early development of this process in Spain.

Vaughn *et al.* (1943) considered the possibility of accelerating the fermentation by incubation. They performed a series of investigations to determine the optimal temperatures for the different lactic acid bacteria present in this fermentation. *Leuconostoc mesenteroides* grew best at 20–35°C, *Lact. plantarum*, around 35°C, although even at 20°C it produced good acidification. Similar results were found for *L. brevis*. Spoilage bacteria (Gram-negatives, butyric bacteria, etc.) also grew well in olive brines at 25–40°C. Thus, any increase of temperature will enhance the activity of all the microorganisms and incubation would not seem particularly to favour lactic fermentation over spoilage. In some cases, temperatures early in the season (September and October) may exceed 20°C in many of the producing areas and no additional heating is necessary. However, in late season or with late maturation cultivars the temperature may be under 20°C and fermentation may be slow or entirely absent.

Table 7.20 Spanish-style green olives in brine. Synthesis of compositional changes during processing (Adapted from Fernández Díez, 1990)

Operation	Change in composition
Debittering process	
Alkaline treatment	Hydrolysis of oleuropein Formation of organic acids Skin permeabilization Texture conditioning
Washing	Excess of alkali elimination Loss of sugars and organic acids Formation of organic acids Partial texture recuperation Partial polyphenol removal
Brining	
Brine addition	Diffusion of fermentable substrates, minerals, vitamins, etc. into the brine NaCl equilibration with the flesh
Fermentation phase	
Lactic acid fermentation	Lactic formation from sugars and other fermentable substrates Formation of acetic and other organic acids Degradation of chlorophylls and some carotenoids
Yeast growth	Formation of aromatic substance
Storage and final packing	
Brine reconditioning	NaCl increase
Growth of propionibacteria	Consumption of lactic acid and formation of propionic and other volatile compounds
Pitting and stuffing	Loss of a small proportion of fat Incorporation of the stuffing material (pimento, almond, etc.)
Bottling	Partial lost of acids, volatile compounds, NaCl, etc.
Product shelf-life	None under normal conditions

Borbolla y Alcalá and Cordón Casanueva (1969) showed that conducting the fermentation in wooden barrels in 'green-houses' of plastic material improved the process, especially when this operation was done at the beginning of the fermentation. However, with the introduction of bulk fermentation (Borbolla y Alcalá *et al.*, 1975) the fermentation containers were no longer portable. In fact, bulk fermenters tended to be underground (see Figure 7.5), because of the lower construction costs and

Fig. 7.5 Fermentation yard with underground 10-tonnes capacity fibre-glass fermenters. (Illustration courtesy of Dr Rejano Navarro.)

better temperature control. In these cases, heat treatments are given to the surrounding brine by means of a heat exchanger. The maximum brine temperature should not exceed 30°C in order to avoid damage to the fruits. The brine is recirculated by an air-tight pump to avoid introduction of air, until the whole mass of fruits and brine reaches equilibrium. Normally, one or two treatments are enough to accelerate the fermentation, as the insulation and the heat of the fermentation process help to maintain the temperature level. Fermentation by thermophilic lactic acid bacteria (*Streptococcus thermophilus* and *Lactobacillus bulgaricus*) at temperatures of 37–40°C was tried by Balatsouras *et al.* (1971) and gave a good final product from the Conservolea Greek cultivar. Fermentation was complete in about 20 days. The procedure was suggested for early delivery of the product and as appropriate for warm regions. However, it has never been applied in practice.

The introduction of bulk fermentation (10 tonnes capacity) has helped enhance fermentability because the fruit : brine ratio is higher than in the traditional barrels (500 kg capacity). Also, bulk fermenters do not leak – unlike barrels, which lose fermentable substrates through leakage.

The main corrections applied are to combined and titratable acidity. The first should be corrected when the value is over 0.14 N, as already noted. This can be done by addition of sufficient food-grade HCl to eliminate the excess combined acidity (which also has the effect of increasing the titratable acidity), or by substituting part of the brine with fresh material with the same salt and lactic acid concentration.

Correction of deficient acidity can be achieved in different ways. With olives that were almost correctly fermented, a small acidity increase can be achieved by adding food-grade lactic acid. However, when the lack of acid is due to deficient fermentation, the best option is to supplement the brine to enhance fermentability. The first step is to investigate the cause of the problem. Any lack of fermentable substrates or lactic acid bacteria should be corrected. Then, the fermenter can be heated if the temperature is inappropriate. In the case of stuck fermentations due to the presence of high levels of inhibitors in the brine, the brine must be diluted before following the procedures outlined above.

7.10 STORAGE OF THE FERMENTED PRODUCT

7.10.1 CHARACTERISTICS OF STORAGE

Under normal conditions with good control of all the factors discussed, a complete lactic fermentation may be reached after approximately 1 month. A low pH value and high acidity, together with the salt contribute to the organoleptic character of the product and its preservation (Fernández Díez, 1983). However, it is important to recognize that both pH and acidity are related to a third factor, combined organic acidity or 'residual lye' (Borbolla y Alcalá and González Pellissó, 1964), which remains in the medium. Relatively high values of 'residual lye' counteract the effect of high acidity, maintaining a pH value too high for good preservation. Excessively low values of 'residual lye' in the fermented product may lead to spoilage if a reduction or accidental loss of acidity causes a rapid rise in pH.

Under normal conditions, the final product of the fermentation should have a pH value in the neighbourhood of 3.8–4.2, a free acidity, expressed as lactic acid and consisting mostly of lactic acid of 0.8–1.2% and a combined acidity or residual lye of about 0.09–0.11 N. The final concentration of salt should be about 7% (w/v) if the product is expected to be sold within a short period, if the sanitary conditions are normal, and if the temperature is not too high. The microbial population should consist of lactobacilli in their declining phase and some fermentative yeasts. For longer periods of preservation or for extremely high summer temperatures, it is advisable to raise the salt concentration to 8% or higher, in order to avoid the growth of Propionibacteria, which may deplete the lactic acid. This so-called 'fourth phase' of the Spanish-style green fermentation was first described by Borbolla y Alcalá *et al.* (1975), who demonstrated that when temperatures were high during the storage period, propionic acid production was common, although its quantity was dependent on physicochemical characteristics of the brine such as pH, titratable acidity and mainly salt concentration (Rejano Navarro *et al.*, 1978b). Later, they found that the amount of propionic acid formed was equivalent to the lactic acid used and this seems to be the only transformation that occurs at this stage. *Propionibacterium acnes* identified as responsible for these changes requires the relatively high temperatures of summer for its activity (González Cancho *et al.*, 1980). In other studies related to the same fourth phase, *Propionibacterium acidi-propioni* has also been found (Castro and Rejano Navarro, 1990). In addition to the influence of pH and salt on the development of this phase, the re-use of lye and elimination of washing during Spanish-style green olive processing limited production of propionic acid. This was in spite of their having pH values slightly higher than those which were normally processed and in which production of propionic acid was marked. This partial inhibition could have been due to higher levels of polyphenol in the olives processed without washing. However, this hypothesis is not yet proven.

7.10.2 GRADING SPANISH-STYLE GREEN OLIVES BY SIZE

In general, table olives are sold according to size. Such grading is carried out after fermentation and at the end of the storage period. These sizes are established in the COI/CODEX Standards (Codex STAN 66–1981), in the Spanish table olive legislation and in the United Standards for Grades of Green Olives, among others. Sizes established by CODEX Standards are shown in Table 7.21. Reasons for this classification are:

- Improvement of appearance; homogeneity gives confidence to consumers.
- Enhancement of profitability by segregating the sizes with higher demand (and prices) or by diminishing losses (better pitting and stuffing characteristics). Large sizes are sold at higher prices because they are scarcer and are more attractive to consumers.
- Different sizes are related to different maturity stages. Even in fruits collected at the same time there is a gradation. Some of them can enjoy the optimum texture, flavour, etc. These obviously would command higher prices.
- Homogeneous sizes are essential for the effective pitting and stuffing operations as well as packing (to obtain the appropriate average weight or the suitable distribution in the bottle when packed by hand).

Normally, the olive size is expressed as the number of fruits per kg. However, as Borbolla y Alcalá (1972) pointed out, this classification has the disadvantage of being an indirect measure since, in practice, olive size-grading is accomplished by separating olives according to their

Table 7.21 Different sizes established by CODEX standards. These are normally expressed as number of fruits/kg (From Codex Standard 66–1981)

60–70	161–180
71–80	181–200
81–90	221–250
91–100	251–280
101–110	281–310
121–130	311–340
131–140	341–370
141–150	371–400
151–160	401–450

transverse diameter. Fruits with the same diameter may have different specific weight or length. This introduces a variability in the olive size when expressed in fruits/kg. Tolerances in the COI/CODEX standards are: 10%, when the size differs in 10 fruits, for example 70–80 fruits/kg; 5%, when such difference is 20 fruits ; and 2% in the case of 30 fruits. Percentages refer to the number of fruits which belong to the immediately higher or lower size. Borbolla y Alcalá, after a detailed study of compliance with the different standards, suggested the use of the fruit diameter as the most appropriate system. However, his suggestion has not been considered until now.

Another problem arises with pitted and stuffed olives which must be graded according to the size of the whole or plain olives from which they were obtained. When the industry knows the origin of each lot size assignment is easy, but when packed olives must be controlled, the normal procedure to calculate the equivalent size is by weighing 1 kg of pitted or stuffed olives, counting them and multiplying the resulting figure by 0.7 or 0.75, respectively, according to the presentation form. The standards use only these two transformation coefficients. A detailed study carried out by Rejano Navarro and Sánchez Gómez (1990) has, however, shown that they change with cultivar and that the current system can lead to difficulties in fulfilling the standards requirements when olive packages are controlled by the authorities. These authors have proposed alternative equations for Manzanilla and Hojiblanca as well as a new method based on the transverse diameter to establish the appropriate size. However, to date there has been no modification of the regulations.

The same researchers have used the differences in shape between Manzanilla and Hojiblanca to differentiate between them. The procedure is based on the difference between the length and diameter of fruits. They form two perfectly separate statistical populations within each cultivar. The procedure can serve to control fraud (use of one cultivar in place of the other).

7.11 PACKING

7.11.1 STABILIZATION BY PHYSICOCHEMICAL CONDITIONS

The final product, sorted and size-graded, is occasionally pitted and stuffed as indicated above, and packed in small hermetically sealed, tin or glass containers, with a new brine. The packing medium must be kept within the following ranges: pH 3.2–4.1; free acidity 0.4–0.6% as lactic acid; NaCl between 5 and 7%; residual lye between 0.02 and 0.07 N.

This means that the normal variability among the different batches has to be adjusted to within a narrow range. Methods used in the correction of chemical values for fermented products used for bottling have been studied by Borbolla y Alcalá and González Pellissó (1980a, b) and Rejano Navarro and González Pellissó (1985). Basically, they consist of the correction of the combined acidity to values between 0.04 and 0.140 by the methods already described.

The levels of the various physicochemical characteristics for bottling determine the product stability. Traditionally, this has been obtained by setting high levels of acidity and salt. Otherwise, the formation of sediment was unavoidable with possible gaseous spoilage and loss of vacuum. The influence of pH, salt content and free acidity was studied by Borbolla y Alcalá and González Pellissó (1972) who found that the sediment was avoided by a complete absence of fermentable substrates, lactic acidity of 0.4% or over and 5% of salt or over. When olives are not completely cured and retain some fermentable sugars, these percentages should be increased to 0.6% of acidity and 7% of salt, although even then a long shelf-life is not completely assured. Low storage stability was strongly related to the combined acidity. When this is higher than 0.025 N the conditions described were not enough to preserve the product.

In spite of its constraints, this work provided packers with an appropriate preservation system that was largely responsible for the expansion of the Spanish-type olive bottling industry and the new markets that were opened up during the two decades following this work.

In addition to lactic acid, packed green olives brines may also contain ascorbic acid (vitamin C) as an antioxidant (Corral Andrade, 1961) as well as citric acid, which exerts a synergistic effect with ascorbate. Acetic acid is also used on occasion.

Table 7.22 shows the main characteristics of the final product and their commercial presentation forms, once they are packed in retail containers.

Packed green olives must fulfil some minimum quality characteristics established in the COI/CODEX Standards (lower pH and higher NaCl percentages than 4.0 and 4.0 in hermetically sealed containers and 4.5 and 6.0 in those which are not). Apparently, most of the green olives in the market are within these limits, but when Marsilio and de Angelis

Table 7.22 Spanish-style green olives in brine. General characteristics of the final product (Adapted from Fernández Díez, 1990)

Parameter	Bulk-fermented product	Packed in small containers
pH value	3.8–4.2	3.2–4.1
Free acidity (as lactic acid, %)	0.8–1.2	0.4–0.6
Combined acidity (as normality, N)	0.09–0.11	0.02–0.07
Sodium chloride (as weight/volume, %)	7.0–8.0	5.0–7.0*

* Salt content may be reduced up to 2% if the small containers are to be pasteurized.

(1993) studied the characteristics of commercial table-packed green olives the results were rather disappointing. Some 20% failed to meet the salt requirement and 7% the prescribed pH level. There were also deficiencies in fulfilling size standards (about 20%), acidity (around 20%) and other deficiencies (of the order of 15%). As such minimum levels are essential for effective preservation, a strong call for the strict observation of these requirements is necessary.

Initially, it was thought that olive spoilage in the pack was due to the growth of *Lactobacillus*. However, Borbolla y Alcalá *et al.* (1975), Rejano Navarro *et al.* (1978b), González Cancho *et al.* (1980) and Castro and Rejano Navarro (1990) demonstrated that usually the microorganisms responsible for spoilage during the shelf-life of the packed product were the same species of *Propionibacterium* which had been isolated and identified during the storage of the fermented olives in bulk.

7.11.2 STABILIZATION BY PASTEURIZATION

The physicochemical conditions which fermented olives must meet for adequate stability after packing are difficult to achieve. The high levels of acidity and salt required are sometimes not appropriate for many markets and the need to reduce pollution by re-using lye and reducing the number of washings is currently making it difficult to obtain the appropriate levels, particularly of combined acidity, in the packing solutions. Hence milder conditions (low salt and acid percentages) and combined acidities higher than 0.025 N are frequently found at final bottling. In these cases the stability of the product is not guaranteed and alternative preservation systems must be used.

From the very early days, pasteurization was thought to be the best option. After a careful study of the thermal resistance of the different microorganisms (yeast, lactobacilli and propionibacteria) present in the packing brines, it was found that *Propionibacterium* was the most heat-resistant and the pasteurization treatment should be established accordingly (González Pellissó *et al.*, 1982; González Pellissó and Rejano Navarro, 1984). After several studies on bottled olives under different physicochemical conditions, it was established that a value of accumulated lethality of 15 pasteurization units (15 minutes at 62.4°C) was enough to assure correct pasteurization (Sánchez Gómez *et al.*, 1989). Specific heating times will depend on the container characteristics and heat penetration curves.

Heat treatment must be applied carefully, since texture can be affected and olives rapidly lose lightness, while red tones increase. However, correct pasteurization conditions affect neither colour nor texture of bottled Spanish-type olives, provided that they are not exceeded (Sánchez Gómez *et al.*, 1991).

Due to this sensitivity of fermented olives to heat, hot bottling with hot brine has also been studied (González Pellissó and Rejano Navarro, 1984; Sánchez Gómez *et al.*, 1989). The procedure has been used successfully when the same accumulated thermal lethality as proposed for bottled olives was applied. The bottling equipment is simpler, the costs are lower and the fruits suffer a less severe heat-shock. Hot fruit contact with air must be reduced to a minimum to avoid browning.

Packing in acidic conditions leads to a texture loss over time. Brenes *et al.* (1994) have shown that the softening rate was more than 0.08 N/h within the pH range of 2.5 to 4. The softening rate was always higher with lactic than with acetic acid, and increased slightly with concentration. Addition of calcium diminished the softening rate at pH values over 3.0, which are those normally observed in green olive packs. On the other hand, the presence of salt up to 1.2 mM had no influence.

When a pack of green olives is opened, a layer of oxidative yeast can appear on the air–brine contact surface after a few days. Its growth can in a few days reduce the acidity of the medium (Table 7.23) and provoke the appearance of malodorous spoilage. This can be avoided by including sorbic acid or its salts as an antifungal agent (Borbolla y Alcalá *et al.*, 1961). Concentrations of 0.025% are effective for only 5–6 days, while higher levels (0.05–0.1%) can extend this period to several months. Used at these concentrations, the presence of sorbic acid does not affect the physicochemical characteristics of the brine such as pH and acidity and does not alter the olive flavour. However, a slight browning may appear, especially when transparent containers are exposed to light. This depends on the surface : capacity ratio of the containers. Lactobacilli are unaffected by the sorbate and should be inhibited in parallel by other mechanisms (physicochemical characteristics or pasteurization).

Table 7.23 Lactic acid consumption by oxidative yeasts. (Reproduced with permission from Borbolla y Alcalá, J.M.R. de *et al.*, 1961)

Sample	Initial characteristics		After 6 days' exposure to air	
	pH	Acidity (g lactic acid/l)	pH	Acidity (g lactic acid/l)
1	4.00	11.8	5.50	1.2
2	4.00	11.8	5.50	1.0
3	4.00	11.8	5.70	0.8
4	4.00	11.8	5.40	1.2
Anaerobic control	4.00	11.8	4.00	11.8

7.11.3 PACKING WITH REGENERATED FERMENTATION BRINES

In general, olives are packed in fresh new brine and the fermentation and storage brines are discarded. These contribute to most of the pollution found in the general wastewater of the packing factory, about 6–12 l/kg of fruit, and make the whole liquid wastes unsuitable for conventional treatments at both plant and municipal treatment facilities. The segregation of these brines which account for 80–85% of the final pollution, but represent only about 20% of the total volume of liquid residues, could help control a serious polluting charge. In addition, their re-use could benefit from the value of the lactic acid and salt they contain.

A good recycling option is their utilization, once regenerated, for packing. Ideally, a good regenerating procedure should eliminate from the fermentation brine suspended solids and colour. Two approaches are possible: adsorption and ultrafiltration. Garrido Fernández (1979) showed that the first option was possible, although the high prices of the adsorbents and the retention of a high proportion of lactic acid by the adsorbent prevented the industrial application of the procedure. However, more recently, Brenes and Garrido (1988) discovered a special activated carbon that adsorbs colour very selectively with the elimination of only a small percentage of lactic acid.

Brenes *et al.* (1990) reported the successful application of ultrafiltration through a polysulphone membrane to recover green table-spent brines. Both systems were tested at industrial level with good results (Garrido Fernández *et al.*, 1992a,b). Complete re-utilization is not possible, since the physicochemical conditions of the packing brines are weaker in both titratable acidity and salt than the storage brines. As an example, Table 7.24 shows the values of the main physicochemical characteristics of Hojiblanca olives packed with a final brine made of a 70% regenerated brine (and 30% tap water). The regenerating procedure was adsorption with activated carbon followed by microfiltration through a ceramic filter. The effect of re-use is apparent in a slight increase of the pH, combined acidity and colour, which is more pronounced when the brine is regenerated by microfiltration in the absence of treatment with the adsorbent. Using lower percentages of regenerated brine in the final solutions, the differences diminish. Because of this increase in pH, product stability requires pasteurization, unless the combined acidity is corrected to lower values.

The use of ultrafiltration gave similar results. Table 7.25 shows the physicochemical characteristics of packed olives in which the proportion of recycled brine in the packing solution was 45% (v/v). Differences were lower than in the case of regeneration with activated carbon, but this effect was due to the lower proportion of recycled brine used. In both cases, the recycled brine was practically colourless due to the decrease of polyphenol polymers. Also observed with ultrafiltration was a marked diminution of the polyphenol content, 55% to 80% in the case of caffeic and 4-hydroxybenzoic acids and 10–30% in most of the other compounds (Brenes *et al.*, 1990). Available experience seems to indicate that ultrafiltration maybe the best option because of the lower labour costs, the easy equipment handling and the possible lower running costs. However, data are still required on the life of the membranes because the pilot and industrial trials carried out until now have been limited.

Independent of the regeneration method used, brine recycling is a potentially important tool for pollution reduction in green table olive packing factories. It can represent a variable proportion of the final packing solution depending on the physicochemical characteristics. But with either of the regeneration procedures used, it is difficult to reach the same pH as fresh brine. Thus, to guarantee the shelf-life stability of the

Table 7.24 Effect on the main physicochemical characteristics of the re-use of regenerated fermentation brines in the final packing* (Reproduced with permission from Garrido Fernández *et al.*, 1992b)

	Re-used brine		
Physicochemical characteristic	0.6%†	0%†	Fresh brine
pH	4.55	4.57	4.20
NaCl (%, w/v)	5.90	5.90	5.70
Titratable acidity (g lactic acid/l)	5.80	6.00	5.70
Combined acidity (mM)	82.00	99.00	76.00
Colour ($A_{440}-A_{700}$)	0.36	0.59	0.34

* Regeneration treatment consisted of adsorption with activated carbon followed by microfiltration through a 0.22 μ ceramic filter. Proportion of re-used brine in the final brine was 70% (v/v). Olive used was Hojiblanca cultivar.

† Proportion of activated carbon used in regeneration.

Table 7.25 Effect of the re-use of regenerated fermentation brines* on the main physicochemical characteristics of the final packing. Hojiblanca cultivar (Reproduced with permission from Garrido, A. *et al.*, 1992b)

Physicochemical characteristic	Recycled brine	Fresh brine
pH	4.05	3.67
NaCl (g/l)	45.00	48.00
Titratable acidity (g lactic acid/l)	6.90	7.60
Combined acidity (mM)	100.00	65.00
Colour ($A_{440}-A_{700}$)	0.09	0.007

* Regeneration was accomplished by ultrafiltration through a 1000-Da pore size membrane. Proportion of re-used brine in the final packing solution was 45% (v/v).

product, the combined acidity must be previously corrected by the procedures already described or the product will require pasteurization.

In order to diminish the combined acidity in the fermentation brines, lye treatment at low NaOH concentrations and closely controlled temperature have also been examined (Sánchez Gómez *et al.*, 1990). Fermentation processes were normal, but the application of this procedure requires a strict control of temperature, which even at 30°C could produce a sensible increase of sloughing in the more sensitive cultivars (Manzanilla).

7.12 COMPOSITION AND NUTRITIONAL ASPECTS OF THE FINAL PRODUCT

In addition to the characteristics of the fresh olives used as raw material, other factors (Fernández Díez, 1983) can, in principle, affect the composition of the final product. The most important are the treatment of the fruit before fermentation or conditioning, fermentation and storage in brine, bottling and storage to the time of consumption.

Until quite recently, few facts were known about the composition of the processed Spanish-style pickled green olives in brine. These concerned the gross composition such as moisture, oil, protein, residual soluble sugars, total carbohydrates, ash, fibre, organic acids produced during the fermentation and limited data on certain vitamins, such as vitamin A, thiamin and riboflavin (Borbolla y Alcalá *et al.*, 1956). Therefore, a detailed study of the composition of the final product and its dependence on the processing conditions was required to establish the biological and nutritional value of the product (Castro Ramos *et al.*, 1979, 1980a, b; Vázquez Ladrón *et al.*, 1979; Nosti Vega *et al.*, 1982, 1984; Fernández Díez *et al.*, 1985).

Table 7.26 shows the main components (Fernández Díez, 1991) for

Table 7.26 The composition of Spanish-style green olives in brine (From Fernández Díez *et al.*, 1985)

Composition	Concentration Ranges
Major component (weight %)	
Moisture	61.00–80.56
Lipids	9.05–28.19
Protein	1.00–1.45
Fibre	1.40–2.06
Ash	4.19–5.46
Vitamins	
Carotene (mg/100 g)	0.02–0.23
Vitamin C (mg/100 g)	1.44–2.87
Thiamin (mg/100 g)	0.40–3.37
Essential amino acids (mg/100 g)	
Valine	55–157
Isoleucine	43–121
Leucine	82–227
Threonine	5–64
Methionine	13–79
Phenylalanine	39–111
Lysine	5–31
Tryptophan	13–18
Minerals (mg/100 g)	
Phosphorus	7–21
Potassium	34–109
Calcium	35–86
Magnesium	6–40
Sodium	1313–1753
Sulphur	14–38
Iron	0.58–1.16
Manganese	0.06–0.12
Zinc	0.25–0.41
Copper	0.42–0.82
Calorific value (cal/100 g)	102–280

some Spanish cultivars which were processed according to the Spanish-style. Indicated ranges include Gordal, Manzanilla, Hojiblanca and Verdial cultivars.

The moisture and oil contents in the flesh show an inverse relationship, as in the fresh fruit. The wide range of these values is influenced by cultivar, maturity of the raw material, and processing conditions.

The percentage of ash increases as a consequence of the alkaline treatment, fermentation, and storage in brine. The majority of essential amino acids as well as appreciable quantities of vitamins and mineral elements, are also present. There are no significant differences in the protein, fibre or ash content among cultivars of olives.

The calorific value is quite variable, depending mostly on the lipid content of the different cultivars. The quality of the oil fraction (Gracian Tous, 1968; Fernández Díez, 1971; Heredia Moreno and Fernández Díez, 1982; Nosti Vega *et al.*, 1982) is good because of its low content of saturated fatty acids (12–19%) and the presence of linoleic acid (5–8%). Overall, it can be said that Spanish-style green olives in brine provide good food value.

Soluble sugars disappear during the fermentation and are absent from the final product. The same happens to the bitter principle, oleuropein, which is mostly hydrolysed during the alkaline treatment. Organic acids and their salts represent about 1.5% of the weight of the pulp of the final product and comprise the organic acids present in the fresh fruit and those formed during the lye treatment and in fermentation.

Dietary fibre in olives is about 3–4%, depending on cultivar. Heredia Moreno and Fernández Bolaños (1982) reported the following contents on a fresh basis: Manzanilla, 3.1%, Gordal, 3.1%, Hojiblanca, 3.6%, and Verdial, 3.3%. Most of this is composed of cellulose, which is over 50% of the total content. Constituents of the fibre fraction, expressed as percentages of the pulp, are: cellulose, 1–3%; lignin, 0.5–0.9%; and hemicellulose, 0.3–0.9% (Heredia Moreno and Fernández Díez, 1982). The lignin : cellulose ratio, which is considered as an index of digestibility, ranges from 0.32 to 0.57, indicating that olive dietary fibre has good digestibility, a property which increases as the figures decrease. Heredia Moreno *et al.* (1993) studied the apparent digestibility of the main components of the olives. As indicated by the above data, digestibility of fibre was marked (about 31% of the neutral detergent fibre). There was also a high digestibility of fats (higher than 60%) and sugars (100% for glucose and 76% for fructose). Some 41% of the protein was also digested, although this is a relatively low value when compared with other products. This behaviour may be related to interference due to the presence of polyphenols, as occurs with other foods rich in these compounds. If the diet was previously desalted, the digestibility of the components diminished. The acid detergent fibre (ADF) digestibility was lower (only about 23%) since this does not include hemicellulose, which is the most easily digested polysaccharide. Separating the ADF into its components, cellulose presented a high apparent digestibility, of the order of 55%, while lignin showed a digestibility of the order of 40%, much higher than found in most foodstuffs. It is probable that, with the lignin partially digested, this reduces the barriers, thus enhancing digestibility for the other cell polysaccharides.

Fermented surface colour of olive fruits is due mainly to carotenoids

(Mínguez Mosquera *et al.*, 1990) since chlorophylls are degraded during the fermentation process. Some brownish coloration may also be due to phenol oxidation.

7.13 MICROBIAL SPOILAGE

Fruits may suffer different kinds of spoilage if the microbial population is not well controlled during the different phases of the fermentation (Fernández Díez *et al.*, 1985; Unidad Estructural de Biotecnología de Alimentos, 1990). Although they pose no risk to health, such microbes do depreciate the quality of the final product, due to deleterious changes in the appearance and organoleptic characteristics of the olives. Several different types of spoilage are recognized.

7.13.1 WHITE OR 'YEASTS' SPOTS

This used to be one of the most common abnormalities associated with olives processed in California and Israel. In Spain, Gordal cultivar may also show them. In general, Sevillano and other easily fermented cultivars most frequently show this blemish, although the spots may be found in all cultivars and types of olives which undergo lactic fermentation. The olives are considered unsightly and may lose value as a result, but are perfectly normal and healthy in other respects. The small white spots actually are colonies of microorganisms which have developed during fermentation. Vaughn *et al.* (1943) reported that most of the pimples examined in the laboratory contained lactobacilli, some cultures of which were identified as *Lact. plantarum*. Later, Vaughn *et al.* (1953) saw bacteria but no yeasts on microscopic examination of spotted olives and cross-sections showed that they were established in stomatal openings. Normally, fleshy Sevillano, Gordal, etc. olives are provided with enlarged stomatal openings. The prominent white areas disappear when the air has been displaced and reappear only when bacteria have developed during fermentation. It is assumed that the bacteria which finally develop into subcutaneous colonies gain entry through those enlarged pores. In this research all isolates were identified as *Lact. plantarum*. Similar bacterial blemishes are also observed in cucumbers, Italian peppers, pickled green tomatoes, etc.

Attempts by Vaughn *et al.* (1953) to avoid or reduce the appearance of these spots failed, and subcutaneous colonies developed during the first few weeks after the olives had been placed in brine. Samish and Dimant (1963) also carried out a series of experiments to diminish the incidence of this problem. They found that conditions which tend to increase the rate of lactic acid fermentation in brined olives (increase in temperature, increase in population of lactobacilli, addition of either yeast extract or tomato juice, lower NaCl concentration, or a lower initial acidity in the brine, etc.), also tend frequently to increase the formation of white spots. It may well be that during active fermentation, when more lactobacilli are concentrated in the brine, more bacteria come into contact with the olive peel and thus more may penetrate below the peel, forming white spots. For the same reason the larger olive cultivars (because of their larger fruit surface) may be more susceptible. In fact, there is a good correlation between the size and number of open lenticels of the fresh olives and the number of white spots in the fermented fruits.

The number of spots was reduced when higher initial acidity, sorbic acid or crushed olive pulp was added to the brine or when fruits well fertilized with nitrogen were used for fermentation. A complete absence was observed when olives were pitted before brining. This could be due to a more rapid diffusion of nutrients into the brine, which assured more satisfactory growing conditions for the lactobacilli in the solution, or to a reduction of the turgor of the fruits' surface.

However, it is also very likely that the appearance of white spots is related to the residual concentration of phenols in olives. Both in California and in Israel, lye treatment and subsequent washes almost completely remove oleuropein and its derivatives and white spots are more frequent. Furthermore, larger cultivars (Sevillano, Gordal, Sede-Eliyahn, etc.) are known to have a lower content of inhibitors (Amiot *et al.*, 1990) and in this case the bacteria can find an appropriate culture medium even in the flesh. Small cultivars though, are higher in phenols and exert stronger inhibitory effect in their flesh, preventing this problem.

7.13.2 SOFTENING

Olive texture is well maintained in anaerobic conditions. Nortje and Vaughn (1953) associated softening with the presence of *Bacillus*. They isolated the species *B. subtilis* and *B. pumilus* which elaborated softening enzymes under suitable conditions. The range of pH for activity was 6.0–11.0 and activity was maximal at 8.0. Optimum temperature was around 45°C. The enzyme complex was reasonably stable in alkaline solutions but rapidly lost its activity under acid conditions. In addition, the enzyme from *B. pumilus* was produced only when the medium contained galacturonic acid, pectic acid, or other pectic substances. Tests carried out with olives at pH of 5.7 proved their softening activity.

These bacteria can provoke softening of green Spanish-type olives only under certain conditions, especially during the first days after brining when pH is still appropriate for their activity.

From the mycelial growth of the surface of brines Balatsouras and Vaughn (1958) isolated 36 pure cultures of moulds. The majority were allocated to the genera *Fusarium* (23) and *Penicillium* (8). *Aspergillus, Geotrichum, Paecilomyces* and *Verticillium* were only represented by single species. From the different tests carried out with respect to pH tolerance, salt resistance, ability to decompose lactic acid, and to degrade pectin and cellulose it was concluded that most of such microorganisms could be regarded as potential agents of spoilage of olives.

King and Vaughn (1961) ascribed softening to five genera of pectolytic bacteria (*Aerobacter, Escherichia, Paracolobactrum, Aeromonas* and *Achromobacter*), since the production of pectin esterase and other pectin-degrading enzymes were detected.

This spoilage can be prevented by a rapid pH decrease during the first days of brining, maintenance of anaerobic conditions on the surface of brines and good manufacturing practices.

7.13.3 FISH-EYES AND GAS-POCKETS

This is characterized by the formation of pockets or blisters in the olive surface caused by the accumulation of gas inside. On the outside of the flesh the olives look as if they have been pressed by a wire, hence the Spanish name 'alambrado'. Cutting the fruits, it can be observed that cracks sometimes reach through the flesh to the pit. This spoilage is more frequent in certain cultivars such as Gordal, Ascolana and Morona, although it may affect other cultivars to a lesser extent. The defect is characteristic of the first stage of the fermentation and is due to the action of Gram-negative bacteria of the genera *Enterobacter, Citrobacter, Klebsiella, Escherichia* and *Aeromonas* which produce CO_2 and hydrogen (Borbolla y Alcalá *et al.*, 1959, 1960).

The ways to prevent this spoilage include: (i) fundamental hygiene measures such as good bacteriological quality water used for washing and brining, cleaning all the utensils and containers; and (ii) correct control to reduce the time for the first stage of the fermentation, in which Gram-negative bacteria can develop. This last point is achieved by a rapid lowering of the pH value of the brine, since the development of Gram-negative bacteria is largely inhibited when the pH is 4.5 or less. Borbolla y Alcalá *et al.* (1959) showed that the use of HCl or of lactic or acetic acids was useful for adjusting the pH, without any negative influence on lactobacilli growth. Acetic acid was the most effective at controlling Gram-negative bacteria (and gas-pockets), especially when initial acidification was followed by subsequent pH control with HCl. The same authors (1960) showed later that providing a CO_2-saturated atmosphere was also inhibitory for the growth of gas-pocket-producing bacteria and this method has been used since then by the industry.

7.13.4 PUTRID AND BUTYRIC FERMENTATIONS AND 'ZAPATERÍA'

These three different kinds of malodorous fermentations have in common the fact that microorganisms belonging to the genus *Clostridium* are implicated, producing undesirable changes in the sensory characteristics of the fruit. Levin and Vaughn (1966) associated the putrid fermentation with the presence in brine of *Desulfovibrio aestuarii*, which was found in concentrations as great as 2×10^6 cfu/ml in fermenting malodorous sulphurous olive brines. The most likely source would appear to be solar salt, since the same organism was obtained from mud along the banks of solar salt ponds.

Gililland and Vaughn (1943) implicated the *Clostridium butyricum* group as the cause of the butyric acid fermentation of olives. The spoilage is always accompanied by an increasing pH value and volatile acidity level in the brine. Putrid and butyric fermentations are generally produced during the first stages of fermentation just a few days after brining and are mainly due to the use of contaminated water for the preparation of the brine and a failure to eliminate the alkaline grounds remaining at the bottom of the fermenters at the end of the washing period. The relatively high pH value and anaerobic conditions favour the growth of the causative microorganisms.

'Zapatería' may develop at the end of the lactic fermentation if the salt concentration is not well controlled. In this case, the development of *Propionibacterium* may cause a depletion of the lactic acid and an increasing pH allowing growth of *Clostridium*, which accidentally may be present if good sanitary conditions are not applied. Kawatomari and Vaughn (1956) were the first to associate species of *Clostridium* with zapatera spoilage of olives and found *C. sporogenes* and *C. bifermentans* to be most abundant. *C. butyricum*, *C. multifermentans*, *C. beijerinckii*, *C. pasteurianum* and *C. spheroides* were also identified. Many others, which were of saccharolytic type, remained unidentified. However, because of the unexplained presence of propionic acid in spoilt brines they deduced that other genera were also involved in the spoilage. Thus, Plastourgos and Vaughn (1957) also isolated species of *Propionibacterium* from zapatera-spoilt olives, particularly *P. pentosaceum* and *P. zeae*. A detailed study of the microorganisms responsible for this spoilage in green Spanish-style olives was carried out by González Cancho and Fernández Díez (1957) and González Cancho *et al.* (1968, 1970, 1973). The main species responsible are shown in Table 7.27.

The same authors have studied the effect of salt and pH on the growth of the microorganisms responsible for the spoilage. In the absence of salt all the *Clostridium* species are able to grow at pH 4.75 or over, while for *Propionibacterium* the lower limit was 4.3. The inhibitory effect of salt at pH 7 was low and relatively high concentrations of 5 to 8% (w/v) for

Table 7.27 Relevant species of *Propionibacterium* and *Clostridium* associated with malodorous fermentations (From Borbolla y Alcalá and González Cancho, 1975)

Species	Relative presence (%)
Propionibacterium zeae	68
Propionibacterium pentosaceum	24
Propionibacterium sp.	7
Clostridium sporogenes	38
Clostridium lituseburense	34
Clostridium limosum	10
Clostridium aerofoetidum	7
Clostridium nauseum	7
Clostridium multifermentans	1
Clostridium innominatum	1

Clostridium and over 9% for *Propionibacterium* were necessary to prevent growth. Fortunately, the combined effect of pH and salt are higher because at pH of 4.75 all species of *Clostridium* were inhibited by 3% NaCl and at pH 4.5, *Propionibacterium* was inhibited by 5%. In the industrial processes, where a certain safety margin is advisable, suggested levels of salt and pH are 7.5–8.0% and 4.0 respectively.

The mixture of metabolites produced by the two genera of microorganisms involved in the alteration are responsible for the off-flavour. Delmouzos *et al.* (1953) recovered formic, propionic, butyric and succinic acids from zapatera brines, while only lactic and acetic acid were present in normal fermentations. Bocklet and Sticella (1972) found acetic (more abundant) and propionic, as well as small proportions of iso-butyric, *n*-butyric and iso-valeric acids in slightly zapatera olives. In strongly zapatera-spoilt olives, the proportion of propionic acid was higher than acetic. Montaño Asquerino *et al.* (1992) have reported the presence of acetic, propionic (which was relatively high in all cases) *n*-butyric, iso-valeric, *n*-valeric and *n*-caproic acids as well as acetaldehyde, methanol, ethanol 2-butanol and *n*-propanol as the major headspace components of the brines. But their most outstanding contribution has been the identification of cyclohexanecarboxylic acid as the main feature of zapatera olives. When the compound was added to normal olives, they acquired a zapatera-like flavour. The presence of this compound, with the other acids and volatiles identified, seems to be responsible for the unpleasant typical smell of zapatera olives. The origin of this compound is not yet known.

Inigo and Bravo Abad (1983) reported the presence of very low amounts of histamine in fermented olives (below 2 p.p.m.) and found a slightly higher level in zapatera olives (below 5 p.p.m.). Hornero Méndéz and Garrido Fernández (1994) have found putrescine (12.64 p.p.m.), cadaverine (8.29 p.p.m.) and tyramine (16.08 p.p.m.) in zapatera olives, but no biogenic amines were present in normal fermentations. Although the amounts of biogenic amines found are far from significant, from a safety point of view their association with zapatera spoilage may serve as an additional tool for the identification of this spoilage.

7.14 INDUSTRIAL EQUIPMENT

Traditionally, Spanish-style green olives were processed in three main stages. Lye treatment and washing waters were given in 1000-kg capacity cement tanks. Later, olives were transferred to wooden barrels similar to those used for wine making in which the fruits underwent fermentation and storage. Finally, they underwent a series of operations (grading, pitting, etc.) and then packing. Factories needed large areas for the storage of barrels and the whole process used many workers because most of the operations were manual.

Modernization of this industry began with the work of Ball *et al.* (1950) who initiated fermentation in bulk in redwood tanks of about 7000 kg capacity. The system was seen as promising if applied under the supervision of trained personnel since it gave greater uniformity of products and saved labour. In Argentina at around the same time, parallelepipedic cement tanks of 1200 to 14 000 kg capacity were introduced. In Greece (Alygisakis, 1971), similar cement tanks covered with paraffin wax and wooden conical tanks were used to ferment naturally black olives and green. In Spain research to apply bulk fermentation began in 1955 and has been applied practically since 1964. Borbolla y Alcalá *et al.* (1969a,b) were the driving force who led this complete transformation.

At the present time, the containers (Fernández Díez et *al.*, 1985; Fernández Díez, 1991; Unidad Estructural de Biotecnología de Alimentos, 1990) used for all the fundamental phases of this process in the majority of countries producing Spanish-style green olives are made of polyester and fibre-glass. They can be fully closed to create anaerobic conditions and exclude the growth of film yeasts on the surface of the brine. The containers have a large opening on the top to load the fruit and valves at the top and the bottom which facilitate unloading of the fruit and the circulation of the brine when necessary. The most common capacity is about 10–15 tonnes of fruit. When the fermenters are situated in the factory above ground and supported on three or four legs, the valve at the bottom allows unloading of the fruit by gravity.

The same material is also used for fermenters placed underground (Figure 7.5). In this case, the opening at the top is used for all the operations (loading and unloading of the fruit and change and circulation of the brine) using suitable pumps. A heat exchanger is used in connection with the pump if necessary to improve the temperature of the product during fermentation.

A complete review of bulk fermentation technology can be found in Borbolla y Alcalá *et al.* (1969a,b) and Borbola y Alcalá and Rejano Navarro (1979), as well as in Fernández Díez *et al.* (1985). In general, large fermenters require rapid loading and unloading to prevent fruit deformation while the lye, washing waters or brine are poured out. Modern installations have a series of above-ground fermenters where the debittering process is applied and a large number of underground ones where fermentation and storage take place. In these, appropriate neck-sealing devices prevent formation of surface layers of yeasts and moulds. Underground vessels also maintain better temperature in both cold and warm periods.

Transportation of olives from these fermenters to other parts of the factory is nowadays made by appropriate pumps or vacuum systems. Fermented or stored olives are transported between factories in plastic drums which are easily loaded and unloaded from trucks as well as through the interior of the factory.

Before packing, the olives are sorted, either manually or by means of some electronic devices, by running them along electric conveyor belts. Then, they are size-graded using machines with diverging steel cables and collected in plastic containers of variable capacity until they are pitted, stuffed and packed into the small containers. Very efficient automatic machines are available for all of these later operations.

7.15 OTHER COMMERCIAL PREPARATIONS OF THE GREEN OLIVE TYPE

As indicated in section 7.1, lye-treated green olives, also called pickled olives, may also be prepared without undergoing a complete lactic fermentation in brine. In that case, their subsequent preservation at a pH value within the limits specified in the Standards (International Olive Oil Council, 1980; Spanish Government Presidency, 1983) may be ensured by sterilization or pasteurization; addition of preserving agents; refrigeration; or inert gas packing without brine.

Generally for all these preparations, after a lye treatment which is quite similar to that of Spanish-style, olives are washed with water to eliminate most of the residual lye, conditioned in brine, and acidified with acetic acid, lactic acid, citric acid or a mixture to give an equilibrium pH value safe enough to pasteurize the product packed in small containers. According to the standards, this pH value must be 4.3 or below if the salt concentration is 2% (w/v) or above. For salt concentrations lower than 2%, the pH value must be reduced at least to 4.0.

Before packing and pasteurization, treated green olives in brine may undergo a lactic, or yeast, or lactic and yeast partial fermentation, depending upon the physical, chemical and microbiological control conditions of the process.

The taste of these products may be relatively close to that of Spanish-style olives, though they are much more green. Chemical composition is also quite similar, but generally the fruit keeps a certain amount of soluble fermentable sugars, which must be considered when fixing the correct pasteurization conditions. These preparations are very popular in several producer countries, but their international market is very limited.

In some cases fermentation is completely eliminated as in the case of the system developed by Montaño Asquerino *et al.* (1986, 1988b). The procedure maintains the lye treatment and washings but substitutes the phase of fermentation with a rapid modification of the physicochemical characteristics of the green olives by an initial acidification and a more or less long period of heat treatment. The olives are then packed. Preservation is achieved by pasteurization. Since there is no time for diffusion of the bitter principle from the untreated part of the fruit into the brine because of the rapid packing, lye treatment must reach the whole flesh.

Concentrations of NaOH must be similar to those used in the traditional process. Washing waters must also be applied generously and the number of washes should also be higher than in the normal Spanish-style. However, with more than five washings the final product loses flavour. Brining is achieved in a 10.5–11.5% (w/v) NaCl solution with the required amounts of acids added. Acids used must be food-grade HCl or a mixture of lactic and citric. The final product should ideally have a pH of 4.0. It is advisable to determine in advance that the olives are not shrivelled when held in such acidified brines. In general, no negative effects of these acids on the organoleptic characteristics are observed. The treatment is completed by an incubation for a few days at 30–40°C. After this, the colour and texture of the olives resemble those found in the traditional process and fruits can be packed similarly to fermented olives.

In spite of the washing waters, significant amounts of reducing sugars remain in the fruits on packing. With correct heat treatment, the stability of the product is assured, but any error can lead to spoilage due to the presence of fermentable substrate. When the concentration of acids is insufficient, the growth of lactobacilli is possible during incubation.

In spite of its similar appearance, the flavour of the product can easily be distinguished from the classical version. This product thus must be considered as an alternative product, rather than a substitute for Spanish-style green olives.

The Castelvetrano system ('dolcificata') from Sicily is also known world-wide, although the product is only sold in the region. It is applied to Nocellara del Bellice cultivar olives with a transverse diameter greater than 19 mm. Fruits are graded by size, placed in 200-litre plastic drums, and treated with lye (2.5–3.5 °Bé). After 1 hour, 5–8 kg of salt is added. In 12–24 hours the brine is formed and the drums are tightly closed and stored. After 15 days of storage, the product is ready. Olives are washed to remove excess alkali and sold. Due to its physicochemical characteristics, the product must be sold before April or May; otherwise, the olives lose their characteristic green colour and anomalous tastes appear as the temperature increases. Other zones of Greece and Spain are also trying to produce similar products.

Finally, some Argentinian agricultural engineers have developed another product in which lye treatment and washing stages were somewhat similar. The final packed product had optimum acceptability when the pH was 4.2–4.3 and the NaCl content 4.0–4.5%. The combinations of acids that obtained the higher organoleptic scores were the mixture of lactic and acetic, followed by acetic and malic acids in equal proportions. When bottled the product had good stability. If canned, however, the researchers suggested a short sterilization which maintained the organoleptic characteristics.

Untreated green olives in brine (Borbolla y Alcalá *et al.*, 1971; Fernández Díez *et al.*, 1985) – also called naturally green olives – are placed directly into a brine with a salt concentration around 6% (w/v) at equilibrium, and preserved by a natural fermentation involving a mixture of lactic acid bacteria and yeasts. This fermentation is similar to that of turning colour or ripe olives (during the storage phase). Detailed information about this is presented in Chapter 8.

This product is generally sold in plastic containers of 5–10 kg capacity, or may be pasteurized and packed in smaller containers. Frequently the product is prepared finally as aromatized green olives in a way similar to that described for certain preparations of turning colour olives in Chapter 8. Salt concentration and pH value for the pasteurized product are the same as for treated green olives. International Standards fix the following limits for non-pasteurized final products:

	Minimum sodium chloride concentration (%)	Maximum pH limit
Hermetically sealed containers	5	4.0
Non-hermetically sealed containers	6	4.5

A comparison between the composition and nutritive value of treated and untreated green olives (Castro Ramos *et al.*, 1980a,b) indicate that treated olives have higher values of moisture, ash and carotene than untreated fruits. However, oil content, protein, thiamin and calorific value show higher values in naturally green olives. Treated olives present also a lower content of potassium and copper and higher values of calcium and sodium.

The taste of untreated green olives is completely different from that of the lye-treated counterpart, due mainly to the residual bitterness which they retain, even after a long period of storage in brine. For that reason they are highly appreciated among consumers of Mediterranean origin. However, the international market for these products is very small.

7.16 REFERENCES

Albi, T. (1981) Residuos de insecticidas en aceitunas. I. Método de determinación de carbamatos en aceitunas. Residuos de carbaryl en los procesos de aderezo de aceitunas. *Grasas y Aceites*, **32**, 381–6.

Albi, T. and Rejano Navarro, L. (1982) Residuos de insecticidas en aceitunas. II. Eliminación o degradación de los insecticidas fosforados dimetoato, diazinone e imidan durante el aderezo de aceitunas. *Grasas y Aceites*, **33**, 146–9.

Alygisakis, M.E. (1971) Fermentation et conservation des olives de table dans des citernes en beton armé enduites d'un vernis plastique special. Proceedings II. International Congress of Olive Technics, Torremolinos, Junio.

Amiot, M.J., Tacchini, M., Fleuriet, A. and Macheix, J.J. (1990) The technological debittering process of olives: characterization of fruits before and during alkaline treatment. *Sciences des Aliments*, **10**, 619–31.

Balatsouras, G. (1985) Taxonomic and physiological characteristics of the facultative rod type lactic acid bacteria isolated from fermenting green and black olives. *Grasas y Aceites*, **36**, 239–49.

Balatsouras, G.D. and Vaughn, R.H. (1958) Some fungi that might cause softening of storage olives. *Food Research*, **23**, 235–43.

Balatsouras, G., Papamichael, V. and El Din, N. (1971) Fermentation of green olives by thermophilic lactic acid bacteria at temperature of 37 40°C. *Grasas y Aceites*, **22**, 357–66.

Balatsouras, G., Tsibri, A., Dalles, T. and Doutsias, G. (1983) Effects of fermentation and its control on the sensory characteristics of Conservolea cultivar green olives. *Appl. Environm. Microbiol.*, **46**, 68–74.

Ball, R.N., Dellen, E., Jaquith, J.B., Vaughn, R.H., Tabachnick, J. and Wedding, G.T. (1950) Experimental bulk fermentation of Californian green olives. *Food Technology*, **IV**, 30–2.

Barret, A. and Bidan, P. (1964) Quelques recherches recentes et leur application des olives vertes de table. *Inf. Oleicole Internationale*, **25**, 53–63.

Bocklet, M.F. and Sticella, M.A. (1972) Etude sur les substances qui conferent aux olives vertes fermentées des saveurs anomales. *Inf. Oleicoles Internationales*, **60–61**, 115–31.

Borbolla y Alcalá, J.M. R. de la. (1972) Aceitunas de mesa. Su clasificación por tamaños. *Grasas y Aceites*, **23**, 237–48.

Borbolla y Alcalá, J.M.R de la. (1975) Sobre la preparación de aceitunas estilo sevillano en grandes fermentadores. *Grasas y Aceites*, **26**, 161–6.

Borbolla y Alcalá, J.M.R. de la. (1979) Sobre la preparación de la aceituna estilo sevillano. La aceituna en salmuera. *Grasas y Aceites*, **30**, 31–6.

Borbolla y Alcalá, J.M.R. de la. (1981) Sobre la preparación de las aceitunas estilo sevillano. El tratamiento con lejía. *Grasas y Aceites*, **32**, 181–9.

Borbolla y Alcalá, J.M.R. de la and Cordón Casanueva, J.L. (1969) La utilización de invernaderos de material plástico en la fermentación de aceitunas aderezadas estilo sevillano. *Grasas y Aceites*, **20**, 171–8.

Borbolla y Alcalá, J.M.R. de la and González Pellissó, F. (1964) Estudios sobre el aderezo de aceitunas verdes. XXII. Correlación acidez-pH. *Grasas y Aceites*, **15**, 233–6.

Borbolla y Alcalá, J.M.R. de la and González Pellissó, F. (1972) Estudio sobre aceitunas envasadas. XI. La inhibición del sedimento. *Grasas y Aceites*, **23**, 107–17.

Borbolla y Alcalá, J.M.R. de la and González Cancho, F. (1975) Preparación de aceitunas verdes. Ponencia al II Seminario Oleícola Internacional, Córdoba, Spain.

Borbolla y Alcalá, J.M.R. de la and González Pellissó, F. (1980a) Corrección de las diferentes características químicas en las aceitunas aderezadas estilo sevillano. Conclusión. *Grasas y Aceites*, **31**, 197–203.

Borbolla y Alcalá, J.M.R. de la and González Pellissó, F. (1980b) Correlación de las diferentes características químicas en las aceitunas aderezadas estilo sevillano. *Grasas y Aceites*, **31**, 111–20.

Borbolla y Alcalá, J.M.R. de la and Rejano Navarro, L. (1978) Sobre la preparación de las aceitunas estilo sevillano. El lavado de los frutos tratados con lejía. *Grasas y Aceites*, **29**, 281–91.

Borbolla y Alcalá, J.M.R. de la and Rejano Navarro, L. (1979) Sobre la preparación de la aceituna estilo sevillano. La fermentación. I. *Grasas y Aceites*, **30**, 175–85.

Borbolla y Alcalá, J.M.R. de la and Rejano Navarro, L. (1981) Sobre la preparación de la aceituna estilo sevillano. La fermentación II. *Grasas y Aceites*, **32**, 103–13.

Borbolla y Alcalá, J.M.R de, Gómez Herrera, C., Guzmán, R. and Vázquez, R. (1952a) Estudio sobre el aderezo de aceitunas verdes. VII. Efecto del tratamiento con lejía. *An. R. Soc. Esp. Física y Química*, **48**, 427–30.

Borbolla y Alcalá, J.M.R. de la., Gómez Herrera, C. and Izquierdo Tamayo. (1952b) El uso de cultivos puros de lactobacilos en el aderezo de aceitunas. *Grasas y Aceites*, **3**, 91–4.

Borbolla y Alcalá, J.M.R. de la, Gómez Herrera, C. and Izquierdo Tamayo, A. (1953) Estudio sobre el aderezo de aceitunas verdes XI. Empleo de cultivos puros de lactobacilos. Comunicación Técnica, **27**, Patronato Juan de la Cierva, Instituto de la Grasa, Sevilla, Spain.

Borbolla y Alcalá, J.M.R. de la, Gómez Herrera, C., González Cancho, F., Fernández Díez, M.J., Gutiérrez González-Quijano, R., Izquierdo Tamayo, A., González Pellissó, F., Vázquez Ladrón, R., and Guzmán García, R. (1956) *El aderezo de aceitunas verdes*. Consejo Superior de Investigaciones Científicas, Madrid, Spain.

Borbolla y Alcalá, J.M.R. de la, Fernández Díez, M.J. and González Cancho, F. (1959) Estudios sobre el aderezo de aceitunas verdes. XVI. Experiencias sobre el alambrado. *Grasas y Aceites*, **10**, 221–34.

Borbolla y Alcalá, J.M.R. de la, Fernández Díez, M.J. and González Cancho, F. (1960) Estudios sobre el aderezo de aceitunas verdes. XIX. Nuevas experiencias sobre el alambrado. *Grasas y Aceites*, **11**, 256–60.

Borbolla y Alcalá, J.M.R. de la, Fernández Díez, M.J. and González Cancho, F. (1961) Empleo del ácido sórbico, o sus sales, en las aceitunas aderezadas. *Grasas y Aceites*, **12**, 10–15.

Borbolla y Alcalá, J.M.R. de la, Fernández Díez, M.J., González Cancho, F. and Cordón Casanueva, J.L. (1969a) Estudio sobre el aderezo de aceitunas verdes. XXIII La fermentación de aceitunas en recipientes de gran capacidad. *Grasas y Aceites*, **20**, 55–66.

Borbolla y Alcalá, J.M.R., Fernández Díez, M.J. and Cordón Casanueva, J.L. (1969b) Estudios sobre el aderezo de aceitunas verdes. XXIII. La preparación y fermentación de aceitunas en recipientes de gran capacidad. *Grasas y Aceites*, **20**, 55–66.

Borbolla y Alcalá, J.M.R de la, Fernández Díez, M.J. and González Cancho, F. (1969c) Influence of pasteurization and lye treatment on the fermentation of Spanish-style Manzanilla olives. *Appl. Microbiol.*, **17**, 734–6.

Borbolla y Alcalá, J.M.R. de la, González Pellissó, F. and González Cancho, F. (1971) Aceitunas verdes y de color cambiante en salmuera. I. *Grasas y Aceites*, **22**, 455–50.

Borbolla y Alcalá, J.M.R. de la, Rejano Navarro, L. and Nosti Vega, M. (1975) La formación de ácido propiónico durante la conservación de las aceitunas verdes de mesa. *Grasas y Aceites*, **26**, 153–60.

Bravo Abad, F. and Iñigo, R.M. (1987) Población bacteriana y evolución de la composición química de la aceituna en el transcurso de la fermentación láctica. *Alimentaria*, **189**, 87–94.

Brenes, M. and Garrido, A. (1988) Regeneración de salmueras de aceitunas verdes estilo sevillano con carbón activo y tierras decolorantes. *Grasas y Aceites*, **39**, 96–101.

Brenes, M., Montaño, A. and Garrido, A. (1990) Ultrafiltration of green table olive brines: Influence of some operating parameters and effect on polyphenol composition. *J. Food Sci.*, **55**, 214–17.

Brenes, M., García, P. and Garrido, A. (1994) Influence of salt and pH on the firness of olives in acid conditions. *J. Food Quality*, **17**, 335–46.

Brighigna, A. and Pelagatti, O. (1980) La preparazione mediante essiccazione spray di fermenti lattici per le olive da tavola. *Industrie Alimentari*, June, 499–502.

Castro, A. and Rejano Navarro, L. (1990) Incidencia de la fermentación propiónica en la conservación de aceitunas verdes de mesa. Seminario Ultimos avances en la tecnología de preparación de encurtidos. Aplicación a pepinillos y otros cultivos de interés en Andalucía, Sevilla, Mayo.

Castro Gómez-Millán, A., Montaño Asquerino, A., Sánchez Gómez, A.H., Rejano Navarro, L. and Garrido Fernández, A. (1989) Influencia de la adición de ácido clorhídrico y de la inoculación con levaduras en la fermentación y características organolépticas de las aceitunas verdes estilo sevillano. *Grasas y Aceites*, **40**, 376–81.

Castro Ramos, R. and Vázquez Ladrón, R. (1976a) Variación del color de las aceitunas según el proceso de fermentación. I. Variedad Verdial. *Grasas y Aceites*, **27**, 151–9.

Castro Ramos, R. and Vázquez Ladrón, R. (1976b) Variación del color de las aceitunas según el proceso de fermentación. II. Variedad Gordal. *Grasas y Aceites*, **27**, 229–31.

Castro Ramos, R., Nosti Vega, M. and Vázquez Ladrón, R. (1979) Composición y valor nutritivo de algunas variedades españolas de aceitunas de mesa. I. Aceitunas verdes aderezadas al estilo sevillano. *Grasas y Aceites*, **30**, 83–91.

Castro Ramos, R., Nosti Vega, M. and Vázquez Ladrón, R. (1980a) Estudio comparativo del valor nutritivo de la aceituna de mesa. Efecto de la variedad del fruto, del tipo de elaboración y del envasado. *Alimentaria*, XVII, **115**, 21–4.

Castro Ramos, R., Nosti Vega, M. and Vázquez Ladrón, R. (1980b) Composición y valor nutritivo de algunas variedades españolas de aceitunas de mesa. IV. Efectos de la reutilización de la lejía de cocido y aguas de lavado. *Grasas y Aceites*, **31**, 91–5.

Corral Andrade, A. (1961) L'Acido ascorbico nella conservazione delle olive in salomoia. *Conserve e Derivati Agrumari*, **10**, 7–9.

Cruess, W.V. (1930) *Pickling green olives*, Univ. Calif., Agr. Expt. Sta. Bull. 490.

Cucurachi, A., Buonocore, V., Marsilio, V. and Brighigna, A. (1974) Osservationi anatomo-istologiche sulle olive da mensa in relazione alle caratteristiche di consistenza della polpa. *Ann. Ist. Sper. Elaiot.*, **IV**, 133–64.

Delmonzos, J.G., Stadmaun, F.M. and Vaughn, R.H. (1953) Malodorous fermentation. Acidic constituents of zapatera olives. *J. Agri. Food Chem.*, **1**, 333–4.

Etchells, J.L., Borg, A.F., Kittle, I.D., Bell, T.A. and Fleming, H.P. (1966) Pure culture fermentation of green olives. *Appl. Microbiol.*, **14**, 1027–41.

FDA (Food and Drug Administration) (1989) Residues in Foods – 1988. *J. Assoc. Off. Anal. Chem.*, **75**, 133A–151A.

Fernández Díez, M.J. (1971) The Olive, in *The Biochemistry of Fruits and Their Products*, Vol. **2** (ed. A.C. Hulme), Academic Press, London, pp. 255–79

Fernández Díez, M.J. (1983) Olives, in *Biotechnology*, Vol. **5** (eds H.J. Rehm and G. Reed), Verlag Chemie, Weinheim, Germany, pp. 379–97

Fernández Díez, M.J. (1990) Biotecnología de las aceitunas de mesa aderezadas al estilo sevillano o español. Proceedings *II Seminario Internazionale 'Olio d'oliva e olive da tavola: Tecnologia e qualita*, Pescara, Italy.

Fernández Díez, M.J. (1991) Processing table olives, in *Encyclopedia of Food Science and Technology*, Vol. **3** (ed. Y.H. Hui) Wiley & Sons, New York, pp. 1910–25.

Fernández Díez, M.J. and Vidal Sigler, A. (1968) Determinaciones de textura en aceitunas.II. Las pruebas de compresión-cizallamiento, *Grasas y Aceites*, **19**, 199–204.

Fernández Díez, M.J., Castro Ramos, R., Garrido Fernández, A., González Cancho, F., González Pellissó, F., Nosti Vega, M., Heredia Moreno, A., Mínguez Mosquera, M.I., Rejano Navarro, L., Durán Quintana, M.C., Sánchez Roldán, F., García García, P., and Castro Gómez Millán, A. (1985) *Biotecnología de la aceituna de mesa*. Consejo Superior de Investigaciones Científicas, Instituto de la Grasa, Madrid-Sevilla, Spain.

Fleming, H.P. and Etchells, J.L (1967) Occurrence of an inhibitor of lactic acid bacteria in green olives. *Appl. Microbiol.*, **15**, 1178–84.

Fleming, H.P., Etchells, J.L. and Bell, T.A. (1969a) Vapor analysis of fermented Spanish-type green olives. *J. Food Sci.*, **34**, 419–22.

Fleming, H.P., Walter, W.M. and Etchells, J.L. (1969b) Isolation of a bacterial inhibitor from green olives. *Appl. Microbiol.*, **18**, 856–60.

Fleming, H.P., Walter,W.M. and Etchells, J.L. (1973) Antimicrobial properties of oleuropein and products of its hydrolysis from green olives. *Appl. Microbiol.*, **26**, 777–82.

Garrido Fernández, A. (1979) Regeneración de salmueras de aceitunas verdes al estilo sevillano o español mediante la adsorción con carbón activo. *Grasas y Aceites*, **30**, 5–10.

Garrido Fernández, A. and Vaughn, R. H. (1976) Utilization of oleuropein by microorganisms associated with olive fermentations. *Can.J. Microbiol.*, **24**, 580–4.

Garrido Fernández, A., González Pellissó, F., González Cancho, F., Sánchez Roldán, F., Rejano Navarro, L., Cordón Casanueva, J.L. and Fernández Díez, M.J. (1977) Changes in the processing and packing of Spanish-style green table olive to reduce waste waters. *Grasas y Aceites*, **28**, 267–85.

Garrido Fernández, A., Cordón Casanueva, J.L., Rejano Navarro, L., González Cancho, F. and Sánchez Roldán, F. (1979) Green Spanish-style olive pickling with lye re-use and washing suppression. *Grasas y Aceites*, **30**, 227–34.

Garrido Fernández, A., García, P. and Brenes, M. (1992a). The recycling of table olive brine using ultrafiltration and activated carbón adsorption. *J. Food Eng.*, **17**, 291–305.

Garrido Fernández, A., Brenes Balbuena, M. and García García, P. (1992b) Tratamiento de salmueras de fermentación de aceitunas verdes. *Grasas y Aceites*, **43**, 291–8

Gililland, J.R. and Vaughn, R.H. (1943) Characteristics of butyric acid bacteria from olives. *J. Bacteriol.*, **46**, 315–21.

González Cancho, F. (1960a) Estudio sobre el aderezo de aceitunas verdes. XVII. Levaduras presentes en la fermentación. *Grasas y Aceites*, **11**, 9–14.

González Cancho, F. (1960b) Estudios sobre el aderezo de aceitunas verdes. XVIII. Bacilos Gram-negativos no esporulados en las salmueras de aceitunas. *Grasas y Aceites*, **11**, 125–31.

González Cancho, F. (1963) Microorganismos que se desarrollan en el aderezo de aceitunas verdes estilo español. *Microbiología Española*, **16**, 221–30.

González Cancho, F. (1965) Levaduras en la fermentación de aceitunas verdes estilo español. Estudio cuantitativo. *Grasas y Aceites*, **16**, 230–4.

González Cancho, F. and Durán Quintana, M.C. (1981) Bacterias cocáceas del ácido láctico en el aderezo de aceitunas verdes. *Grasas y Aceites*, **32**, 373–9.

González Cancho, F. and Fernández Díez, M.J. (1957) Resistencia térmica de diferentes especies de *Clostridium* aislados de salmueras de aceitunas verdes aderezadas alteradas. *Microbiología Española*, **20**, 73–81.

González Cancho, F. and Fernández Díez, M.J. (1968) Especies de *Clostridium* aisladas de salmueras de aceitunas aderezadas. Influencia del pH y cloruro sódico sobre el desarrollo. *Microbiología Española*, **21**, 129–41.

González Cancho, F., Nosti Vega, M., Fernández Díez, M.J. and Buzcu, N. (1970) Especies de *Propionibacterium* relacionadas con la zapatería. Factores que influyen en su desarrollo. *Microbiología Española*, **23**, 233–52.

González Cancho, F., Fernández Díez, M.J., Nosti Vega, M., and Buzcu, N. (1973) Characteristiques des principaux microorganismes responsables de la zapatería. *Información Oleicola Internacional*, **60–61**, 133–44.

González Cancho, F., Rejano Navarro, L. and Borbolla y Alcalá, J.M.R. (1980) La formación de ácido propiónico durante la conservación de aceitunas verdes de mesa. III Microorganismos responsables. *Grasas y Aceitas*, **31**, 245–50.

González Cancho, F., Rejano Navarro, L., Durán Quintana, M.C., Sánchez Roldán, F.,

García García, P., Castro Gómez Millán, A. and Garrido Fernández, A. (1983) Influence of the HCl addition on the fermentation of Spanish-style green olives. *Grasas y Aceites*, **34**, 375–9.

González-Cancho, F., Rejano Navarro, L., Durán Quintana, M.C., Sánchez Roldán, F., Castro-Gómez Millán, A., García García, P. and Garrido Fernández, A. (1984) Spanish-style green olive elaboration without washing waters. Solutions to the problems caused by initial HCl addition. *Grasas y Aceites*, **35**, 155–9.

González Pellissó, F. and Rejano Navarro, L. (1984) La pasterización de aceitunas estilo sevillano. II. *Grasas y Aceites*, **35**, 235–9.

González Pellissó, F., Rejano Navarro, L. and González Cancho, F. (1982) La pasterización de aceitunas estilo sevillano. *Grasas y Aceites*, **33**, 201–7.

Gourama, H. and Bullerman, LL. B. (1987) Effects of oleuropein on growth and aflatoxin production by *Aspergillus parasiticus*. *Lebensm.-Wiss. u.-Technol.*, **20**, 226–8.

Gourama, H., Letutour, B., Tantoui-Elaraki, A., Benbya, M. and Bullerman, L.B. (1989) Effects of oleuropein, tyrosol, and caffeic acid on the growth of mold isolated from olives. *J. Food Protection*, **52**, 264–6.

Gracián Tous, J. (1968) Olive oil, in *Analysis and Characterization of Oils, Fats, and Fat Products.*, Vol. 2 (ed. H.A. Bockenoogen), Interscience, London, pp. 315–606.

Hassapidou, M.N., Balatsouras, G.D. and Manoukas, A.G. (1994) Effect of processing upon the tocopherol and tocotrienol composition of table olives. *Food Chemistry*, **50**, 111–14.

Heredia Moreno, A. and Fernández Díez, M.J. (1982a) Composición de fibra. V. Hemicelulosas e índice de digestibilidad en aceitunas de mesa. *Grasas y Aceites*, **33**, 197–200.

Heredia Moreno, A. and Fernández Díez, M.J. (1982b) Composición de fibra. V. Hemicelulosas y digestibilidad de distintas variedades de aceitunas. *Grasas y Aceites*, **33**, 197–200.

Heredia Moreno, A., Ruiz-Gutiérrez, V., Felizón, B., Guillén, R., Jiménez, A. and Fernández Bolaños, J. (1993) Apparent digestibility of dietary fibre and other components in table olives. *Die Nahrung*, **37**, 226–33.

Hornero Méndez, D. and Garrido Fernández, A. (1994) Biogenic amines in table olives. Analyses by high performance liquid chromatography. *Analyst*, **119**, 2037–41.

Iñigo, R.M. and Bravo Abad, F. (1983) Fermentación y nivel de histamina en aceitunas verdes. *Alimentaria*, **141**, 33–40.

International Olive Oil Council (COI) (1980) *Unified qualitative standard applying to Table Olives in International Trade*, COI, Madrid.

Jiménez Díaz, R., Rios-Sánchez, R.M., Desmazeaud, M., Ruiz-Barba, J.L., and Piard, J.C. (1993) Plantaricins S and T, two new bacteriocins produced by *Lactobacillus plantarum* LPCO10 isolated from a green table olive fermentation. *Appl. Environ. Microbiol.*, **59**, 1416–24.

Juven, B. and Henis, Y. (1970) Studies on the antimicrobial activity of olive phenolic compounds. *J. Appl. Bacteriol.*, **33**, 721–32.

Juven, B., Samish, Z., Henis, Y. and Jacoby, B. (1968a) Mechanism of enhancement of lactic acid fermentation of green olives by alkali and heat treatments. *J. Appl. Bacteriol.*, **31**, 200–7.

Juven, B., Samish, Z. and Henis, Y. (1968b) Identification of oleuropein as a natural inhibitor of lactic fermentation of green olives. *Israel J. Agric. Res.*, **18**, 137–8.

Juven, B., Henis, Y. and Jacoby, B. (1972) Studies on the mechanism of the antimicrobial action of oleuropein. *J. Appl. Bacteriol.*, **35**, 559–67.

Kawatomari, T. and Vaughn, R.H. (1956) Species of *Clostridium* associated with zapatera spoilage of olives. *Food Research*, **4**, 481–90.

King, A.D. and Vaughn, R.H. (1961) Media for detecting pectolytic Gram-negative bacteria associated with the softening of cucumbers, olives and other plant tissues. *J. Food Sci.*, **26**, 635–43.

Levin, R.E. and Vaughn, R.H. (1966) *Desulfovibrio aestuarii*, the causative agent of hydrogen sulfide spoilage of fermenting olive brines. *J. Food Sci.*, **31**, 768–72.

Mahjoub, A. and Bullerman, Ll.B. (1987) Effects of nutrients and inhibitors in olives on aflatoxigenic molds. *J. Food Protection*, **50**, 959–63.

Marsilio, V. and de Angelis, M. (1993) Indagine su alcuni parametri di qualità di campioni di olive da tavola del commercio. *Industrie Alimentari*, **XXXII**, 1093–5.

Mínguez Mosquera, M.I. and Gandul Rojas, B. (1994) Mechanism and kinetics of carotenoid degradation during the processing of green table olives. *J. Agric. Food Chem.*, **42**, 1551–4.

Mínguez Mosquera, M.I., Garrido Fernández, J. and Gandul Rojas, B. (1989) Pigment changes in olives during fermentation and brine storage. *J. Agr. Food Chem.*, **37**, 8–11.

Mínguez Mosquera, M.I., Garrido Fernández, J. and Gandul Rojas, B. (1990) Quantification of pigments in fermented Manzanilla and Hojiblanca olives. *J. Agr. Food Chem.*, **38**, 1662–6.

Mínguez Mosquera, M.I., Gandul Rojas, B., Montaño Asquerino, A. and Garrido Fernández, J. (1991) Determination of chlorophylls and carotenoids by high-performance chromatography during olive lactic fermentation. *J. Chromatogr.*, **585**, 259–66.

Mínguez Mosquera, M.I., Gandul Rojas, B. and Mínguez Mosquera, J. (1994) Mechanism and kinetics of the degradation of chlorophylls during the processing of green table olives. *J. Agric. Food Chem.*, **42**, 1089–95.

Mondher, K. and Chakib, B. (1993) Elaboracíon de las aceitunas verdes aderezadas estilo sevillano de la variedad 'Meski'. *Olivae*, **48**, 54–7.

Montaño Asquerino, A., Rejano Navarro, L. and Sánchez Gómez, A.H. (1985) Determination of total polyphenol in extracted olive pulp and fermentation brines. *Grasas y Aceites*, **36**, 274–85.

Montaño Asquerino, A., Rejano Navarro, L. and Sánchez Gómez, A.H. (1986) Proceso rápido de elaboración de aceitunas verdes aderezadas. *Grasas y Aceites*, **37**, 141–7.

Montaño Asquerino, A., Sánchez Gómez, A.H. and Rejano Navarro, L. (1988a) Método para la determinación del color en salmueras de aceitunas verdes de mesa. *Alimentaria*, **193**, 78–82.

Montaño Asquerino, A., Rejano Navarro, L. and Sánchez Gómez, A.H. (1988b) Proceso rápido de elaboración de aceitunas verdes aderezadas. II. *Grasas y Aceites*, **39**, 174–9.

Montaño Asquerino, A., Sánchez, A.H. and Rejano, L. (1990) Rapid quantitative analysis of headspace components of green olive brine. *J. Chromatogr.*, **521**, 153–7.

Montaño Asquerino, A., Castro, A., Rejano, L. and Sánchez, A.H. (1992) Analysis of zapatera olives by gas and high-performance liquid chromatography. *J. Chromatogr.*, **594**, 259–67.

Montaño Asquerino, A., Sánchez, A.H. and Castro, A. (1993) Controlled fermentation of Spanish-type olives. *J. Food Sci.*, **58**, 842–52.

Montero Marín, M. and Fernández Díez, M.J. (1976) Comportamiento de los parámetros de textura y evolución del tamaño en la variedad Gordal. *Grasas y Aceites*, **27**, 169–74.

Nortje, B.K and Vaughn, R.M. (1953) The pectolytic activity of species of the genus *Bacillus*: Qualitative studies with *Bacillus subtilis* and *Bacillus pumilis* in relation to the solftening of olives and pickles. *Food Research*, **18**, 57–69.

Nosti Vega, M., Castro Ramos, R. and Vázquez Ladrón, R. (1982) Composición y valor nutritivo de algunas variedades españolas de aceitunas de mesa. V. Influencia de la reutilización de las lejías y de la supresión de los lavados en las aceitunas verdes aderezadas. *Grasas y Aceites*, **33**, 5–8.

Nosti Vega, M., Castro Ramos, R., Vázquez Ladrón, R. (1984) Composición y valor nutritivo de algunas variedades españolas de aceitunas de mesa. VI Cambios debidos a los procesos de elaboración. *Grasas y Aceitas*, **35**, 11–14.

Nychas, G.J.E., Tassou, S.C. and Board, R.G. (1990) Phenolic extracts from olives: inhibition of *Staphylococcus aureus*. *Letters in Appl. Microbiol.*, **10**, 217–20.

Paster, N., Juven, B.J. and Harshemesh, H. (1988) Actimicrobial activity and inhibition of aflatoxin B_1 formation by olive plant tissue constituents. *J. Appl. Bacteriol.*, **64**, 293–7.

Pelagatti, O. and Brighigna, A. (1981) Revitalizzatione e impiego di fermenti lattici essiccati per la preparazione delle olive da tavola col sistema Sivigliano. *Industrie Alimentari*, March, pp. 182–4.

Plastourgos, S. and Vaughn, R.H. (1957) Species of *Propionibacterium* associated with zapateria spoilage of olives. *Appl. Microbiol*, **5**, 267–71.

Reig, J.M. and Hernández, J.M. (1991) El uso de microorganismos iniciadores (starters) en la fermentación de la aceituna. *Olivae*, **37**, 20–8.

Rejano Navarro, L. and González Pellissó, F. (1985) Corrección de las características químicas en aceitunas verdes aderezadas. Nuevos procedimientos de cálculo. *Grasas y Aceites*, **36**, 207–16.

Rejano Navarro, L. and Sánchez Gómez, A.H. (1990) Clasificación por tamaños y diferenciación de variedades de las aceitunas verdes de mesa. *Alimentación, Equipos y Tecnología*, Junio, pp. 133–8.

Rejano Navarro, L., González Cancho, F. and Borbolla y Alcalá, J.M.R. de la. (1977) Estudios sobre el aderezo de aceitunas verdes. XXIV. Nuevos ensayos sobre el control de la fermentación. *Grasas y Aceites*, **28**, 255–65.

Rejano Navarro, L., Fernández Villasante, J., Márquez González, J., Albarreal López, J., González Cancho, F. and Fernández Díez, M.J. (1978a) Efecto del tiempo de almacenamiento del fruto fresco sobre el proceso de elaboración de aceitunas verdes estilo sevillano. XIV Asamblea Miembros Instituto de la Grasa, Sevilla, Mayo.

Rejano Navarro, L., González Cancho, F. and Borbolla y Alcalá, J.M.R. de la. (1978b) La formación de ácido propiónico durante la conservación de las aceitunas verdes de mesa. II. *Grasas y Aceites* **29**, 203–10.

Rodríguez Quiñones, F., Nieto, J.J. and Megias, M. (1983) Estudios fisiológicos y genéticos en *Lactobacillus plantarum*, in *Proceedings of the II Reunión de Microbiología Industrial*, Sevilla, Spain, pp. 36–46.

Ruiz-Barba, J.L. and Jiménez Díaz, R. (1989) Actividad antibacteriana de las salmueras de aceitunas verdes sobre *Lactobacillus plantarum*. *Grasas y Aceites*, **40**, 102–8.

Ruiz-Barba, J.L. and Jiménez Díaz, R. (1994) Vitamin and amino acids requirements of *Lactobacillus plantarum* strains isolated from green olive fermentations. *J. Appl. Bacteriol.*, **76**, 350–5.

Ruiz-Barba, J.L., Rios Sánchez, J.M., Fedriani-Iriso, C., Olias, J.M., Rios, J.L. and Jiménez Díaz, R. (1990) Bactericidal effect of phenolic compounds from green olives on *Lactobacillus plantarum*. *System. Appl. Microbiol.*,**13**, 199–205.

Ruiz-Barba, J.L., Garrido-Fernández, A. and Jiménez Díaz, R. (1991) Bactericidal action of oleuropein extracted from green olives against *Lactobacillus plantarum*. *Letters in Appl. Microbiol.*, **12**, 65–8.

Ruiz-Barba, J.L., Brenes-Balbuena, M., Jiménez Díaz, R., García García, P. and Garrido-Fernández, A. (1993) Inhibition of *Lactobacillus plantarum* by two different kinds of olive brines. *J. Appl. Bacteriol.*, **74**, 15–19.

Ruiz-Barba, J.L., Cathcarth, D.P., Warner, P.J. and Jiménez Díaz, J. (1994) Use of *Lactobacillus plantarum* LPCOIO, a bacteriocin producer, as starter culture for green Spanish-style olives. *Appl. Environ. Microbiol.*, **60**, 2059–64.

Ruiz Cruz, J. and González Cancho, F. (1969) Metabolismo de levaduras aisladas de salmueras de aceitunas aderezadas al estilo español. I. Asimilación de los ácidos lácticos, acético y cítrico. *Grasas y Aceites*, **20**, 6–11.

Russo, C., Lanza, C.M., Tomaselli, F.G. and Nicolosi-Asmundo, C. (1987) Influenza della conservazione delle olive in salamoia sui residui dei pesticidi usati nella lotta antidacica. *Industrie Alimentari*, December, pp. 1162–4.

Samish, Z. (1954) Olive pickling research in Israel. Proceedings of the Xth International Congress of Agricultural and Food Industries, Madrid.

Samish, Z. and Dimant, D. (1963) White spots on green olives. *Israel J. Agric. Res.*, **13**, 37–45.

Samish, Z., Cohen, Sh. and Sudin, A. (1968) Progress of lactic acid fermentation of green olives as affected by peel. *Food Technology*, **22**, 77–80.

Sánchez Gómez, A.H., Rejano Navarro, A.H. and Montaño Asquerino, A. (1985) Determinaciones del color en las aceitunas verdes aderezadas de la variedad Manzanilla. *Grasas y Aceites*, **36**, 258–61.

Sánchez Gómez, A.H., Rejano Navarro, L. and Montaño Asquerino, A. (1986) Evolución del color durante el proceso de aderezo de aceitunas verdes. Efecto de la adición de HCl y de la supresión de los lavados. *Grasas y Aceites*, **37**, 86–91.

Sánchez Gómez, A.H., Montañon Asquerino, A. and Rejano Navarro, L. (1989) Optimización del proceso de pasterización de aceitunas verdes. Asamblea de Miembros del Instituto de la Grasa, Noviembre.

Sánchez Gómez, A.H., Rejano Navarro, L., Durán Quintana, M.C., Castro Gómez Millán, A. de, Montaño Asquerino, A., García García, P. and Garrido Fernández, A. (1990) Elaboración de aceitunas verdes con tratamiento alcalino a temperatura controlada. *Grasas y Aceites*, **41**, 218–23.

Sánchez Gómez, A.H., Rejano Navarro, L. and Montaño, A. (1991) Kinetics of the destruction by heat of colour and texture of pickled green olives. *J. Sci. Food Agric.*, **54**, 379–85.

Sciancalapore, V. (1984) La temperatura di deamerizzazione nella preparazione delle olive verdi col sistema Sivigliano. *Industrie Alimentari*, **23**, 941–4.

Spanish Government Presidency (1983) Reglamentación técnico sanitario para la elaboración, circulación, y venta de las aceitunas de mesa. *Spanish State Official Bull.*, **108**, 12619–25.

Tassou, C.C. and Nychas, G.J.E. (1994) Inhibition of *Staphylococcus aureus* by olive phenolics in broth and in model food systems. *J. Food Protection*, **57**, 120–4.

Tassou, C.C., Nychas, G.J.E. and Board R.G. (1991) Effect of phenolic compounds and oleuropein on the germination of *Bacillus cereus* T spores. *Biotechnol. Appl. Biochem.*, **13**, 231–7.

Tranter, H.S., Tassou, S.C. and Nychas, G.J. (1993) The effect of the olive phenolic compound, oleuropein, on growth and enterotoxin B production by *Staphylococcus aureus*. *J. Appl. Bacteriol.*,**74**, 253–9.

Tunçel, G. and Nergiz, C. (1993) Antimicrobial effect of some olive phenols in a laboratory medium. *Letters in Appl. Microbiol.*, **17**, 300–2.

Unidad Estructural de Biotecnología de Alimentos. Instituto de la Grasa. (1990) *Table Olives Processing*, International Olive Oil Council, Madrid.

Vaughn, R.H., Douglas, H.C. and Gililland, J.R. (1943) *Production of Spanish-type green olives.* Bulletin 78, University of California, College of Agriculture, Agricultural Experiment Station, Berkeley, California, USA.

Vaughn, R.H., Won, W.D., Spencer, F.B., Pappagranis, D., Foda, I.O. and Krumperman, P.M. (1953) *Lactobacillus plantarum*, the cause of yeast spots on olives. *Appl. Microbiol.*, **1**, 82–5.

Vázquez Ladrón, R. and Castro Ramos, R. (1976) Variación del color de las aceitunas según el proceso de fermentación. III. Variedad Manzanilla. *Grasas y Aceites*, **27**, 299–302.

Vázquez Ladrón, R., Castro Ramos, R. and Nosti Vega, M. (1979) Composición y valor nutritivo de algunas variedades españolas de aceitunas de mesa. III. Aceitunas verdes aderezadas envasadas. *Grasas y Aceites*, **30**, 221–6.

Vázquez Roncero, A. and Janer del Valle, M.L. (1977) Evolución de los polifenoles durante el aderezo de aceitunas verdes. I. Estudio cualitativo. *Grasas y Aceites*, **28**, 421–6.

Walter, W. M., Fleming, H. P. and Etchells, J.L. (1973) Preparation of antimicrobial compounds by hydrolysis of oleuropein from green olives. *Appl. Microbiol.*, **26**, 773–6.

Turning colour olives type

8

8.1 INTRODUCTION

According to the Unified Qualitative Standard Applying to Table Olives in International Trade (International Olive Oil Council, 1980) and the Spanish Standard (Spanish Government Presidency, 1983), **turning colour olives** type are obtained from rose, wine-rose or brown-coloured fruits harvested before the stage of complete ripeness is attained, which may or may not have been subjected to alkaline treatment and which are ready for consumption. Two groups of commercial or trade preparations have to be considered within this type of olives:

1. Untreated turning colour olives in brine. Directly placed in brine, preserved by fermentation and ready for consumption. These preparations may frequently be pasteurized.
2. Treated turning colour olives in brine. Obtained after alkaline treatment and preserved by natural fermentation in brine or by a heat treatment after brining. Californian green-ripe and black-ripe olives are included in this group.

Untreated turning colour olives have a small international market. Even including their internal markets, they represent a small percentage when compared with green, black-ripe and naturally black olives.

Among untreated turning colour olives in brine, aromatized (also called seasoned olives in brine) are presently becoming more popular among consumers, but their commercial importance is confined exclusively to the internal market.

Untreated turning colour olives in brine are customarily stored in brine until most of the bitterness is lost. Californian ripe olives in brine are sometimes stored in brine for a certain period until they are subjected to the oxidation process. The fermentation that takes place during the olive immersion in brine is similar in both cases and will thus be considered together. However, specific storage systems applied only to Californian ripe olives will be considered separately.

As indicated in Chapter 2, the Spanish Standard considers treated black-ripe olives in brine obtained by alkaline oxidation of turning colour fruits as a different type, known as **black olives**, within which California ripe olives are the most important. This division is correct from a technological standpoint. However, in this chapter we will follow the Unified Qualitative Standards Applying to Table Olives in the International Trade because of its wider coverage. The black-ripe olives (Californian-style) will thus be included here. This also accords with the Table Olive Statistics published by the International Olive Oil Council which include this trade presentation within the turning colour category.

Treated turning colour olives in brine were developed mainly in California. However, during the 1980s production of black-ripe olives expanded greatly in many table olive-producing countries (Spain, Morocco, Italy, etc.) and currently accounts for about 30% of the total production and an important share of the international trade. On the contrary, green-ripe olives remain restricted almost exclusively to California.

8.2 HARVESTING AND TRANSPORTATION

The picking optimum is, according to the definition of this type, rather vague and includes all fruits collected after the green type and before the naturally black type. However, in the case of black-ripe olives, the interval of maturity is more strictly defined and must be carefully followed to obtain the best quality. As with green olives, unmarketable sizes, leaves, stones, etc. must be removed by appropriate divergent cable machines, preferably in the orchard and always before brining. This operation must be carried out carefully to avoid fruit damage.

Transportation and handling of these olives must also follow the recommendations given in Chapter 6. They must be brined as soon as possible, since respiration and transpiration are rapid in these fruits and appreciable weight losses can occur if they are kept in the boxes or bins.

The epiphytic flora carried by fruits can have an important influence on the subsequent fermentation in the brine since it is not destroyed by the lye treatment as occurs with green olives.

8.3 MICROFLORA OF THE FRESH TURNING COLOUR OLIVES

One of the first to study the microorganisms associated with olives was Florenzano *et al.* (1973) who found an abundant population of bacteria, yeasts and moulds. This population included organisms that are not generally favourable for fermentation such as Gram-negative bacteria, *Clostridium* and *Bacillus* spp. as well as the more desirable yeasts and lactic acid bacteria (LAB). In the normal fermentative process, the presence of moulds does not have any significance since they are unable to grow in the environmental conditions prevailing.

Pelagatti (1978–1980a,b) studied the LAB and yeast microflora on the surface of 12 Italian cultivars, identifying 13 pure cultures of lactobacilli and 56 species of yeasts. Among the lactobacilli, *Lactobacillus plantarum* and *Lactobacillus brevis* were isolated from almost all the cultivars. *Lact. delbrueckii, Lact. coryniformis* subsp. *coryniformis* and *Lact. fermentum* were present in only four. *Lact. helveticus, Lact. casei* subsp. *alactosus, Lact. curvatus* and *Lact. hilgardii* were also isolated from two cultivars. Finally, *Lact. leichmannii* and *Lact. acidophilus, Lact. casei* subsp. *rhamnosus* and *Lact. xylosus* were isolated from only one cultivar. The author considered that only *Lact. plantarum, Lact. brevis* and *Lact. fermentum* were actually associated with the olives generally while the others might be considered as associated with specific cultivars and climatic conditions.

A great diversity of yeast species was found, with a noticeable influence of climatic and plant conditions. Among the 56 isolated species, *Candida utilis, Hansenula anomala,* var. *anomala* and var. *schneggii, Kloeckera apiculata, Metschnikowia pulcherrima, Pichia fermentans* and *Rhodotorula glutinis* var. *glutinis* were present in all cultivars. Other species which could also be related to the olive surface were: *Candida guilliermondi, Candida parasilopsis, Candida parasilopsis* var. *intermedia, Debaryomyces hansenii, Pichia membranaefaciens, Saccharomyces bayanus, Saccharomyces cerevisiae* var. *ellipsoideus, Saccharomyces italicus* and *Torulopsis glabrata.*

There was an apparent evolution with maturation since there were species that only appeared in more mature fruits. These were of genus *Pichia* (*P. pinus, P. polymorpha, P. terricola, P. vini* var *vini*), *Saccharomyces* (*S. fermentati, S. globosus, S. kluyveryi, S. oleaceus, S. rosei, S. rouxii, S. uvarum*), and *Torulopsis* (*T. candida, T. holmi, T. magnoliae, T. stellata*). Farris *et al.* (1989) isolated 384 pure yeast cultures belonging to the following species: *Candida boidinii, Candida diddensiae, Candida krusei, Cryptococcus melibiosum, Rhodotorula glutinis, Rhodotorula minuta, Debaryomyces hansenii, Hansenula anomala, Metschinikowia pulcherrima* and *Saccharomyces cerevisiae.* Deiana *et al.* (1992) paid special attention to the identification of LAB present on the olive surface and in the paste prepared from them. The species identified were *Lactobacillus casei* subsp. *casei, Lact. plantarum, Lact. brevis* and *Streptococcus faecium (Enterococcus faecium).*

As will emerge later, some of these species are normally found in fermentation brines. So, the most probable source of inoculation in the spontaneous fermentation of directly brined olives is the fruits themselves.

8.4 WASHING

Due to their direct immersion in brine, turning colour olives must be thoroughly washed to remove dust and other contaminants before brining. In addition, this operation has also been used as a step for removing part of the oleuropein since the partial debittering of these olives is only accomplished by dilution of their bitter principle.

One of the usual methods consists of placing the fruits in the fermentation vessels containing water or a weak brine (2–3% NaCl) which is changed frequently over a period of 2–7 days. Torre *et al.* (1993) followed the effect of these washings in the Spanish cultivar Arbequina and failed to find any debittering, but noticed that the Gram-negative population grew markedly during this operation. In parallel, there was an accumulation of gases produced by fruit respiration and from bacterial activity which increased the volume and caused the liquid surrounding the olives to overflow. The lost liquid must be replaced immediately to prevent shrivelling.

Slow diffusion of the bitter principle into the brine must be expected since the olive skin is intact, presenting an important barrier to osmotic exchange. Untreated fruit cells offer a greater barrier than NaOH-treated fruits. In view of these drawbacks, static washings are not recommended.

However, olive fruits can be washed by any other system. Ideally the method should permit a final contact between fresh water and fruits immediately before their introduction into the fermenters. Thus, counter-current washing machines or showers placed over the conveyors that carry the olives to the fermentation vessels are effective. It is desirable to remove drops of water adhering to the olives since these might be heavily contaminated; this can be achieved with fans.

8.5 BRINING

Well-washed olives are then introduced in the same fermenters used for preparing Spanish-style green olives, cement tanks, wooden vats, etc., according to tradition or available technology. These containers must be one-third to one-half full with water or brine to cushion the fruit's fall. As the fermenter becomes filled the falling distance decreases, as does the risk of damage. If the liquid is tap water or a weak brine then it is removed once the container is full and replaced by the fermentation brine. Containers should be filled with brine within 10–20 minutes to avoid excessive exposure of olives at the bottom to the full weight of the fruits above them.

Brine concentrations usually vary between 5 and 10% according to cultivar and tradition. In this aspect, the operation is less standardized than with Spanish-style olives. More sensitive cultivars should be brined with weaker solutions, otherwise the osmotic pressure may be enough to extract olive juice and give rise to a surface shrivelling that is difficult to remove. In California, Sevillano and Ascolano can be subject to salt shrivel, so it is customary to use a lower concentration of salt (4–5%) if either cultivar is to be sold as Sicilian-type olives. Otherwise, the brine strength will range between 5 and 8% salt because, in the processing of black-ripe olives, shrivelling is reduced to a minimum by puncturing the fruits. This facilitates osmotic exchange between the needled olives and the processing solution (Vaughn, 1982). In Spain, Borbolla y Alcalá *et al.* (1971) used concentrations ranging from 3 to 8% NaCl for Gordal, Manzanilla and Hojiblanca and found no shrivelling.

Diffusion of salt in untreated olives is slow. Drusas *et al.* (1988) gave values ranging from 3.2×10^{-11} to 4.3×10^{-11} m^2/s, which are about one-fifth those obtained for olives pretreated with 1.8% NaOH. This value was independent of initial concentration in the range 7–16% NaCl. Consequently, it may take 3 months or more to achieve equilibrium.

Depending on the kind of fermentation expected, salt additions are necessary to replace the salt absorbed by the fruits. The best system is to dissolve solid salt in a small volume of brine removed from the fermenter. After it is added back, all the liquid in the fermenter is circulated to ensure a homogeneous distribution of the added NaCl as well as vitamins, amino acids and other nutrients.

Fermentable substrates and other organic and inorganic (Mn, Fe, etc.) solutes from the fruit also dissolve into the brine with difficulty. This slow diffusion (in contrast to the rapid process with NaOH-treated olives) is responsible for the rather deficient medium that brines of untreated olives offer microorganisms. As a result, differences in fermentation between both types of olives are evident.

Traditionally, brines used for directly brined olives were just water and salt. However, there is currently a tendency to make a preliminary adjustment to the pH using hydrochloric, acetic or lactic acid. This can help prevent the growth of spoilage organisms such as *Clostridium*, *Bacillus* and Gram-negative bacteria that produce large volumes of gas.

8.6 TRADITIONAL FERMENTATION

The slow diffusion of olive compounds into brine is a constraint to the development of any kind of fermentative process. In turning colour olives this stage may last from 3 to 6 months in contrast to Spanish-style green olives in which fermentation can be complete in only 1 month.

There is considerable production of gas during the first days of fermentation, but there is little foaming and the gas is released to the atmosphere without major losses of brine. This contrasts with the persistent foam seen in the Spanish-style process.

8.6.1 INFLUENCE OF CULTIVAR AND SALT CONCENTRATION

Chemical composition, mainly the polyphenol content, has a marked effect on the fermentation. In general, those cultivars with a high proportion of inhibitors make LAB growth difficult. In the spontaneous process carried out industrially it is common to find very variable behaviour, especially in fermentations carried out in plastic drums. Some containers follow a lactic process while others fail without apparent reason. Borbolla y Alcalá *et al.* (1971) sampled a series of plastic drums which had been packed from the same lot of olives and brine and found that only some followed a lactic fermentation reaching high levels of acidity (Table 8.1). The cause of such differences was unexplained.

Normally, the fermentation is carried out by yeasts and a moderate acidity level is developed, probably due to organic acids from the fruits, Gram-negative bacteria and possibly some fermentative yeasts. Consequently, pH remains rather high. Combined acidity is always low because of the lack of NaOH treatment, remaining around 60 mEq/l. In the case reported by Borbolla y Alcalá *et al.* (1971), the salt level was also low, but this will be discussed in more detail later.

Borbolla y Alcalá *et al.* (1971) compared the fermentations of Gordal, Manzanilla, Hojiblanca and Verdial using different salt levels (Table 8.2). They found that Gordal always followed a lactic process, while with Manzanilla and Hojiblanca, lactic acid production was observed only at the lowest salt concentrations (Table 8.2). However, the authors pointed out that this behaviour could not be considered usual as they had found many drums with similar levels of salt that had not supported lactic acid bacteria (LAB) (Table 8.1). Verdial, on the other hand, always showed a typical fermentation by yeasts and limited acid pro-

Table 8.1 Spontaneous fermentation of turning colour olives in brine, Hojiblanca cultivar. Physicochemical characteristics at the end of the process (Reproduced with permission from Borbolla y Alcalá *et al.* 1971)

Fermenter	Titratable acidity*	pH	NaCl content†	Combined acidity‡
6	0.46	4.35	3.1	–
7	0.46	4.40	3.0	–
8	0.48	4.45	3.2	0.062
9	0.50	4.40	3.2	–
16§	1.00	3.95	3.5	0.060
18	0.46	4.45	3.1	–
22	0.44	4.30	3.7	–
23	0.46	4.50	4.5	0.060
24	0.44	4.45	4.0	–
25§	0.88	4.15	3.5	0.064

* Expressed as g of lactic acid/100 ml of brine.
† Expressed as g NaCl/100 ml brine.
‡ Expressed as Eq/l.
§ Fermenters that followed a lactic process.

Table 8.2 Effect of cultivar and salt content on the fermentation process of turning colour olives measured by the production of acidity (g lactic acid/100 ml brine) (Reproduced with permission from Borbolla y Alcalá, 1971)

NaCl initial content (%, w/v)	Time after brining (days)			
	14	67	90	214
Gordal				
3	0.24	0.50	0.78	1.25
5	0.20	0.54	0.88	1.20
8	0.22	0.40	0.74	1.04
Manzanilla				
3	0.28	0.44	0.62	0.94
5	0.28	0.30	0.38	0.47
8	0.26	0.26	0.36	0.45
Hojiblanca				
3	0.24	0.46	0.66	1.19
5	0.16	0.30	0.36	0.41
8	0.11	0.30	0.46	0.55

duction. In any case, when there is formation of acid, the acidification process takes a long time due to the slow diffusion of fermentable substrates into the brine. This was also observed by Durán Quintana *et al.* (1991) in Gordal, which in the presence of 6% initial NaCl developed final acidities around 1%. The Spanish Aloreña cultivar also readily supports LAB growth. García García *et al.* (1992b) reported the occurrence of a lactic fermentation with either 9 or 5% initial NaCl. Apparently, most of the Californian cultivars are also able to support lactic fermentation to give 0.2–0.7% acidity (Vaughn, 1982). This wide range indicates that while some cultivars were able to develop a typical lactic process (Sevillano is cited as reaching the highest acidities) others or at least some lots of them cannot.

Verdial and Arbequina always produce low acidity (Torre *et al.*, 1993) and do not support growth of *Lact. plantarum*, even in the case of inoculation or with treatments to facilitate its colonization. Hojiblanca is another cultivar in which yeast fermentation normally occurs (Borbolla y Alcalá *et al.*, 1971). Fernández González *et al.* (1992) studied this fermentation with initial salt concentrations of 6 and 3% and found that LAB were absent in both cases.

In some circumstances, when the presence of LAB is detected sporadically (as in Hojiblanca) the initial NaCl content could play an important role to facilitate their implantation. Borbolla y Alcalá *et al.* (1971) reported that in the presence of only 3% initial salt, the lactic process was established in both Hojiblanca and Manzanilla. However, in big fermenters the growth of *Lact. plantarum* occurred only in the total absence of salt (Fernández González *et al.*, 1993). Thus, it seems that a low salt content may help the survival and growth of lactic bacteria in cases where they encounter difficulties in colonizing.

8.6.2 INITIAL pH CORRECTION

It is well established that Gram-negative bacteria, which are less nutritionally fastidious than LAB, can grow from the very beginning of olive brining. They are associated with gas-pocket spoilage in green olives (Fernández Díez *et al.*, 1985) but have not been implicated in the same problem with turning colour or naturally black olives. However, their production of gas, in conjunction with the CO_2 released by fruit respiration can increase the volume of the brines, causing them to overflow. To reduce this and to inhibit the growth of *Bacillus* spp., it is becoming customary to adjust the initial pH of the brine to levels around 4.0–4.5 using hydrochloric, lactic or, most commonly, acetic acid. Its efficacy has been established in the Hojiblanca cultivar (Fernández González *et al.*, 1993), Cacereña (Durán Quintana *et al.*, 1994b), Aloreña (García García *et al.*, 1992a) and Gordal (Durán Quintana *et al.*, 1991). No negative effect of pH adjustment on further development of LAB has been observed. A decrease in the yeast population may be observed with high additions of organic acid, but this does not represent a drawback for the normal process.

8.6.3 MICROBIOLOGICAL ASPECTS OF THE TRADITIONAL FERMENTATION

According to Vaughn (1982), there is a sequence of microorganisms. The initial stage is dominated by organisms other than LAB. After 4 or 5 days, if the brine strength is less than 5% NaCl, LAB begin to appear, *Leuconostoc mesenteroides*, *Pediococcus cerevisiae* and *Lact. plantarum* have been found, but *Leuc. mesenteroides* never dominates the fermentation, as it does during the initial stage of cabbage fermentation. *Lact. plantarum* is always found, often with *P. cerevisiae*. The initial population of organisms, mainly coliform bacteria and bacilli, disappear gradually and in normal fermentations are no longer detectable after 10–14 days. Fermenting yeasts appear in the first 1–2 weeks and persist throughout the fermentation. Some are acid-formers, but their contribution to the total acidity of the olive fermentation may be offset by acid decomposition by oxidative yeasts which may also become established if there is a brine surface in contact with air.

This microbial sequence has been reported by other authors, although the specific species may differ, mainly with respect to the LAB present. Farris *et al.* (1989) found only yeasts in the natural fermentation of Tonda, Maiorca, Bosana, Nera di Gonos, Terza Grande, Olieddu, Semidana and Manna Italian cultivars. The most representative species were: *Saccharomyces cerevisiae* (present in 92% of the samples); *Debaryomyces hansenii, Candida boidinii* (42%), *Candida krusei* (42%) and *Hansenula anomala* (33%). All of them were also present in the fresh olive. Of special relevance are the first two species due to their good fermentative qualities and the production of aromatic compounds.

The lactic microorganisms in directly brined Italian cultivars were studied by Deiana *et al.* (1992) who found the species shown in Table 8.3. Some of them (*Lact. casei* subsp. *casei, Lact. plantarum, Lact. brevis* and *Streptococcus faecium*) were present also in the fresh olives.

Pelagatti (1978–1980a,b) also found in the natural fermentation of Nocellara Etnea and Intesso D'Abruzo an initial aerobic microflora (*Pseudomonas fluorescens, Micrococcus* sp., *Serratia marcescens, Sarcina aurantiaca, Flavobacterium diffusum* and *Achromobacter superficialis*) and some moulds (*Geotrichum* sp., in Nocellara and *Aspergillus niger* and *Penicillium* sp., in Intesso), which were in low counts and diminished with the fermentation time. There were clear differences with respect to the LAB in these fermentations, which were abundant in Nocellara but practically absent in Intesso. There were also differences with the yeasts: *Candida krusei, Pichia fermentans, Torulopsis glabrata, Candida valida* and *Rhodotorula mucilaginosa* were found in Nocellara, while *Pichia membranaefaciens, Torulopsis holmi, Debaryomyces nicotianae* and *Trichosporon* sp. were found in Intesso.

Apparently, the species of yeasts that dominate specific fermentations depend on factors such as the olive growing area and cultivar, although certain species are widely found in all the directly brined turning colour olive fermentations and thus must be well adapted to the utilization of

Table 8.3 Species of lactic acid bacilli identified in directly brined Italian cultivars (Tanda, Maiorca, Bozana, Torza Grande, Nera di Gonnos, Semidana, Manna and Olia Nidda) (Adapted from Deiana *et al.*, 1992)

Species	Frequency in brine (%)
Lactobacillus casei subsp. *casei*	5.4
Lactobacillus casei subsp. *pseudoplantarum*	5.4
Lactobacillus plantarum	42.3
Lactobacillus brevis	5.4
Streptococcus (Enterococcus) faecium	33.8
Leuconostoc paramesenteroides	5.4
Leuconostoc mesenteroides subsp. *cremoris*	2.3

Table 8.4 Species of yeasts isolated from spontaneous fermentations of directly brined olives (Adapted from Marquina *et al.*, 1992)

Candida boidinii
Candida hellenica
Candida oleophila
Candida parasilopsis
Candida rugosa
Candida tenuis
Candida valida
Debaryomyces hansenii
Kluyveromyces lactis
Pichia anomala
Pichia membranaefaciens
Pichia pseudocactophila
Rhodotorula mucilaginosa
Saccharomyces cerevisiae

this specific substrate. Species isolated in Portugal by Marquina *et al.* (1992) included those already mentioned plus others which could be more related to that specific environment (Table 8.4). They did not find any LAB.

8.6.4 PHYSICOCHEMICAL CHANGES DURING FERMENTATION

Acidity and pH

Total acidity and pH depend on the kind of fermentation that takes place. In low polyphenol cultivars such as Gordal, Sevillano, Aloreña, when there is LAB growth, the acidity may reach values of 1% or over. Vaughn (1982) considers that in Sevillano olives held for a minimum of 30 days the lactic acid content is about 0.1%, but if they are stored for longer periods and oxidative yeasts and moulds are controlled, the total acidity may be as high as 0.6%. Other examples of acidity levels in lactic processes are shown in Tables 8.1 and 8.2. However, in the absence of LAB, the acidity reaches a maximum of 0.5% during storage for 6 months or longer.

The pH value is always lower than 4.0 in the case of LAB growth and around 4.4 when only yeasts are present (Table 8.1). The strong pH decrease with increasing acidity is due to the low combined acidity of these olives (which have not been lye-treated) and their respective brines.

Fermentable substrates

Total fermentable substrates in these olives may be 2.5% or higher, according to the cultivar. Their distribution can be seen in Table 8.5. As in

Table 8.5 Proportion of the main fermentable substrates in the flesh of turning colour olives. Hojiblanca cultivar. (Reproduced with permission from Garrido Fernández *et al.*, 1993)

Substrate	Percentage of total
Glucose	71.5
Fructose	9.2
Sucrose	2.5
Mannitol	11.6
Malic acid	5.0

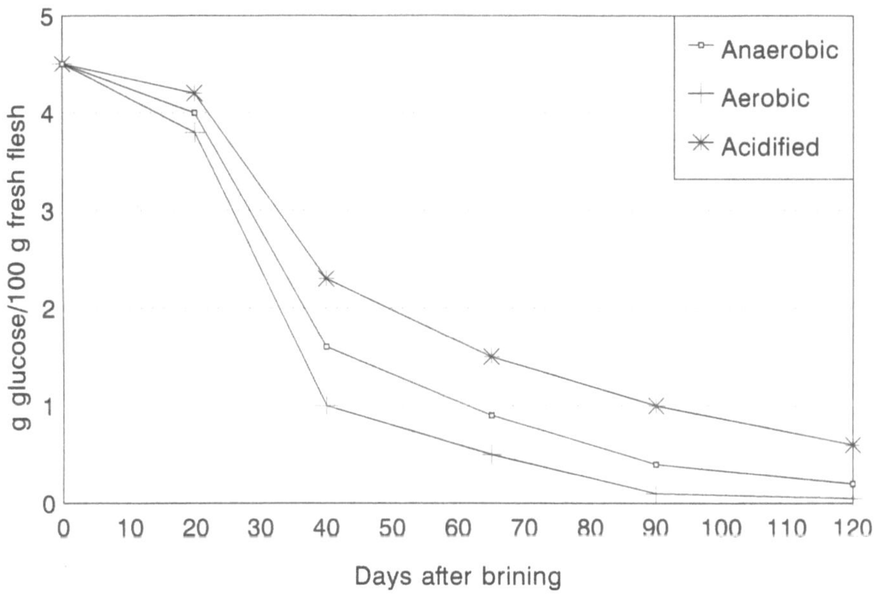

Fig. 8.1 Changes in the reducing sugar content (expressed as glucose) in flesh throughout storage in anaerobic (□), aerobic (–) and highly acidified (×) medium. Gordal cultivar. Initial NaCl, 6% (w/v).The initial pH was corrected to 4.2 in anaerobic and aerobic media. Concentration of acetic acid was 1.5% in acidified medium. (Redrawn from Durán Quintana *et al.*, 1991.)

the case of green Spanish-style olives, the main fermentable substrate is glucose, followed by mannitol (11–12%), malic acid (4–6%) and sucrose (2–3%). Their diffusion into the brine is very slow and may continue for six months or more. Mannitol is the slowest to diffuse due to is practically negligible utilization in yeast-only fermentations and consequent accumulation in the brine. Figure 8.1 shows glucose diffusion from Gordal cultivar prepared using three different fermentation systems.

Fermentable substrates are normally scarce in the brine throughout fermentation and accumulate only in the case of dormant (or stuck) fermentations. Glucose, fructose and sucrose are easily used, but mannitol and malic acid may reach relatively high concentrations. This means that the usual yeast microflora is unable to use these compounds. However, mannitol is metabolized when LAB are present (Garrido Fernández *et al.*, 1993).

Phenolic compounds

The phenolic content of turning colour olives is slightly lower than green olives as they decrease during maturation. The main component is oleuropein, which may reach 0.5–1.0 g/kg, depending on cultivar and maturation stage. Other phenols occur in markedly lower concentrations. Hydroxytyrosol level ranges from 70–350 mg/kg, tyrosol 0.2–0.5 mg/kg, verbascoside 0.10–1.30 mg/kg and luteolin 7-glucoside 0.4–0.6 mg/kg. Differences among cultivars are of interest both for a varietal classification and for the detection of fraud.

Brining causes a drastic drop in oleuropein content and a parallel hydroxytyrosol increase due to its formation by acidic hydrolysis of the glucoside. On the other hand, verbascoside is more stable and a greater proportion of it remains in the olive, even at the end of the storage period. Oleuropein diffuses into the brine at a relatively high rate (apparently higher than fermentable substrates). There is an initial accumulation followed by a period of continuous decrease due to hydrolysis by the acidic medium or the microbial action. In general, the brine oleuropein content is always lower than that found to be inhibitory for *Lact. plantarum* (Ruíz Barba *et al.*, 1991).

There is a progressive enrichment of the brines in hydroxytyrosol, mostly from oleuropein hydrolysis. Its final concentration may reach 10 mmol/l and the inhibitory effect of brines of directly brined turning colour olives must be mainly due to this rather than oleuropein (Ruíz Barba *et al.*, 1993). Other phenols are slowly diffused into the brine but their low concentrations suggest they play little, if any, role in the fermentation process.

Differences in the ability to support LAB growth among cultivars may be explained by differences in oleuropein (and its derivative hydroxytyrosol) content and, to a lesser extent, by the presence of verbascoside.

It has been shown that the diffusion of phenols and the presence of hydroxytyrosol in brines increase markedly with temperature (Brenes Balbuena *et al.*, 1993a). Thus, heating the fermentation of turning colour olive will have a negative effect on the process. Final concentrations are similar to those obtained at room temperatures (20°C) and the hydrolysis rate seems to be more dependent on concentration than on temperature.

The effect of the initial salt concentration on the polyphenol content in brine is slight (Brenes Balbuena *et al.*, 1993a). Its main influence on the type of fermentation must be mainly related to its inhibitory action on LAB.

Metabolic end-products

The main metabolic products in fermented turning colour olives are acetic acid, which may reach 0.4%, ethanol (up to 120 mmol/l) and methanol (about 15 mmol/l). They accumulate during the first 2 months and then stabilize or decline slightly due to their evaporation. In the event of LAB growth, lactic acid is also present, rising rapidly at the start and then more slowly during the rest of the fermentation. When surface growth of oxidative yeasts and moulds occurs, at least part of the acid is used and the lactic acid stabilizes or even diminishes.

8.7 MODIFICATIONS TO THE TRADITIONAL FERMENTATION

The traditional fermentation method for colour turning olives has some drawbacks: (i) the uncertain development of LAB in the case of high polyphenol cultivars; (ii) the difficulties in reaching physicochemical characteristics for safe product storage; and (iii) the accumulation of excess CO_2 in both olives and brines that may cause overflow in fermenters as well as irreversible changes in the olive surface when the fruits are removed from the fermenters and the gas is released.

These problems could be avoided by using a lactic fermentation with all cultivars, similar to that possible in Gordal (Durán Quintana *et al.*, 1991) and Aloreña (García García *et al.*, 1992b), and by removing most of the CO_2 produced during the process.

8.7.1 FERMENTATION IN AEROBIC CONDITIONS

The aerobic fermentation process was developed initially at Instituto de la Grasa (Sevilla) by Garrido Fernández *et al.* (1985) to solve gas-pocket spoilage of naturally black olives. However, since this spoilage also occurs in turning colour olives, it has also been applied here. A detailed description is given in the chapter on naturally black olives in brine, so only those aspects directly related to turning colour olives will be mentioned here.

A complete and detailed account of the use of this method with Gordal, Manzanilla and Hojiblanca can be found in Durán Quintana *et al.* (1991, 1994b), Garrido Fernández *et al.* (1986) and Fernández González *et al.* (1992). In essence, the application of this method to turning colour olives consists of the introduction of air at a rate of 0.1–0.5 l/h per litre of fermenter capacity through a column introduced into the conventional 10-tonne vessels, cement tanks or other fermentation vessel. The process is maintained under these conditions for at least 3 months. Aeration is then stopped or sporadically applied when accumulation of CO_2 in the brines is evident through an increase of brine volume or gas evolution.

Physicochemical changes

Aerobic fermentation substantially diminishes the CO_2 dissolved in the medium. In the traditional process CO_2 levels could reach 125–150 mg/ 100 ml brine, while aerobically the concentration is maintained below 30 mg/100 ml, except during the first days when production by fruit respiration coincides with microbial activity. At this stage, levels may reach about 40–60 mg CO_2/100 ml brine, but soon decrease to the equilibrium content after 1–2 weeks (Figure 8.2).

Aerobic fermentation also accelerates diffusion of fermentable substrate into the brine (see Figure 8.1) and salt penetration into the flesh. Thus, equilibrium is established more rapidly than in the traditional process.

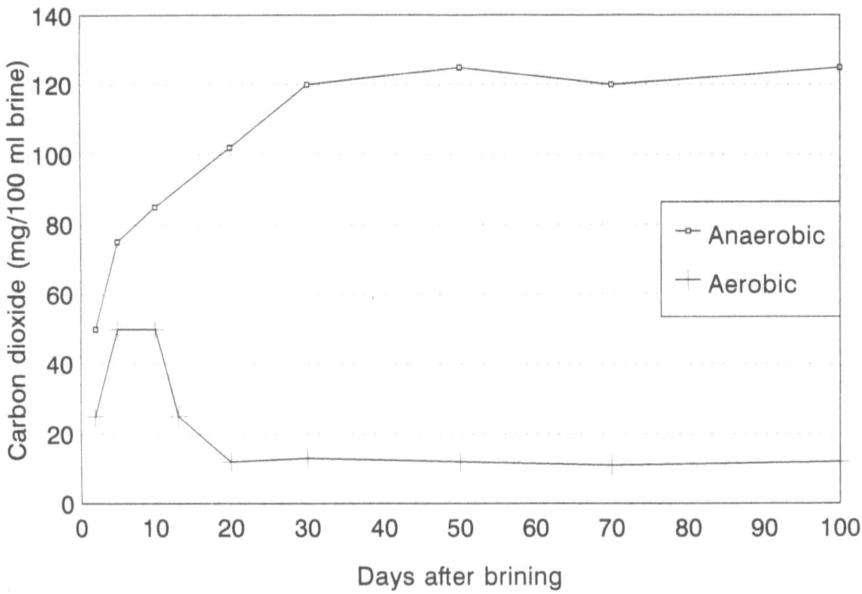

Fig. 8.2 Changes in the carbon dioxide content during anaerobic (□) and aerobic (+) storage in brine of Gordal cultivar. Initial conditions were similar to that of Figure 8. 1. (Redrawn from Durán Quintana *et al.*, 1991.)

The level of fermentable substrates in the brine usually increases only slightly compared with anaerobic conditions, but differences are appreciable during the first 20 days. However, in cold conditions, when the salt content is high and the brines are initially acidified, there is some accumulation of fermentable substrates, reaching a maximum around 50–70 days. But then, and possibly when levels of other nutrients are high enough and temperatures appropriate, it is possible to overcome the combined inhibitory effect of salt, acetic acid and polyphenols and the fermentation sugars disappear rapidly from the brine. However, it is interesting to note that this accumulation only happens in those cases where there is no LAB growth.

The most readily fermented substrates are glucose, fructose and sucrose; the last two disappear from the flesh after 30–50 days and are rarely found in brine. Mannitol is slowly utilized in the case of yeast-only fermentations, so relatively high amounts can be found in these brines (Garrido Fernández *et al.*, 1993). In LAB fermentations its disappearance parallels the growth of these microorganisms. The use of mannitol as a carbon source by *Lact. plantarum* when electron acceptors such as acetic and malic acids are present, has been demonstrated by McFeeters and Chen (1986). Aerobic conditions enhance the ability of LAB to use mannitol and permit its utilization by facultative yeasts. However, its accumulation in anaerobic conditions where only fermentative yeasts are present indicates that they cannot use it anaerobically, as occurs also with cucumber fermentation (Etchells *et al.*, 1975).

Finally, malic acid usually decreases at a rate intermediate between that for glucose and mannitol. Some accumulation does occur, unless LAB are present when it is completely exhausted in brines regardless of whether conditions are anaerobic or aerobic.

Generally, in anaerobic fermentations of turning colour olives relatively high amounts of residual fermentable substrates may be found, while in those carried out aerobically or with LAB growth, they are practically exhausted.

Aerobiosis only increases the pH slightly in relation to the anaerobic system (Figure 8.3). The increase can be of approximately 0.5 units, although an excess of air can increase the difference markedly and lead to unacceptable pH levels, which must be avoided.

Where LAB growth occurs, the lack of buffering capacity in the brine means that the pH reaches very low values (about 3.6) even in aerobic conditions. Consequently, aerobiosis does not interfere with achieving appropriate physicochemical conditions for prolonged safe storage of the olives.

Contrary to what might be expected, the use of air during the process leads to a relatively low oxygen content in brines (Durán Quintana *et al.*, 1991), especially during the first 60–80 days. This must be due to a higher

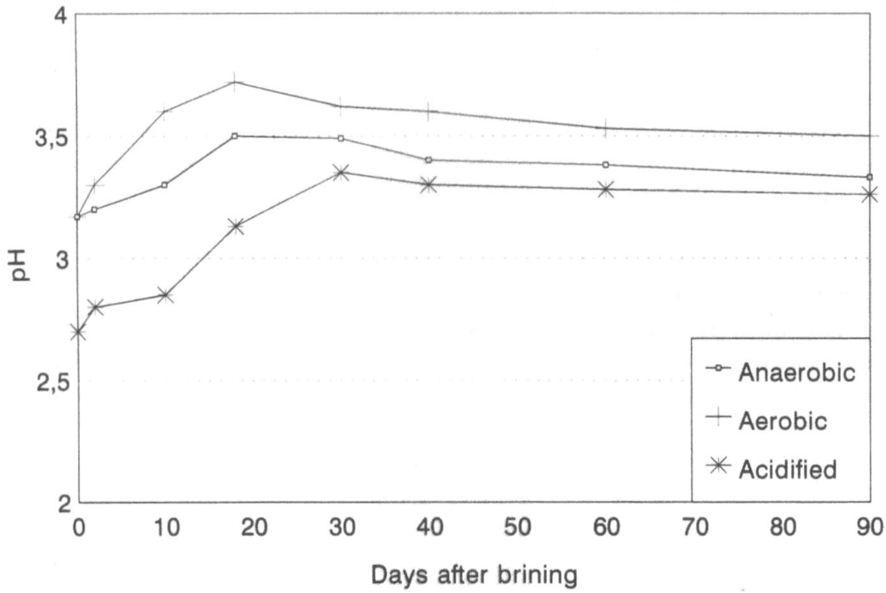

Fig. 8.3 Changes in pH during the anaerobic (□), aerobic (–) and acidified (×) storage in brine of Gordal cultivar. Initial conditions were similar to those of Figure 8.1. (Redrawn from Durán Quintana *et al.*, 1991.)

consumption rate by the induced facultative microflora that predominates in aerated fermenters compared with the fermenting organisms present in the anaerobic process.

The higher rate of brine circulation in aerobic fermentation gives rise to higher levels of both lactic and acetic acid in the olive flesh. With an induced lactic process there is also substantial production of acetic acid, possibly due to a heterofermentative lactic process or the conversion of lactate to acetate in the presence of citrate. Lactic acid is only formed in the lactic process.

Accumulation of the volatiles, ethanol and methanol, occurs only during the first 50 days when the fermentation is more active. After this period, in the case of aerobiosis they are progressively eliminated throughout the rest of the fermentation process (Garrido Fernández *et al.*, 1993).

The aerobic process, when used at the recommended aeration rate, does not interfere with LAB growth in those cultivars to which they are suited. (Durán Quintana *et al.*, 1991). It has also been confirmed that this system, when properly applied, has no negative effect on texture. Since aerobically processed olives do show a slightly brownish surface colour, however, its utilization with olives in which a light colour must be retained is not advisable.

Microbiological aspects

The presence of air in the brine induces the growth of a specific microflora. Fernández González *et al.* (1993) and Durán Quintana *et al.* (1994a) have studied the main species of yeasts found in the aerobic fermentation of turning colour Hojiblanca olives, and found that they were mainly composed of *Pichia membranaeficiens* (44% of the pure strains isolated), *Pichia vini* (10.5%), *Pichia fermentans* (9%) and *Hansenula polymorpha* (9%). Other species found are shown in Table 8.6. As in the case of the traditional process, Cacereña cultivar showed a slightly different yeast population with the main species isolated being *Pichia minuta* (47%), *Pichia fermentans* (21%), *Pichia membranaefaciens* (17%) and *Rhodotorula* spp. (9%). The same species of yeasts were found regardless of the presence or absence of LAB, although the total counts were always lower when LAB were present. Figure 8.4 shows typical profiles of yeast growth in the three most common olive storage systems.

LAB growth was very reduced in the case of Aloreña. LAB isolated belonged predominantly to the genus *Pediococcus* (García García *et al.*, 1992b). In Hojiblanca, *P. inopinatus* was the only *Pediococcus* species identified (Fernández Gonzalez *et al.*, 1993). *Lact. plantarum* was the only lactobacillus in both cultivars. Obviously LAB are only present in those

Table 8.6 Species of yeasts identified in the fermentation of directly brined olives in the presence of air. Hojiblanca cultivar (Adapted from Fernández González *et al.*, 1992)

Species	Frequency (%)*
Pichia membranaefaciens	44.0
Pichia vini	10.5
Pichia fermentans	9.0
Hansenula polymorpha	9.0
Pichia etchelsii	3.0
Hansenula dimenae	3.0
Debaryomyces hansenii	3.0
Pichia media	3.0
Saccharomyces oleaceus	1.0
Saccharomyces ovarum	1.0
Torulopsis candida	1.0
Candida lambica	1.0
Saccharomyces oleaginosus†	10.5

* With respect to the total number of isolated pure cultures.
† Isolated only from the flesh of the stored olives.

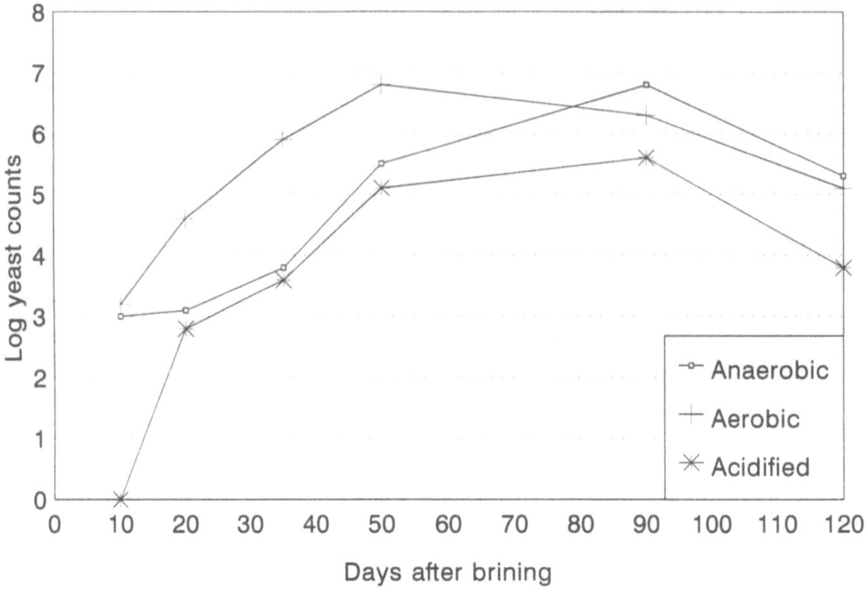

Fig. 8.4 Changes in the population of yeasts during the anaerobic (□), aerobic (+) and acidified (×) storage in brine of Gordal cultivar. Initial conditions were similar to that of Figure 8.1. (Redrawn from Durán Quintana *et al.*, 1991.)

cultivars which normally support their growth and, with Hojiblanca, Cacereña and other difficult cultivars, where a starter culture is used or conditions are suitable for inducing a spontaneous fermentation. Figure 8.5 shows a typical LAB profile in the three most common olive brine storage systems.

In general, *Lact. plantarum* is the first species to appear, 2–5 days after brining, while cocci are detected later, between 7 and 30 days after brining. This sequence is different from the behaviour of these species in the fermentation of other vegetables and in green Spanish-style olives. Normally, cocci initiate the fermentation and disappear when the pH is low and acidity high. *Lact. plantarum* generally finishes the process since it has higher acid resistance due to its low growth limiting internal pH and the ability of this genus to maintain a pH gradient with respect to the surrounding medium (McDonald *et al.*, 1990).

Maximum growth of *Lact. plantarum* is normally reached between 30 and 60 days after brining. In the presence of *Pediococcus* this was delayed to between 80 and 100 days, after which numbers diminished faster than those of *Pediococcus* which thus outnumbered *Lact. plantarum* from about the 70th day to up the end of the process.

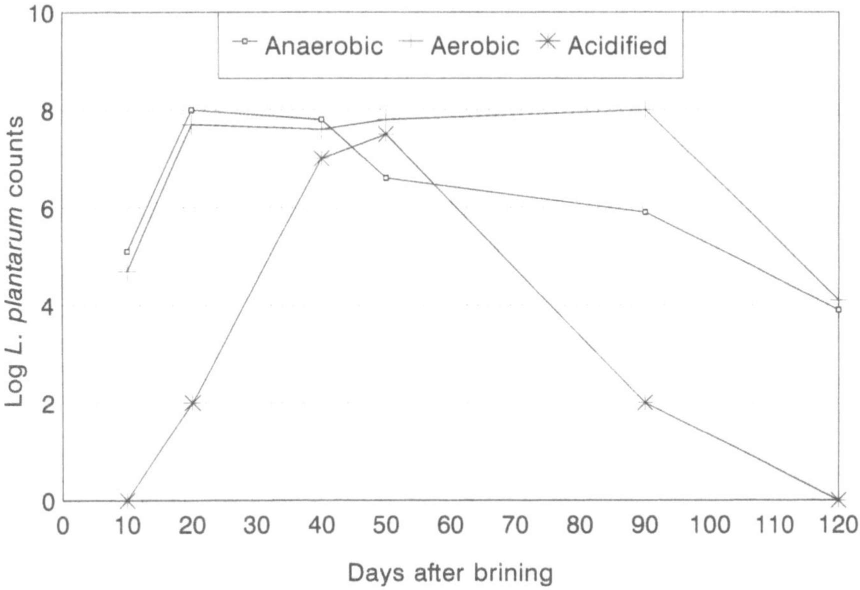

Fig. 8.5 Changes in the population of *Lact. plantarum* during the anaerobic (□), aerobic (–) and acidified (×) storage in brine of Gordal cultivar. Initial conditions were similar to that of Figure 8.1. (Redrawn from Durán Quintana *et al.*, 1991.)

This behaviour in turning colour olives was unexpected and remains unexplained. It might be related to certain characteristics of the fermentation such as the presence of air (despite the actually lower oxygen content of aerobic brines), greater availability of vitamins and amino acids at the end of the process, better adaptability of *Pediococcus* to the prevailing environmental conditions, relative resistance to the inhibitors present in the brine which change from being mainly oleuropein at the beginning to hydroxytyrosol from the 50–70th day.

8.7.2 INDUCTION OF A LACTIC FERMENTATION PROCESS

As is evident from previous paragraphs, a lactic fermentation in turning colour olives may or may not occur depending on cultivar and brining conditions. From the point of view of debittering, lactic fermentation will have little significance since the elimination of oleuropein in this case is only due to the diffusion into the brine. However, lactic fermentation is more appropriate for obtaining a stable product for storage. A lactic process is possible in the case of Gordal (Durán Quintana *et al.*, 1991) and Aloreña (García García *et al.*, 1992b), but fails frequently in Manzanilla

(Borbolla y Alcalá *et al.*, 1971), Hojiblanca (Borbolla y Alcalá *et al.*, 1971; Fernández Gonzalez *et al.*, 1992, 1993). The development of a reliable lactic process for those last cases would be of some advantage. It would be of undoubted help for preventing malodorous and softening spoilage by other microorganisms and would provide adequate acidity and pH to guarantee safe storage. To overcome difficulties with LAB growth the inhibitory effect of the polyphenols must be decreased, nutrients must be more rapidly available and physicochemical characteristics such as salt and initial pH must be more suitable for LAB colonization.

Etchells *et al.* (1966) showed that heated lye and washed Manzanilla cultivar olives fermented rapidly. Borbolla y Alcalá *et al.* (1971) applied the same procedure to untreated olives (74°C for 3 min) and obtained a good spontaneous lactic fermentation with a final acidity of about 1.3% (w/v). However, the procedure was never used industrially due to the risk of softening and gas-pocket spoilage from the heat treatment. Lower temperatures did not give the same results. Furthermore, some cultivars (Arbequina) did not support LAB growth even when this treatment was applied (Torre *et al.*, 1993).

The aerobic process in naturally ripe olives shows more rapid fermentable substrate diffusion. Its utilization in turning olive fermentations with cultivars which support LAB growth did not show any negative effect on *Lact. plantarum* growth or acid production (Durán Quintana *et al.*, 1991) and efficiently removed the CO_2 produced in the process, thus helping eliminate shrivelling. Thus, its utilization in enhancing the lactic process has been widely used.

The salt content had been a decisive factor in promoting spontaneous lactic processes in difficult cultivars (Manzanilla and Hojiblanca) according to Borbolla y Alcalá *et al.* (1971). These authors found that only with 3% NaCl was LAB growth possible. In naturally black olives salt concentrations higher than 8% were inhibitory for LAB (Gonzalez Cancho *et al.*, 1975). This is due to the synergistic action of NaCl with other natural inhibitors present, since all of the LAB isolated from directly brined olives by Balatsouras (1985) were able to grow at concentrations below 5% NaCl and most at 7%. In these fermentations, the salt concentration used has been always higher than that suggested for green Spanish-style olives and has ranged from 6 to 8% NaCl. Bearing in mind that even 6% could be inhibitory when used with Hojiblanca cultivar, successful trials to enhance fermentability have been carried out lowering the salt content to below this level (Fernández Gonzalez *et al.*, 1992).

The other factor which can be modified is the initial pH. Its use in controlling the undesirable microflora initially present has been widely observed (Durán Quintana *et al.*, 1991; Fernández Gonzalez *et al.*, 1992), but presence of acid could also negatively affect LAB development in the extreme conditions of turning olive fermentations.

Systematic studies using combinations of these factors have determined appropriate conditions for LAB development in different cultivars. Spontaneous growth of LAB in Hojiblanca cultivar has been obtained when just tap water, pH-corrected with 0.3% of acetic acid, was added and aerobic conditions used. In this case, *Lact. plantarum* colonized the solution 5 days after fermentation and *Pediococcus* was also detected after 30 days. Maximum numbers were reached at the 100–120th and 70–80th day, respectively. After 70–80 days, the population of *Pediococcus* was higher than that of *Lactobacillus* (Fernández Gonzalez *et al.*, 1993). In recent experiments, the use of inoculum 5 days after brining with Hojiblanca and at the first day with Cacereña cultivar enhanced LAB growth. There was substantial survival of *Lact. plantarum* in brines of up to 3% NaCl and pH corrected with acetic acid. However, the organism only persisted throughout the whole fermentation and storage time in Hojiblanca cultivar olives initially immersed in just tap water. In other cases there was a progressive diminution of the *Lact. plantarum* counts which disappeared after the 30th and 40th days in 3% NaCl with Cacereña and Hojiblanca respectively. In Cacereña, treatments in water gave longer *Lact. plantarum* survival up to the 80th day. In both cases, but especially in Hojiblanca, the acidity produced was enough to guarantee good product preservation during the rest of the olive storage (Durán Quintana *et al.*, 1994b). The greater difficulties found with Cacereña can be explained by its higher phenols (mainly oleuropein) content and, later, the hydroxytyrosol produced by the hydrolysis of oleuropein.

The addition of a starter culture can thus enhance LAB colonization even in the presence of 0.1–0.3% acetic acid and 3% NaCl – conditions necessary to avoid excessive growth of spoilage microorganisms in the initial stages. The time of inoculation seems to be critical. A detailed study of this factor in zero and 3% NaCl has been carried out by Durán Quintana *et al.* (1993). They found that with inoculation after 24 hours, *Lact. plantarum* survived in just tap water in the case of Cacereña. However, when the starter culture was added after the third day, survival in tap water was observed in both cultivars, although it was always lower in Cacereña. Similar results were obtained with inoculations up to 9 days after brining.

In the presence of salt, survival was observed in inoculations after 5 days (very limited) and after 7 days. The higher inhibitory effect of Cacereña was stronger in the presence of salt. Addition of glucose to the brine always slightly improved survival.

Thus, this experiment showed that to improve LAB colonization, a delay of at least 7 days is preferable before *Lact. plantarum* inoculation. Initial salt content in the brine must be as low as possible.

Deiana *et al.* (1992) have also studied the use of a mixture cultures of LAB alone and with *Saccharomyces cerevisiae* as starters for preparing directly brined olives. They selected *Lact. plantarum* and *Streptococcus (Enterococcus) faecium* because they were the most abundant species in both fresh olives and in the subsequent immersion brines. The addition of *S. cerevisiae* was suggested because of its physiological characteristics and its capacity for producing aromatic compounds. Use of the different starter cultures in anaerobic conditions and with olives previously washed for 4 days and immersed in a 8% NaCl brine, increased the LAB population which showed a tendency to diminish with time. The authors also observed an increase in *Lact. plantarum* growth due to the presence of *Strep. faecium* and more rapid formation of lactic acid, mainly of L(+) lactic acid. However, the presence of *S. cerevisiae* gave rise to lower lactic acid levels (and higher pH) which could be a serious drawback to obtaining appropriate final physicochemical storage conditions. On the other hand, the presence of the yeast slightly improved the texture and gave the product a fruity taste. The procedure gives appreciable amounts of acetic acid.

8.7.3 MICROBIAL ENHANCEMENT OF DEBITTERING

Yeasts are normal microorganisms present in directly brined olives and have not been shown to be susceptible to inhibition by olive polyphenols. Some of them may use phenols as a carbon source. Materassi *et al.* (1975) tested a series of 104 strains from 19 genera and 37 species from the collection of the 'Istituto di Microbiologia Agraria e Tecnica di Fiorenze' (Italy), finding that those shown in Table 8.7 had β-glucosidase activity and produced oleuropein hydrolysis.

Marquina *et al.* (1992) tested strains isolated from Portuguese olives in brine for their ability to grow in oleuropein as carbon source, finding that the five shown in Table 8.8 gave positive results. Some of them also had killer activity.

Table 8.7 Species of yeasts able to produce oleuropein hydrolysis in solid and liquid media (Adapted from Materassi *et al.*, 1975)

Candida guilliermondii
Candida parasilopsis
Candida utilis
Candida veronae
Cryptococcus albidus
Hansenula anomala
Lipomyces starkeyi
Rhodotorula glutinis
Trigonopsis variabilis

Table 8.8 Species of yeasts able to grow using oleuropein as carbohydrate substrate (Adapted from Marquina *et al.*, 1992)

Candida hellenica
Candida oleophila
Debaryomyces hansenii
Pichia anomala
Rhodotorula mucilaginosa

Recently, Ciafardini *et al.* (1994) have reported the isolation and characterization of a species of *Lact. plantarum* with oleuropeinolytic activity. The β-glucosidase produced by those strains was active on oleuropein and 7 days after inoculation the authors could detect oleuropein hydrolysis products: aglycone and β-3–4-dihydroxyphenyl ethanol (no elenolic acid was detected). With time the effect increased. The production of β-glucosidase decreased sharply (40–50%) with increased glucose concentration, even with 2% glucose. However, enzyme activity would rarely be inhibited by this in real fermentation brines due to the low glucose content and its use by the bacterium. In any event, activity was noticed mainly at the end of the fermentation when other fermentable substrates are exhausted.

Some suggestions about the commercial use of yeasts (because of their constant presence in the directly brined olive fermentations) have been reported. Balloni *et al.* (1977) studied the utilization of the oleuropein-splitting yeast *Candida veronae*. Its action was used to produce a microbiological debittering of table olives which might replace the chemical debittering by NaOH and thus prepare olives by an entirely microbiological process. The procedure consists of inoculation of brined olives with the yeast, which destroys the oleuropein, followed by a subsequent inoculation with *Lact. plantarum* to produce the necessary acidity. The new process was denominated 'fiorentino-pescarese' since the two Institutes involved in the work were situated in Fiorenze and Pescara.

Plain olives are, however, slowly debittered because of the slow diffusion of oleuropein. A more rapid transformation can be obtained with picked and pitted olives, although the process can hardly be applied in fresh olives because of its negative influence on colour and texture. From the nutritive point of view, the amino acid content of olives processed by this method was similar to that of fresh fruits, while NaOH-treated olives had lower concentrations of most nutrients.

The method was also applied by Pelagatti (1978–1980a,b) to Nocellara etnea and Intosso d'Abruzzo Italian cultivars, with similar results.

8.8 SPOILAGE OF TURNING COLOUR OLIVES DURING FERMENTATION IN BRINE

This type of olive can potentially suffer the same malodorous fermentations already noted in green Spanish-style olive and its control must be accomplished by the same methods described in Chapter 7. Their incidence in turning colour olives is, however, lower due to the limited nutrient availability in these brines.

Other types of spoilage are more specific to this type of olives or have clear differences with respect to green or naturally black olives. Most of them are due to the presence of undesirable microorganisms.

8.8.1 SHRIVELLING

This spoilage is frequent in turning colour olives, especially in less mature fruits. It is produced either by the use of high salt concentrations (when it is apparent from the very beginning of brining), or by an excessive gas accumulation in the interior of fruits. In this case, the problem is observed when olives are removed from the fermenters for grading and sorting and is caused by the rapid release of the gases. The deformations are difficult to repair even when fruits are repeatedly treated with NaOH (black-ripe olives). The problem can be avoided using appropriate initial salt concentrations, adding limited amounts in further corrections, circulating the brine carefully after each addition to obtain a homogeneous solution, and removing the CO_2 produced during storage. The application, when possible, of the aerobic processing system can efficiently achieve the last two effects and it is particularly good at maintaining the CO_2 concentration at low levels throughout the whole storage period.

8.8.2 GASSY, 'FLOATER' OR 'FISH-EYE' SPOILAGE

This problem is characterized in turning colour olives by the development of blisters caused by the accumulation of gases. These can cause separation of the olive skin from the flesh and the formation of fissures or gas-pockets which may extend to the fruit pits. Slightly pink olives are more prone to this than those with green or green yellow tones.

Apparently, the same bacteria associated with the formation of gas-pockets in green Spanish-style olives can also produce the spoilage in these olives. In fact, all the species of coliform bacteria, except *Escherichia coli*, have been implicated in gassy spoilage by Foda and Vaughn (1950).

Other species related to gassy deterioration of this type of olives were *Bacillus polymyxa*, *Bacillus macerans* and *Aeromonas liquefaciens*. Saccharolytic species of the anaerobes of the genus *Clostridium* can also cause a violent gassy fermentation (Vaughn, 1982). All of them can simultaneously cause softening and the clostridia are involved in butyric fermentations and zapatería. However, these microorganisms are rarely found in olive brines if adequate control of NaCl and pH is achieved.

Of greater concern is the association of yeasts of the genera *Saccharomyces* and *Hansenula* with gassy fermentations. These yeasts caused typical gas blisters rather than fissures. Two of the species, *S. kluyveromyces* and *S. oleaginosus* also produced softening, whereas *Hansenula* species were not pectinolytic. Since yeasts are normally present throughout the natural spontaneous fermentation of turning colour olives, their control is more difficult. In fruits destined for Californian ripe olives alternative procedures are available (salt free-storage, application of the aerobic system, etc.). However, in the case of other final uses, the only method to control excessive growth of these undesirable yeasts is initial pH correction with acetic acid and the induction of a lactic process, which, in general, diminishes the yeast population.

8.8.3 SOFTENING

Apart from the problems caused by physicochemical treatments, softening in turning colour olives is usually caused by the activity of pectolytic microorganisms. One of the most common types of spoilage is soft stem end, which is produced by a variety of bacteria and yeasts that eventually enter the olive by the unprotected olive flesh remaining when fruits are destemmed. Their pectolytic activity softens the stem area and causes a puckering of the stem end. The attack continues until the entire flesh is affected. Fruits then become soft and often mushy. The bacteria known to be involved include species of *Bacillus* and *Clostridium* and species of Gram-negative bacteria mentioned in Chapter 7. Moulds responsible for stem end softening include *Aspergillus*, *Fusarium*, *Geotrichum*, *Paecilomyces* and *Penicillium* (Balatsouras and Vaughn, 1958) and oxidative yeasts of the genus *Rhodotorula* (Vaughn *et al.*, 1969a) as well as fermentative yeasts of the genus *Saccharomyces* (Vaughn *et al.*, 1972).

Since the bacterial pectolytic enzymes are active only at pH 6.0–8.0, they can only affect turning colour olives in cases of abnormal fermentation. The problem can thus be avoided by initial brine acidification or by lactic acid fermentation. The fungal and yeast enzymes, however, are active at acidic pH and the best control for them is providing conditions that prevent growth of the corresponding microorganisms, as already noted.

Additionally, species of *Fusarium* (Balatsouras and Vaughn, 1958) were also found to be cellulolytic and are thus able to attack the olive skin. Sometimes, the effect is only observed after further treatments, e.g. the darkening process of ripe-olives.

8.9 FINAL PRODUCTS

As noted in the Introduction, fresh or stored turning colour olives can be sweetened by dilution of the bitter principle (untreated turning colour olives) or by an alkaline treatment (treated turning colour olives). Previously stored olives are used as raw material, although sometimes fresh olives may be used, in which case the operations described up to now are omitted.

8.9.1 UNTREATED TURNING COLOUR OLIVES

Turning colour olives without any lye treatment can be marketed in brine. A great proportion of them are, however, used as raw material for the preparation of 'Seasoned or Aromatized Olives' (Unidad Estructural de Biotecnología de Alimentos, 1990) which are prepared differently depending on regional customs.

Untreated turning colour olives in brine

Preparation of this final product is very simple. It consists of sorting, to remove shrivelled and damaged fruits, and grading. After washing, they are packed mainly in plastic bags or PVC containers. Other materials are not used because the product does not command high prices in the markets.

A minimum equilibrium salt concentration of 6% is specified by the International Olive Oil Council Standard. Although a maximum pH limit is not indicated for this product, values of 4.0 or below and a minimum salt content of 4% must be recommended for non-hermetically sealed containers. Both lactic and acetic acids are used to adjust the pH. To reduce browning, it is customary to add ascorbic and citric acids. In the case of pasteurized products, the maximum pH limit is 4.3 for salt concentrations as low as 2%, and 4.0 if the sodium chloride content is less than 2%.

Seasoned or aromatized olives

This term applies to olives that have undergone a special treatment, involving the addition of aromatic plant substances and possibly wine vinegar to the covering liquid. They are known in Spain as 'aceitunas aliñadas'.

Olives used for this can be subjected to the debittering treatments immediately after harvesting but, more generally, they are placed in a 6–8% brine in which they undergo a partial or total fermentation. As has already been established, a mixture of yeasts and sometimes LAB are generally responsible for the fermentation process (Borbolla y Alcalá *et al.*, 1971). This second procedure is most frequently used and permits storage of the olives for a long time until they have to be seasoned and packed.

Before seasoning and packing, fresh or brine-stored turning colour olives are washed with water, two or three times, changing the water every 10–12 hours, to eliminate residual fermentable material and as much as possible residual bitterness which could be present. The fruits are then conditioned in a brine containing the aromatic products and acids.

As indicated above, the preparation of seasoned olives varies depending on regional consumer preferences, but there are a certain number of aromatic substances (Fernández Díez *et al.*, 1985) which are common to most of them: garlic, thyme, bitter orange, lemon, green pepper, paprika, oregano, bayleaf, fennel, cumin, hot chilli pepper, sweet pepper, clove, coriander, and black pepper. At least three of these substances are employed together. Among the aromatic mixtures which are more popular are:

- garlic, green pepper, cumin and oregano
- garlic, thyme and fennel
- garlic, oregano, paprika and lemon peel
- garlic, oregano, paprika, thyme, cumin and coriander

Frequently, the aromatic natural products are substituted either totally or partially by extracts, essential oils or oil-resins which have the advantage of giving a more uniform flavour and having less risk of deterioration. However, in these cases a little of the natural substances is generally added to make it more attractive to the consumer. Wine vinegar (sometimes with lactic acid) is added to adjust the pH value and free acidity to improve the keeping quality of the packed product. When the product is not pasteurized, a mixture of sorbic and benzoic acids, 500 and 1000 p.p.m. respectively, is added to the covering brine. The shelf-life of these preparations is not long, never more than 2–3 months, mainly due to the enzymatic action of some of the aromatic substances used as natural products.

Pasteurization is the only method that completely stabilizes these preparations, destroying any vegetative microorganisms which may be present and inactivating enzymes. This way the product may be kept unaltered for years, even when the olives have not been fully fermented before packing. The following guide data were obtained at the Instituto de la Grasa (Unidad Estructural de Biotecnología de Alimentos, 1990) for different sizes of glass containers:

Pasteurization bath temperature	80°C
Added brine temperature	70°C
Sealing temperature	35°C

Size of container	*Pasteurization time (min at 80°C)*
5 Cylinder	6.0
8 Par	7.5
20 Par	9.5
16 Ref	10.5
Half-a-gallon	15.0
Gallon	19.0

The sealed containers should be immersed previously for 1 minute in a bath at 45°C to prevent any breakage as a result of thermal shock. For tin containers, with equivalent diameters to those above, shorter times between 1 and 6 minutes are sufficient.

Although International Standards do not specify the limits for hermetically sealed containers of turning colour olives, those indicated for seasoned green olives may be applied: a minimum salt concentration of 4% and a maximum pH of 4.0 for non-pasteurized products. In the case of pasteurized containers, the maximum pH limit may be 4.3 for salt contents as low as 2%, and 4.0 if the sodium chloride content is less than 2%.

A free acidity between 0.6 and 0.9%, expressed as lactic acid, is generally accepted by consumers and sufficient to keep the pH within the limits indicated above.

8.9.2 TREATED TURNING COLOUR OLIVES IN BRINE

This section includes Californian green-ripe olives and Californian ripe olives. There are three main differences (Fernández Díez, 1969) between these two processes:

1. The maturity stage of the fresh olives selected for Californian green-ripe olives is, in general, earlier than that of fruits destined to be oxidized and sold as Californian ripe olives.
2. The raw material in green-ripe olives is always processed immediately after the fruits reach the factory, without any storage in brine.
3. No oxidation of the fruits is performed with green-ripe olives between lye treatments, to keep the colour of the olives close to that of the fresh fruit.

In Algeria, fresh or previously brine-stored olives are oxidized by an unusual method. The product is called 'confites' and they can also be included within the treated turning colour olives.

Californian green-ripe olives

These are produced almost exclusively in the United States, using a relatively small percentage of the turning colour fruits of the Manzanilla,

Mission and Sevillano or Gordal cultivars. They are used either in the preparation of salads or consumed directly as an appetizer.

According to the United States Standards for Grades of Canned Ripe Olives (United States Department of Agriculture, 1983) green-ripe olives are 'those which have not been oxidized in processing; which range in colour from yellow-green, green-yellow or other greenish casts; and which may be mottled'. The general scheme of the process is shown in Figure 8.6.

Fresh olives, once sorted and size-graded, are lye-treated to remove the bitterness, washed with water to eliminate most of the residual lye and then brined. They are packed in cans and sterilized. A bulk pasteurization after washing and preceding brining is often used to avoid the risk of bacterial spoilage during brining and packing operations. The alkaline treatment is generally made using two successive lyes; the first, a little stronger to penetrate the skin, has a concentration of 1.5–1.7% NaOH (w/v), and the second (1.2–1.3%) is allowed to penetrate up to the pit to ensure total removal of the bitterness.

The number of washings to eliminate the residual lye is variable but it is important that each does not exceed 8–10 hours, to avoid the growth of Gram-negative bacteria which may affect the product, particularly its texture. The pH value at the end of the washing period is 7.5–7.8. The final

Fresh olives
|
Harvesting
|
Transport to the factory
|
Sorting and size grading
|
Lye treatment
|
Washing
|
Bulk pasteurization
|
Brining
|
Packing
|
Sterilization

Fig. 8.6 Scheme for the Processing of Californian Green-Ripe Olives. (Reprinted from Fernández Díez, 1969 Elaboración de aceitunas de mesa en California. *Grasas y Aceites*, **20**, 194–205.)

sterilized canned product has a pH of 5.8–6.7 and a sodium chloride content in the equilibrated brine of 1.98–2.80% (w/v).

Sterilization conditions for green-ripe and ripe olives have been revised periodically in the United States by the National Canners Association (1976), and may be summarized as follows:

Sterilization temperature	*Sterilization time (min)*
115.6°C (240°F)	60
121.1°C (250°F)	50

The indicated sterilization times are valid for initial temperatures between 21 and 71°C and for metal containers with a net capacity (fruit and brine) up to 3 kg.

Black olives by alkaline oxidation

Californian-style pickled black olives in brine – generally known in United States as Californian ripe olives or black-ripe olives – are the most representative commercial preparation of this type. International Olive Oil Council Standards (International Olive Oil Council, 1980) and International Olive Oil Council (1960–1990) commercial statistics include this style within the turning colour olives type because of the maturity stage of the raw material used in their preparation.

Some other presentations which may also be included within this type will be described briefly at the end of this chapter, though they are of little importance in international trade.

Definition of Californian ripe olives

As indicated in Chapter 2, the Spanish Standard (Spanish Government Presidency, 1983) defines this type of table olives as obtained from fruits harvested when not fully ripe, which have been blackened by oxidation and have been debittered by alkaline treatment. They must be canned in brine and preserved by heat sterilization.

Although their specific technology indicates them to be a well defined different type, the International Olive Council does not have a different section for them.

This preparation, Spanish-style green olives in brine (Chapter 7) and naturally black olives in brine (Chapter 9) are the three most important commercial preparations, which collectively account for about 70% of the international trade (Chapter 3). The United States and Spain are the main producers of this type, with about 50% and 30% respectively of world production.

General overview of Californian ripe olive processing

Processing of Californian ripe olives may be carried out in two ways, summarized in Figure 8.7 (Cruess, 1958; Fernández Díez, 1969, 1983). In the first process, the fresh fruit is treated successively with sodium hydroxide solutions for varying periods of time to achieve a progressive penetration of the lye into the pulp. After each alkaline treatment the olives are placed into water and oxidized by injecting air under pressure into the water. This oxidation of the polyphenolic compounds permits a complete blackening of the fruit skin and an uniform coloration of the pulp.

The number of lye treatments is generally between three and five. Penetration into the fruit is adjusted so that the first sodium hydroxide treatment merely passes through the skin. Subsequent treatments are chosen so that they penetrate deeper into the pulp and so that the lye from the last treatment reaches the stone. The concentration of sodium hydroxide in the lye solution depends on the maturity of the fruit, its cultivar, the temperature of the treatment and the desired penetration. Generally, it varies between 1 and 2%, the highest concentration being used for the first treatment.

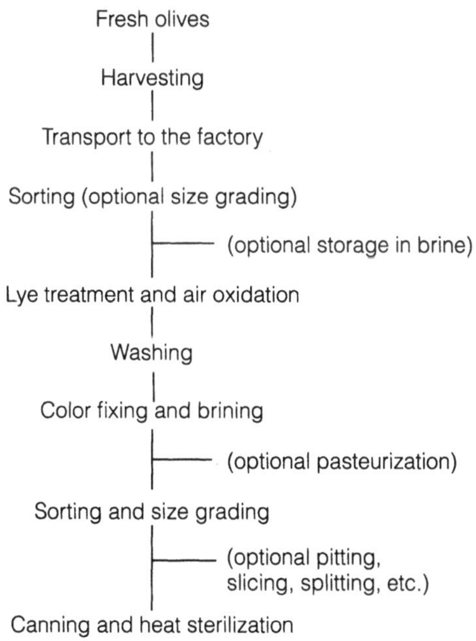

Fig. 8.7 General scheme for the processing of Californian-style black-ripe olives in brine. (Reprinted from Fernández Díez, 1969 Elaboración de aceitunas de mesa en California. *Grasas y Aceites*, 20, 194–205).

After the lye treatments and oxidation, the olives are washed several times with water to remove most of the residual lye and 0.1% of iron gluconate is added to the final wash water to stabilize the colour achieved by oxidation. The olives are then placed into a 3% NaCl brine, bottled in glass jars or, most frequently, tin cans, and sterilized by heating.

Since not all the fruit available at the time of harvest can be processed at the same time and because it is not economic to keep large stocks of canned merchandise in stock, a large portion of the olives harvested at the correct maturity is stored in brine or other solutions until market demand calls for their processing. This also permits, if necessary, the use of the brined olives for other processes, such as untreated aromatized turning colour olives in brine. The time for storage generally varies from 2 to 6 months. During this period, when stored in brine, a fermentation takes place (Vaughn, 1982), most relevant aspects of which have already been described.

Specific characteristics of harvesting, transport to the factory, sorting and size-grading

General rules for harvesting and transport of the raw material to the factory have already been described in Chapter 6 and in section 8.2 above. Hence, only some specific details concerning this olive style will be added here.

At the time of harvest, the colour of the fruit skin varies considerably (Fernández Díez, 1983) from a yellowish-green to a more or less pronounced purple, even though the pulp may still be light in colour. The taste, texture, and colour of the final product may vary greatly depending on the maturity of the fruit and the cultivar.

There are varying criteria for the optimum maturity range at harvest (Fernández Díez *et al.*, 1985). In the United States, the optimum stage is considered when the fruits have a straw yellow surface colour with certain variations, depending upon cultivar and harvesting time. The relationship between raw material and final product quality has been studied by several authors. Hilts and Hollingshead (1920) proposed some maturity standards based on oil content, establishing a minimum of 17% for Mission and 15% for Manzanillo. However, Hartmann *et al.* (1959) found that the most mature fruits of such cultivars hardly reach those oil contents and that, in any case, such olives would result in a product too soft to be palatable. They concluded that this criterion would not be reliable from year to year.

Other criteria studied by these authors are shown in Table 8.9. They observed that flavour and off-flavours in the final product were only slightly related to the maturity stage at harvest. The percentage of shrivelled fruits after processing and canning was apparently unrelated to

Table 8.9 Panel scores of black-ripe olives processed in different maturation stages. Manzanillo cultivar. Scores range, 0–10 (Adapted from Hartmann *et al.*, 1959)

Date	Flavour	Off-flavour	Colour
11 Sept.	4.50	1.00	6.84
24 Sept.	4.58	1.50	6.53
1 Oct.	4.58	1.17	5.47
8 Oct.	4.67	1.92	2.79
15 Oct.	5.00	1.50	4.32
22 Oct.	3.67	2.42	4.68
30 Oct.	4.50	2.67	2.89

harvest date, but differences were noticeable among cultivars: 2% average for Manzanillo, 5% for Mission and 25% for Sevillano. Oil content increased with maturation and did not change with processing. No relationship with flavour was evident. The colour obtained after processing was observed to change from a dark to a lighter tone as the harvest date advanced. This can be related to the continuous decrease of polyphenols observed by Amiot *et al.* (1986) during fruit development.

During maturation fruit size (as measured both by weight and size indices) and flesh-to-pit ratio both increased. This effect is due exclusively to flesh enlargement, since the pit remained unchanged from mid-September. In general, fruits with the greatest colour development were larger with a higher flesh-to-pit ratio, lower density and softer flesh texture. The product from red and black fruits was undesirably soft. When olives were classified by colour, olives in the same group gave the same flavour and off-flavour scores, independent of cultivar.

Measurements of changing flesh texture seem to be the best method for judging changes in olive fruit maturity, since flesh texture at harvest was closely related to the texture of the processed fruit (Figure 8.8). Of the devices to measure texture, they proposed the Christel texture meter as the most suitable since it is easy to use, portable, relatively inexpensive, gave consistent results and gave good correlations with panel test and shear-press results. Optimum values for Manzanillo were around 110 (with a maturity distribution of 50% green fruits, 40% straw, 10% red and 0% black), 75 for Mission (50, 35, 15, 0) and 60 for Sevillano (25, 70, 5, 0).

Fruit size and texture of the processed fruit are the only two factors that should determine the optimum time to harvest. However, it is to the grower's advantage that the fruits remain on the tree for as long as possible in order to obtain higher profits. Both points of view have been optimized and reconciled by Ferguson *et al.* (1991) as already mentioned in Chapter 6. In essence, the criteria given are: take daily samples of 100 olives from three trees of the orchard, determine the proportion of

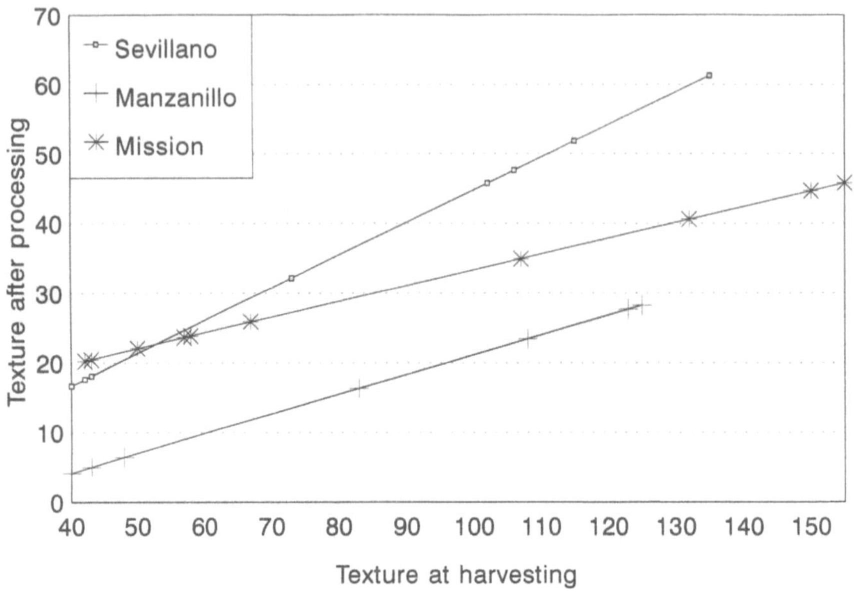

Fig. 8.8 Relationship between texture (Christel texture meter values) of the raw material and the consistency of black-ripe olives obtained as a final product. Regression was automatically adjusted. (Redrawn from Hartmann *et al.*, 1959.)

standard, medium, large and extra large olives, and begin to pick the olives one week after 50% of olives are within these categories and this percentage is increasing between 3 and 5% every week.

From the point of view of surface colour, the raw material must follow the rules established in the Californian Fresh Olives Inspection Manual (Federal State Inspection Service, 1993) that establishes that bigger cultivars (Sevillano and Ascolano) must be picked when they have no more than 15% immature fruits. In the case of smaller olives (Manzanillo and Mission), this percentage is increased up to 20%. There are also limits for over-ripe fruits and picking must cease when the proportion of these reaches 10%.

In Spain, the optimum maturity to get a better colour during processing is generally considered to be when a purple colour shows on the skin in approximately 30% of the fruits, but no more than 50%. However, considering the increasing demand of the market for a better texture, there is now a tendency to pick the olives sooner.

In general, Manzanillo is collected after the Spanish-style green olive harvesting, while Hojiblanca is picked when still completely green. Cacereña, on the other hand, is collected when there is a mixture of pale green and blush olives. Usually end-products from this cultivar are softer. Consequently, Cacereña cultivar is mainly used for plain olives, though it develops a distinctive flavour which is esteemed in certain markets.

Objective analytical methods for controlling the colour of the raw material have been established to study the different cultivars during ripening and fix the best harvesting period. The study was carried out with samples of Hojiblanca, Lechín and Verdial cultivars. Reflectance through all the visible region was studied and the reflectance at specific wavelengths selected as adequate to describe colour evolution (Fernández Díez and Garrido Fernández, 1971). Changes for the same subjective colour were the same, irrespective of the cultivars, indicating that the method is generally applicable. Figures 8.9 and 8.10 show the average reflectance for the different subjective colour classifications for 535 and 595 nm. It is apparent that there is a maximum when surface colour changes from green to straw yellow. Limits for the different

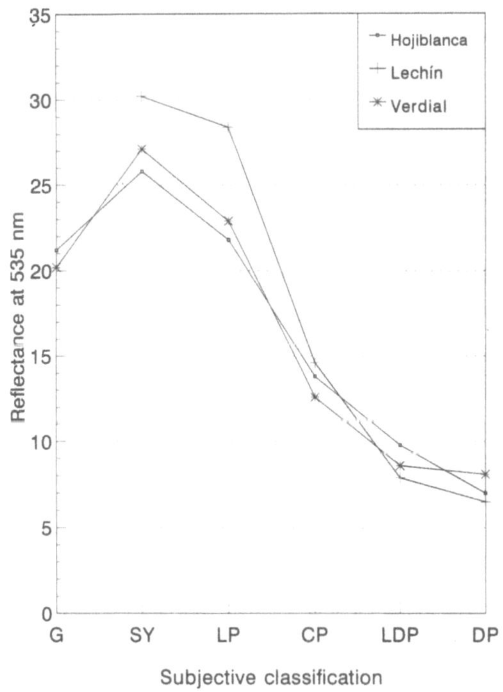

Fig. 8.9 Reflectance (R_{535}) associated with different turning olive colour homogeneous groups obtained by a subjective colour classification. G, green; SY, straw-yellow; LP, light purple; CP, clear purple; LDP, light dark purple; DP, dark purple. (Redrawn from Fernández Díez and Garrido Fernández, 1971.)

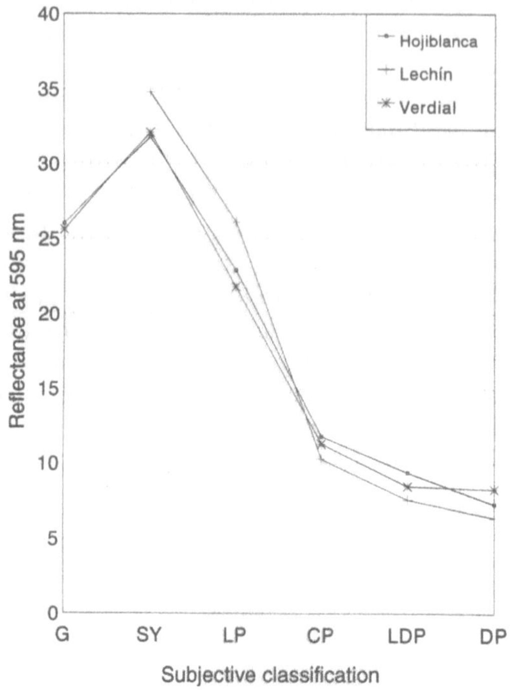

Fig. 8.10 Reflectance (R_{595}) associated with different turning olive colour homogeneous groups obtained by a subjective colour classification. G, green; SY, straw-yellow; LP, light purple; CP, clear purple; LDP, light dark purple; DP, dark purple. (Redrawn from Fernández Díez and Garrido Fernández, 1971.)

colours can be deduced statistically and are summarized in Table 8.10. The approximate optimum for picking to prepare ripe olives for Spanish tastes coincides with the reflectance maximum; R_{535} ≥29.00 and R_{595} ≥23.00.

Texture of the fresh fruits has been monitored in Spain with an Allo-Kramer Shear Press or IG Texturometer using both shear-compression and compression cells (Garrido Fernández and Fernández Díez, 1972; Garrido Fernández *et al.*, 1975; Fernández Díez, 1979). The evolution of this attribute during maturation was described in Chapter 6. The texture in the fresh Hojiblanca olives may range between 2.000 and 3.000 N/100 g of pitted olives. It is more difficult to establish limits in the case of Cacereña due to the tendency to pick a mixture of maturation stages. Spanish cultivars gave similar results to those of Hartmann *et al.* (1959) showing that the optimum date for picking olives for black-ripe type must be a compromise between texture (strongly related to the

Table 8.10 Relationship between subjective evaluation of the superficial colour of fresh olives of Hojiblanca, Lechin and Verdial cultivars, and reflectance at 535 and 595 nm (Reproduced with permission from Fernández Díez and Garrido Fernández, 1971)

Subjective evaluation	R_{535}*	R_{595}*
Green	≤ 28.5	≤ 22.5
Straw yellow	≥ 29.0	≥ 23.0
Light purple	28.5–17.0	26.0–19.0
Clear purple	16.5–10.0	18.5–11.0
Light dark purple	9.5–8.0	10.5–8.0
Dark purple	≤ 7.5	≤ 7.5

* R, reflectance

maturity stage and, consequently, olive colour) and size (which determines farmers' profits). In general, the greatest risk is from excessively mature rather than immature fruits since texture recovery is practically impossible. So, rules to avoid soft raw material at the end of the season are usually more strict than those used at the start of picking. Sometimes, picking is adapted to the availability of pickers and their work is distributed in such a way that green, turning colour and naturally black olives may be picked progressively and without interruption.

Other biochemical procedures have also been used as an indirect estimation of the best picking moment. Among these are measurement of pectinesterase and polygalacturonase activities, changes in which always accompany textural changes (Fernández Díez *et al.*, 1974; Mínguez Mosquera *et al.*, 1976; Castillo Gómez *et al.*, 1978; Fernández Díez, 1979; Mínguez Mosquera, 1982). Tables 8.11 and 8.12 show typical results.

The Californian cultivars used for this type of processing are Mission, Manzanilla, Ascolano and Sevillano, the first two producing the best product. In Spain, Cacereña and Hojiblanca cultivars supply the bulk of

Table 8.11 Pectinesterase activity in the extracts of fresh olives, Hojiblanca cultivar, at different stages of maturity (Reproduced with permission from Fernández Díez *et al.* 1974)

Stage of maturity	Date	Pectinesterase activity (U/g)
Green	November 20	0.02
Turning colour	December 21	0.12
Black	January 26	1.40
Black	February 3	2.47
Black	February 20	1.06
Black (over-ripe)	March 16	0.57

Table 8.12 Polygalacturonase activity (pH 4.0), in the extracts of fresh olives, Hojiblanca cultivar, at different stages of maturity (Reproduced with permission from Fernández Díez *et al.* 1974)

Stage of maturity	Decrease in viscosity (%)
Green	0.0
Turning colour	0.0
Ripe	16.4
Overripe	39.4

production, although there are small amounts produced from the Manzanilla and Gordal cultivars. Others like Lechín (small size) and Verdial (high oil yield) were initially considered but their use was discontinued due to the excellent results obtained with Hojiblanca and Cacereña which are grown on a scale sufficient to cover any demand for black-ripe processing.

Sorting in factories is made either manually or with electronic sorters. This permits separation of excessively green olives for processing as Spanish-style green olives in brine or green-ripe olives, and over-ripe fruits for use as naturally black olives, as described in Chapter 9.

In some cases, olives are also destemmed by automatic machines. This leaves a sensitive part of the flesh exposed and some moulds, yeasts or bacteria can grow on this area very easily even before brining. Soft stem ends are due to microbial penetration through this weak surface. Stem removal after preservation in brine appreciably diminishes the risk of this sort of spoilage.

Size-grading with divergent cables graders is optional for fresh fruits, although it has the advantage of allowing a more homogeneous penetration of the lye during the alkaline treatment, and, as a consequence, a better, more uniform final colour during the air oxidation.

Any olive handling performed before brining must be done with great care because any damage to the skin and adjacent cells will result in liberation of a series of enzymes that may cause a severe attack on the surrounding tissues. Damage increases with time and brining does not necessarily stop these biochemical transformations since the salt content and pH in the flesh only reach inhibitory levels after prolonged immersion times. This, and the large volumes of product arriving during picking, have discouraged many processors from any previous sorting and size-grading. Only those operations essential for removing inappropriate (size, maturation degree, etc.) fruits and those related to the price (size distribution) are normally applied at this stage. Destemming, size-grading, etc. are left for the period after fermentation or before processing.

During harvest, processing plants usually do not have enough capacity to process all the olives available. Therefore, most of the crop is stored until the darkening process can be applied. Different storage procedures can be used, but the most common are anaerobic or aerobic brining and immersion in a relatively dilute acid solution.

Storage before processing

Holding in brine In general, storage in brine follows the same criteria already described in section 8.5. However, some special conditions relate to the Californian industry where the black-ripe process has traditionally been applied. Containers are paraffin- or plastic-coated concrete tanks or wooden tanks of 20 tonnes capacity. Holding tanks are located out doors and usually covered by plastic sheet or another device to provide anaerobic conditions. Initial brines range from 5 to 7.7% NaCl, which is progressively raised to 7.5–9.0% over the first 3 months. When warm weather approaches, salt is added to reach about 10%. A mild acid fermentation (0.40–0.50% lactic acid) may occur.

In Spain, the aerobic system already outlined (section 8.7.1) has also been used. Table 8.13 compares the effect of the storage system on the main quality characteristics: texture and colour. Differences were minimal and both the anaerobic traditional brining storage and the aerobic method gave comparable final products.

Normally, no other additives are added to the holding brine but since relatively soft fruits may come to the factory numerous trials of calcium addition to this brine have been conducted. Table 8.14 shows the results obtained with Manzanilla cultivar to which 1 and 2% $CaCl_2$ had been added to improve texture. Apparently, its use did not disturb the fermentation process, although in aerobic conditions it produced a diminution of the population of LAB. The effect of Ca^{++} on olive texture was scarcely noticed at the end of the fermentation process, as well as in the final product in both anaerobic and aerobic storage conditions. This behaviour could be related to cultivar, but it is more reasonable to suppose that Ca^{++} cannot improve texture when the flesh structure is inadequate.

Table 8.13 Relationship between storage system and texture and colour of final product. Gordal cultivar (Adapted from Durán Quintana *et al.*, 1992)

	Texture (N/100 g pitted olives)		
Storage system	After 1 year storage	After processing	Colour*
Anaerobic†	195	93	6.9
Aerobic†	187	89	7.4
Highly acidified‡	188	94	6.8

* Reflectance at 700 nm; initial conditions: †, NaCl 6%, pH 4.2; ‡, NaCl 6%, 1.5% acetic acid.

Table 8.14 Effect of $CaCl_2$ addition to the storage brine* on the texture and colour of the stored fruits and on the final product. Manzanilla cultivar (Reproduced with permission from Garrido Fernández *et al.* 1986)

	At the end of storage (120 days)		After processing	
Storage system	Texture†	Gas-pocket (%)	Texture†	Colour‡
Aerobic§				
Control	131.9	10.5	51.9	6.7
With 1% $CaCl_2$	137.3	14.5	54.3	6.8
With 2% $CaCl_2$	135.6	9.5	56.6	6.7
Anaerobic				
Control	136.4	20.3	55.5	6.7
With 1% $CaCl_2$	142.3	30.5	–	–
With 2% $CaCl_2$	145.3	14.5	60.2	6.8

* Initial conditions: NaCl, 8% (w/v); pH, 4.2.
† N/100 g pitted olives.
‡ Reflectance (%) at 700 nm.
§ 0.2 l/h in continuous.

The storage systems, on the other hand, had an important influence on gas-pocket formation since this type of spoilage was reduced by 50% with the air bubbling. Presence of Ca^{++} had contradictory effects on such deterioration.

Salt-free storage This procedure is used mainly in California because of the strict local legislation in relation to the disposal of brines. One of the simpler methods to solve the problem would be the complete elimination of the salt brines used for storage. The use of a combination of acid, chemical food preservatives, temperature and anaerobiosis can efficiently prevent the growth of microorganisms in an environment free of salt. Research on this method was developed by Vaughn and co-workers (1969c). They found the best results using a solution containing 0.67% lactic acid, 1.00% acetic acid and 0.3% of both sodium benzoate and sorbate. Microbiological analysis showed that the system had reduced markedly the total counts, although there was evidence of some microbial growth. The viable counts decreased throughout the entire period of salt-free storage. However, there was no evidence of the fermentation found in olives stored in brine. Pectinolytic yeasts, which are highly undesirable due to their gas-pocket and softening activity, were found only sporadically. Because of the low pH, the coliform count was never significant. The procedure has advantages with cultivars which are prone to shrivel, such as Sevillano and Ascolano, where salt-free storage eliminated the problem.

This has the added advantage that it avoids the need to pass olives over the needle board to fracture the skin and release trapped CO_2. This effect must be related to the lower amount of CO_2 produced during storage since microbial activity is practically eliminated. Also, the flavour of fruits stored in these conditions apparently improved compared with those stored (and fermented) in brine. In particular, the flavour of Sevillano decidedly improved and lost the characteristic strong flavour and woody texture of this cultivar. These effects on flavour can be related to the lack of the natural flavour in brined olives due to microbial activity during the fermentation. When the storage period lasts 8–12 months, the influence of the fermentation in brine is very evident and undesirable for the Californian industry and possibly for American consumers. Due to the absence of fermentation throughout salt-free storage, there are no adverse flavour changes.

The method must, however, be applied under strict anaerobic conditions, otherwise vinegar flies and dried-fruit beetles are attracted and colonize the olives. Another problem is the surface growth of *Acetobacter* and oxidative yeasts or moulds. Anaerobiosis is obtained in California by immersing the sorted, size-graded olives in solution inside open-top redwood tanks (approximately 1.5 m high, 1.8 m diameter) of 2.5 tonnes capacity. The slatted, false lid is then fixed and covered with the acidic-preservative solution, after which the plastic cover is fitted in place and secured with pliable slats nailed to the inside of the tank. A wax compound, Sealfite, provides an airtight seal between the plastic and the inside of the tank.

The system has also been applied in other places. In Spain, it is used sporadically and has frequently been modified by using a highly acidified solution without any other preservative. The method efficiently reduces the Gram-negative bacteria and yeast population, providing a reduced fermentative medium sufficient to preserve the olives of most cultivars. However, with those cultivars (Gordal, Sevillano, etc.) where LAB are present during the holding in brine, lactobacilli and cocci growth is still observed after a period of latency. The accumulation of fermentable substrates and other nutrients, and the lack of other competing microorganisms, provide a very rich medium in which LAB finally grow in spite of the high acidity. Fortuitously, they help to increase the acidity just when the warmer days of the spring are approaching. However eventually, the hostile environment inhibits their growth and LAB may even disappear (see Figure 8.2). The effect of acid storage on texture and colour of the final product seems to be minimal (Table 8.13). In most of the Mediterranean countries the system can be applied in the 10-tonnes fibre-glass fermenters in which the head space is very small in relation to the volume. Additionally, there are different devices to provide anaerobic conditions in such containers.

The procedure is more expensive than the traditional brining system, but it also reduces the weight losses during the holding period – a feature that can compensate (probably in excess) for the supplementary costs.

According to the Californian researchers, acidification may produce a slow softening of the raw material. The effect is different for each cultivar and type of acid. Thus, the most appropriate mixture of acids must be determined before being applied at industrial level. Control of this effect by different chemicals was tried by Vaughn (1970). Ammonium or potassium alum had a negative effect while the Ca^{++} salts gave inconclusive results. However, Vaughn observed a clear negative effect with an excess of Ca^{++} when fruits were darkened because its presence increased the lye treatment periods and caused a rather heterogeneous penetration into the flesh.

Other organic or inorganic acids (citric, malic, HCl, H_4PO_3) may also give good results (Vaughn, 1975), although the concentrations of inorganic acids must be lower (0.50–0.74%,w/v) to decrease the softening effect of the acid in which case the yeast population is inhibited less well.

Other alternative storage methods Other chemical salts have also been used or suggested as substitutes for NaCl in holding solutions. Those used as fertilizers were partially successful in holding olives for several months without spoilage. Ammonium nitrate brines could be used for up to 15 weeks, although the salt was difficult to remove completely during olive preparation for canning. Brines containing ammonium sulphate (with the pH controlled by addition of H_2SO_4) held olives for 6 months. The difficulties in completely removing the ammonium sulphate from the olives limited the commercial potential for this type of storage.

Food-grade glycols containing bacteriostats can be used to store olives for over 6 months. Olives submerged in propylene glycol–diethyl pyrocarbonate mixtures showed severe shrivelling. The subsequent recovery from shrivelling was satisfactory only with smaller sizes of olives. Similar results were obtained for olives stored in a mixture of glycerol–benzoic acid–sorbic acid. The most interesting storage mixture was water–propylene glycol–benzoic acid–potassium sorbate–HCl. By using KOH in place of NaOH in the preparation steps, it was possible to prepare canned ripe olives without generating liquid wastes containing sodium ions (Ralls *et al.*, 1971).

A great research effort has been devoted to using brines again the following year. When brines are re-used without any treatment, the quality of the final products is poorer, probably due to the accumulation of metabolites from the previous fermentation (Rose, 1966). However, regeneration by activated charcoal before re-use led to products with characteristics similar to those from fresh brine storage (Ralls, 1969). The regeneration must be accomplished immediately before re-use to avoid off-flavour accumulation after carbon treatment. The procedure is technically feasible, although it introduces greater technological complexity as well as higher costs.

Other procedures recover the salt by evaporating the water. Darkey *et al.* (1973) tried submerged combustion in which a burner is placed under the brine surface. The slurry obtained was then calcined at a slightly lower temperature than NaCl fusion. Salt is recovered as a solid (with some carbon impurities). Olives stored in brine prepared from this salt were similar in quality to those from the traditional system.

Finally, dry storage of freshly harvested olives has also been studied. Storage at temperatures below 5°C produces internal browning which begins around the pit and progresses to the olive surface. Slightly higher temperatures (5–10°C) can be used for periods ranging from 4 to 8 weeks before chill injury or excessive ripening occurs. Obviously as the temperature rises the storage period possible diminishes. This can be increased slightly by the use of a controlled atmosphere with a reduced O_2 content (Kader *et al.*, 1990). More information about cold storage is included in Chapter 6.

Darkening processes

Lye treatments Lye treatment and air oxidation are the two most important steps of the Californian ripe olive process and are essential to get a good final product (Fernández Díez, 1969; Unidad Estructural de Biotecnología de Alimentos, 1990). Once the fresh or stored fruits are sorted and occasionally graded, they are treated with a series of dilute sodium hydroxide solutions and exposed to air between treatments. In the United States, three to five lyes are generally used. During the first alkaline treatment, the lye penetrates only the skin of the fruit (one quarter of the flesh); the remaining lyes are permitted to penetrate the pulp progressively until the last one reaches the stone. In other cases, the first four lye treatments last only long enough for the lye to barely penetrate the skins of all olives; the last then reaches the pit. All the lyes are known as 'colour lye', except the last, which is called 'pit lye'.

Sodium hydroxide concentration is variable, mainly depending on cultivar, stage of maturity and temperature, ranging generally between 1 and 2% (w/v). The highest concentration is used in the first lye used to penetrate the skin. For example, the five lyes used during a typical industrial process were 1.53%; 1.17%; 1.20%; 1.27%; and 1.33% (Fernández Díez *et al.*, 1985).

A practical way to check the lye penetration in the fruit during the process is to cut a sample of 5–10 olives longitudinally, close to the pit and observe the lye-penetrated surface of the flesh, which darkens in air within a few seconds. The same method is used to check lye penetration in the other treated olives, such as in the Spanish-style green olives procedure. As indicated in Chapter 7, lye temperature during treatment must not exceed 20°C.

In Spain, the number of lye treatments is normally restricted to two or three. The concentration used for the first treatment is generally between 1.5 and 2.0% of sodium hydroxide (w/v) and the olives are left in the alkaline solution until the lye penetrates the skin. The concentration of the two successive lyes ranges from 1.0 to 1.5%, leaving the first to penetrate 1–2 mm of the flesh, and the second to reach the pit.

In California, lye treatment (and the whole oxidation process) is carried out in tanks of approximately 3 × 1.5 × 0.75 m, built of cement and coated in the interior by paraffin or plastic (Luh and Ferguson, 1994). Each tank is supplied with four pipes for supplying water, lye, brine and air and has outlets at the bottom to permit the removal of either the solutions or the solutions and suspended olives. Air is normally supplied through the olives in suspension by horizontal perforated pipes on the tank bottom. Aeration supplies the O_2 for oxidation, as well as stirring the olives, which are circulated continuously in the tanks. When fruits are not well stirred, problems of under-oxidation can occur. Figure 8.11 shows a series of tanks used in a Spanish plant to darken black-ripe olives.

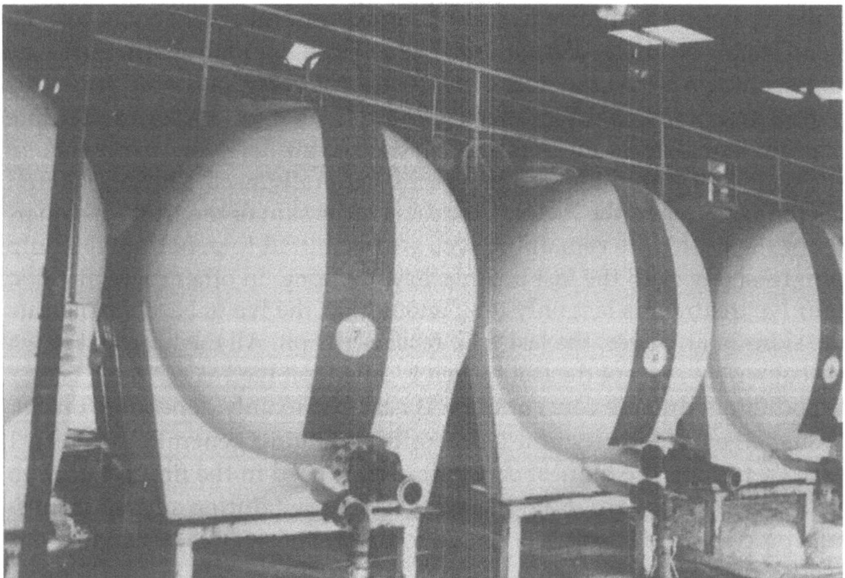

Fig. 8.11 Modern fibre-glass tanks used in a Spanish processing plant to darken black-ripe olives. (Courtesy of Internacional Envasadora S.A. Dos Hermanas, Sevilla; the photograph was taken by Dr Rejano.)

Air oxidation, washing, colour fixing and brining After each colour lye treatment fruits are blackened by oxidation, either by exposing them to air for 24 hours or by blowing pressurized air into the mass of olives immersed in water or low-strength brine. This second procedure is recommended because the simple air exposure may produce deterioration of the fruit texture. Air oxidation periods between lyes are variable, frequently around 12 hours.

The number and duration of washings (Fernández Díez *et al.*, 1985; Fernández Díez, 1991) must be regulated so that the washed olives reach a final pH value around 7, generally between 6.0 and 7.8. Values higher than 8 produce less final colour. The washing water must be changed frequently to avoid growth of spoilage aerobes. Sometimes food-grade hydrochloric acid (0.01 to 0.1%) is added to the water to speed neutralization of the residual lye. Lactic acid and CO_2 injection are also sometimes used for this purpose.

Generally, 0.1% (w/v) of ferrous gluconate is added to the last wash to stabilize the colour produced by oxidation of polyphenols during air injection. Ferrous lactate may also be used, at a lower concentration of 0.05% (García García *et al.*, 1986). Immersion time in the iron solution is variable, from 1 to 12 hours, depending on the cultivar and the effectiveness of the previous process stages. Procedures for controlling the iron content of the final product have been established (Garrido Fernández *et al.*, 1973; Albi Romero and Garrido Fernández, 1975).

The olives are then placed in a 3% brine for between 1 and 3 days to equilibrate the sodium chloride content in the flesh. For certain cultivars it is customary before this to perforate the skin of the fruits with a rotating needle board in order to speed up the process and prevent shrivelling of olives.

A bulk pasteurization, at 60°C for 45 minutes, to minimize the action of some aerobic bacteria during brining is optional. It may also be achieved by injecting steam into the olives and brine, cutting the steam off as soon as the temperature reaches 90–95°C, and allowing the mass to cool down (Unidad Estructural de Biotecnología de Alimentos, 1990).

Factors affecting the darkening process and fruit quality

Lye concentration, temperature and maturity Higher NaOH concentrations lead to faster penetration but the influence of temperature is decisive. In places with low winter temperatures, lye concentration of 3.5% (w/v) or higher may be necessary to achieve a similar effect to a 1% lye used in spring or autumn. Previous experience is necessary to decide the appropriate concentration, since practices elsewhere are not always relevant. It must also be adjusted later according to results and temperature changes.

Fresh olives generally need higher lye concentrations than stored fruits to penetrate the skin at the same rate. It seems that brine holding affects the skin making it more permeable so some processors diminish the lye concentration depending on how long the olives have been stored. Penetration also depends on the olive size and the maturation stage of each olive. Smaller fruits which are less developed always offer higher resistance to NaOH penetration. As an example, Table 8.15 shows time required for 1 mm, half-flesh and pit penetration of different lyes in two maturation stages of fruits from the same lot.

Aeration system, NaOH concentration and number of lye treatments
Lye treatments produce a sharp drop in the fruit consistency. The effect of lye concentration on fruits of different maturation stage, subjected to dry or wet oxidation is shown in Table 8.16. Final texture clearly diminishes as NaOH concentration increases (for the same temperature). The effect of maturation stage (within the same lot) is also evident. Differences in olives picked the same day can vary appreciably. If they are processed all together it is difficult, if not impossible, to obtain a high degree of homogeneity.

Table 8.15 Time required for 1 mm, half-flesh and pit lye penetration, using different concentrations and fruit maturation stages. Lechin cultivar. Temperature 22°C (Reprinted with permission from Garrido Fernández *et al.*, 1973)

Lye concentration and order	Duration (h : min)	
	Large (blush)	Standard (pale green)
1% (w/v)		
1st lye	5 : 00	8 : 15
2nd lye	3 : 45	5 : 50
3rd lye	2 : 10	3 : 15
1.5%		
1st lye	3 : 45	4 : 20
2nd lye	1 : 45	2 : 00
3rd lye	1 : 15	1 : 00
2.0%		
1st lye	2 : 10	2 : 20
2nd lye	1 : 15	1 : 30
3rd lye	0 : 45	0 : 45

Table 8.16 Effect of lye concentration, fruit maturity and aeration system on the final product texture (N/100 g pitted olives). Lechin cultivar (Reproduced with permission from Garrido Fernández *et al.*, 1973)

Aeration system and lye concentration (%, w/v)	Maturity Stage*				
	LDP	CP	LP	SY	G
Oxidation by exposure to air					
1.0	800	960	1060	1120	3940
1.5	710	960	1000	1010	3680
2.0	680	800	1070	950	3500
Oxidation in an aqueous (or brine) suspension					
1.0	1090	1140	1200	1250	4770
1.5	900	1080	1180	1160	4320
2.0	850	910	1030	1020	3810

* Maturity stages: LDP, Light-dark purple; CP, clear purple; LP, light purple; SY, straw yellow; G, Green.

Less-mature fruits have more initial texture, which is also better retained through processing. Excessively mature olives are initially deficient and the lye treatment renders them softer. These differences must be considered to adapt consistency to the specific demands of each market. In general, the final product may retain between 40 and 60% of the original texture.

Table 8.16 also shows the effect of the oxidation system on texture. In general, wet oxidation gives a more homogeneous product with higher texture than dry oxidation. Differences seem to be less important as NaOH concentrations increase. Thus, oxidation with the fruits suspended in water or in a light brine is preferable and is used in most developed and developing countries. However, it has the significant drawback of using large volumes of water, up to 30–35 l/kg of fruits, especially during the washing periods. This causes a huge pollution problem that the industries in the USA and Europe need to overcome in the near future if the business is to be continued. Alternative procedures for olive oxidation must be seriously considered.

An excessive number of lye treatments may negatively affect texture while causing no improvement in colour (Table 8.17). The effect of the number of lye treatments can be decreased by using lower NaOH concentrations. In this case, differences obtained in both texture and colour are slight.

Sometimes, especially when olive fruits have been stored for longer than 3 months, only one lye treatment is frequently applied. Garrido Fernández (1981) found that this procedure could be used successfully, although the oxidation rate was slower and the final colour was lighter

Table 8.17 Average effect of the number of lye treatments and oxidation system on texture (N/100 g pitted olives, in the Allo Kramer shear press) and colour (reflectance at 700 nm). Lechin cultivar. Olives prepared with a constant lye concentration of 1% (w/v) NaOH (Reproduced with permission from Garrido Fernández *et al.*, 1973)

	Fruit texture		Colour*	
Oxidation system	3 treatments	5 treatments	3 treatments	5 treatments
By exposure to air	1270	1130	8.0	10.5
In aqueous suspension	1510	1350	7.5	9.5

* Darker colour indicates lower reflectance values.

when fresh fruits were used. This meant that the washing periods should be enlarged, so the duration of the whole process is practically the same as when a higher number of lye treatments is applied. Best results were obtained when the lye immersion was followed by a washing with fresh water or a light brine, in which the pH had been corrected with food-grade HCl.

Results showed that a good final colour and adequate characteristics were obtained with 1% NaOH. Different timings of the immersion in the ferrous gluconate solution did not influence the final results (Table 8.18). Particularly interesting is the fact that the incorporation of ferrous gluconate did not enhanced the colour obtained. This observation is in agreement with the hypothesis that the cation acts only as a colour fixer and does not directly promote colour formation.

Adequate combinations of NaOH concentrations, number of treatments, maturation stage and oxidation system can lead to the same final product – hence the very diverse industrial practices observed throughout the different black-ripe olive-producing countries.

Additives Olive fruits harvested at the proper maturity stage and processed adequately do not need the use of any additives, other than ferrous gluconate or lactate, to reach the optimum quality. However, sometimes the industry has tried using some additives to improve colour or texture. The two most common have been tannins and $CaCl_2$, although the first is not authorized by the COI/CODEX. The effect on surface colour of the addition of two different types of tannins (soluble and insoluble in ethanol) used in the wine industry is summarized in Table 8.19. It shows that the addition of tannin before colour fixation has only a slight improving effect before packing and the influence on colour after this operation is varied. When tannin was added after the immersion in the ferrous solution, the effect was negligible after the darkening

Table 8.18 Darkening process with only one lye treatment. Effect of different modifications. Hojiblanca cultivar (Reproduced with permission from Garrido Fernández 1981)

Lye concentration (%, w/v) and modifications	Before packing	Packed in fresh brine	Packed in RFG* solution
1% NaOH			
Aeration throughout the whole process	14.8	12.9	14.5
Ferrous gluconate addition after lye	14.6	14.0	13.8
Ferrous gluconate after 4 days of oxidation	15.8	13.9	14.9
0.75% NaOH			
Aeration throughout the whole process	16.7	14.5	14.8
Ferrous gluconate addition after lye	17.3	14.3	14.8
Ferrous gluconate after 4 days of oxidation	17.4	13.9	14.5

* RFG, Re-used ferrous gluconate.

Table 8.19 Effect of addition of tannins on black-ripe olive colour (reflectance at 700 nm). Hojiblanca cultivar. Oxidation treatment consisted of three successive lyes of 1.5, 1.2 and 1.2% NaOH (Reproduced with permission from Garrido Fernández, 1980b)

	Tannin type					
Processing stage	Soluble in ethanol*			Non soluble*		
	NP	TBFe	FeBT	NP	TBFe	FeBT
Before packing	15.5	14.0	14.5	15.5	14.2	15.5
After packing	19.0	18.0	25.0	14.0	14.0	19.0

* Insoluble or soluble in ethanol. Concentrations: 0.2% with respect to the solution volume. NP, Normal process; TBFe, Addition of tannin before colour fixation with ferrous salts; FeBT, Addition of tannin after colour fixation with ferrous salts.

process, but had an important negative impact on the final product, which was discoloured (Garrido Fernández, 1980b). The same products always had a negative effect on texture (Table 8.20). However, in the presence of $CaCl_2$, the effects of the tannins was not appreciable (Garrido Fernández, 1980a).

In summary, it seems that tannin may give a slightly improved colour before packing, a fact that can encourage some processors to use it. However, any previous effect is lost in the final product and the result can be detrimental. In these conditions, transgressing the good manufacturing practices established by COI is an unnecessary risk.

In the case of $CaCl_2$ addition is different since this additive is authorized for table olives and its utilization depends only on raw material, industrial practices and market demand in relation to texture. The effect of Ca^{++} on the texture of black-ripe olives processed in different conditions is shown in Table 8.20. In all cases, there is significant texture enhancement in relation to the control. However, it is of interest to study its effect in detail. Figure 8.12 shows the relationship between initial texture and texture obtained after different treatments. The first point apparent is that the effect is processing-dependent. Another important observation is that the effect is strongly dependent on the initial texture. When texture was high at the beginning, the presence of Ca^{++} enhanced the texture more than when the initial texture was low. This relationship is well described by a straight line. This means that Ca^{++} is able to work better with fruits that initially have a good consistency, but when the initial raw material is poor recuperation of the consistency is limited. Again, the use of the proper maturity of fruits and a good preservation system is the best way of assuring an adequate final texture.

Table 8.20 Effect of $CaCl_2$ on black-ripe olives texture (N/100 g pitted olives), using different darkening conditions. Hojiblanca cultivar. Oxidation process consisted of three successive treatments with 1.5, 1.2 and 1.2 NaOH solutions (Reproduced with permission from Garrido Fernández 1980a)

	Soluble tannin*			Unsoluble tannin		
	NP	TBFe	FeBT	NP	TBFe	FeBT
Before packing						
Control	140	126	130	155	136	141
With $CaCl_2$†	146	166	150	177	162	176
After packing						
Control	91	89	89	94	93	91
With $CaCl_2$†	106	119	119	119	119	117

* Insoluble or soluble in ethanol. Concentrations: 0.2% with respect to the solution volume.
† Added at a proportion of 1.5% (w/v) with respect to the olive flesh (pulp).
NP, normal process; TBFe and FeTB, addition of tannin before or after the ferrous fixing colour solution.

According to Luh and Ferguson (1994) colour formation is most rapid at pH of 8 to 9.5 and its retention is improved when hard water is used. This effect is attributed to a favourable influence of calcium salts in fixing the colour. Apparently, calcium chloride greatly aids colour retention when added in low concentration (0.1–0.5%) to the lye solutions or the washing waters between lye treatments.

In general, sterilization of the final product always produces texture losses. This is dependent on processing and canning conditions. Table 8.21 shows the average changes for different cases. The presence of Ca^{++} in the washing waters improves the olive texture with respect to the control by about 24%. However, packing and sterilization reduce the texture in all kinds of fruits. In fruits elaborated with Ca^{++}, the decrease ranges from 16%, when olives were packed re-using the ferrous gluconate solution, to 26%, when packed in fresh brine. In the absence of the cation, the texture degradation was still more important and reached values between 33% and 37%, respectively. The re-use of ferrous gluconate solution was less important on this occasion.

Previous aeration and desalting of fruits Some industries precede their darkening step with an aeration of the fruits, apparently with the objective of removing salt and acidity. There is controversy over the usefulness of this step. Those in favour argue that during aeration there is a slight darkening of the fruits that enhances further oxidation. Table 8.22 shows the colour retained before and after packing with and without previous desalting. In general, the operation leads to a decrease of about 1 or 2% of

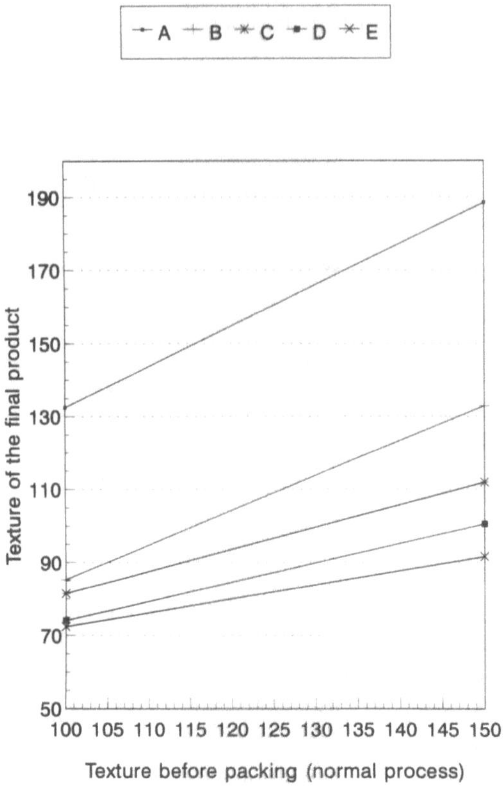

Fig. 8.12 Effect on texture (Allo Kramer shear press, in kg/100 g pitted olives) of different processing and packing conditions. Symbols correspond to the following treatments: A, texture before packing of olives with $CaCl_2$ addition during the last washing of the darkening process; B, texture after packing of olives prepared with $CaCl_2$ addition during the darkening process and packed with re-used ferrous gluconate solution; C, as for B, but packed in a fresh brine; D, texture of ripe olives prepared without $CaCl_2$ and packed in re-used ferrous gluconate solution; E, as for D, but packed in fresh brine. (Redrawn from Garrido Fernández, 1980a.)

Table 8.21 Effect of $CaCl_2$ addition during processing and different packing conditions (Reproduced with permission from Garrido Fernández, 1980a)

	Texture (N/100 g pitted olives)		
Processing stage and conditions	With $CaCl_2$ added	Without $CaCl_2$ added	Increment due to $CaCl_2$
Before packing	159.8	129.7	30.1
Packed with re-used gluconate solution	108.3	86.8	21.5
Packed in fresh brine	96.7	81.5	15.2

Table 8.22 Effect of previous aeration and desalting on black-ripe olive colour (reflectance at 700 nm). Hojiblanca cultivar. Oxidation process consisted of three successive treatments with 1.5, 1.2 and 1.2% NaOH (Reproduced with permission from Garrido Fernández, 1980b)

Experiment no.	Desalted	Non-desalted
Before packing		
1	14.5	15.5
2	14.5	19.0
After packing		
1	14.0	15.5
2	12.0	14.5

the reflectances (a slight colour improvement). Thus, its use at an industrial level is simply a matter of the balance between costs and profits obtained as consequence of the improvement of quality.

On the other hand, the presence of a certain higher acidity in olives when they are not desalted may be easily overcome by an increase in the NaOH concentration. In the case of preservation in a highly acidified medium however, the acidity in fruits may be high enough to require previous desalting (or deacidification) to avoid any possible interference with the lye treatment and excessive consumption of NaOH.

Previous storage and re-use of lyes and washing waters There is no single criterion governing the best moment for applying the darkening process. In California, there is a tendency to prepare the fruits as soon as possible. This reduction of the storage period to the minimum diminishes texture deterioration and the appearance of off-flavours. In other countries, processors prefer to elaborate the final product throughout the year in line with demand. This needs less initial investment in cans, labels, etc. and thus distributes costs over a longer period. In general, fresh fruits are more difficult to darken, as can be deduced from Table 8.23, while fruits stored for 8 months darken better (Garrido Fernández, 1983).

Influence of storage on texture is less evident, although there is a tendency for this to improve after a certain holding time. A relationship between calcium presence and olive flesh texture was observed by Chung *et al.* (1974), who found that the texture of stored olives, independent of their maturation degree, improved when stored for 2 months in brine. This improved firmness of olives stored in salt brine can be explained by the action of pectin esterase present in olive flesh which demethoxylates the pectin, causing an increase in the free carboxyl groups. These can then combine with calcium ions in the brine to form

Table 8.23 Effect of the previous olive fruits preservation and re-use of lyes and washing waters on the colour (reflectance at 700 nm) and texture (N/100 g pitted fruits) of the final product. Hojiblanca cultivar. Darkening process consisted of successive lyes of 1.5, 1.0 and 1.0% (w/v) NaOH. (Reproduced with permission from Garrido Fernández, 1983)

	Normal process		Lye re-use		Lye + water re-use	
Type of fruit	Fresh brine	Holding brine + Fe gluconate	Fresh brine	Holding brine + Fe gluconate	Fresh brine	Holding brine + Fe gluconate
Colour						
Fresh fruit	26	23	26	20	32	23
4 months' storage	15	13	18	14	18	14
Texture						
Fresh fruit	76	75	75	77	73	75
4 months' storage	78	81	81	80	81	73

insoluble calcium pectate. This is supported by the observation that the calcium content of canned ripe olives which had been stored in brine was much higher than in directly cured olives.

Long storage periods lead to fermentation in the case of brines or possibly limited microbial development and slight softening when storage is in an acidic medium. Some of the aromatic components produced during this period will remain after the darkening process and sterilization. The special flavour produced in the final product can be judged as an off-flavour in certain markets (e.g. USA), but is appreciated in others (Europe, North Africa, etc.).

Long storage periods (over a year) cause excessive leaching of black-colour precursors and the olives are difficult to darken. In many cases texture can be also deficient.

The preservation system has little influence on product colour and texture (Brenes Balbuena *et al.*, 1986a,b) (Table 8.24).

The use of Spanish-style green olives as a raw material for producing black-ripe olives has been tried. This could be of interest in case of short harvests or demand distortion. Since oxidation is based on the presence of polyphenols in the olive flesh and these are deliberately reduced in Spanish-style fermented fruits, application of the darkening process is difficult. García García *et al.* (1987) developed an approach for Hojiblanca and Manzanilla cultivars. Hojiblanca gave the best results due to its higher polyphenol content. However, a good colour, similar to that obtained from directly brined olives, was only obtained when polyphenols were added to the washing waters. Apparently this compensates for the losses during the processing of Spanish-style green olives. Due to the

Table 8.24 Average effect of preservation system on colour (reflectance at 700 nm) and texture (N/100 g pitted olives). Hojiblanca cultivar. Darkening process consisted of three successive immersions in 1.5, 1.0 and 1.0% lyes (Reproduced with permission from Brenes Balbuena *et al.* 1986b)

	Preservation system					
	Aerobic		Anaerobic		Acidic solution	
Period after packing	Fresh brine	Re-used gluconate	Fresh brine	Re-used gluconate	Fresh brine	Re-used gluconate
Colour						
15 days	13	10	14	10	13	8
12 months	12	8	14	7	12	8
Texture						
15 days	76	79	75	76	72	73
12 months	70	70	70	71	70	70

reduced texture of these olives, the lye concentration must be lower (0.3–0.6%). The best results were obtained with three lye treatments, aeration during all the darkening steps, addition of food-grade polyphenols (in all washing waters with Manzanilla, but only in the final one with Hojiblanca) and immersion in 0.06% ferrous gluconate or its equivalent, 0.03% lactate. The final immersion in ferrous salts and the subsequent processing steps were similar to those applied with the normal raw material. The process includes utilization of non-authorized additives which will have to be approved before industrial application.

The product prepared from green olives has a very characteristic flavour which was preferred by most of the taste panel members but is radically different from the traditional black-ripe olives. Product substitution will then be difficult.

Lastly, black-ripe darkening requires large amounts of water, up to about 35 l/kg. Some control of the volume of wastewaters produced can be achieved by re-using the lyes, washing waters and the ferrous gluconate or lactate solution. As was seen in Tables 8.19–8.21 and Figure 8.12, re-use of the colour-fixing gluconate solution for packing always produces an improvement in colour and texture. Its re-use at industrial level must face the drawback arising from the higher coloration (darker tone) acquired by the holding solution. If this aspect proves a serious drawback for the market acceptability of the product, the practice could be discouraged. If not, it could be a good approach to reduce this liquid waste.

Other solutions able to be re-used are the lyes and washing waters. The effect of the application of such practices is shown in Table 8.23. Lye re-use has no negative effects when the NaOH content is readjusted adequately after each treatment. However, re-use of washing waters has a negative effect on colour development. More importantly, an excessively long period of washing water use may lead to development of Gram-negative bacteria, yeasts, etc. and to product deterioration. Consequently, lye re-use can be suggested in those cases in which the industry has water disposal problems, but washing water re-use is not recommended.

Oxygen and temperature The presence of **oxygen** in the oxidation medium can enhance the oxidation rate during the first lye immersions. However, it seems that the stability of the polymers formed, or their fixation in the skin or outer parts of the flesh, is weaker than that obtained more slowly at low oxygen concentrations. This effect is independent of the step in the process (lyes or water rinses) in which the excess of oxygen is used. Presence of O_2 also increases the brown colour of the flesh, which can be a quality drawback for certain markets that are accustomed to just a brown ring around the olive surface and light flesh. Thus, at room temperature there is no advantage in increasing the O_2 concentration above that obtained by bubbling air alone (García García *et al.*, 1991).

Temperature increase when lye is applied is discouraged because of the possibility of severe texture deterioration. However, when heat is used during washing, moderate levels (30–70°C) can speed up darkening. The Californian industry has taken advantage of this and often heats the oxidation tanks to shorten the darkening period.

Washing waters and acids and flesh neutralization The aim of washing is to reduce the excess alkali in the flesh while it, and especially the surface, darkens. When the appropriate colour is obtained, neutralization must be achieved as soon as possible. This will shorten the whole process and diminish the volume of the residual wastewater. Traditionally, neutralization of the excess NaOH in flesh has been achieved by frequent changes of the waters and some processors still believe that only with this method can they reach the highest quality. The problem is the disposal of these used waters. Neutralization with acids will decrease the problem. Commonly used acids are HCl, H_2SO_4 or CO_2. The strong mineral acids have the drawback that only a small excess of them can cause a dangerously low pH and subsequent olive discoloration. However, CO_2 is a weak acid that hardly can decrease the pH below 6.5 and consequently its addition has no negative effects. In fact, both types of acid are used in practice. A comparative study of the behaviour of HCl and CO_2 is shown in Figure 8.13. Neutralization rate depends on the acid; CO_2 produces a quicker decrease in flesh pH. The effect also depends on the pH of the surrounding solution; the lower this can be maintained (within

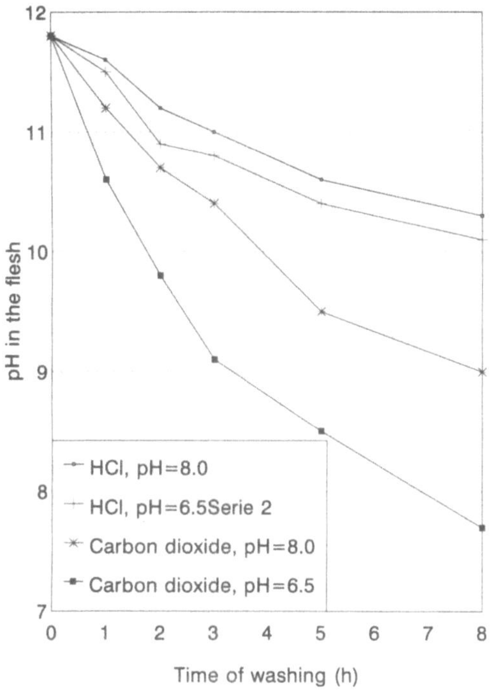

Fig. 8.13 Changes in the pH of flesh during washing period of the darkening stage of ripe olive processing, using different acids and pHs of equilibrium in the surrounding solutions. (Redrawn from Brenes Balbuena *et al.*, 1993)

reasonable limits), the faster the neutralization. This behaviour of CO_2 during neutralization of the flesh must be related to its easier diffusion, as a gas, through the skin and flesh. This increases as the concentration increases and its presence as gas increases with low pH. Brenes Balbuena *et al.* (1993b) also demonstrated that this effect was dependent on Ca^{++} concentration in the solution; flesh neutralization with HCl or CO_2 was improved as the Ca^{++} content diminished. Calcium may thus have favourable effects on texture but can also slow the neutralization rate. A compromise can be achieved by adding the Ca^{++} when the neutralization process has finished.

As with other osmotic and equilibration processes, neutralization is more rapid as temperature increases.

Recent modifications of the darkening process in California

In the past 15 years the Californian industry has worked hard developing procedures to reduce the volume of waste water. McCorkle *et al.* (1984)

developed a procedure and apparatus for producing black-ripe olives which consisted of periodic addition of alkaline solution (0.45– 0.7% NaOH for fresh cure olives and 0.9–1.2% for stored olives) into a tank containing the olives to be processed. The olives are agitated in the solution and the solution is drained from the tank leaving the olives to dry in the tank for a period before the alkaline solution is re-introduced. After repeating the wet–dry cycle a number of times the blackened ripe olives are rinsed with fresh water. The lye is retained in a separate storage tank at the completion of each period of wetting the olives and is periodically strengthened to the desired concentration. The agitation results from bubbling high pressure air into the processing tank below the olives. The bottom of the tank is corrugated and air is introduced along the valleys of the corrugations to provide improved agitation. Carbon dioxide is added to the rinse water to neutralize the alkali in the blackened ripe olives at the end of the process.

Scrimshire (1987) patented a new method and apparatus for producing black-ripe olives which mainly improved the neutralization phase. Olives are immersed in lye until the alkali penetrates to the pit, eliminating the bitter principle and then neutralizing the lye by immersing the olives in a carbonic acid solution which is initially supersaturated with carbon dioxide. There is no air agitation or heat applied and this step is followed by treatment with a ferrous gluconate solution.

Changes in some physicochemical characteristics of olives during darkening

Industrial darkening treatments are diverse and so changes in the flesh and surrounding solutions are also varied. We will describe some changes in several of the common sequences of lye and washing operations, following mainly the work of Garrido Fernández *et al.* (1973). Fruit size and weight both change as the darkening process advances, due mainly to water retention caused by expansion of the flesh tissues following the NaOH treatment.

Figure 8.14 shows the change in olive fruit weight during different darkening processes expressed as percentages of the initial weight. There is a marked increase with the first lye treatment. This continues with the second alkaline immersion to a lesser extent. With the exception of the first washing, there is always moderate loss on washing. The highest weight losses occur during brining and subsequent pasteurization due to the osmotic effect of the 4–6% NaCl in the brine. If the heat treatment is applied after equilibrium, weight changes are small. During the whole darkening process there is a net weight increase that depends on the procedure used, cultivar and the initial storage system. In the experiments cited, the higher increases were obtained with only two lye treatments (2% and 1% NaOH, respectively).

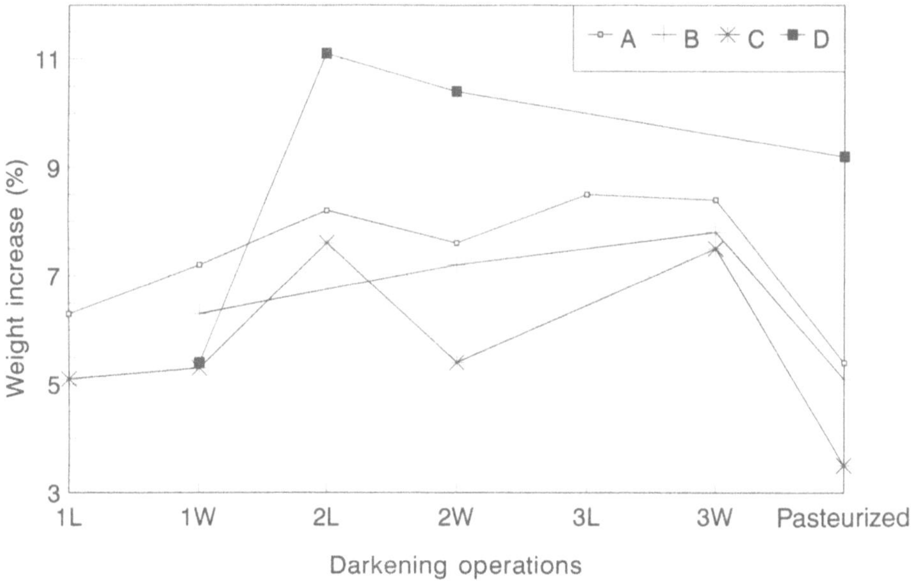

Fig. 8.14 Changes in olive fruit weights (expressed as percentages of initial weight) during the darkening stage of ripe olive processing. Lechin cultivar. Treatments were as follows: A, three successive lye treatments of 2, 2 and 2% NaOH; B, three successive lye treatments of 2, 1 and 1% NaOH; C, three successive treatments of 1, 1 and 1% NaOH; D, two successive lye treatments of 2 and 1% NaOH. L, lye treatment; W, wash. (Redrawn from Garrido Fernández *et al.*, 1973.)

Texture changes are also important during this step. A general overview of changes during the whole darkening process has already been given, but more detailed information will be reported here. Figure 8.15 shows the evolution of this quality attribute in different darkening processes, expressed as percentages of the initial consistency. In general, each lye treatment produces a deterioration of the consistency which is more marked as the concentration of NaOH increases. As is obvious from Figure 8.15, significant differences in this attribute can be obtained according to the lye treatment (NaOH content and number of immersions). Some processors take advantage of this to adapt the texture to the market demand. Good results (lower texture diminution) are obtained with low NaOH concentrations or a reduced number of treatments. Some recuperation of this characteristic is observed during the curing in brine, which increases the values, even after the heat pasteurization.

Jiménez Araujo *et al.* (1994) reported that when olive fruits were processed, substantial amounts of cell wall polysaccharides were solubilized. While all non-cellulosic components appear to be involved, the

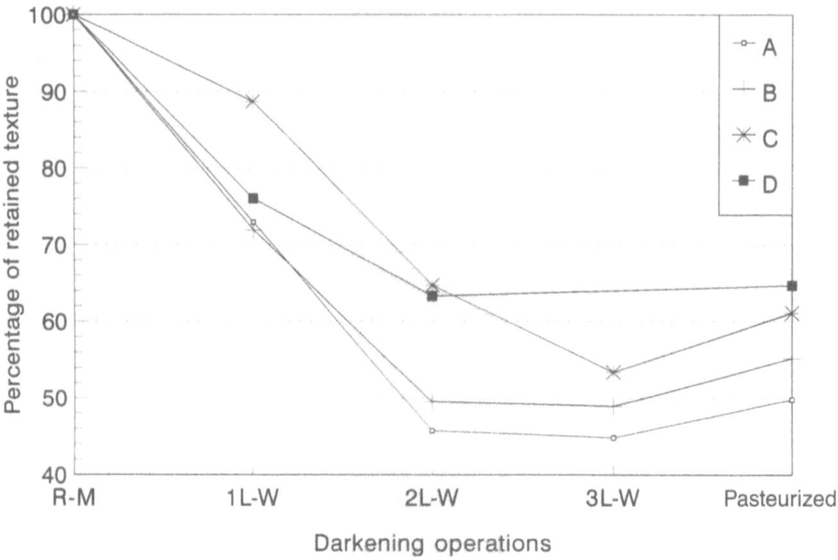

Fig. 8.15 Changes in olive fruit texture (expressed as percentages of initial texture) during the darkening stage of ripe olive processing. Lechin cultivar. Treatments were as follows: A, three successive lye treatments of 2, 2 and 2% NaOH; B, three successive lye treatments of 2, 1 and 1% NaOH; C, three successive treatments of 1, 1 and 1% NaOH; D, two successive lye treatments of 2 and 1% NaOH. L, lye treatment; W, wash. (Drawn from Garrido Fernández *et al.*, 1973.)

pectic polysaccharides appear to be most directly affected. Apparently, processing has its greatest impact on neutral sugar-poor less branched polymers. The most important pectic polymer solubilized was uronic acid. This leaves behind a pectin fraction that is relatively low in uronic acid.

Although quantitatively, the pectins are the wall components most affected by processing, in relative terms wall glucan is also greatly affected. This causes an accumulation of more glucose than uronic acids despite the fact that there is 25 times as much uronic acid as non-cellulosic glucose in the walls of unprocessed fruits. The apparent loss of olive cell wall uronide from processing liquids can be explained by a depolymerization or disaggregation of pectins in the presence of NaCl. However, the cellulose content is fairly constant during processing.

Most of the fruits used for darkening have been previously stored in brine. During the first stages of lye treatment and washing, salt in the flesh is washed out. It is of interest to know how this occurs because later

Table 8.25 Changes in the NaCl content in the olive flesh during processing. (Reproduced with permission from Garrido Fernández *et al.*, 1973)

	Treatments*			
Operation	A	B	C	D
1st lye and washing	–	2.5	5.1	3.8
2nd lye and washing	1.3	0.9	2.0	1.8
3rd lye and washing	0.3	0.3	1.3	–
Brining and pasteurization	2.3	2.0	2.4	2.0

* Sequence and concentration of lyes: A 2%, 2%, 2%; B 2%, 1%, 1%; C 1%, 1%, 1%; and D 2%, 1%.

it is necessary to add salt to establish a final equilibrium in the product normally around 2–3%. Table 8.25 shows the salt content of the olive flesh after the different lye and washing cycles.

Salt elimination depends on the treatment energy (Table 8.25). Those treatments with high NaOH levels produce, as in the case of the Spanish-style green olives, a higher skin permeability and thus more rapid NaCl leaching. The lowest NaCl content was obtained with 1% NaOH, although three treatments were applied. In any case, by the end of the third cycle the NaCl content of the olive flesh is very low and this diminishes still further during the subsequent washing period for the elimination of the excess NaOH. This period lasts several days by which time fruit must have lost practically all the NaCl. This is different if, as is done in some factories, a light brine is used during this period. Then the flesh NaCl content equilibrates with that in the brine.

The darkened olives reach equilibrium very quickly, but are also sensitive to osmotic pressure differences which may cause water losses in the olive flesh. This effect is increased with heat. Thus, any heat application should be applied only after equilibrium between olives and any holding solution is reached.

Finally, it is of interest to show the evolution of the lye concentration during each treatment (Figure 8.16). As is evident, only a small amount of the initial NaOH content is used during lye treatment.

Traditionally, fresh solutions were used for all treatments; however, as shown, this practice wastes much of the NaOH. In addition, its disposal causes a major pollution problem. Bearing in mind both circumstances, re-use of these solutions is advisable. Due to their high pH, these solutions normally remain free of undesirable microorganisms, but elimination of suspended solids and sediments in the storage tanks could help to improve the re-utilization conditions. After each cycle the concentration must be readjusted by adding back the amount of NaOH absorbed by the fruits. The NaOH concentrations required are calculated from titration

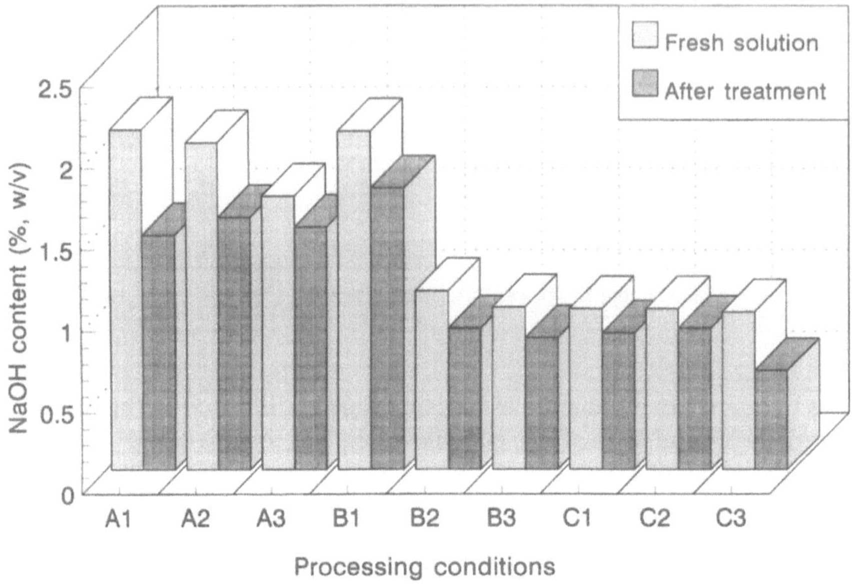

Fig. 8.16 Changes in NaOH content in solution during lye treatment. Treatments were as follows: A1–A3, three successive lye treatments of 2, 2 and 2% NaOH; B1–B3, three successive lye treatments of 2, 1 and 1% NaOH; C1–C3, three successive treatments of 1, 1 and 1% NaOH. (Redrawn from Garrido Fernández *et al.*, 1973.)

data because the density of these solutions not only reflects the NaOH content but also all other organic and inorganic components that progressively accumulate. The number of cycles of re-use can vary and is dependent on the disposal facilities of the factory and the recuperation treatments applied. Re-utilization of lyes is nowadays a common practice.

Type and concentration of the colour fixing solution

It was observed from the very beginning of black-ripe olive elaboration that colour faded after the darkening stage. So, when the consumer demanded a permanent black colour it was necessary to develop a method by which the colour could be fixed.

Cruess (1962) studied the addition of iron as a colour fixer, comparing the results obtained with different salts (gluconate, sulphate, ammonium sulphate, ferric chloride) as well as different application procedures. Vaughn (1966) completed those experiments with other salts. Both

Californian researchers established the basis for the use of iron in black-ripe olive preparation. Among their conclusions were that:

- the olives have to be processed to get good colour or the iron will not fix the colour;
- the amount used must not be excessive or the interior of the flesh will turn an unnatural grey colour;
- colour fixation is more rapid at moderately high temperatures than at room temperature;
- iron salts in light brines give better results than in plain water;
- iron salts added to the brine used in canning ripe olives give good colour but impart an unnatural purplish tint to the brine after retorting;
- application of a pit lye after colour fixation does not appreciably affect the colour;
- iron salts have the highest effect when applied after the final colour lye and after the pit lye.

These studies were used for the authorization of ferrous gluconate as an additive for colour fixation in black-ripe olives. Later, García García *et al.* (1986) compared the effectiveness of this compound with ferrous lactate in the most important Spanish cultivars (Hojiblanca and Cacereña). Results showed that the effect was the same when used in concentrations such that the ferrous cation contents were the same. Based on such research the COI/CODEX and COI has authorized the indiscriminate use of either of them in the fixing phase of darkening.

The absorption of iron increases with the immersion time (Tables 8.26 and 8.27). Vlahov *et al.* (1977) found that absorption was strongly cultivar-dependent (Table 8.26). Provenzale Italian cultivar absorbed a significantly higher amount than Cucco. Similar behaviour can be expected in other cultivars.

When packed, the olive-holding brine acquires a dark colour due to the solubilization of part of the iron retained in the flesh. Iron found in the brine was proportional to the flesh content but was only a small proportion of the total. This means that the cation is fixed in the flesh not only by combination with the polyphenol polymers but also possibly with other flesh components, since the total iron absorbed is markedly higher than that attributable to the polyphenol retention.

Table 8.27 shows iron absorption in both pasteurized and non-heat-treated olives darkened under different conditions. In addition to the increase in iron with time, it increased with heat treatment. This explains the greater darkening effect when immersion in iron salts was followed by a heating period. The absorption is clearly dependent on NaOH concentration, being higher as the alkali content increased. Apparently, the alkali is able to hydrolyse some compounds, possibly from the cell wall, which may serve as iron complexers.

Table 8.26 Iron content (p.p.m.) in flesh and brine as a function of cultivar and duration of the immersion* (Adapted from Vlahov *et al.*, 1977)

Cultivar and time from packing	Flesh			Brine		
	3 hours†	9 hours†	24 hours†	3 hours†	9 hours†	24 hours†
Cucco						
After pasteurization	24	70	112	–	–	–
30 days*	23	60	90	5	12	45
60 days	21	55	71	6	15	56
90 days	20	35	90	7	16	48
Provenzale						
After pasteurization	132	291	321	–	–	–
30 days	122	280	302	16	32	50
60 days	114	270	280	16	34	47
90 days	114	269	250	35	31	55

* Containing $2^0/_{00}$ of ferrous gluconate.
† Immersion time in the gluconate solution.

Table 8.27 Effect of the treatment on iron absorption by olive flesh in different darkening treatments and immersion times*. (Reproduced with permission from Garrido Fernández *et al.*, 1973)

Darkening treatment	Immersion time			
	6 hours		21 hours	
(% NaOH)	Pasteurized	Non-pasteurized	Pasteurized	Non-pasteurized
2, 2, 2%	37.9	22.7	40.8	37.4
2, 1, 1%	29.3	19.8	56.7	27.8
1, 1, 1%	14.6	11.2	34.4	21.9
2 and 1%	31.2	–	34.3	–

* Gluconate solution contained 0.1% (w/v) of the compound.

Compounds responsible for the darkening process

Simpson *et al.* (1961) hypothesized that the black colour developed during the black-ripe processing was due to the polymerization of polyphenol compounds. They found up to three compounds that increased their colour in presence of alkali.

Chambionat (1961) held the view that the black colour was obtained by polymerization of *ortho*-diphenols in the fruits and explained the colour fixation as due to the reaction of these substances with NH_2 groups of the amino acids present in the skin and flesh. The phenomenon thus might occur in two steps:

1. The formation of polymers from the oxyflavones by bonding through the polyphenolic groups.
2. The subsequent reaction of the polymers with amino acids.

The aptitude of each cultivar for darkening was thus related to its phenolic content. Those low in polyphenols could be processed to obtain a good colour when adequately supplemented with polyphenols. Based on this hypothesis, Chambionat found that the reaction could be catalysed by cobalt, a metal that enhances the oxidation process and induces a good colour in fruits.

However, it is only relatively recently that the compounds associated with darkening have been identified. Brenes Balbuena *et al.* (1992) followed the polyphenol changes during the lye and washing treatments. The most important polyphenols present before processing and after storage for a moderately long period are hydroxytyrosol and verbascoside. As already commented, the oleuropein content is low by this time because of its progressive hydrolysis during storage at relatively low pH (about 4.2 or lower).

After the NaOH treatment, the most important components were hydroxytyrosol and caffeic acid. The first must come from the material originally present plus that produced by oleuropein hydrolysis by the alkali; the caffeic acid is derived from the hydrolysis of verbascoside. Other compounds also present, such as tyrosol (which increased slightly after the alkali treatment) and vanillic acid were relatively resistant to the action of NaOH. The presence of coumaric acid after lye treatment could be from the destruction of certain *p*-coumaric esters. There were other phenolics in small amounts which apparently do not play any role in darkening.

The behaviour of polyphenols in the flesh and the different solutions in which olives are immersed was strongly related to their structure. Those without *o*-diphenol character (tyrosol, *p*-coumaric and vanillic acid) remain constant during oxidation (apart from the obvious dilution into the surrounding liquid). However, the two most representative phenolic substances of the olive flesh after lye treatment, hydroxytyrosol and caffeic acid (which are *o*-diphenols) diminish markedly through the oxidation process. The oxidation rates for both compounds were similar, but caffeic acid disappeared more rapidly possibly due to its higher water solubility.

The changes were the same with fruits oxidized in air or suspended in water. The virtual disappearance of both *o*-diphenols can be explained on the basis of their oxidation in the presence of alkali and air.

There was a good relationship between the hydroxytyrosol and caffeic acid disappearance and the colour formed. As a consequence, it can now be stated that the brown to black colour formed during the oxidation process is due mainly to the oxidative polymerization of hydroxytyrosol and caffeic acid.

The effect of pH on the kinetics of the reaction was studied by García García *et al.* (1992a) in both liquid model systems and with olive surface colour development. Darkening of model solutions and hydroxytyrosol and caffeic acid disappearance obeyed first-order kinetics. The constants diminished as the pH decreased within the range studied of 6 to 10. The rate of hydroxytyrosol decrease was always higher than the rates of caffeic acid decline and darkening. The surface colour was thus more related to caffeic acid oxidation–polymerization.

Surface colour development was similar across the pH range used. Colour formation is rapid during the first 12 hours and then becomes more gradual. The first part of this behaviour can be adapted to second-order kinetics; rates are very similar between pH 6 to 8, but with significantly lower parameters than those obtained in the model solutions. This may be due to the slower oxygen diffusion into the flesh, bearing in mind the low permeability of the fruit skin to chemical substances and gases. A procedure which increased the oxygen diffusion into the flesh could possibly enhance and accelerate the darkening process. In practice, it is apparent that oxidation always occurs with the flesh alkaline, since the pH in the flesh decreases only very slowly after lye treatment. Evolution of colour in the flesh will therefore depend on the pH in the surrounding solution but at least for the 24 hours following alkaline treatment, the pulp pH remains over 9.0.

Sorting, size-grading, packing and heat sterilization

After brining and before packing, the olives are sorted to separate defective units such as broken or uncoloured fruits and then size-graded with diverging cable graders. Before packing, an important fraction of the product is prepared for presentation forms other than whole olives, such as pitted, sliced, divided, broken or olive paste, by means of different mechanical devices (Fernández Díez *et al.*, 1985; Unidad Estructural de Biotecnología de Alimentos, 1990).

Once these operations are completed, the fruits are generally placed in varnished tin cans and a hot brine at about 3% NaCl (w/v) is added before sealing. After filling and brine addition, the cans are exhausted at 93–95°C for 5 minutes to reach 77°C or higher. They may then be re-brined and sealed at 77°C in a double seamer. Sometimes, brine may be replaced by the addition of solid salt and hot water followed by the double seaming with steam injection. Glass jars are also used but to a much lesser extent. Olives packed in glass jars must be sterilized in a retort with superimposed compressed air, otherwise the pressure that develops in the jar during processing forces the lids off. In the United

States, heat sterilization is conducted according to rules established by the National Canners Association, currently the National Food Processors Association (1976): 60 minutes at 115.6°C (240°F) or 50 minutes at 121.1°C (250°F).

An initial temperature between 21 and 71°C is assumed. These values can be applied to containers with a net content (fruit and brine) of up to 3 kg. In Spain (Instituto de la Grasa y sus Derrivados, 1970) relatively similar temperatures and times were established for this product:

Net content	*Temperature (°C)*	*Time (min)*
1 kg or smaller	115–116	60
3 kg	115–116	70
1 kg or smaller	121	45
3 kg	121	50

It is strongly recommended that for each new sterilization plant or new presentation form of the product, the sterilization process be studied by heat processing specialists. Figure 8.17 shows a modern, computer-controlled retort used in a Spanish black-ripe olive factory.

The California State Board of Health requires that a temperature-chart record of every retort load of canned olives be made available to the

Fig. 8.17 Modern computer-controlled retort used for black-ripe olive sterilization. (Courtesy of Internacional Envasadora S.A., Dos Hermanas, Sevilla; the photograph was taken by Dr Rejano.)

inspection services and that records of the double-sealing inspection, pH and weight from each lot are also kept and made available to the inspection service.

The equilibrated final product in the market has a pH between 5.8 and 7.9, and a sodium chloride concentration which ranges from 1 to 3% (Fernández Díez, 1969, 1984), though it may be as high as 5%, depending on the demand and taste preferences of consumers. Ripe olives have very different organoleptic properties from the fermented fruits obtained by other processes and are an excellent complement for green salads, pizzas and other foods.

The effect of processing on different quality characteristics in the final product has been studied (Garrido Fernández *et al.*, 1973; Garrido Fernández, 1980a,b, 1981). In particular, the canning conditions have been studied with respect to the level of additional ferrous gluconate, acids and the re-use of the ferrous gluconate. García García *et al.* (1991) showed that the addition of a certain level of ferrous gluconate or lactate to the final holding solution always improved the colour (Table 8.28). It is worth pointing out that, in spite of the marked improvement in some cases, those olives with high values of reflectance (poor colour) hardly reached acceptable colour even in presence of high amounts of iron, corroborating the first observations by Cruess (1962) and Vaughn (1966).

Sometimes the industry needs to neutralize the alkali remaining in processed fruits by adding acid to the packing brine. Cruess (1962) demonstrated that when the ferrous gluconate solution had an excessively low pH, the effect on darkened olives was one of discoloration instead of fixation. Thus, any addition must be made with caution. Concentrations are normally of around 0.1–0.2% for organic acids such as lactic or acetic and a little lower for strong mineral acids (HCl or H_2SO_4).

Table 8.28 Effect on colour (reflectance at 700nm) of the use of different concentrations of ferrous gluconate in the packing solution (Adapted from García *et al.*, 1991)

	Percentage ferrous gluconate		
Number of experiment	0.03	0.06	0.09
1	11.4	9.5	8.2
2	9.3	8.2	7.7
3	10.8	9.1	7.8
4	8.0	7.1	7.0
5	19.5	16.7	12.5
6	17.0	13.9	12.1
7	19.0	16.7	12.1
8	16.9	13.3	11.8

Table 8.29 Effect of acids on colour (reflectance at 700 nm) and texture (N/100 g pitted olives) of canned ripe olives (Reproduced with permission from Brenes Balbuena *et al.*, 1986b)

	Colour		Texture	
Characteristics of holding brine	Aerobic storage	Anaerobic storage	Aerobic storage	Anaerobic storage
Ferrous lactate*	9.0	10.6	900	1000
Ferrous lactate + 0.1% acetic acid	8.4	10.0	1160	1250
Ferrous lactate + 0.05% HCl	8.2	9	1050	1170

* Concentration of 0.015% (w/v).

Following these specifications, their utilization can even be favourable (Table 8.29). Surface olive colour mostly improves and texture is decidedly enhanced. This must be related to the re-adjusted pH that the acid addition provokes.

Re-use of the ferrous gluconate or lactate solution as packing brine has also been proposed. This means that the holding solution can have a higher content of iron and in fact its colour is markedly browner than those of fresh brine-packed olives. Concentrations in the olive flesh also reach higher levels (Table 8.30). The data reveal that re-use of lyes and washing waters can lead to increased iron absorption possibly due to the higher levels of substances able to complex with the cation. However, although rather high, the iron levels are far from the 150 p.p.m. limit for iron in the flesh fixed by the COI and COI/CODEX standards.

Lastly, olives can also be canned in an acidic medium. In this case, the product is an acid product and does not require full sterilization, only pasteurization. In general, the use of acids gives a discoloration (Table 8.31). This effect cannot be counteracted even in the presence of re-used ferrous gluconate solution. The final colour of acidified fruits is only slightly worse than those sterilized, but at these reflectance levels 4–5 units of difference in reflectance signify important colour changes. The concentration of acid needed differs with the acid. Lactic acid requires lower concentrations as it is a stronger acid than acetic.

Organoleptic characteristics of the acidified product are completely different from the traditional black-ripe olive and, in absence of a strong flavour from olives, the acidic flavour predominates. The commercial presentation could be supplemented with the addition of natural spices or their concentrates which do not influence any other physicochemical characteristics. In the opinion of some consumers and researchers, this is not the ideal form of preparing black-ripe olives.

Table 8.30 Effect of the re-use of ferrous gluconate solution in canning, on the iron (p.p.m.) retained in the fruits processed with and without lye and washing waters re-use (Reproduced with permission from Garrido Fernández 1983)

	Processing system		
Canning type	Normal process	With lyes re-used	With lyes + wash water re-used
Fresh brine	21.6	20	26.7
With re-use of ferrous gluconate solution	59.5	83.5	72.3

Table 8.31 Effect on colour of the holding brine acidification, in both fresh brine and re-used ferrous gluconate solution (Reproduced with permission from Garrido Fernández and Sánchez Roldán 1982)

	Sterilized		Acidified and pasteurized	
Treatment	pH	Colour	pH	Colour
Fresh brine				
1	5.7	8.1	4.25	12.1
2	6.50	8.2	4.30	11.6
3	6.50	8.1	4.25	13.1
4	6.60	7.2	4.30	10.0
5	6.30	8.4	4.25	10.7
Re-used solution				
1	5.90	7.2	4.10	11.9
2	6.50	7.0	4.05	11.0
3	6.50	8.1	4.05	11.5
4	6.40	7.1	4.10	12.0
5	6.30	7.0	4.15	10.9

Changes related to texture of black-ripe olives during packing were studied by García García *et al.* (1994), who reported the effect of Ca^{++} and Na^{+} on the consistency of this product in pH range 6–10. The presence of Ca^{++} always increased the ripe olive firmness, the effect being more intense at the lower pH. When the Ca^{++} content is higher than 10–20 mM there is no, or only slight, improvement. The effect of Na^{+} on olive firmness is very slight.

With respect to the softening rate, there are important differences between cultivars (e.g. Cacereña cultivar softens more rapidly than Hojiblanca). In general, the presence of calcium diminishes the softening rate, but at pH 8.5 or over there is a rapid increase in the softening rate. This could be a serious inconvenience for preserving adequate texture during the sterilization of canned black-ripe olives when their pH is excessively high. However, this is not the case since during the sterilization process there is a rapid (and not until now well explained) decrease in the pH of the flesh and the brine that stabilizes the product at values around 6.5–7.5, where the softening rate is relatively low. However, some differences between the values in the brine and flesh are maintained during the heat treatment. In the presence of Ca^{++} the differences may even reach 1 pH unit. In spite of this, as has already been commented, sterilization always involves a serious texture degradation.

As with iron, there is a great capacity for calcium retention by the olive flesh. This leads to a substantial difference in calcium concentrations between flesh and brine. Calcium is strongly bound to the flesh and is not released by the heat treatment applied during sterilization.

Microbial spoilage

If the microbial population is not well controlled during storage of the fruits in holding brines, and the medium is not maintained within the suitable ranges of pH, salt content, and free acidity, different types of deterioration may develop. These are similar to those described previously for Spanish-style green olives in brine, such as fish-eyes and gas-pockets, putrid and butyric fermentations, and 'zapatería'.

There are some other changes which are more specific for ripe olives and have been studied by Vaughn and co-workers (Vaughn, 1954, 1982).

Sloughing

This is mainly produced during the last stages of washing after the lye treatments and air oxidation. It is characterized by a softening of the flesh and loss and degradation of the skin of the fruits, and is attributed to the action of certain genera of Gram-negative bacteria, such as *Aerobacter*, *Aeromonas*, *Paracolobactrum*, *Achromobacter* and *Escherichia* (Vaughn *et al.*, 1969b).

However, in spite of being able of producing a rapid softening, these bacteria do not rupture the skin or produce sloughing of the flesh. Because the last two symptoms (as well as a generalized tissue softening) are typical of this spoilage, it was postulated that cellulase-producing microbes might also be associated with the deterioration. A bacterium identified as *Cellulomonas flavigena* was isolated (Patel and Vaughn, 1973). This was highly cellulolytic, attacking filter-paper, carboxymethyl cellulose gel and olive tissue. Its activity was enhanced by other bacteria represented by the genera *Xanthomonas*, *Aerobacter* and *Escherichia*, all of which were able to degrade carboxymethylcellulose and olive tissue, but not filter-paper. Apparently, the association of both types of microorganisms is required for this type of spoilage. The deterioration is usually observed by the fourth day of washing (darkening stage). A good method for its control is reducing this phase to only 2–3 days. Mild pasteurization can also destroy rapidly the microorganisms. A few minutes at 40°C is enough.

Nailhead

This is a defect which appears under the skin and is attributed to spore-forming microorganisms and pectinolytic bacteria and yeasts. It is also produced during the final stages before canning operations and may be avoided if the product is pasteurized before canning.

Soapy centres

This is characterized by a disintegration of the flesh close to the pit in processed olives and is accompanied by a special off-flavour. It has been found in processed fruits of the Ascolano cultivar and its origin is unknown. It is attributed to certain circumstances occurring before processing, such as an overheating of the fresh fruits in the field or the action of fertilizers.

Fishy fermentation

This is mainly produced in Gordal and Manzanilla cultivars. It is an off-flavour in processed ripe olives of unknown origin, although it has been attributed to the action of certain microorganisms during the washing and brining stages.

Other off-flavours

The quality of black-ripe olives is also dependent on the presence of so-called off-flavours. This rather ambiguous term includes diverse sensations. Only the term metallic is well identified by panellists when the concentration of iron is over 50 p.p.m. Other off-flavours have been related to the fermentation process during holding in brine. Their intensity increases with time. The first to appear are described as 'sea water' and 'wet'. Later, about 9 months after brining, the most outstanding flavour seems to be 'rancid'. After 1 year's storage the 'wet' flavour re-appears, along with others like 'medicinal'. The profile is strongly related to the population of microorganisms since there is a progressive increase in these sensations up to 9 months after brining and then a decrease in parallel with the microbial counts. The presence of these off-flavours could not be related to lipid oxidation, although there is slight tendency for unsaturated fatty acids to decrease. Apparently, in spite of the NaOH treatment and strong aeration, lipids remains rather stable during the process possibly protected by the presence of the high level of polyphenols. A close association of these off-flavours was found with the presence of volatile carbonyl compounds (i.e. ketones). A concentration over 2 p.p.m. coincides with the detection of off-flavours. The maximum content of these compounds is detected after 2 months, a moment at which the off-flavours are detected by panellists. Chromatographic profiles for normal and off-flavour olives has demonstrated marked differences in headspace volatiles. However, interpretation of the role played by the numerous compounds found and the even higher number of peaks still unidentified is complex and no significant progress has been seen in this field in the past two decades.

Composition and nutritional aspects of the final product

The most complete data available on the composition of ripe olives were obtained by Nosti Vega *et al.* (1984) and later by Nosti Vega and Castro Ramos (1985) with processed olives of Hojiblanca cultivar. Table 8.32 shows some typical values which are in agreement with those obtained for Mission and other Californian cultivars (Cruess *et al.*, 1939; United States Department of Agriculture, 1977; Californian Olive Industry, 1990).

A detailed study of the changes in the natural pigments during the blackening operation is presently missing in the literature and is a challenge for technologists. Table 8.33 summarizes the compositional changes during processing and the general characteristics of the final product.

8.9.3 OTHER COMMERCIAL PREPARATIONS OF THIS TYPE

Californian-style pickled black olives in brine, also known as ripe olives, or black-ripe olives, are the most representative commercial preparation of the turning colour olive type. For that reason, it is the only one which has been studied and is currently manufactured with sufficient technological and scientific support.

The rest of the commercial products of this type are prepared on the basis of routine and tradition and have a reduced, generally local, demand with very little significance on the international market. Among them, three types of preparation must be mentioned (Fernández Díez *et al.* 1985):

1. 'Confites' olives. This is a procedure popular in Algeria. To produce this commercial presentation, fruits are harvested when they are in the turning colour stage. Fruits are usually preserved in 1000-kg cement tanks (generally the same used for the lye treatment of the Spanish-

Table 8.32 Composition of Californian-style black-ripe olives in brine (Adapted from Fernández Díez *et al.*, 1985)

Composition	Typical values in wet pulp of Hojiblanca cultivar fruit
Major component (weight %)	
Moisture	68.98
Lipids	21.24
Protein	1.12
Fibre	2.35
Ash	2.04
Vitamins	
Carotene (mg/100g)	0.09
Thiamin (μg/100g)	0.21
Riboflavin (μg/100g)	96.30
Essential amino acids (mg/100 g)	
Valine	119
Isoleucine	118
Leucine	147
Methionine	32
Phenylalanine	89
Tryptophan	15
Minerals (mg/100 g)	
Phosphorus	15
Potassium	12
Calcium	68
Magnesium	18
Sodium	634
Sulphur	8
Iron	7
Manganese	0.14
Zinc	0.49
Copper	0.29
Caloric value (cal/100 g)	222

style green olives) or in bigger cement tanks situated underground. Grading and sorting are accomplished as in green or any other table olive preparation. The darkening process is carried out according to demand. Stored olives are lye-treated using NaOH concentrations slightly lower than those used for green olives and lightly washed. Olives are then put in boxes of 20–25 kg capacity for 3–4 days. Every day, olives are changed from one box to another to produce a homogeneous darkening on the surface. When olives have darkened, they are lye-treated again to achieve a complete debittering of the fruits and thoroughly washed to remove excess alkali. Finally, processed

Table 8.33 Californian-style black-ripe olives in brine. Composition changes during processing and general characteristics of the final product (Adapted from Fernández Díez 1984)

Operation	Change in composition
Prior phase (optional)	
Fermentation in brine, by lactic acid bacteria and yeasts	slow loss of sugars, tannins, and oleuropein
	Formation of organic acids, and possibly ethanol and other aromatic compounds
Fundamental phase	
Alkaline treatment	Hydrolysis of oleuropein
Washing	Loss of sugars and organic acids
Oxidation by air	Organic acid formation from sugars
Brining	Pigment degradation
Final phase	
Storage of the sealed and sterilized product	None, under normal conditions
General characteristics of the final product	
pH value	5.8 – 8.0
Sodium chloride (w/v, %)	1–5 according to commercial preparations

olives are put in wooden barrels of 160–180 kg (230–250 l) capacity in a 8–10°Bé brine in which they are sold. Malodorous and gas-pocket spoilage are frequent because the high pH favours the growth of spoilage organisms (Gram-negative bacteria, yeasts and *Clostridia*), the salt content being insufficiently high to inhibit such growth.

2. Treated (or pickled) black olives in dry salt. These are obtained from lye-treated and air-oxidized olives which are packed in alternative layers of fruit and dry salt, or by sprinkling salt on the fruits. Because of the conditions of preparation and storage, the final product frequently has a relatively high pH and is only kept safely by using high salt levels or an elevated degree of dehydration (humidity less than 20%). This naturally restricts consumer demand.
3. Shrivelled treated (or pickled) black olives. These are generally obtained from fruits harvested just before complete ripeness has been attained. After a light immersion in a diluted lye, the fruits are kept by being sprinkled with dry salt or being mixed with salt, being placed in wooden barrels, and then rotated daily to improve the blackening of the olives. Eventually the fruits become shrivelled by the action of the salt and naturally dehydrated, presenting the same difficulties as with

the procedure indicated above. The market importance of this product is also very limited.

8.10 INDUSTRIAL EQUIPMENT

At present, the optional storage of the fresh fruits in brine, before the lye treatment and the remainder of the processing, is generally made in paraffin- or plastic-coated concrete tanks or wooden tanks of 20 tonnes capacity (in the USA) or in the same kind of fermenters (Fernández Díez *et al.*, 1985) used for Spanish-style green olives in most other countries. They are made of polyester and fibre-glass, the average capacity being about 10 tonnes of fruit. They are designed to keep close to anaerobic conditions to avoid the growth of film yeasts and moulds on the surface which may spoil the texture of the olives. As indicated above, an alternative aerobic procedure has been developed to be used with the same fermenters in order to avoid 'fish eyes' produced by some species of fermentative yeasts, which may affect certain cultivars of olives (Manzanilla and Hojiblanca). By means of a simple aeration device placed in the fermenter, an air flow of 0.3 1/h per litre capacity of the fermenter is supplied for 8 hours each day, or less if the levels of dissolved CO_2 and O_2 permit. Excellent results have been obtained when processing olives previously stored using this method (Unidad Estructural de Biotecnología de Alimentos, 1990).

For the lye treatments, washing and air oxidation (Fernández Díez, 1984) open tanks of different shapes and sizes and varying from 2 to 10 tonnes fruit capacity, have been used. These are made of concrete, stainless steel, polyester and fibre-glass. The same fermenters used to store the olives in brine may even be used. In all cases, they must have a similar network of pipes for the distribution of pressurized air so that the oxidation phase is uniform (Fernández Díez, 1969; Fernández Díez *et al.*, 1972).

Frequently, specific oxidation vats are used only for this operation (Unidad Estructural de Biotecnología de Alimentos, 1990). These are cylindrical or six-sided in shape, their bases consisting of two slanted planes which contain a perforated tube through which pressurized air is bubbled at their point of intersection.

Electronic sorters, diverging steel cables graders and specific automatic machines for pitting, slicing, dividing, and stuffing olives, complete the necessary equipment.

8.11 REFERENCES

Albi Romero, M.A. and Garrido Fernández, A. (1975) Aceitunas negras por oxidación en medio alcalino. III. Estudio comparativo de métodos de determinación de hierro. *Grasas y Aceites*, **26**, 133–5.

Amiot, M.J., Fleuret, A. and Macheix, J.J. (1986) Importance and evolution of phenolic compounds in olive during growth and maturation. *J. Agric. Food Chem.*, **34**, 823–6.

Balatsouras, G. (1985) Taxonomic and physiological characteristics of the facultative rod type lactic acid bacteria isolated from fermenting green and black olives. *Grasas y Aceites*, **36**, 239–49.

Balatsouras, G.D. and Vaughn, R.H. (1958) Some fungi that might cause softening of storage olives. *Food Res*, **23**, 235–43.

Balloni, W., Paoletti, C., Fiorenzano, G., Pelagatti, O. and Cucurachi, A. (1977) Un nuovo metodo di preparazione delle olive da mensa per via microbiologica. *Rev. Ital. Sostanze Grasse*, **LIV**, 351–5.

Borbolla y Alcalá, J.M.R. de la, Gonzalez Pellissó, F. and Gonzalez Cancho, F. (1971) Aceitunas verdes y de color cambiante en salmuera. I. *Grasas y Aceites*, **22**, 255–60.

Brenes Balbuena, M., Durán Quintana, M.C. and Garrido Fernández, A. (1986a) Estudio comparativo de sistemas de conservación de aceitunas tipo negras. *Grasas y Aceites*, **37**, 123–8.

Brenes Balbuena, M., García García, P. and Garrido Fernández, A. (1986b) Estudio comparativo de sistemas de conservación de aceitunas tipo negras. II. Efecto sobre las características del producto final. *Grasas y Aceites*, **37**, 301–6.

Brenes Balbuena, M., García García, P. and Garrido Fernández, A. (1992) Phenolic compounds related to black colour formed during the processing of ripe olives. *J. Agric. Food Chem.*, **40**, 1192–6.

Brenes Balbuena, M., García García, P., Durán Quintana, M.C. and Garrido Fernández, A. (1993a) Concentration of phenolic compounds change in storage brines of ripe olives. *J. Food Sci*, **56**, 347–50.

Brenes Balbuena, M., García García, P., Romero Barragán, C. and Garrido Fernández, A. (1993b) Estudio de los factores que afectan a la velocidad de neutralización de la pulpa durante la elaboración de aceitunas tipo negras. *Grasas y Aceites*, **44**, 190–4.

California Olive Industry (1990) California ripe olive. Nutritional Facts and Caloric Information. Lindsay, California.

Castillo Gómez, J., Mínguez Mosquera, M.I. and Fernández Díez, M.J. (1978) Presencia de poligalacturonasa (PG) y su relación con el ablandamiento en algunos productos empleados en la industria del aderezo (pimientos y aceitunas). *Grasas y Aceites*, **29**, 97–101.

Chambionat, A. (1961) Contribution a l'étude du noircissement artificiel des olives vertes. *Cahiers de la Recherche Agronomique*, **11**, 61–8.

Chung, J.I., Sakamura, S. and Luh, B.S. (1974) Effect of harvest maturity on pectin and texture of canned black olives (*Olea europaea*, L.). *Confructa*, **19**, 227–35.

Ciafardini, G., Marsilio, V., Lanza, B. and Pozzi, N. (1994) Hydrolysis of oleuropein by *Lactobacillus plantarum* strains associated with olive fermentation. *Appl. Environ. Microbiol.*, **60**, 4142–7.

Cruess, W.V. (1958) *Commercial Fruits and Vegetable Products*, 4th edn, McGraw-Hill, New York.

Cruess, W.V. (1962) *Experiments on Colour Fixation with Iron Salts*, 41st Annual Technical Report, California Olive Association. Sacramento, California, USA.

Cruess, W.V., el Saifi, A. and Develter, E. (1939) Changes in olive composition during processing. *Ind. Eng. Chem.*, **31**, 1012–14.

Darkey, E.L., Lowe, E., Baker, K.A. and Burgers, J.W. (1973) Field tests of salt recovery system for spent pickle brine. *J. Food Sci.*, **38**, 507–11.

Deiana, P., Farris, G.A., Catzeddu, P. and Madan, G. (1992) Impiego di fermenti lattici e lieviti nella preparazione delle olive da mensa. *Industrie Alimentarie*, **XXXI**, 1011–23.

Drusas, A., Vanegas, G.K. and Saravacos, G.D. (1988) Diffusion of sodium chloride in green olives. *J. Food Eng.*, **7**, 211–22.

Durán Quintana, M.C., Brenes Balbuena, M., García García, P. and Fernández González, M.J. (1991) Aceitunas tipo negras. Estudio comparativo de tres procedimientos para la conservación previa de frutos de la variedad Gordal (*O. europaea regalis*). *Grasas y Aceites*, **42**, 106–13.

Durán Quintana, M.C., García García, P., Brenes Balbuena, M. and Garrido Fernández, A. (1993) *Lactobacillus plantarum* survival during the first days of ripe olive brining. *System. Appl. Microbiol.*, **16**, 153–8.

Durán Quintana, M.C., García García, P., Brenes Balbuena, M. and Garrido Fernández, A. (1994a) *Lactobacillus plantarum* survival in aerobic, directly brined olives. *J. Food Sci.*, **59**, 1197–201.

Durán Quintana, M.C., García García, P., Brenes Balbuena, M. and Garrido Fernández, A. (1994b) Induced lactic acid fermentation during the preservation stage of ripe olives from Hojiblanca cultivar. *J. Appl. Bacteriol.*, **76**, 377–82.

Etchells, J.L., Borg, A.F., Kittel, I.D., Bell, T.A. and Fleming, H.P. (1966) Pure culture fermentation of green olives. *Appl. Microbiol.*, **14**, 1027–41.

Etchells, J.L., Fleming, H.P., Hontz, L.H., Bell, T.A. and Monroe, R.J. (1975) Factors influencing bloater formation in brined cucumbers during controlled fermentation. *J. Food Sci.* **40**, 569–75

Farris, G.A., Deiana, P. and Budroni, M. (1989) La microflora blastomicetica delle drupe e delle salamoia delle olive da mensa. *Industrie Alimentari*, **XXVIII**, 263–6.

Federal State Inspection Service (1993) *California Fresh Olives Inspection Manual*. United States Department of Food and Agriculture, Sacramento, CA.

Ferguson, L., Gawad, H., Sibbet, G.S., Freeman, M. and Hatakeda, J.J. (1991) Using fruit size to predict harvest timing of California Manzanilla olives for ripe-black processing. *HortScience*, **26**, 26–9.

Fernández Díez, M.J. (1969) Elaboración de aceitunas de mesa en California. *Grasas y Aceites*, **20**, 194–205.

Fernández Díez, M.J. (1979) Texture of table olives and pimientos. *Journal of Texture Studies*, **10**, 103–16.

Fernández Díez, M.J. (1983) Olives, in *Biotechnology*, Vol. 5 (eds H.J. Rehm and G. Reed), Verlag Chemie, Weinheim, Germany, pp. 379–97.

Fernández Díez, M.J. (1984) Changes in the chemical components during the processing of table olives and their relation to the quality, in *Proceedings of M.O.C.A.*, Vol.I, Valencia, Spain, pp. 301–18.

Fernández Díez, M.J. (1991) Procesing table olives, in *Encyclopedia of Food Science and Technology*, Vol. III (ed. Y.H. Hui), Wiley & Sons, New York, pp. 1910–25.

Fernández Díez, M.J. and Garrido Fernández, A. (1971) Aceitunas negras por oxidación en medio alcalino. I. El color como criterio de madurez y calidad en el producto elaborado. *Grasas y Aceites*, **22**, 193–9.

Fernández Díez, M.J., Garrido Fernández, A., González Cancho, F., Durán Quintana, M.C. and Cordón Casanueva, J.L. (1972) Elaboración de aceitunas negras de mesa. Instituto de la Grasa, Sevilla, Spain.

Fernández Díez, M.J., Mínguez Mosquera, M.I. and Castillo Gómez, J. (1974) Composición péctica y presencia de pectinesterasa en algunos productos del aderezo. *Grasas y Aceites*, **25**, 331–40.

Fernández Díez, M.J., Castro Ramos, R., Garrido Fernández, A., González Cancho, F., González Pellissó, F., Nosti Vega, M., Heredia Moreno, A., Mínguez Mosquera, M.I., Rejano Navarro, L., Durán Quintana, M.C., Sánchez Roldán, F., García García, P. and Castro Gómez Millán, A. (1985) *Biotecnología de la aceituna de mesa*. Consejo Superior de Investigaciones Científicas, Instituto de la Grasa, Madrid, Sevilla, Spain.

Fernández González, M.J., Garrido Fernández, A., García García, P., Brenes Balbuena, M. and Durán Quintana, M.C. (1992) Características del proceso fermentativo durante la conservación de aceitunas de la variedad Hojiblanca, destinadas a la elaboración del tipo negras. *Grasas y Aceites*, **43**, 212–18.

Fernández González, M.J., García García, P., Garrido Fernández, A. and Durán Quintana, M.C. (1993) Microflora of the aerobic preservation of directly brined green olives from Hojiblanca cultivar. *J. Appl. Bacteriol.*, **75**, 226–33.

Florenzano, G., Margheri, M.C. and Pelasi, E. (1973) Recherches sur la microflore des olives, pates grignons et sur l'activité lipolitique des especes predominantes. *Inf. Oleicoles Internationales*, **60/61**, 145–51.

Foda, I.O. and Vaughn, R.H. (1950) Salt tolerance in the genus *Aerobacter*. *Food Technol.*, **4**, 182–8.

García García, P., Brenes Balbuena, M. and Garrido Fernández, A. (1986) Uso de lactato ferroso en la elaboración de aceitunas tipo negras. *Grasas y Aceites*, **37**, 33–8.

García García, P., Brenes Balbuena, M. and Garrido Fernández, A. (1987) Elaboración de aceitunas tipo negras a partir de verdes estilo español. *Grasas y Aceites*, **38**, 228–32.

García García, P., Brenes Balbuena, M. and Garrido Fernández, A. (1991) Effect of oxygen and temperature on the oxidation rate during darkening of ripe olives. *J. Food Eng.*, **13**, 259–71.

García García, P., Brenes Balbuena, M., Vattan, P. and Garrido Fernández, A. (1992a) Kinetic study at different pH values of the oxidation process to produce ripe olives. *J. Sci. Agric.*, **60**, 327–31.

García García, P., Durán Quintana, M.C., Brenes Balbuena, M. and Garrido Fernández, A. (1992b) Lactic fermentation during the storage of Aloreña cultivar untreated green table olives. *J. Appl. Bacteriol.*, **73**, 324–30.

García García, P., Brenes Balbuena, M. and Garrido Fernández, A. (1994) Effect of pH and salts on the firmness of canned ripe olives. *Sciences des Aliments*, **14**, 159–72.

Garrido Fernández, A. (1980a) Aceitunas negras por oxidación en medio alcalino. IV. Estudio de la influencia de algunos aditivos y formas de envasado sobre la textura. *Grasas y Aceites*, **31**, 323–9.

Garrido Fernández, A. (1980b) Aceitunas negras por oxidación en medio alcalino. V. Estudio de la influencia de algunos aditivos y formas de envasado sobre el color. *Grasas y Aceites*, **31**, 385–90.

Garrido Fernández, A. (1981) Elaboración de aceitunas negras por oxidación en medio alcalino con un solo tratamiento de lejía. *Grasas y Aceites*, **32**, 219–25.

Garrido Fernández, A. (1983) Estudio de las aguas residuales del proceso de elaboración de aceitunas negras por oxidación y sus reutilizaciones (y II).

Influencia sobre la calidad del producto final y sus salmueras de envasado. *Grasas y Aceites*, **35**, 165–71.

Garrido Fernández, A. and Fernández Díez, M.J. (1972) Aceitunas negras de mesa. Evolución de las principales características del fruto durante el desarrollo y maduración. *Grasas y Aceites*, **23**, 309–18.

Garrido Fernández, A. and Sánchez Roldán, F. (1982) Envasado de aceitunas negras por oxidación en medio ácido y con esencias aromáticas. *Grasas y Aceites*, **33**, 340–5.

Garrido Fernández, A., Albi, M.A. and Fernández Díez. (1973) Aceitunas negras por oxidación en medio alcalino. II. Determinación del contenido en hierro y evolución de diversos factores durante el proceso de elaboración. *Grasas y Aceites*, **24**, 287–92.

Garrido Fernández, A., Cordón Casanueva, J.L. and Fernández Díez, M.J. (1975) Diseño y estudio de un texturómetro experimental para su empleo en diversos tipos de aceitunas aderezadas. *Grasas y Aceites*, **26**, 195–202.

Garrido Fernández, A., García García, P. and Sánchez Roldán, F. (1985) Nuevo proceso aerobico de fermentación de aceitunas negras al natural. Optimizacion de la columna de aireación. *Alimentación, Equipos y Tecnología*, May–June, 73–81.

Garrido Fernández, A., Durán Quintana, M.C. and García García, P. (1986) Estudio de diferentes formas de conservación de aceitunas de color cambiante de la variedad Manzanilla. *Grasas y Aceites*, **37**, 1–7.

Garrido Fernández, A., García García, P., Montaño Asquerino, A., Brenes Balbuena, M. and Durán Quintana, M.C. (1993) Biochemical changes during the preservation stage of ripe olive processing. *Die Nahrung*, **37**, 583–91.

González Cancho, F., Nosti Vega, M., Durán Quintana, M.C., Garrido Fernández, A. and Fernández Diez, M.J. (1975) El processo de fermentación de aceitunas negras maduras en salmuera. *Grasas y Aceites*, **26**, 297–309.

Hartmann, H.T., Simone, M., Vaughn, R.H. and Maxie, E.C. (1959) Relation of fruit maturity at harvest to quality of black-ripe processed olives. *Am. Soc.Hort. Sci.*, **73**, 213–28.

Hilts, R.W. and Hollingshead, R.S. (1920) A chemical study of the ripening and pickling of California Olive. US Department of Agriculture Bureau of Chemistry, Bulletin 308.

Instituto de la Grasa and Instituto de Agroquímica y Tecnología de Alimentos (1970) *Elaboración de aceitunas negras por oxidación en medio alcalino*, Valencia, Spain.

International Olive Oil Council (COI) (1960–1990) *Yearly Reports on Production of Table Olives*, Madrid, Spain.

International Olive Oil Council (1980) *Unified Qualitative Standard Applying to Table Olives in International Trade*, Madrid, Spain.

Jimenez Araujo, A., Labavitch, J.M. and Heredia Moreno, A. (1994) Changes in the cell wall of olive fruit during processing. *J. Agric. Food Chemistry*, **42**, 1194–9.

Kader, A.A., Nanos, G.D. and Kerbel, E.L. (1990) Storage potential of fresh Manzanillo olives. *California Agriculture*, **44**, 23–4.

Luh, B.S. and Ferguson, L. (1994) Processing California olives, in *Olive Production Manual*, University of California, Division of Agriculture and Natural Resources, Oakland, California, Publication 3353, pp. 133–42.

Marquina, D., Peres, C., Caldas, F.V., Marques, J.F., Peinado, J.M. and Spencer-

Martin, J. (1992) Characterization of the yeast population in olive brines. *Letters in Appl. Microbiol*, **14**, 279–83.
Materassi, R., Midans, N. and Pelagatti, O. (1975) Sulla idrolisi della oleuropeina nei lieviti. *Ann. Ist. Sper. Elaiot.*, **V**, 53–68.
McCorkle, K.C., Wells, H.M., Scrimshire, H.D. and Beckwith, R.N. (1984) Method for producing black-ripe olives. *United States Patent Number* 4, 463, 023.
McDonald, L.C., Fleming, H.P and Hassan, H.M. (1990) Acid tolerance of *Leuconostoc mesenteroides* and *Lactobacillus plantarum*. *Appl. Environ. Microbiol.*, **56**, 2120–4.
McFeeters, R.F. and Chen, M. (1986) Utilization of electron acceptors for anaerobic mannitol metabolism by *Lactobacillus plantarum*. Compounds which serve as electron acceptors. *Food Microbiology*, **3**, 73–9
Mínguez Mosquera, M.I. (1982) Evolución de los constituyentes pécticos y de las enzimas pectinolíticas durante el proceso de maduración y almacenamiento de la aceituna Hojiblanca. *Grasas y Aceites*, **33**, 327–33.
Mínguez Mosquera, M.I., Castillo Gómez, J. and Fernández Díez, M.J. (1976) Variaciones en la composición péctica durante la elaboración, fermentación y conservación de productos aderezados (pimiento y aceituna). *Grasas y Aceites*, **27**, 27–32.
National Canners Association (1976) *Process for low-acid canned foods in metal containers*. Bulletin 26-I, 11th edn, Washington, DC.
Nosti Vega, M. and Castro Ramos, R. de. (1985) Composición y valor nutritivo de algunas variedades españolas de aceituna de mesa. VII. Aceitunas negras oxidadas. *Grasas y Aceites*, **36**, 203–6.
Nosti Vega, M., Castro Ramos, R. de and Vázquez Ladrón, R. (1984) Composición y valor nutritivo de algunas variedades españolas de aceitunas de mesa. VI. Cambios debidos a los procesos de elaboración. *Grasas y Aceites*, **35**, 11–14.
Patel, J.B. and Vaughn, R.H. (1973) Cellulolitic bacteria associated with sloughing spoilage of California ripe olives. *Appl. Microbiol.*, **25**, 62–9.
Pelagatti, O. (1978–80a) La preparazioni di olive da tavola verdi al naturale delle varietá Nocellara etnea e Intosso d'Abruzzo. *Ann. Ist. Sper. Elaiot.*, **VIII**, 177–92.
Pelagatti, O. (1978–80b) Sulla microflora lactica e blastomicetica associata alle drupe di alcune cultivars di *Olea europaea*. *L. Ann. Ist. Sper. Elaiot.*, **VIII**, 195–206.
Ralls, J.W. (1969) The use of activated carbon to renovate brines. 48th Annual Technical Report, California Olive Association, Sacramento, California, USA.
Ralls, J.W., Maagdenberg, H.J., Lemoine, G. and Mercer, W.A. (1971) Alternative storage systems for the production of canned black-ripe olives. *J. Food Sci.*, **36**, 408–12.
Rose, W.W. (1966) Research on low-salt brines and olive storage problems. 45th Annual Technical Report, California Olive Association, Sacramento, California, USA.
Ruíz Barba, J.L., Garrido Fernández, A. and Jiménez Díaz, R. (1991) Bactericidal action of oleuropein extracted from green olives against *Lactobacillus plantarum*. *Letters in Appl. Microbiol.*, **12**, 65–8.
Ruíz Barba, J.L., Brenes Balbuena, M., Jiménez Díaz, R., García, P. and Garrido Fernández, A. (1993) Inhibition of *Lactobacillus plantarum* by polyphenols extracted from two different kinds of olive brines. *J. Appl. Biotechnol.*, **74**, 15–19.

Scrimshire, H.D. (1987) Method for processing black-ripe olives. *United States Patent Number 4, 664, 926.*

Simpson, K.L., Chichester, C.O. and Vaughn, R.H. (1961) Compounds responsible for the colour of black-ripe olives. *J. Food Sci.*, **26**, 227–33.

Spanish Government Presidency (1983) Reglamentación técnico sanitaria para la elaboración, circulación, y venta de las aceitunas de mesa. *Spanish State Official Bulletin*, **108**, 12619–25.

Torre, J.E. de la, Moya, E.R., Bota, E. and Sancho, J. (1993) Estudio físico-químico y microbiológico de la fermentación de aceitunas verdes arbequinas. *Grasas y Aceites*, **44**, 274–8.

Unidad Estructural de Biotecnología de Alimentos (Instituto de la Grasa) (1990) Table olive processing, International Olive Oil Council, Madrid, Spain.

United States Department of Agriculture (USDA) (1977) *Home and Garden*, Bull. **72**, Washington, DC.

United States Department of Agriculture (1983) United States Standards for Grades of Canned Ripe Olives, Washington, DC.

Vaughn, R.H. (1954) Lactic acid fermentation of cucumbers, sauerkraut and olives, in *Industrial Fermentations*, Vol. II (eds L.A. Underkofle and R.J. Hickey), Chemical Publishing Co. Inc., New York, pp. 417–78

Vaughn, R.H. (1966) Use of iron compounds for fixation of colour in olives. 45th Annual Technical Report, California Olive Association, Sacramento, California, USA.

Vaughn, R.H. (1970) Progress report on the year's research with olives. 49th Annual Report, California Olive Association, Sacramento, California, USA.

Vaughn, R.H. (1975) Research progress report. 53rd Annual Technical Report, California Olive Association, Sacramento, California, USA.

Vaughn, R.H. (1982) The fermentation of olives, in *Industrial Microbiology*, 4th edn (ed. G. Reed), AVI Publishing Co., Westport, Connecticut, pp. 207–36.

Vaughn, R.H., Jakubczyk, T., McMillan, J.D., Higgins, T.E., Dave, B.A. and Crampton, V.M. (1969a) Some pink yeasts associated with softening of olives. *Appl. Microbiol.*, **18**, 771–5.

Vaughn, R.H., King, A.D., Nagel, C.W., Ng, H., Levin, R.E., McMillan, J.D. and York, G.K. II (1969b) Gram-negative bacteria associated with sloughing, a softening of California ripe olives. *J. Food Sci.*, **34**, 224–7.

Vaughn, R.H., Martin, M.H., Stevenson, K.E., Johnson, M.G. and Crampton, V.M. (1969c) Salt-free storage of olives and other produce for future processing. *Food Technology*, **23**, 124–6.

Vaughn, R.H., Stevenson, K.E., Dave, B.A. and Park, H.C. (1972) Fermenting yeasts associated with softening and gas-pocket formation in olives. *Appl. Microbiol*, **23**, 316–20.

Vlahov, G., Marsilio, V., Brighigna, A. and Cucurachi, A. (1977) Sull'impiego di sali di ferro per l'ennerimento delle olive. *Ann. Ist. Sper. Elaiot.*, **VII**, 3–16.

Naturally black olives type 9

9.1 INTRODUCTION

According to the Unified Qualitative Standard Applying to Table Olives in International Trade (International Olive Oil Council, 1980) and Spanish Standard (Spanish Government Presidency, 1983), naturally black olives type are those 'obtained from fruits harvested when fully ripe or slightly before full ripeness is reached; they may, according to the production region and time of harvesting be reddish black, violet black, deep violet, greenish black or deep chestnut'. It must be added that cultivar is also an important factor determining colour.

Within this type, the most important commercial preparation or trade preparation, covering about 30% of the international market, is denominated naturally black olives in brine. It is included within the group of commercial preparations known as untreated black olives; that is, those which are placed directly in brine without any lye treatment. They have a fruity flavour, which is more marked than that of the treated black olives, and the product usually retains a slight bitter taste. They are preserved by natural fermentation in brine.

9.2 NATURALLY BLACK OLIVES IN BRINE

This process has been practised traditionally in Greece where it remains of great importance (Fernández Díez, 1969, 1983; Fernández Díez et *al.*, 1972). The most important production region is the central part of Continental Greece starting from the Egoumenitsa region and terminating at Volos by the Aegean Sea. The peninsula of Halkidiki, south of Salonica, is dedicated almost exclusively to production of the cultivar of the same name. Other regions producing naturally black table olives in Greece are Egoumenitsa and the Temby valley (between Larissa and Salonica). The wide international distribution of naturally black olives and their importance for local markets has been studied and described by Balatsouras (1967, 1990).

Production of naturally black olives in brine is an old tradition in Turkey. The well-known black olive pickling area is the Marmara Sea district (north-west Turkey), especially the province of Bursa (Borcakli *et al.*, 1993).

Some characteristics of this type are:

- Natural processing, with a minimum of chemical treatment other than the presence of NaCl.
- Use of practically completely mature fruits, with the maximum possible profits for farmers (counteracted by the possibility of losses due to early frosts).
- Simple processing procedure with low energy consumption compared with black-ripe olives (California-style) or Spanish-style green olives.

In Spain, the use of this process has increased greatly during the past 20 years (Fernández Díez *et al.*, 1985), mainly in the province of Cordoba, although it has been partially substituted by an increasing production of treated black olives (ripe olives).

At present, the main producing country is Turkey with over 100 000 tons/year (about 25% of the world's naturally black olives in brine). Most of these olives are destined for the internal market. However, Greece, with 18–21% of world production, is responsible for more than 50% of total exports of this product.

The Greek Conservolea and the Spanish Hojiblanca are the two more well-known and appreciated cultivars used to prepare naturally black olives. Both have particular characteristics which make them the best raw material for the preparation of this type of olive and both give a fairly good final product flavour, colour and texture. Other countries of the Middle East such as Syria, Jordan, Lebanon and Cyprus are also important producers, although their crops are sold in their own interior markets.

The general scheme of the process is shown in Figure 9.1. Olives are transported to the factory (Fernández Díez, 1991), sorted to separate damaged fruits, occasionally roughly graded to eliminate the small sizes, and quickly washed with water to remove superficial dirt. They are then placed into a brine with a salt concentration ranging from 8 to 10°Bé, or slightly lower.

A natural fermentation takes place for which a complex microflora of bacteria and yeasts are responsible. Because of the slow diffusion of sugars and other solubles from the fruits to the brine in untreated olives and the presence of oleuropein, the process is frequently very slow and incomplete if suitable physical, chemical and microbiological controls are not applied.

Afterwards, the fermented olives are exposed to air to improve the skin colour, classified by size and packed with an appropriate new brine in barrels, cans or plastic bags.

Fresh olives
|
Harvesting
|
Transport to the factory
|
Sorting (rough size grading)
|
Washing
|
Brining
|
Fermentation
|
Sorting and size grading (air exposure)
|
Packing

Fig. 9.1 General scheme for the processing of Greek-style naturally black olives in brine. (Source: Fernández Díez *et al.*, 1985.)

9.2.1 HARVESTING, TRANSPORT TO THE FACTORY, SORTING AND SIZE-GRADING

As indicated in section 9.1, fruits must be harvested at, or slightly before, full ripeness. For this reason their superficial colour is not a suitable criterion to fix the harvest time. To get a good colour in the fermented final product it is necessary to take a representative sample of olives, cut them close to the pit and observe the flesh colour (Fernández Díez *et al.*, 1985). A good ripeness stage is when a colour between violet and black reaches at least up to 2 mm from the pit. Indeed, skin and flesh colours are very variable, depending upon production region, time of harvesting, and cultivar used. This criterion, shared by Greek (Balatsouras, 1967) and Spanish (Fernández Díez *et al.*, 1972) workers, is now changing and with newly acquired knowledge it is possible to bring forward harvesting to when the desired colouring is only half-way through the flesh (Unidad Estructural de Biotecnología de Alimentos, 1990).

There is a big difference between fruit picked at the beginning and at the end of the season. Olives harvested early have a good texture but the colour after processing is not good. Fruit picked at the end of the season retains excellent colour after processing but the texture is less firm (Fernández Díez, 1983).

Harvesting for this type of olives must be manual. The fruits should be hand-picked and carefully transported to the factory in holed plastic bins with an approximate capacity of 20–22 kg fruit, and the time interval between harvesting and the start of processing should be as short as possible.

In Greece, olives intended to be processed in this way are harvested between 15th October and 15th December. When there is a heavy crop, the harvest begins early and is completed by three or four successive pickings, but with a light crop this operation is performed in only one picking when the fruit reaches maximum colour.

The olives must be picked before the winter frosts, since these cause shrivelling and loss of crop. Early frost is always a menace to naturally ripe olives and has caused many total losses of the crop in Greece and has been a serious problem in Spain where the main cultivar devoted to this preparation (Hojiblanca) usually reaches the appropriate ripeness by January or February.

In spite of its desirability, harvest by hand is not always applied and sometimes a substantial part of the crop is picked by knocking the fruits from the trees using slender poles or rakes. Another technique is to strip the fruit off by hand and leave them to fall on canvas or nylon nets spread under the tree. With both procedures, an important proportion of fruits is damaged, leading to low yields of first-class final product.

In spite of the surface colour of naturally black olives – which can hide slight damage better than in green olives – their advanced stage of maturity makes the flesh softer and very sensitive to bruising and scratching. Thus, damage during picking must be reduced as much as possible. The hand-picking system is therefore strongly recommended. In Greece, harvested fruits are also put into large cane baskets with capacity up to 45–50 kg where they can sometimes be slightly damaged by the pressure of fruits.

Similarly, transportation of these fruits over long distances is not advisable. In Greece, factories are located mainly within a few miles from the orchards.

In Turkey, olive picking begins in Bursa in October or at the beginning of November and can last until February. The fruit is picked carefully into small woven baskets. Only the ripe, violet-coloured fruit is picked and brought to fermentation plants, where the bruised, insect-damaged, undersized and unripe fruit is sorted out by hand. Extreme care must be taken to avoid bruising.

After a manual sorting to separate damaged fruits, the olives are occasionally subjected to a rough grading to eliminate small sizes which will be used for oil extraction.

9.2.2 WASHING AND BRINING

After sorting and optional size-grading, olives must be washed with water to remove superficial dirt from the fruits. Some years ago a static washing system was used in Greece (Fernández Díez, 1969). Olives were placed in water in the fermentation tanks and the water changed several times over a period of 2–3 days before brining. The effects claimed for this operation were: (i) cleaning the fruits; (ii) partial elimination of the bitter principle by diffusion into the water; and (iii) a gradual increase in NaCl in the olive flesh when a light brine was used instead of just water.

Fernández Díez *et al.* (1972) and Durán Quintana *et al.* (1973) studied the changes during these washings. Table 9.1 shows data from their experiments with light brines (similar results were obtained with water, except those related to salt content). The washings remove a negligible proportion of polyphenols (data not shown) and changes in NaCl were also minimal. Change in pH seems to be more rapid due to the lack of any buffering capacity in the solution. The microbial population is composed mainly of Gram-negative bacteria, although lower counts of yeasts were also found. The microflora is responsible for the pH diminution and the abundant gas production during this phase. Counts for both were always higher in the second washing. No lactic acid bacteria have been found in washings. The dominance of Gram-negative bacteria appears related to their low nutrient requirements. Such nutrients are scarce in solution during this period due to the low skin permeability of fruits untreated with lye.

The effect of Gram-negative bacteria on naturally black olives is apparently not the same as that described in Spanish-style green olives. They have never been related to gas-pocket formation or other types of spoilage of these olives, although they may be potential spoilage organisms. The most evident effect of their presence is the frequent overflowing of

Table 9.1 Characteristics of static washing (2% NaCl brine for 24 hours) of naturally black olives in brine (Adapted with permission from Fernández Díez *et al.*, 1972, and Durán Quintana *et al.*, 1973)

Operation	Physicochemical changes				Final microbial population	
	pH		NaCl			
	Initial	Final	Initial	Final	Gram-negatives/ml	Yeasts/ml
Lechín cultivar						
1st washing	6.8	5.8	2.5	2.5	80	20
2nd washing	6.8	5.5	2.5	2.1	4500	30
Hojiblanca cultivar						
1st washing	6.7	5.2	2.5	2.4	1.27×10^5	–
2nd washing	6.7	4.7	2.5	2.3	1.27×10^6	2.2×10^3
Verdial cultivar						
1st washing	7.0	5.1	2.2	2.1	6×10^5	1.2×10^3
2nd washing	7.0	5.0	2.2	2.2	1.2×10^6	1.9×10^3

the brines out of the fermenters due to CO_2 production. The contribution of fermenting yeasts to the gas production is low during this period. Differences among cultivars are evident and are probably related to the skin permeability and polyphenol content.

Due to the disadvantages of the presence of Gram-negatives, there is a tendency to control their growth by adjusting the initial pH with HCl, acetic or lactic acid. This correction is also useful to avoid or reduce the growth of other undesirable organisms such as *Bacillus* and *Clostridium* species and is thus an advisable practice. When this is done, a slight pH increase is observed during the first hours of washing due to the solubilization of compounds from the flesh; this then diminishes again. The number of bacteria can also be reduced by frequent changes of the washing brine or water, reducing the contact time between olives and the solution to 18 hours or less.

Apart from the cleaning effect, it seems that static washings have little positive effect on other characteristics. On the contrary, they represent an appreciable volume of wastewater and a labour-intensive operation that may interfere seriously with the proper handling of the olives which are received every day. Efficient handling of the large amounts of fruits arriving at the factory requires simpler washing procedures. In Greece, it has been found that maintaining the olives for several days in water renders fruits more resistant to shrivelling, gives the best colour, a better consistency and, sometimes, a rapid lactic fermentation.

Olives are now mainly washed using water sprayers. Other methods use revolving drums (through which the olives are immersed in running water), quickly drained, and then placed as soon as possible into brine. It is important that fermentation tanks or fermenters are one-third filled with brine in advance to prevent damage to the fruit hitting the bottom or knocking against each other (Unidad Estructural de Biotecnología de Alimentos, 1990).

Initial brine concentration generally ranges between 8 and 10°Bé. Sometimes to avoid shrivelling, the salt content is lower, around 6% sodium chloride. This is mainly in the case of rainy weather just before harvesting or when using certain delicate cultivars (Fernández Díez *et al.*, 1985).

Processing conditions vary widely according to local traditions. In places where temperatures are low when fruits are picked, they are put in just tap water. This is maintained during the cold winter and when spring approaches the liquid is supplemented with salt to avoid malodorous spoilage. Some traditions use high initial salt contents (14% or higher), although this is limited by the sensitivity of the olives to salt.

In Turkey, for fermentation, cement tanks (1.5–3.0 m in height) are used, containing alternate layers of olives and salt, which are finally covered with mattings or boards (Fernández Díez, 1972a). Large stones are placed on the boards to keep the fruits submerged in the brine. Once the containers are prepared, water is added which dissolves the salt and forms the brine. The final brine salt content may reach 10–12% (w/v) (Fernández Díez, 1972a). However, recently fruits harvested when fully ripe are put into concrete basins containing 12–14% NaCl, a procedure which, although it still maintains a high salt content, is more similar to those used in other countries.

Other traditions use part of previous brines to initiate the fermentation. This procedure has been used in Greece with the aim of also preserving a better colour, thinking that a previously used brine can dissolve less colouring matter than a freshly prepared one. This practice is being abandoned because there is no scientific confirmation of its putative advantages. On the contrary, it may represent a spoilage risk in cases of previous contamination (Fernández Díez *et al.*, 1972; Balatsouras, 1990).

Barret and Bidan (1972) described a similar brine re-use in the production of naturally black olives according to the Nyons (France) system. In this case fruits are permitted to stay on the trees until over-ripe and after cold weather. This provokes a certain shrivelling and sweetening of the fruits. New olives are brined in previously used brines. The high organic matter concentration prevents weight losses and, at the same time, supplies appropriate strains of microorganisms for the fermentation. However, the process is not widely used.

Traditionally, the fresh brine used for naturally black olives was simply a solution of salt in water. However, due to the possibility of initial contamination with extraneous microorganisms carried by the olives and the abundant production of gases during the initial brining period, it is becoming customary to correct the initial pH using the same acids as used with the washing waters, acetic acid being the most common.

Throughout the period olives are in brine there are two important effects: (i) a progressive debittering due to the diffusion of oleuropein into the brine; and (ii) a fermentative process of different characteristics depending on the cultivar and brining conditions but which always leads to a pleasant flavour in the final product.

A detailed explanation of the characteristics of the Greek traditional process can be found in Balatsouras (1964, 1966a,b, 1967, 1969) and in Balatsouras and Polymenakos (1964). The production of naturally black olives from Spanish cultivars is well described by González Cancho *et al.* (1975) and Fernández Díez *et al.* (1985).

Currently, two different fermentation systems can be applied: (i) the traditional one, which is carried out under more or less strict anaerobic conditions; and (ii) an aerobic process recently developed by Spanish researchers to avoid the formation of gas-pockets and shrivelling. These will be described separately.

9.2.3 TRADITIONAL ANAEROBIC FERMENTATION AND STORAGE

In Greece, naturally black olives had been mainly prepared by the growers (Balatsouras, 1967, 1969, 1990), who put the olives into wooden vats with capacities ranging from 1–10 tonnes and some 2–3 m in height. They were fitted with a wooden or plastic tap via which the aqueous solutions can be drawn out. Due to the limited production per grower, it could take some time to fill one vat. Eventually, the olives were covered with a 10% NaCl brine and submerged by covering them with old bags and wooden screens weighted down with stones to counteract the pressure of gases produced.

Salt concentration decreased to 6–7% or lower. In spring, to prevent spoilage, the salt content was increased to about 14% by addition of solid salt. Frequently the olives became salty and shrivelled. A thick leathery membrane was formed which was later studied and found to be formed by a pad of fungal mycelium interspersed with bacteria and yeasts. To empty the vats, olives were removed with perforated buckets, the workers occasionally having to descend into the vat to do this. The olives were then sent to the packers who sorted, graded by size and canned the product, using rudimentary machinery.

Deficiencies in the process were evident. Growers had no grading and washing machines, no running water or draining system. Under these conditions olives were stored without even a hand sorting to eliminate heavily bruised, insect-damaged or highly off-coloured fruits. In the 1950s a systematic study of the Greek traditional naturally black olive process was initiated by Balatsouras whose researches contributed to the transformation of such industry in his country. Growers were convinced progressively of the need to process in modern plants to enhance quality. The proportion of olives prepared by growers has decreased dramatically during the past 20 years and may represent in 1995 only about 20% of total production. These changes were led by Eleourgiki, an organization constituted by the olive producers, which is currently the main industrial producer of table olives in this country.

In these large industrial units, containers are made of concrete and lined internally with plastic material resistant to corrosion. More recently, fibre-glass has been used successfully as an alternative covering for the inner surface of tanks. Tanks are 2–2.35 m high with a capacity of 10–15 tonnes and are built with their bases at floor level. Exceptionally the bottom of the tanks is sunk 1–1.5 m into the ground, in which case loading and unloading become easier, provided that an underground drainage system exists for the removal of the wastewater. To seal the tanks, close-fitting covers comprising two or three parts are used. All tubes and valves that come in contact with brine are made out of polyvinyl chloride (PVC) or stainless steel. The recent trend has been to decrease the height of tanks to 1.85 m to avoid squashing the fruit and to facilitate discharge. Unloading the tanks is still done by hand or with cane baskets and is very labour intensive. A complete comparison of fermentation in cement tanks and wooden containers was made by Aligisakis (1971) who found no differences with respect to the fermentation, nor in relation to the organoleptic characteristics. Furthermore, mould growth in old wooden tanks could give olives wet odours which are impossible to remove by customary washing procedures.

During the first period of fermentation, the component parts of the cover are left loose to provide an easy escape for the gas evolved. The wooden covers are also provided with two or more circular openings through which solid table salt in cheesecloth sacks is hung into the brine from time to time to increase its strength to the desired level. The same openings serve for withdrawing samples of either brine or olives and for inspecting the tank contents. To close them, processors place circular pieces of plastic material over them which they cover with paraffin.

In general, the spontaneous fermentation process is slow because diffusion of soluble components through the skin is slow when fruits have not been treated with alkali. A complex microbiological population develops. Gram-negative bacteria of the coli–aerogenes type, Gram-positive lactic cocci, diverse yeasts and some species of *Lactobacillus* may be present if the sodium chloride concentration is kept below 6–7%, but generally yeasts are dominant (González Cancho *et al.*, 1975).

If for some reason the dominating organism is a *Lactobacillus*, then the final product will have a low pH value (about 4) and the acidity will be higher than 0.6%, expressed as lactic acid. This product will have a very different taste from the usual product, and its colour will be much lighter, developing a pink tone.

In most instances, when fermentative yeasts predominate, the final product has a pH between 4.5 and 4.8 and a free acidity between 0.1 and 0.6%. The low acidity and relatively high pH require a gradual increase of the salt concentration. At the end of the fermentation the concentration of sodium chloride should be 10% or higher for stable storage. Acidification with acetic acid to maintain a pH value of 4.0–4.2 is frequently used in order to decrease the salt level.

In Greece, as the stormy fermentation subsides, the component parts of the wooden cover are fitted and made air-tight by flooding with paraffin. These covers are heavy and are kept in place by means of two projections in the inner surface of the walls, constructed deliberately a few centimetres below the surface of the tank. Not all olive processors want complete anaerobiosis. Most pay little attention to air-tight sealing or deliberately give oxidative microbes the chance to grow giving a low titratable acidity.

The whole process is considered to be over when the fermentable substrates are exhausted and the fermentation has subsided. In most regions in Greece, this happens late in August or even earlier, depending on the raw material, the brine strength and other factors.

The brine is never supplemented with sugar since high acidity is not desirable as it can change the characteristic taste and give the product a cherry-red colour not typical of naturally ripe olives.

The traditional process in Turkey was described as achieving the following objectives:

- Elimination of the natural bitterness of fruits which disappears in the first two months as a result of oleuropein hydrolysis.
- Establishing different kinds of fermentations, the lactic fermentation being the desirable one.
- Absorption of some olive juice by the brine and absorption by the olives of salt from the brine.

So, the pickled olive keeps well in the market due to its high salt and lactic acid content. The total loss of weight of the olive during the whole pickling process averages about 7%.

Salt concentration

Salt plays an important role in the fermentation and stability of naturally black olives. Concentrations used in the brining range from 8 to 14%. This is not optimal for the growth of most microorganisms, although yeasts often show good salt tolerance and are the most common organisms found in these fermentations. This factor helps control malodorous fermentations since the bacteria responsible (*Clostridium*, *Propionibacterium*, etc.) are more sensitive to the salt level than other microorganisms in the fermentation such as the lactic acid bacteria (LAB). It is well known that the use of low salt content permits the growth of proteolytic bacteria and the generation of unpleasant odours.

In the traditional process the salt level has practically inhibited any lactic fermentation, as demonstrated by the high pH and low free acidity found in most of the fermentation brines of any cultivar (Balatsouras, 1964; Fernández Díez and Garrido Fernández, 1969; Durán Quintana *et al.*, 1971; González Cancho *et al.* 1975). The influence of cultivar is also important. González Cancho *et al.* (1975) showed that lactic acid bacteria only developed in naturally black olives from the Hojiblanca cultivar when the salt level was maintained below 8%. However, there was no LAB growth in fermentations of Lechín and Verdial under the same conditions. The inhibitory effect of salt occurs despite the fact that LAB can be progressively adapted to high levels of salt. Balatsouras (1985) found that LAB that fermented naturally black olives were more acclimatized to higher salt concentrations than those isolated from green olives. About

75% of the black olive isolates grew profusely at 8% salt, but 10% salt prevented the growth of practically all of them.

The salt content also has an important effect on the solubilization of organic and inorganic components of the fruits and chemical changes during fermentation. Higher levels of polyphenols have been found in brine when the salt concentration is lower. This supports the experience of some experts who prefer low brine salt concentrations or even tap water in the natural debittering of olives. The same can also occur with sugars, but the effect of salt on the fermentation rate leads to a higher brine sugar content as the salt level increases. Influences of salt on pH and acidity are mainly due to the effect on LAB growth. In general, high salt leads to a relatively high pH (4.2–4.5) and low acidity.

Cultivar tolerance to initial salt content is diverse, but generally high. Lechín cultivar did not shrivel with salt content up to 16%. Other Spanish cultivars (Hojiblanca and Verdial) were not affected by 14%, but at 16% NaCl showed 10 and 31% shrivelling, respectively (Durán Quintana *et al.*, 1973). As brining progresses, salt passes into the flesh and the brine strength decreases.

Equilibrium between the brine and flesh occurs long after brining, since naturally ripe olives retain for some time their selective permeability. To secure homogeneity the brines are circulated from time to time using portable pumps. In Greece, solid salt is not added during winter time so the salt content falls to about 6% or even lower. This low salt level favours LAB growth although this is limited because of the low temperatures. In spring, solid salt is added and the brine circulated to complete solubilization and homogenization. This is repeated until the initial levels are reached again when LAB can hardly grow. Concentrations at equilibrium are around 14–16% in the case of grower/processors and 10% or even lower in the case of new industries (Balatsouras, 1990). Borcakli *et al.* (1993) followed the NaCl level in Turkish cultivars. Initially the brine salt content was 14% (w/v) and reached an equilibrium between the 75th and 100th day of fermentation for the Edencık cultivar and after 100 days for Gemlik, the final equilibrium salt concentrations being 11.3% and 12%, respectively. However, the flesh salt contents were 7.4% and 6.9%, attained at the end of the 225th day of fermentation. Differences between flesh and brine can be explained by the maintenance of the flesh structure in these olives because of the absence of NaOH treatment and their decreased moisture content. Hojiblanca cultivar (the Spanish cultivar most often used for naturally black olive production) reaches equilibrium very slowly and may also be around 150 days from brining when the initial concentration is 10% and around 90 days when it is 6% (García García *et al.*, 1982). Spanish producers normally add salt periodically, even during winter. In this way the salt content is about 8–9% when spring comes and fruits preservation is assured.

Sugar changes

During the ripening process reducing sugar content diminishes. However, their concentration in the flesh of the raw material for preparing naturally black olives is still high enough to support a vigorous fermentation. Table 9.2 shows the sugar content of the main cultivars dedicated to produce this type. Actual content can fluctuate around these levels according to specific degree of ripeness, climatic conditions, etc. About 80% of the total sugars are reducing sugars, mainly glucose.

Concentrations are lower than those found in green olives. However, bearing in mind the sugar losses during lye treatment and washing of green olives, effective concentrations can be approximately the same. This means that in the event of lactic fermentation, the production of acid can reach similar levels. However, in naturally black olives, the optimum yield of acidity is not pursued since high acidity may change the characteristic taste and colour of naturally black olives. For the same reason, these brines are never supplemented with sugar or other fermentable substrates.

Diffusion of sugars begins when the olives are brined. It is a relatively slow process which produces an accumulation in brine only in the case of unfavourable fermentation conditions (high salt level, low temperature). They are obviously important determinants for the fermentation. Their availability in the brine depends, in addition to the skin permeability, on factors such as the olive-to-brine ratio, temperature and salt concentration. In general, bulk fermentation in large fermenters (10 tonnes) gives a ratio of 2 : 1 and high levels of fermentable substrates that lead to a stormy fermentation, abundant production of gases and, sometimes, the appearance of gas-pocket spoilage. It is advisable to reduce the ratio in these cases by putting less olives into the fermenters. This also prevents olive flattening.

Table 9.2 Reducing sugar content of some cultivars when picked to prepare naturally black olives in brine (Adapted with permission from Durán Quintana *et al*, 1973; Balatsouras and Polymenakos, 1964; and Borcakli *et al*., 1993)

Cultivar	Sugar content (glucose/100 g wet flesh)
Hojiblanca (Spanish)	2.08
Lechín (Spanish)	3.00
Verdial (Spanish)	1.60
Gemlik (Turkish)	4.45
Edencik (Turkish)	5.94
Conservolea (Greece)	2.92

Increase of temperature always favours any diffusion phenomenon and hence the presence of sugars in the brine, especially during the initial period of brining. Sugar diffusion is also favoured by low salt concentrations in the brines, so high temperatures and low salt can accelerate fermentation. In general, when the salt is between 6 and 10% and the fruit-to-brine ratio is high, sugars accumulate during the low temperature period (winter), and rapidly disappear when temperatures increase in spring. With re-used brines according to the Nyon system (Barret and Bidan, 1972), sugar diffusion is retarded and fruits retain most of these compounds during storage due to the high solute content of the brines. Final concentrations, in both brine and flesh, may range between 0.3 and 0.6%, especially when determined by the Lane–Eynon method due to interference of other organic substances also present. These levels are reached only after long periods of brining and it is common to find high levels of sugar in fruits. Thus, Borcakli *et al.* (1993) reported the presence of 0.65% in the flesh of Edencik cultivar and 1.8% in Gemlik and 0.98% and 0.65% in their respective brines after 237 days of fermentation.

In addition to the free sugars, olive flesh also contains compounds such as anthocyanins and polyphenols which have sugar residues that can be slowly hydrolysed by yeasts and other microorganisms present in the brine. These are able to support microbial growth throughout the time olives are in brine and even after packing, making it difficult to establish an end to fermentation.

The whole process is considered to be over when the fermentable materials are exhausted and the fermentation has subsided. In most countries and regions this occurs late in August. Some customers traditionally buy such olives only after 1 year of storage to assure their complete fermentation and equilibration. However, this period may range depending on factors such as the raw material, the salt content of brines and temperature. Differences of great importance are mentioned by Balatsouras (1990), who reported that olives from some localities in the area of Agrinion can reach the foreign markets in approximately 40 days after harvest. Due to the special characteristics of the fruits, the process consists simply of a thorough washing followed by packing in a 10% brine. In contrast, olives of the same cultivar from certain localities around Amphissa lose their bitterness slowly over a period of 1 year or more.

This slow sugar diffusion and the diversity of fermentable substrates cause difficulties in packing, which must use methods that avoid any further secondary fermentation.

Debittering

Natural black olives are debittered only by diffusion of oleuropein into the brine. Oleuropein and other polyphenols diffuse at the same rate as other organic substances. Removal is never complete, since an equilibrium is established where both brine and flesh contain almost equal amounts of oleuropein, salt, acids, proteins and other water-soluble compounds. By changing the brine several times during processing it would be possible to reduce the oleuropein content of the flesh further. This is, however, a costly procedure so that the industry relies upon the leaching effect of what is called 'mother brine' to remove most of the oleuropein (Balatsouras, 1990). The residual bitterness is thus a characteristic of this product which is expected by consumers, although this is partially masked by the product's saltiness. Traditionally, this taste is liked by consumers of Greek naturally ripe olives and has been the basis of the product's world-wide prestige.

In addition to the sugar level, the residual bitterness is another characteristic to be considered when setting the end of the fermentation. Only when most of the oleuropein has diffused into the brine and the period of very active fermentation is over can olives be considered ready for marketing. By this time, the level in the flesh may be between one-half and one-third of the initial concentration.

The concentration of inhibitory polyphenols in naturally black olives is lower than in less mature olives, so the inhibition of LAB is less strong than in turning colour olives. Furthermore, soluble substrates are more readily available in naturally black olives because of the less rigid structure of the flesh, making a lactic fermentation possible in most cultivars. However, this rarely occurs in the traditional processes because of the high level of salt. A good example is the Hojiblanca cultivar which can hardly support LAB growth when picked as turning colour olives. However, when mature, it is able to undergo a very active lactic fermentation if the brine salt content is less than 8% NaCl.

Gram-negative bacteria are abundant in the first days of naturally ripe olives if there is no initial pH correction but are rare after 20–30 days in brine. They may be inhibited by the polyphenols, such as anthocyanins, although there is no scientific evidence for this at present. Yeasts are not affected, but their counts in these olives, very rich in polyphenols, are relatively low.

A final debittering step is achieved when packing, since it is customary to add a fresh brine to the olives. This new dilution makes the fruits sweeter, although a slight bitter taste still remains when they reach the consumers.

Polyphenol changes in brine during fermentation depend on the raw material polyphenol content and the skin permeability. Changes for Lechín, Hojiblanca and Verdial cultivars are shown in Figure 9.2. Equilibrium of polyphenol concentration is reached only about 100 days after brining.

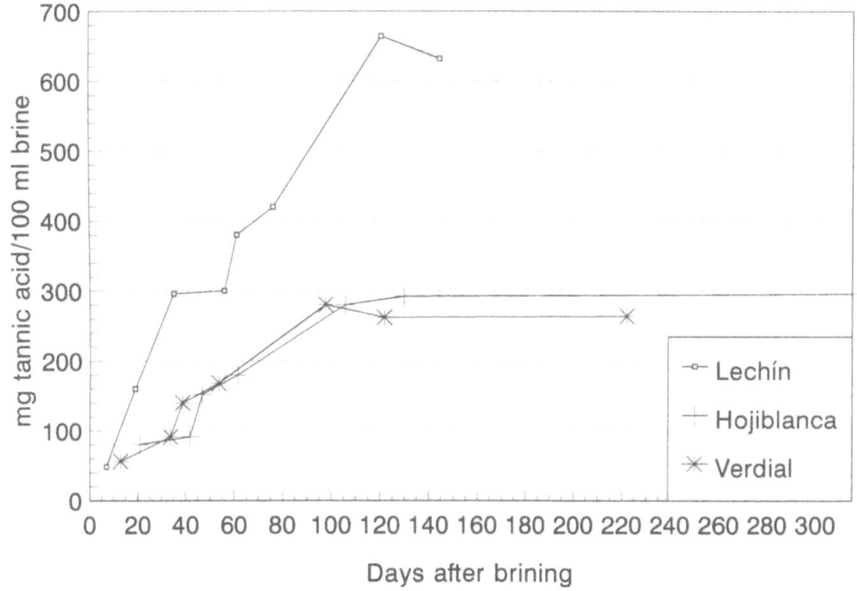

Fig. 9.2 Changes in polyphenol content (mg tannic acid/100 ml brine) during the fermentation of Lechín (□), Hojiblanca (+) and Verdial (×) cultivars. (From Durán Quintana *et al.*, 1973.)

Changes in pH and titratable acidity

In the traditional process, when no initial correction was applied to the brine and salt concentrations were about 12–14%, the pH decreased slowly to stabilize around 4.0–4.5, corresponding to a very weak development of acidity, around 0.5 g lactic/l (Table 9.3). Free acidity formed during the fermentation of Edencik and Gemlik Turkish cultivars reached final values of 0.41% and 0.35%, respectively. The corresponding pH values were also between 4.5 and 5.0 (Borcakli *et al.*, 1993), although the authors did not find correspondence between both values probably due to their differing buffering capacities. According to Balatsouras (1990) in the Greek traditional process, the pH values ranged from 4.5 to 4.8, and the titratable acidity 0.5 to 0.6% (expressed as g lactic acid/100 ml brine). This researcher thought this due to the high salt content, the loss of fermentable substances through growth of oxidative microorganisms, as well as the inherent low sugar content of the Conservolea cultivar. In addition, since many processors paid little attention to air-tight sealing, this gave oxidative microorganisms the chance to grow and lower titratable acidity. In modern fibre-glass vessels, however, brine surface in contact with air is minimal and such losses, even in the event of surface growth, are negligible. Only a small proportion of the acid present is volatile (about 30–40%) while the rest is lactic acid formed during fermentation and other organic acids from the fruit itself (Table 9.3).

If the initial salt content of brines is lower, as currently happens in most processing plants, a complete lactic acid fermentation is possible, at least in some cultivars like Conservolea, Hojiblanca, etc. In these cases, pH values fall below 4.0 and total acidity reaches 1% or higher. The product obtained differs markedly from that produced by the traditional process. The colour becomes deeply cherry-red when olives are rich in colouring substances and the taste is decidedly acid. According to Balatsouras and Polymenakos (1964) the flavour is not objectionable and this type of fermented product is considered excellent by many customers. In his opinion, a complete lactic acid fermentation of naturally black olives is neither necessary nor objectionable.

Lactic fermentation leads to low pH and high acidity that can assure a safe preservation of the product without the need for high levels of salt. If necessary, this characteristic could be corrected before packing. In this case, appropriate preservation procedures must be used since the product would not be preserved by the physicochemical characteristics alone.

Combined acidity

Natural black olives are not lye-treated, so combined acidity is due only to organic acids from the olives (or formed during fermentation) and the existing inorganic substances. Polyphenols may also play an important role in this buffering capacity. Although actual values may depend on cultivars, salt concentration, type of fermentation, etc., typical values are around 5.0 mEq/l or slightly over.

Directly related to this characteristic is the buffering capacity curve of naturally black olive brines. Maximum values as well as the evolution of combined acidity can be observed also in Table 9.3. In the case of lactic fermentation they show a marked maximum around the pK of lactic acid (3.86), reflecting the formation of large amounts of this acid. In the absence of lactic acid, the curve shows a maximum close to the pK of acetic acid (4.75), since 30–40% of the total acidity in these cases is volatile. Balatsouras (1966a) studied in detail the buffering capacity of samples from traditional processing. He found slight differences in the magnitude of buffering capacity, evident at the pH value where the maximum was displayed. Furthermore, the buffering capacity curves of the commercially processed black olives were below the curves of the same product debittered with alkali over the whole pH range. This signifies less buffering strength because of lower total acidity and absence of sodium hydroxide. In spite of this, the buffering capacity is adequate to

Table 9.3 Changes in pH, NaCl, titratable acidity, combined acidity and reducing sugars in brine during fermentation of naturally black olives in brine. Lechín and Picual cultivars (Reproduced with permission from Fernández Díez and Garrido Fernández, 1969)

Days after brining	pH	NaCl	Reducing sugars (g glucose/100 ml brine)	Titratable acidity (g lactic/100 ml brine)	Volatile acidity (g lactic/100 ml brine)	Combined acidity (mEq/l)
4	4.5	–	–	–	–	–
6	–	–	–	0.17	–	–
15	4.4	–	–	–	–	–
18	4.4	10.0	0.11	0.34	0.11	43
25	4.0	11.2	0.11	0.38	0.11	48
37	4.0	11.5	0.21	0.40	0.15	56
51	4.0	11.4	0.17	0.48	0.17	58
66	4.0	11.5	0.21	0.52	–	59
88	4.1	11.2	0.20	0.46	0.19	62
100	4.1	11.0	0.17	0.44	0.18	62

keep the pH value around 3.8 long enough for 0.8–1% of acidity to be formed.

The maximum buffering capacity at pH values around 3.8–4.0, signifies that the decrease of pH to this level is based upon the activity of acid tolerant microbes, such as *Lactobacillus plantarum*. However, in spite of the favourable buffering conditions in the brines, lactobacilli are unable to complete the fermentation in traditional processes because of the strong competition, the inadequacy of fermentable material and the high salt level. Thus, a complete conversion of fermentable substrates into lactic acid in naturally black olives fermentations is prevented.

In cases where only Gram-negative and yeast growth occur, the buffering maximum is around pH 4.6 corresponding approximately to the pK of acetic acid (4.76). Two representative buffer curves are shown in Figure 9.3.

Nitrogen content

Balatsouras (1966a) determined the nitrogen content in traditional Greek naturally black olives. He found that in brines from Conservolea cultivar it varied considerably; from 18.7 to 97.4 mg/100 ml of brine, or 0.115 to

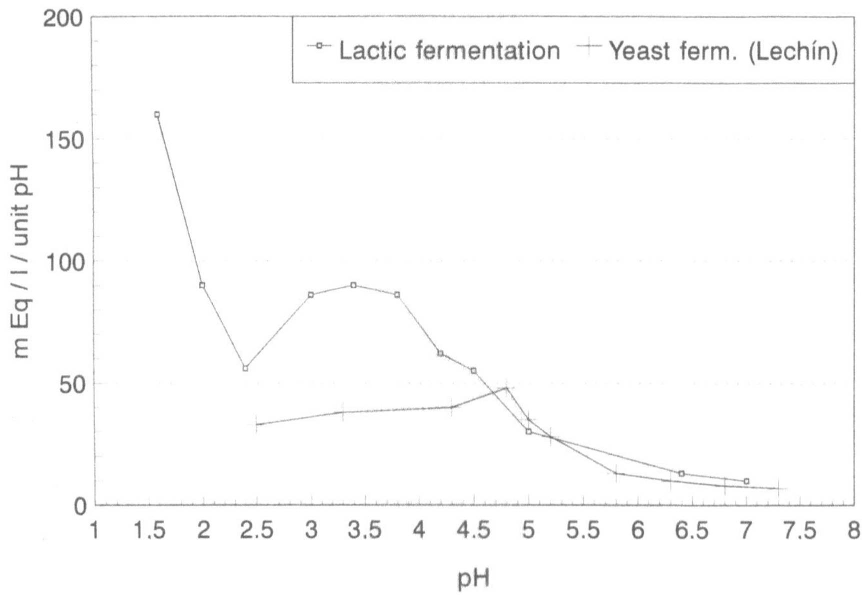

Fig. 9.3 Average buffering capacity curves of traditional (+) and lactic (□) fermentation brines of naturally black olives (From Balatsouras, 1966; Garrido Fernández *et al.*, 1972.)

0.956% when expressed as protein. This was markedly lower than values reported for the flesh of the same cultivar. The amount present at the stage of storage was enough, when combined with the appropriate source of carbon, to support the growth of the microorganisms that commonly grow in brines.

The inorganic nitrogen serves as a nitrogen source for genera such as *Saccharomyces*, *Hansenula* and *Rhodotorula*. In most cases, low nitrogen content was accompanied by high total acidity, possibly due to the proliferation of LAB at the expense of the nitrogenous compounds in brine.

The maximum nitrogen content was found in high-salt brines (17–18% NaCl), perhaps reflecting excessive osmosis or the complete inhibition of microbial growth.

Volatile compounds

A characteristic of fermented naturally black olives is the special flavour they acquire during processing. Like any fermented product, the aroma is very complex. Head space analysis has shown the presence of the compounds shown in Table 9.4. The main component is ethanol, which represents more than 95% of all volatiles. Others also present but in relatively low concentrations were acetaldehyde, acetone and ethyl acetate. These accumulate progressively during the phase of more active fermentation, then remain stable or slowly diminish. When control of aerobiosis is not strict, volatiles are easily lost.

Changes in physicochemical characteristics of fruits

Changes in sugars, polyphenols and other soluble substances have been already mentioned. The main objective of the naturally black olive fermentation is partial elimination of the bitter principle, oleuropein, the

Table 9.4 Volatile compounds in brines. Hojiblanca cultivar (From Fernández Díez *et al.*, 1985)

Compound	Relative presence (%)
Acetaldehyde	0.31–2.85
Acetone	0.08–1.64
Ethyl acetate	0.09–1.35
Ethanol	94.06–98.05
Propanol	0.03–0.08
2-Methyl-propanol-1*	0.01–0.26

* Low proportions of 2-methyl-butanol-1 and 3-methyl-butanol-1 as well as some other unidentified components were also present.

Table 9.5 Changes in the olive flesh during fermentation (Reproduced with permission from Durán Quintana *et al.*, 1973)

Olive cultivar	Polyphenol content (mg tannic acid/100 g flesh)	Texture (N/100 g pitted fruits)						
		Shear compression	Pressure					
			15*	16*	17*	18*	19*	20*
Hojiblanca								
Before processing	1300	2900	–	578	524	467	397	408
After processing	624	1775	–	–	470	452	353	–
Lechín								
Before processing	2500	2200	1027	893	875	–	–	–
After processing	816	1976	1040	840	–	–	–	–
Verdial								
Before processing	1140	2340	–	594	635	599	575	546
After processing	384	1289	–	–	470	452	353	–

* Values are olive sizes expressed as transverse diameter in mm.

main polyphenol of such fruits, and anthocyanins. Diminution in polyphenols can be observed in Table 9.5. Approximately two-thirds of polyphenols are solubilized. This probably gives an excessively bitter taste to olives, which complete their debittering process in the final pack during their shelf-life. Anthocyanins, which are responsible for the black to red colour of naturally black olives, are also solubilized and give the brines their characteristic tone. As pH decreases the brine colour turns progressively to red. This solubilization can discolour fruits with low levels of these pigments.

Other chemical components are also affected. In general, flesh moisture content increases only slightly and protein content remains almost constant or slightly lower, which means that most of the nitrogen is bound protein which is not easily solubilized. In the case of Gemlik cultivar, it changed from 2.01, at the beginning of the process, to 1.86% w/w. In Edencik, the decrease was even lower (Borcakli *et al.*, 1993). In contrast, inorganic substances in flesh increase progressively as fermentation time progresses. Changes of typical fermentations are shown in Figure 9.4. The increase is due mainly to the absorption of salt by the flesh. However, other cations such as K^+ are leached into the brine. Olives are also rich in Ca^{++}, but this cation is usually strongly linked to organic constituents and its solubilization is more difficult.

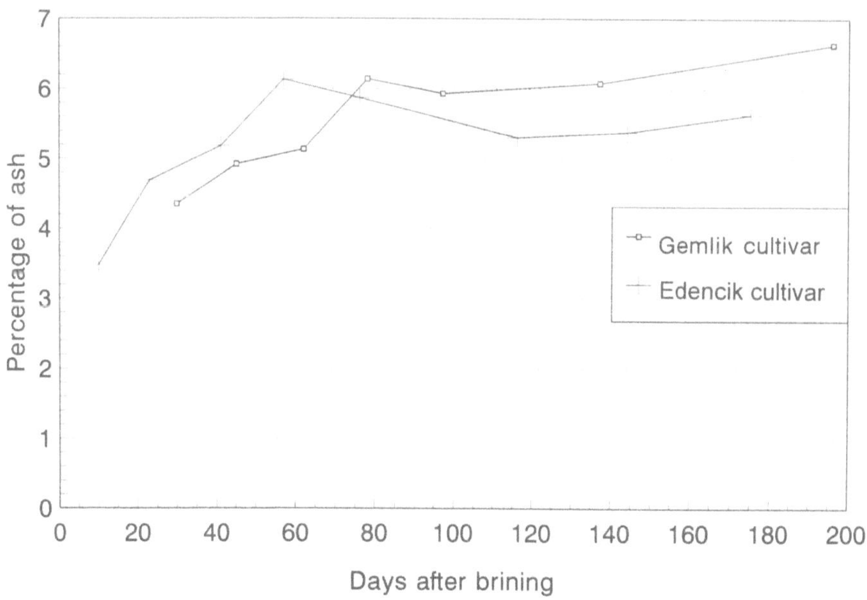

Fig. 9.4 Inorganic content changes (as percentage of ash) in the flesh during the fermentation of naturally black olives. (From Borcakli *et al.*, 1993.)

Fermentation always produces a decrease in olive texture. Changes in shear-compression and compression are shown in Table 9.5. Shear-compression diminishes by up to 40–45% and the effect is more pronounced in those cultivars with a higher flesh-to-pit ratio. The change is more important than that found in turning colour olives, possible due to the fact that soluble substances increase as the fruit matures. The decrease in compression values is less pronounced, which may be explained by the fact that the skin is scarcely affected during storage in brine, so the force needed to destroy it is rather similar. Texture decreases as the size of the olives and the flesh-to-pit ratio increase. In cases of the growth of pectolytic bacteria, texture deterioration may be important.

Microorganism growth during fermentation

There are no detailed studies on the microflora carried by fresh naturally black olives, although it is reasonable to suppose it to be similar to that found in turning colour olives. However, the lower sugar and polyphenol content may cause slight differences. Borcakli *et al.* (1993) found that the adhered microflora was mainly composed of yeasts (10^4–10^5 cfu/ml), Gram-negative bacteria (10^4–10^5 cfu/ml) and coliforms (10^3–10^4 cfu/ml) with negligible numbers of lactobacilli and streptococci.

Traditional naturally black olive fermentations were usually carried out without much attention to whether conditions were aerobic or anaerobic, and the closing devices were used more with the aim of keeping fruits submerged. In most cases the growth of moulds was permitted on the brine surface, the aim being to accelerate the sugar depletion and utilization of organic acids if these were produced in excess during fermentation. This technique was abandoned by the middle of the century when it was decided that anaerobic conditions were essential. This change of attitude was not only due to spoilage problems arising from mould growth (softening, malodorous spoilage, etc.) but also because most of these organisms could produce mycotoxins. Balatsouras (1966b) studied the composition of the leathery surface films and found that the moulds belonged to the genera *Penicillium*, *Aspergillus* and *Rhizopus*. Oxidative yeasts were also found. Barret and Bidan (1972) also found *Aspergillus* and *Penicillium* in the films which grew on the brines when olives are processed according to the system of Nyons.

All cultivars seem to show a similar fermentation under anaerobic conditions. Generally speaking, most of the potential spoilage organisms, moulds, *Bacillus*, *Clostridium* etc., carried by the raw material should be reduced by the washing process and are normally inhibited by the high salt level (and pH correction when this is applied). They are therefore either not detected at all in brines or only for a very short time. The rest of the microflora is composed of Gram-negative bacteria, yeasts and (sometimes) LAB. There is a sequence of genera observed, especially during the first days, but later there is no real succession of organisms, as happens in the case of the fermentation of Spanish-style green olives, but largely a coexistence of various species of yeasts – and LAB when they grow.

Immediately after naturally black olives are brined, a strong fermentation begins. This is characterized by the production of a great volume of gases, mainly CO_2, produced by the Gram-negative bacteria, which are initially present in high numbers (Figure 9.5 (a–c)), and also from the metabolism of the fruits which continues until the anaerobiosis and salt stop their biological activity. As fermentative yeasts increase, so too does their contribution to the production of gas.

Durán Quintana and González Cancho (1973), González Cancho *et al.* (1975) and Fernández Díez *et al.* (1985) have reported in detail on the different genera found and their relative levels. Most abundant were *Citrobacter* (64.8% of the isolated pure cultures), *Klebsiella* (20.3%), *Achromobacter* (9.2%), *Aeromonas* (3.7%) and *Escherichia* (1.8%). Of these, only *Achromobacter* used glucose without production of gas. However, the other genera produce gas, sometimes with acid, and are thus responsible for the abundant CO_2 produced in this phase and also the decrease in pH.

Gram-negative bacteria reach their maximum level during the first 2 days after brining and are present in brine for up to 7–15 days. Counts and period of growth depend on the cultivar. One important factor can be the availability of fermentable substrates and the concentration of polyphenols which can inhibit Gram-negatives. Then is no specific evidence for this effect, but according to Figure 9.5 (a–c), the disappearance of Gram-negatives cannot be completely related to pH or to the activity of yeasts, since they begin to decrease very early when the pH is still relatively high and are completely inhibited at levels of pH of 5.0 or more. At the same time, the yeasts counts are not high enough to affect the Gram-negative population.

Yeast counts rise slowly from the start of the fermentation and reach their maxima approximately 10–25 days after brining. Among the Spanish cultivars, the highest population of yeasts has always been found in Hojiblanca. This effect of cultivar can be due to the fermentable substrate content but apparently is also due to the polyphenol content. The decrease of the yeast population when the sugar concentration in fruits is still high, suggests some inhibitory factors, although this hypothesis has not been checked. The different species of yeasts isolated and identified from naturally black olives brines are shown in Table 9.6. In Spanish cultivars the most significant species are *Saccharomyces oleaginosus* (34.8% of pure cultures isolated), and *Hansenula anomala* (27.3%). Also found in all cultivars, but with lower frequencies are: *Torulopsis candida* (10.8%), *Debaryomyces hansenii* (9.5%), *Candida diddensii* (8.2%), *Pichia*

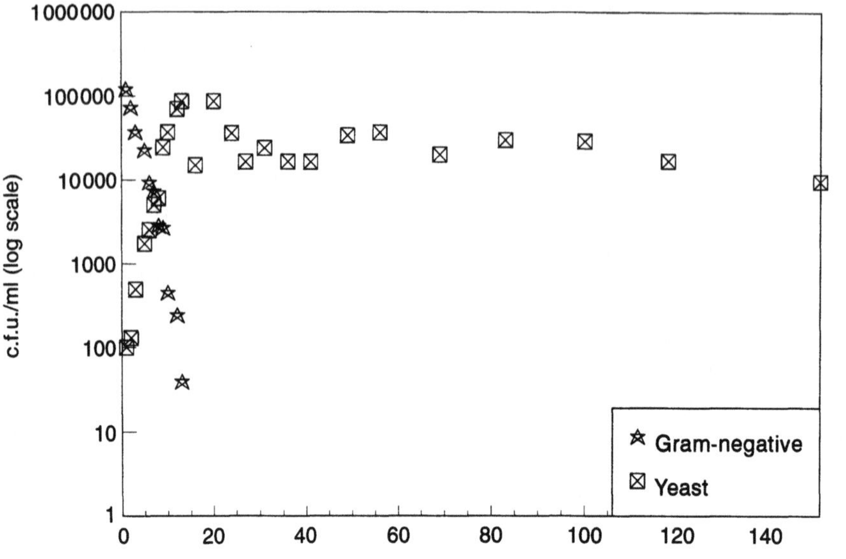
1000000
100000
10000
1000
100
10
1
c.f.u./ml (log scale)
0
20
40
60
80
100
120
140
Time after brining (days)
Gram-negative
Yeast

(a)

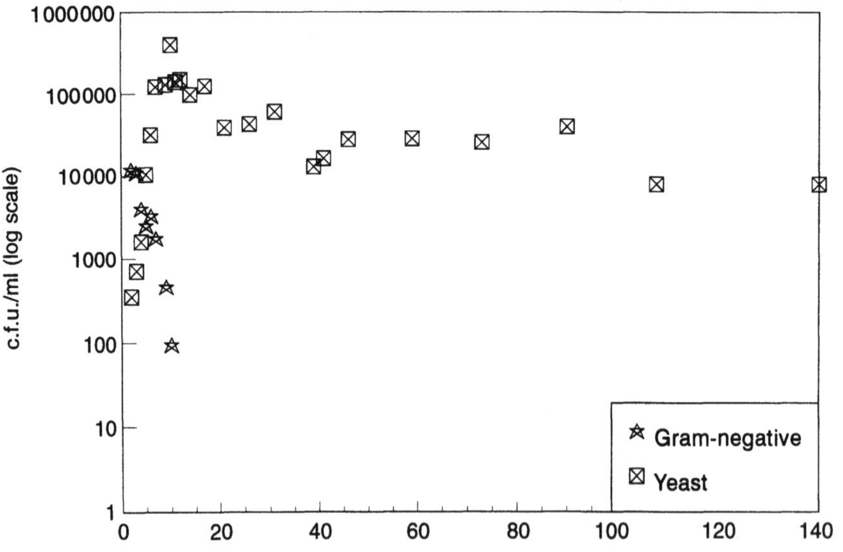
1000000
100000
10000
1000
100
10
1
c.f.u./ml (log scale)
0
20
40
60
80
100
120
140
Time after brining (days)
Gram-negative
Yeast

(b)

Fig. 9.5

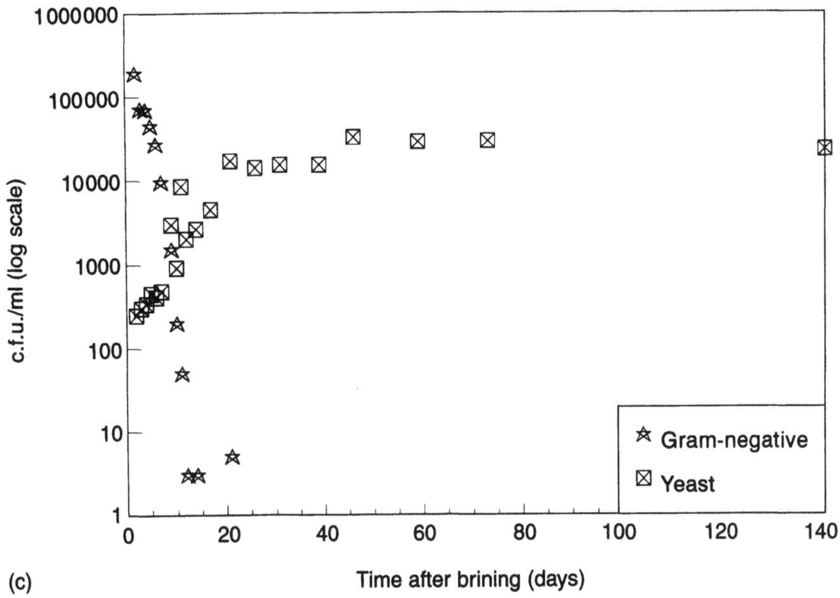

(c)

Fig. 9.5 Changes in Gram-negative bacteria and yeast populations during fermentation of naturally black olives. (a) Lechín cultivar; (b) Hojiblanca cultivar; (c) Verdial cultivar. (Redrawn from Garrido Fernández *et al.*, 1972.)

Table 9.6 Species of yeasts isolated from naturally black olives fermentations. (Adapted from Durán Quintana and González Cancho, 1973; Fernández Díez *et al.*, 1985)

Spore-forming yeasts	Non spore-forming yeasts
Saccharomyces oleaginosus	*Torulopsis candida*
Hansenula anomala	*Torulopsis norvegica*
Hansenula subpelliculosa	*Candida diddensii*
Debaryomyces hansenii	*Candida boidinii*
Pichia membranaefaciens	*Candida krusei*
Pichia farinosa	*Candida valida*
Pichia fermentans	*Cryptococcus ater*
Kluyveromyces veronae	

membranaefaciens (4.9%) and *Candida krusei* (0.4%). Others were isolated sporadically, although all were identified in Hojiblanca cultivar fermentations.

Some characteristics of their growth are shown in Table 9.7. The most representative yeasts are present in the fermentations throughout the period of storage in brine and do not show any succession of species. Only *Sacch. oleaginosus* has a clear maximum level which is observed between 13 and 28 days, depending on the cultivar. This adaptable species has a high fermentative ability for most of the sugars present, so its frequent and permanent presence in naturally black olives is not surprising. It was isolated by Santamaría in 1958 from the wastewaters from oil mills and its presence has often been detected inside olive flesh. It appears well adapted to the olive substrate and this can, in turn, sometimes cause problems, as will be seen later.

In Greek black olive fermentations, Balatsouras (1966b) has also identified different genera of yeasts including *Trichosporon, Candida, Pichia, Kloeckera, Torulopsis* and *Debaryomyces.*

Borcakli *et al.* (1993) also found that the main microorganisms were also yeasts in the traditional process (initial salt content of 14%) using Turkish cultivars. Yeasts developed starting from the 40th day of brining and were present throughout fermentation. The yeast flora consisted of *Debaryomyces hansenii, Rhodotorula mucilaginosa, Rhodotorula glutinis, Candida membranaefaciens* and *Saccharomyces cerevisiae* for the Edencik cultivar. In this case, spore-forming bacteria, Gram-negatives and LAB were not detected. In general, there is a coincidence with respect to the type of organisms (yeasts) and even their species, apparently reflecting the product's selection of the more adapted microorganisms.

The presence of LAB in the fermentation of naturally black olives is sporadic. In fact, neither the salt concentration nor the temperature are appropriate for the growth of these organisms. Consequently, the pH

Table 9.7 Growth characteristics of yeasts in the naturally black olives fermentations. Hojiblanca cultivar (Adapted from Fernández Díez *et al.*, 1985)

Yeast	Presence interval (days)	Average frequency (%)
Saccharomyces oleaginosus	4–159	48
Hansenula anomala	2–141	27
Torulopsis candida	2–141	39
Candida diddensii	4–109	16
Debaryomyces hansenii	2–159	28
Pichia membranaefaciens	4–109	15

reached ≈4.5 and the acid formed (0.1–0.4%) indicate their lack of activity. The growth of LAB in these fermentations depends on cultivar, salt concentration and temperature. Some cultivars still retain, when mature, a relatively high polyphenol content that is enough to inhibit their growth. A clear example is the Spanish Lechín cultivar which never permits LAB growth. In other cultivars the polyphenol content is lower and they can support LAB growth if other conditions are favourable. This is true of the Spanish Hojiblanca and the Greek Conservolea cultivars, where LAB are able to grow if the salt is under 8% NaCl and temperature is not too low. Spanish researchers (Durán Quintana and González Cancho, 1973; Durán Quintana *et al.*, 1971, 1973; Garrido Fernández *et al.*, 1972; González Cancho *et al.*, 1975) have demonstrated the presence of LAB in Hojiblanca cultivar in such conditions. When the salt content is below 8%, the presence of the genera *Pediococcus* (homofermentative) and *Leuconostoc* (heterofermentative) has been observed during the first days of the fermentation. They are followed by the growth of *Lact. plantarum*, which is present practically all the time that olives are in brine because of its higher tolerance to low pH and acidity. Borcakli *et al.* (1993) reported also the sporadic presence of *Pediococcus* in Edencik cultivar between the 76th and the 89th days of fermentation.

Balatsouras (1966a) suggested that the naturally black olives fermentation must be considered a lactic process based on the acidity formed. This definition could possibly be judged as appropriate in current Greek fermentations in cement or fibre-glass vessels, where the salt concentration is lower than that used in the traditional process. According to him, LAB predominate during the phase of moderate fermentative activity. Initially, they are represented by the genus *Streptococcus*, sensitive to high salt and acidity. Later, *Leuconostoc* and *Lactobacillus* are present. The heterofermentative lactobacilli are represented by *Lact. brevis*; *Lact. plantarum*, which is more resistant to salt and acidity, normally concludes the process.

Modifications to the traditional anaerobic process

Some modifications to the traditional naturally black olive fermentation process have been proposed. They have mainly been suggested with the aim of obtaining a better lactic acid yield as well as producing a more rapid debittering.

Balatsouras and Polymenakos (1964) proposed a lye treatment similar to that given to Spanish green olives, but introducing air during the process with the objective of preserving the colour of the raw material. After a series of washings to remove excess sodium hydroxide, olives were put in a 12.5°Bé brine to achieve the same level of salt as when fermenting Spanish-style green olives. Sometimes, the addition of LAB to initiate the fermentative process was also tried. After fermentation, pH values below 4.0 (3.7–3.8) were obtained with a titratable acidity around 0.8% (expressed as lactic acid). Levels of salt after equilibrium ranged from 5.50 to 8.0%.

Combined acidity increased from 28–60 mEq/l in traditional naturally black olives to 90–120 mEq/l when the fermentation of lye-treated naturally black olives was over.

During the first days moulds, yeasts, coliform bacteria and spore-formers were present. All of these, except yeasts, were soon replaced by the LAB. Lactobacilli isolated were homofermentative and were active during the time that the brine had fermentable substrates. Despite the fact that the coliform bacteria disappeared after the 15th day of fermentation, a small proportion (1–2%) of olives were affected by the gas-pocket spoilage.

The taste of such olives was improved with respect to the traditional product due to the combined effect of the high acidity and the moderate salt content. The colour, however, was cherry-red and the consistency was also relatively low, although acceptable.

The final product had a good stability and kept well in anaerobic conditions.

In spite of its good characteristics, the product has never been developed at industrial level because the use of lye to treat naturally black olives is forbidden in Greece. In addition, the product could not be sold as naturally black olives, but as treated black olives. It should thus be oriented to a new market.

In some cultivars, a lactic process is obtained simply by reducing the salt levels as demonstrated by Aligisakis (1971) in Conservolea cultivar. When the process took place in 10–12% NaCl the acidity reached only 0.38% and the pH 4.3. However, in 6.4–8.2% NaCl the respective values were 0.88 and 3.90 (Tables 9.8 and 9.9). Similar results are obtained with the Spanish Hojiblanca cultivar. As temperature is increased, acidity formation also improves.

If the traditional colour and taste are to be maintained, the lactic process must be controlled and limited to the extent compatible with safe preservation of the product. Adequate growth of LAB can be easily achieved by proper NaCl control. Once the desirable acidity has been produced, any further production can be stopped or markedly reduced by raising the level of salt above 8%. Another option can be to permit the complete lactic fermentation and take advantage of the optimum pH reached for the safe preservation of the product during storage. Final appropriate physicochemical characteristics can be corrected just before sorting and grading or during packing.

As shown above, the production of acidity is compatible with concentrations of salt markedly higher than those customary in Spanish-style olives. This is possible because the LAB that grow in naturally black olive

Table 9.8 Changes in acidity and pH during the fermentation of naturally black olives. Conservolea cultivar at high salt concentration (10–12%) (From Aligisakis, 1971)

Days after brining	pH	Acidity (g lactic acid/ 100 ml brine)	Salt (g NaCl/ 100 ml brine)	Temperature (°C)
0	6.5	0.00	10.0	10
15	5.0	0.01	9.8	9
32	4.8	0.10	10.1	8
47	4.8	0.17	10.2	7
63	4.8	0.18	11.0	8
78	4.7	0.19	11.0	8
93	4.7	0.22	11.5	9
108	4.6	0.25	12.2	11
123	4.5	0.30	12.2	15
153	4.4	0.35	12.5	17
183	4.3	0.38	12.5	22
213	4.3	0.38	12.5	27

Table 9.9 Changes in acidity and pH during fermentation of naturally black olives. Conservolea cultivar at low salt concentration (6.4–8.2%) (From Aligisakis, 1971)

Days after brining	pH	Acidity (g lactic acid/ 100 ml brine)	Salt (g NaCl/ 100 ml brine)	Temperature (°C)
0	6.2	0.01	6.4	13
29	5.0	0.20	7.5	9
62	4.8	0.20	8.2	7
91	4.6	0.30	8.0	10
121	4.2	0.45	8.2	15
152	4.1	0.55	7.4	18
182	3.9	0.70	7.5	21
212	3.9	0.88	7.8	28

brines are normally more salt-tolerant. Balatsouras (1985) found that 75% of the isolates from this type of brine grew profusely at 8% NaCl and only NaCl over 10% was able to completely inhibit them.

Some researchers have shown the possibility that some strains of *Lact. plantarum* could hydrolyse oleuropein (Ciafardini *et al.*, 1994). Three strains isolated from naturally black olives grew at pH of 3.5 and could tolerate 1% oleuropein as well as 8% NaCl. The β-glucosidase produced by one of them gave, after 7 days of incubation, all the oleuropein hydrolysis products. After 21 days, oleuropein and its aglycone decreased to trace levels with a simultaneous increase in the concentration of β-3–4-

dihydroxyphenyl ethanol. The use of a starter of such a strain can simultaneously yield a good pH and acidity and a higher degree of debittering. However, its commercial use is not yet established.

The debittering effect of yeasts was studied by Materassi *et al.* (1975). They found that eight of 36 species of yeasts isolated from olive fermentation had the capacity of hydrolysing oleuropein. The two most active species were of the genera *Rhodotorula* and *Candida*. Later, Pelagatti (1981–83) applied the use of *Candida veronae* to the debittering process of naturally black olives. The presence of the yeast caused a more rapid destruction of the bitter principle from the 5th day of brining. Thus, the use of yeasts able to hydrolyse the oleuropein can reduce the processing time and yield sweeter final products. In spite of this, utilization of this ability at industrial level has not been reported.

Controlled fermentations with different species of yeasts isolated from naturally black olive brines were carried out by González Cancho *et al.* (1975). Olives were washed with chlorinated water to eliminate most of the microorganisms adhering to the surface. Brine pH was also corrected to 4.2 units to prevent Gram-negative growth. It was observed that *Saccharomyces oleaginosus* (Table 9.10) was the only species growing in brine for the first 15 days, reaching, as in the spontaneous fermentation, a clear maximum, but this time a little earlier. From the 10th day there was a progressive decrease of the yeast population and the proportion of the inoculated species present fell. *Hansenula anomala* grew rapidly during the first days and lost dominance slightly more rapidly than *Saccharomyces oleaginosus*. After 19 days only 50% of the yeast population came from the inoculated species. Results obtained with *Torulopsis candida* and *Debaryomyces hansenii* are also shown in Table 9.10. There were no differences between the controlled fermentations and the spontaneous ones with respect to texture, colour and flavour. Gas-chromatographic analysis of volatiles showed the same components present as in the spontaneous fermentation and in similar concentrations. In general, 95–99% of the volatiles was ethanol.

Re-use of brines

The only wastewaters produced during the naturally black olive fermentation are the washing waters and the brines. In order to reduce the pollution problems of the packing industry some trials to regenerate the brines and re-use them for further fermentations have been conducted. The regeneration procedures will be described in the following chapter. Only the fermentation behaviour using such reconditioned brines will be described here.

Changes in the sugar content of the different re-use brines are shown in

Table 9.10 Controlled fermentation of naturally black olives in brine changes in the inoculum during fermentation (From González Cancho *et al*, 1975)

Species	Days after brining	Yeasts counts* (c.f.u./ml) × 10^3	Inoculum survival (%)
Saccharomyces oleaginosus	3–10	32–250	100
	11–21	14–100	100–90
	25–59	49–15	90–50
	68–108	13–12	50–10
Hansenula anomala	3–10	440–115	98–80
	12–17	165–125	80–50
	19–38	70–67	50–20
	42–118	53–10	20–0
Torulopsis candida	1–9	2–58	100
	10–29	11–175	90–50
	30–119	44–2	50–10
Debaryomyces hansenii	3–7	0–96	100
	10–14	370–65	85–30
	18–59	31–60	10–0

* c.f.u., colony forming units

Table 9.11. No differences are apparent between treatments and the re-use of brines does not produce any sugar accumulation in the solution. Apparently they are consumed as they pass into the brine. However, yeast counts are affected (Table 9.12). Direct re-use or any kind of dilution causes a one log decrease in microbial count. This is more marked when regenerated brines are used, especially when the regeneration uses treatment with chlorine. Despite this, in all cases the number of organisms in solution was sufficient to use the sugars in brine.

The direct re-use of brine and the different dilutions as well as the regenerated brines removed a similar amount of polyphenols as did fresh brine (treatment A) (Figure 9.6). Accordingly, olives were of similar taste, including bitterness.

Polyphenol concentrations in directly re-used and diluted brines were markedly higher than in those fermentations carried out with fresh brine. This could account for the absence of Gram-negative bacteria and also for the lower population of yeasts in these processes. In the case of the brines regenerated with chlorine, formation of chlorinated phenol compounds could have increased the inhibitory effects since microbial growth was unusually low in such brines.

Table 9.11 Changes in sugar content during fermentation in re-used brines with different dilutions and regeneration treatments (From Durán Quintana and Garrido Fernández, 1978)

Days after brining	Treatments*						
	A	B	C	D	E	F	G
3	–	–	–	–	–	0.14	0.14
9	–	0.26	0.13	0.11	–	–	–
16	–	0.22	0.18	0.11	–	–	–
17	–	–	–	–	0.23	0.19	0.22
31	0.12	–	–	–	0.20	0.21	0.23
68	–	–	–	–	0.29	0.21	0.26
84	0.16	0.44	0.37	0.17	–	–	–
108	–	–	–	–	0.22	0.20	0.17
124	0.23	0.42	0.33	0.31	–	–	–
153	–	–	–	–	0.22	0.24	0.17
170	–	0.37	0.39	0.30	–	–	–
195	0.15	–	–	–	0.23	O.26	0.25
212	0.31	0.26	0.28	0.26	–	–	–

A, traditional anaerobic process with fresh brine; B, direct re-use of previous fermentation brines; C, dilute (1 → 2) previous fermentation brines; D, dilute (1 → 4) previous fermentation brines;E, re-use of brine regenerated with adsorbent earths; F, re-use of brine regenerated by active carbon; G, re-use of brine regenerated with Cl_2.

Table 9.12 Microbial growth of yeasts (c.f.u./ml × 10^3) during the fermentation of naturally black olives in re-used brines, with different dilutions and regeneration treatments (From Durán Quintana and Garrido Fernández, 1978)

Days after brining	Treatments*						
	A	B	C	D	E	F	G
3	200	–	–	–	4	3	2
7	–	–	–	–	2	2	0
9	–	20	20	20	–	–	–
16	–	50	10	20	–	–	–
17	–	–	–	–	1	1	0.1
23	200	10	90	20	–	–	–
31	100	–	–	–	2	3	0.3
68	–	–	–	–	0.1	9	5
84	100	6	3	3	–	–	–
108	–	–	–	–	0.0	3	0.0
124	70	10	4	0.1	–	–	–
153	–	–	–	–	4	4	0.1
170	–	28	14	11	–	–	–
195	20	–	–	–	1	1	5
212	30	3	2	1	–	–	–

* See Table 9.11 for treatment definitions.

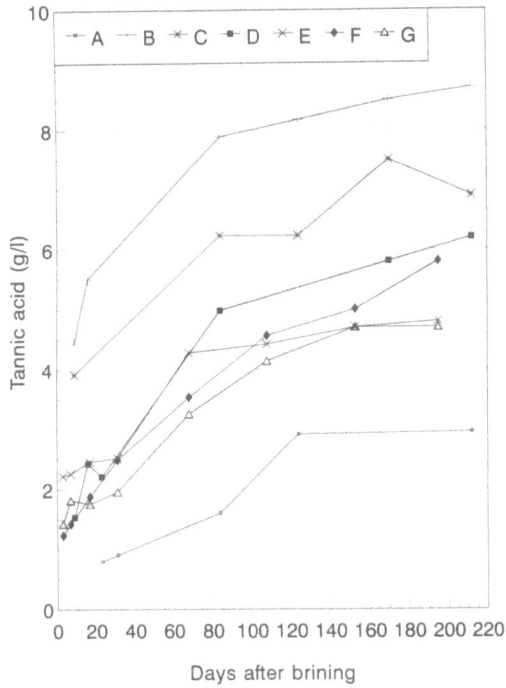

Fig. 9.6 Polyphenol (tannic acid) changes during fermentation of naturally black olives in re-used brines with different dilutions and regeneration treatments. See Table 9.11 for definitions of treatments A–G. (From Durán Quintana and Garrido Fernández, 1978.)

9.2.4 AEROBIC FERMENTATION

The Spanish Hojiblanca cultivar is very sensitive to fish-eye or gas-pocket spoilage which will be discussed later. To overcome this problem, some modifications to the anaerobic, traditional, naturally black olives fermentation have been tried.

Most effective has been fermentation under aerobic conditions. It is based on the fact that this spoilage was caused in naturally black olives by *Saccharomyces oleaginosus* and *Hansenula anomala*, the two most characteristic yeasts of the traditional anaerobic process and on the finding that when olives are put in brine, fruit metabolism is able to produce the same modifications as those caused by the microorganisms, probably due to a change in the normal physiology of the olives in absence of oxygen.

In essence, the process consists of the introduction of air into the fermentation brine at a rate sufficient to induce growth of facultative (preferably) or oxidative microorganisms.

Salt concentration and pH

Initial salt content can be similar to those used in the traditional process and also depends on the combination of initial pH correction and degree of maturation of the fruits, especially when the air supply is moderate. A high level of salt (over 8.5%) leads to a complete inhibition of the LAB and the process is carried out mainly by yeasts. If the initial salt content is low (2–3%) there is a rapid production of acidity with the Hojiblanca cultivar. However, use of low salt may be risky if no other controls are used. In general, pH control of the initial brine with acetic acid is recommended to decrease the population of Gram-negative bacteria and other possible spoilage organisms.

Equilibration of the salt is achieved more rapidly in this process than in the traditional process and the amount of salt required is also lower. Table 9.13 shows that, in anaerobic conditions, about 12 kg salt/100 kg of olives is needed and the time to equilibrium is 140 days, but in aerobic conditions these figures were 10–11 kg and 80–100 days with a progressive increase as aerobic conditions change to anaerobic. Differences were relatively less important when the air rate supplied was lower than 0.1 l/h × litre capacity (l.c.) and the process was discontinuous. This means that bubbling air can accelerate the process and makes it more efficient and more cost effective. When the initial salt content is lower, the equilibrium is reached more slowly, although the total time to equilibrium depends on the frequency of the salt additions. When these are made at short intervals, a similar time to that shown with high-salt contents can be obtained.

Table 9.13 Influence of the aeration rate and type of aeration on the time needed to reach equilibrium. Initial salt content; 10%. (Reproduced with permission from García García *et al.*, 1985)

Type of process	Final salt content (g/100 ml)	Salt used (kg/100 kg olives)	Days to equilibrium
Traditional anaerobic	9.8	12	140
Continuous aeration			
1.00 l/h × lc	10.7	10	80
0.50 l/h × lc	10.9	10	90
0.25 l/h × lc	10.6	10	95
0.10 l/h × lc	10.0	11	120
Discontinuous aeration, 1 l/h × lc			
1 : 2, air : stop	10.8	11	100
8 : 16, air : stop	10.7	11	100

lc, fermenter capacity in litres

Acceleration of the process is due to the frequent removal of the brine layer in contact with the olive surface. This maintains the gradient between brine and flesh, enhancing the exchange of substances between both. When the salt is high, the gradient is also high, as is the osmotic exchange rate. If the salt is low, all of these phenomena are slowed. Working at aeration rates close to the effective limit, the initial salt content can be a very important factor. Figure 9.7 shows the effect of the initial salt content on gas-pocket formation for two different maturation degrees. It is evident that the influence is more marked in naturally black mature olives. When working exclusively with completely mature fruits however, the salt concentration may become absolutely determinant as shown in Table 9.14. In this, it is observed that a rate of 0.25 l/h × litre capacity air injected continuously is able to prevent gas-pocket formation, regardless of the salt level. However, when the process is discontinuous, 9% initial salt leads to the appearance of high percentage spoilage. In contrast, 2% initial salt can efficiently control the formation of the gas-pockets. Thus, as a general rule, it seems that in case of cultivars sensitive to gas-pocket formation, it is advisable to initiate the aerobic process with low salt concentrations (2–3% NaCl). Such brines must have also the pH corrected to values around 4.0–4.2 with acetic acid. Further careful control of fermentation must be established to avoid any malodorous or softening spoilage.

Aeration causes brine circulation and thus a more homogeneous fermentation than occurs with sporadic circulations of brine by pumps in the anaerobic process. Salt additions are also easier since the required amount can simply be added at the top of the fermenter. For the same reason, any other chemical addition or correction needed in the brine is very easily made.

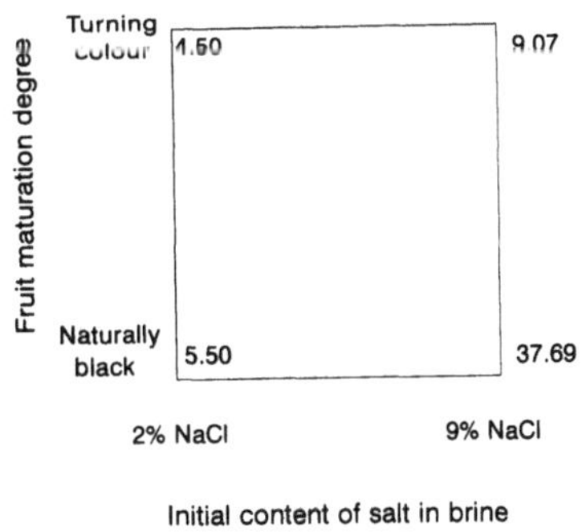

Fig. 9.7 Interaction between initial salt content and maturation degree of fruits. Initial pH control to 4.2 with acetic acid and discontinuous aeration at a rate of 0.3 l/h × lc capacity (l). (From Durán Quintana *et al.*, 1986.)

Table 9.14 Appearance of gas-pocket spoilage (60 days after brining), as a function of the initial salt content and air injection rate. Hojiblanca cultivar. Initial pH corrected to 4.5 with acetic acid (From Durán Quintana *et al.*, 1986)

Fermentation conditions	Gas-pocket (%)
Initial salt content 9%	
Continuous aeration, 0.25 l/h × lc	11
Discontinuous aeration	
1.00 l/h × lc	62
0.75 l/h × lc	73
0.50 l/h × lc	82
0.25 l/h × lc	81
Initial salt content 2%	
Continuous aeration, 0.25 l/h × lc	6
Discontinuous aeration	
1.00 l/h × lc	7
0.75 l/h × lc	2
0.50 l/h × lc	8
0.25 l/h × lc	8

lc, fermenter capacity in litres

Final salt concentration is normally between 8 and 10% NaCl, to ensure safe product storage. This is especially important when fermentation is carried out at high salt levels and LAB are absent. If the pH is below 4.0, salt content can be slightly lower but only to about 7–8% NaCl because the residual bitterness of fermented olives requires this amount of salt for an adequate flavour.

Final pH depends on the aeration rate and fermentation type. Excessive aeration enhances growth of oxidative organisms and pH may increase. In extreme cases this can lead to values inappropriate for product preservation.

If the initial pH is corrected, as suggested, there is a slight increase during the few days after brining, followed by a slight decrease and stabilization at around 4.3–4.5, when high salt levels are used. If the initial salt content is about 2–3% NaCl, the pH decreases rapidly to 3.6–3.8. As mentioned in the case of the anaerobic process, this low pH can be modified and the acidity corrected in the last phase of the storage period or during final packing.

Fermentation rate and debittering

One of the most important advantages of the new aerobic process is the marked increase of the brine circulation. In addition to a more rapid salt equilibrium, it gives a faster solubilization of sugar and other organic substances. Sugar passes progressively to the brine and is exhausted in the flesh by 90–100 days after brining. This coincides approximately with the salt equilibrium. There is no accumulation of sugar in the brine at any moment, since the microbial population also increases substantially. Thus, it can be said that the olives have finished their period of active fermentation after just 3 months of processing. This is a major advantage when compared with the 6 to 9 months needed in the traditional system.

Polyphenol changes in the flesh show a similar pattern to sugars so, consequently, olives are debittered earlier and to a higher degree than in anaerobic conditions. As it is supposed that an equilibrium is reached in this process, polyphenol destruction must occur in aerobic conditions. There is no evidence for this effect, but it is possible that some of the facultative or oxidative yeasts found in this process could play a role in oleuropein destruction.

In general, the period of active fermentation and that required for olive debittering and salt equilibrium coincide. Aerobically processed olives can be sold after about 3 months of brining. This provides the market with a greater flexibility and can decrease the processing costs because of the more rapid return.

Changes in colour and texture

Due to the extended presence of oxygen in the brine, the aerobic process enhances the surface black colour of naturally black mature olives, and its effect is greater as the air flux used increases (Table 9.15). Percentage reflectances obtained in the aerobic process are about 30–35 while those in the anaerobic system can be lower than 20. This blacker colour must be due to polymerization of anthocyanins in the presence of air and their binding to the olive flesh by a still-unknown mechanism. This tone is maintained even at the same pH at which the anaerobic process gives cherry-red olives. The advantage of the aerobic procedure is thus obvious, from this point of view. As shown in Table 9.15, the colour stabilization is achieved earlier than other physicochemical characteristics (salt, reducing sugar, etc.) since it has already been achieved by the 60th day after brining. Interestingly, final colours obtained are similar in both continuous and discontinuous applications.

With respect to texture (Table 9.16), the presence of oxygen gives similar consistency products with a discontinuous process (8 hours of aeration and 16 without it) and lower or very close levels with continuous injection. There is a correlation between the air flux and final texture. After the period of active fermentation (60 days) there is a slight decrease in consistency throughout the time olives are in brine (storage). However, this is slow and has no significance from the quality point of view.

Table 9.15 Influence of different aerobic fermentation conditions on colour (% reflectance at 700 nm). Initial salt content, 10% NaCl (Reproduced with permission from García García *et al.*, 1985)

Type of process	Colour	
	After 60 days	After 124 days
Traditional anaerobic	32.2	36.2
Continuous aeration		
1.00 l/h × lc	19.7	19.9
0.50 l/h × lc	26.0	–
0.25 l/h × lc	29.3	26.9
0.10 l/h × lc	31.7	31.7
Discontinuous aeration, 1 l/h × lc		
1 : 2, air : stop	26.4	19.5
8 : 16, air : stop	24.5	23.1

lc, fermenter capacity in litres.

Table 9.16 Influence of different aerobic fermentation conditions on texture (N/100 g pitted olives). Initial salt content, 10% NaCl (Reproduced with permission from García García *et al.*, 1985)

Type of process	Texture	
	After 60 days	After 124 days
Traditional anaerobic	977	928
Continuous aeration		
1.00 l/h × lc	887	843
0.50 l/h × lc	957	906
0.25 l/h × lc	915	877
0.10 l/h × lc	931	986
Discontinuous aeration, 1 l/h × lc		
1 : 2, air : stop	882	876
1 : 2, air : stop	982	927

lc, fermenter capacity in litres.

The use of divalent cations to improve texture has a limited effect when fermentation is performed at high levels of salt (Table 9.17), regardless of the storage time. However, if the initial salt content is low (2–3%), the texture increase due to the Ca^{++} presence is significant at the 60th day and is still noticeable after 125 days of brining. Bearing in mind the absolute values of texture (Table 9.16) and the magnitude of the calcium effect, the advantage of calcium utilization is debatable, especially considering that its presence can introduce some undesirable flavour. In general, if the process is properly carried out, the consistencies are of the same magnitude as those obtained with the anaerobic process and the presence of any additive to improve texture is unnecessary. This is in agreement with results described for turning colour olives and California-style olives: the effect of additives to enhance texture diminishes when the texture of olives also decreases. Since maturation is paralleled by a rapid decline in texture, it is normal that calcium has only a limited influence on naturally black olives.

Air flux

Since the objective of the new process was the prevention of gas-pocket formation, aeration rates were fixed to yield the lowest percentage of this type of spoilage. Real fluxes are dependent on the volume of the fermentation tank, efficiency of the device used to introduce the air, height of the fermenter, etc. Consequently, although data have been calculated with the intention of the widest possible application, specific fluxes must be established case by case. However, some general ideas and data on the application to Spanish cultivars can be useful to any engineer or processor interested in applying the process.

An efficient process is obtained with modern fibre-glass fermentation vessels of 10 tonnes fruit capacity. The process is performed by introducing a tube of approximately 120 mm exterior diameter. As shown in Tables 9.13–9.17, the fluxes tried have ranged from 0.1 l/h × l.c. to 1.0 l/h × l.c., in the case of the continuous application of air. Good results can be obtained within the range 0.2–0.5 l/h × l.c. Each vessel will thus require approximately 12.5 l/s of air.

The air may come from a compressor or any other device able to give an air pressure sufficient to overcome 2–3 m of brine back-pressure, and the factory must have a pipeline network to distribute the air to each vessel. Air must be free from any inorganic oil and microorganisms. To achieve this, adequate filters must be fitted.

Air injection was initially applied continuously during the fermentation. However, this was possibly excessive and does not match well with the schedule of the naturally black olives processing industry. Most olive factories are not working all day but in 8-hour shifts which, depending on demands, can number one to three. So, the adaptation of the aerobic process to the factory time-table was studied. Results on colour, texture and gas-pocket formation are shown in Tables 9.13–9.17. Nowadays, air injection is usually applied only 8 h/day throughout the period of active fermentation. Excess air, apart from being an unnecessary cost, can cause maturation progression in olives and the growth of oxidative yeast that may form mycelium in the brine. This interferes with brine circulation and impedes air flux through the columns. Additionally, loss of texture

Table 9.17 Influence of $CaCl_2$ addition* on texture (increase of consistency expressed as N/100 g pitted fruits), under different aerobic fermentation conditions. Hojiblanca cultivar. Initial pH corrected to 4.5 (From Durán Quintana *et al.*, 1987)

	Days after brining	
Fermentation conditions	60	125
Initial NaCl content 9%		
Continuous aeration, 0.25 l/h × lc	6.3	4.9
Discontinuous aeration		
1.00 l/h × lc	4.0	3.9
0.50 l/h × lc	5.2	–
Initial NaCl content 2%		
Continuous aeration, 0.25 l/h × lc	10.0	7.7
Discontinuous aeration		
0.75 l/h × lc	15.6	5.2
0.25 l/h × lc	19.6	8.6

* Calcium chloride concentration was 2% (w/v).
lc, fermenter capacity in litres.

resulting from the growth of aerobic pectolytic and cellulolytic microorganisms can be expected.

Changes in the dissolved oxygen in brine

According to experimental results, the percentage of oxygen saturation in brine remains almost constant during the anaerobic process at about 27%. Any initial surplus must be consumed very rapidly by aerobic organisms that contaminate the brines because oxygen saturation levels are about this level from the very beginning. Under these conditions, strictly fermentative yeasts such as *Saccharomyces oleaginosus* can grow. Absence of air can cause physiological disorders in many fruits and some, such as gas-pocket formation in the absence of microorganisms, are observed in olives.

In the aerobic system, the oxygen supply must be enough to prevent the growth of strictly fermentative yeasts and provide, at the same time, a more friendly environment (more oxygen available) to the fruits. According to experimental data these objectives are reached with a limited air supply. Obviously, as more air is injected higher percentages of dissolved oxygen can be obtained. However, this also leads to growth of more oxidative microorganisms which, in turn, cause more rapid depletion of oxygen when bubbling is stopped. Typical percentages of oxygen with respect to saturation are shown in Figure 9.8. The discontinuous aer-

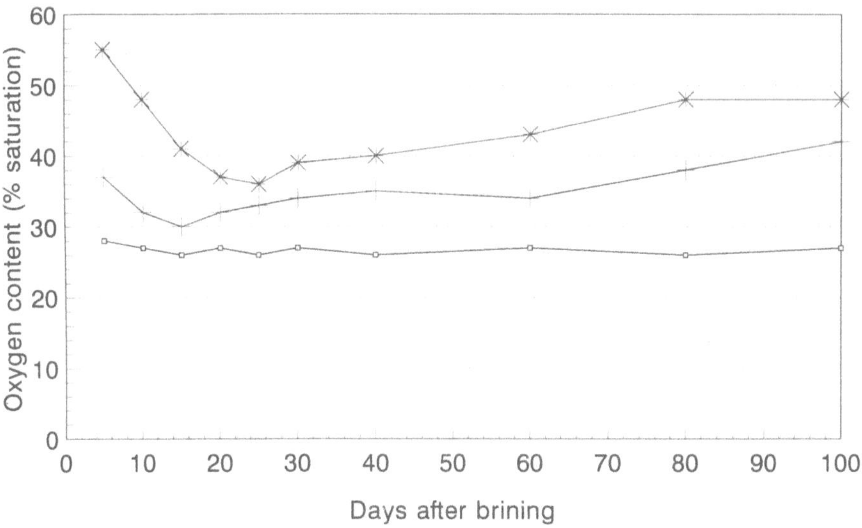

Fig. 9.8 Changes in dissolved oxygen content (percentage with respect to saturation) under anaerobic (□) and discontinuous (+) or continuous (×) aerobic conditions (From García García *et al.*, 1985.)

obic system provides oxygen at between 30 and 40% saturation, while the continuous process increases the percentage to over 40% for most of the fermentation period.

There is a period during which the percentage saturation decreases when the activity of fruits and microorganisms coincides. A minimum is obtained when the combination of both oxygen consumers reaches a maximum. The fruit consumption then progressively decreases and the oxygen level in the brine slowly increases. Behaviour of the continuous and discontinuous processes are the same in this respect.

The oxidative characteristics of the growing microflora can be assessed by measuring the oxygen uptake curve when air injection is stopped. Tangents to the curves can be used as an aerobic index for the flora. It has been observed that continuous air injection leads to higher slopes (growth of more oxidative microorganisms), indicating that the microflora of the discontinuous process is predominantly facultative organisms. Consequently, this is the situation found most widely since, in practice, only alternate air application is used.

Changes in the carbon dioxide content of brine

Brine circulation and air sparging cause release of the dissolved CO_2. This is also favourable for the fermentation and the final quality of fruits. CO_2 elimination prevents overflow from fermenters and brine losses. At the same time, it also favours gas diffusion from the flesh through the skin barrier. This diminishes the internal pressure and possible damage such as liquid exudation and formation of fissures. Other osmotic exchanges are also enhanced. Decreased olive gas content increases the specific weight of fruits and facilitates their immersion in the brine, so no extra pressure (like stones on covers) is needed to keep the fruits submerged.

Concentrations of CO_2 in the anaerobic process are over 100 mg/100 ml of brine and frequently reach higher levels in the large fermenters. Its loss through the scarce brine–air contact surface is shown in Figure 9.9. This problem of CO_2 accumulation has been worsened by

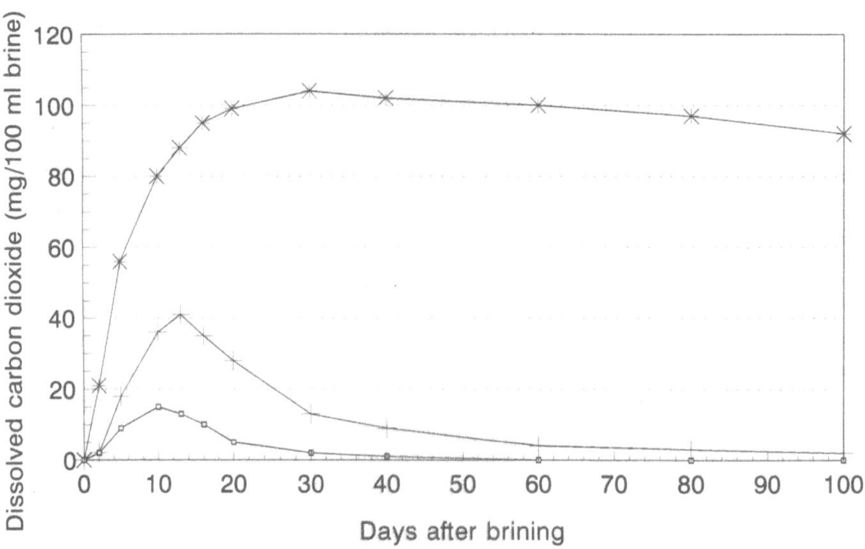

Fig. 9.9 Changes in carbon dioxide content in brine (mg CO_2/100 ml brine) under anaerobic (□) and discontinuous (+) or continuous (×) aerobic conditions. Fluxes of air were: 0.1 l/h × lc in continuous process; 0.25 l/h × lc in discontinuous process. (From García García *et al.*, 1985.)

the introduction of large fibre-glass fermenters where the fruit-to-brine ratio is higher than that found in the old wooden tanks or cement containers. Gas production during fermentation is higher and leads to higher contents in the brine. Gas release in these containers is very slow because of the low surface area of contact with air compared with total volume. There are no data available on the CO_2 concentrations in the old containers, but it is certain that they were markedly lower than in the modern fibre-glass fermenters.

Air injection purges the CO_2 in brine in a proportion which depends on the air flux. Continuous aerobic conditions with a flux over 1 l/h × l.c. completely removes the gas throughout the process. However, normal fluxes are not capable of removing all the CO_2 produced during the first brining days when the maximum respiration of fruits and microbial growth coincide. This gives a characteristic curve that increases to a maximum during the first phase of fermentation, then falls rapidly as fruit metabolism is inhibited by the salt that passes to the flesh. Later, its concentration remains very low and constant. Lower levels are obtained with the continuous aerobic process (0.25–0.10 l/h × l.c.). The alternate or discontinuous system is less efficient in this period, although concentrations are usually below 50 mg/100 ml brine, which could be enough to avoid shrivelling and other problems caused by excess CO_2. If necessary, initial gas accumulation can be avoided by increasing the air flux during this period or by the application of the continuous process. After 30–40 days, aeration can be returned to the alternate system.

Microbial growth

Adequate salt and pH correction prevent the growth of most of the undesirable microorganisms. In addition, aerobic conditions avoid the presence of anaerobic *Clostridium* and *Propionibacterium* species responsible for malodorous fermentations. In spite of the aerobic conditions, microbial films do not form on the brine surface because of the frequent replacement of the liquid on the surface due to circulation.

As previously stated, the aerobic process can enhance the availability of fermentable substrates and, consequently, the growth of microorganisms, especially yeasts. Since most of the LAB are oxygen-tolerant, the system does not interfere with lactic fermentations in appropriate cultivars, provided that levels of salt and pH are suitable.

Growth of microorganisms begins just after brining. As always in olives, the first types to appear are Gram-negative bacteria. Their counts and days of persistence depend on pH. If this is not controlled and the aeration is strong, they may stay in brine for periods of up to 50 days, since this favours the growth of oxidative organisms that maintain the pH at high values. If the pH is initially adjusted to 4.0 or below,

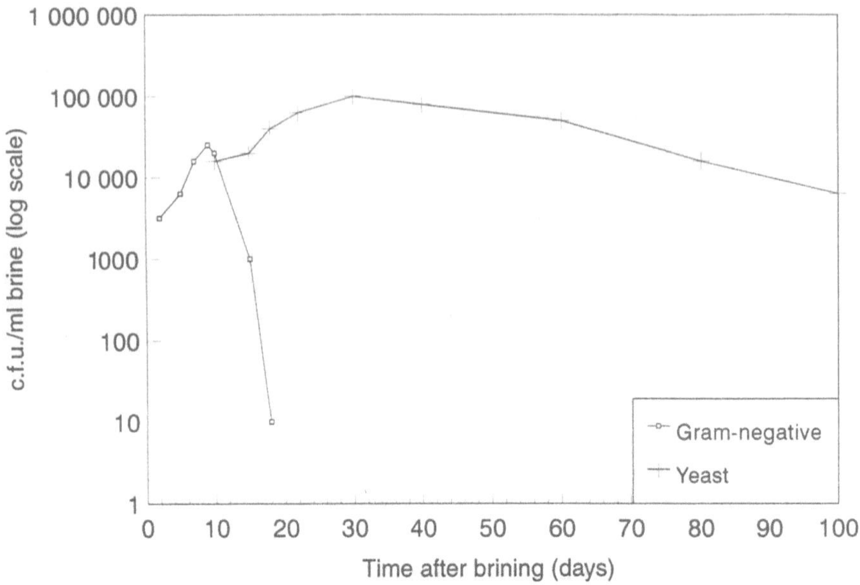

Fig. 9.10 Changes in Gram-negative bacteria (□) and yeast (+) populations during fermentation under discontinuous aerobic conditions at an air flux of 0.3 l/h × lc and no initial pH correction. Salt concentration was 9% and 2% in the case of Gram-negatives and yeasts, respectively. (From Garrido Fernández *et al*., 1987.)

Gram-negative bacteria may be completely inhibited. Growth of LAB also helps eliminate the Gram-negative bacteria. Figure 9.10 shows the moderate growth of Gram-negative bacteria in a typical fermentation at high salt concentration (9% NaCl) and pH controlled at 4.5 with food-grade acetic acid. Their maximum growth occurs usually between the 4th and 7th day after brining and only two species – both from the Enterobacteriaceae family – have been found: *Citrobacter freundii* (96%) and *Proteus mirabilis* (4%).

Yeasts are present during the whole fermentation, their counts being higher than in anaerobic conditions because of the more rapid leaching of nutrients into the brine. Initially, their counts are very low, but they grow rapidly, reaching a significant population by the 10th day of fermentation. A characteristic curve in the absence of LAB is shown in Figure 9.10. The rapid growth during the first stage of fermentation is usually followed by a slower increase that reaches a maximum between the 60th and the 80th day. The population then slowly decreases, but maintains important counts after all the free sugars have been exhausted and throughout the time the olives are brined. Apparently, the yeast population is able to use other carbon sources or survive on the sugars produced by the slow hydrolysis of substances such as anthocyanins.

The yeast population depends on air flux. It is higher and has more oxidative yeasts as air injection increases and decreases with the growth of LAB. The most representative oxidative species (Durán Quintana *et al.*, 1986) are *Pichia membranaefaciens*, *Hansenula mrakii* and *Candida boidinii*, which only utilize glucose (the most abundant sugar in olives) and form a film on liquid malt extract. Other species also important are *Torulopsis candida*, *Debaryomyces hansenii*, *Hansenula anomala*, and *Candida diddensii*, which display facultative metabolism. Other species found are shown in Table 9.18. It is interesting to note the isolation of *Saccharomyces oleaginosus* in aerobic fermentations only from the olive flesh. The aerobic conditions prevented its growth in the brine, but inside the flesh the skin barrier and the yeast's own CO_2 production can provide the required anaerobic conditions for growth. This also indicates the possibility of the organism growing if environmental conditions in the brine become anaerobic, and the need for good control to induce the proper fermentation (absence of fermentative yeasts).

Table 9.18 Main representative species of yeasts identified from fermentation of naturally black olives under aerobic conditions. Hojiblanca cultivar (From Durán Quintana *et al.*, 1986)

Species	Frequency (%)
Torulopsis candida	20
Debaryomyces hansenii	19
Pichia membranaefaciens	14
Hansenula mrakii	11
Hansenula anomala	8
Candida boidinii	7
Candida diddensii	6
Torulopsis norvegica	2
Torulopsis globosa	<1
Torulopsis nitratophila	<1
Hansenula polymorpha	<1
Hansenula silvicola	<1
Candida guilliermondii	<1
Candida rugosa	<1
Saccharomyces uvarum	<1
Rhodotorula marina	<1
Rhodotorula glutinis	<1
*Saccharomyces oleaginosus**	3

*Found in interior of olive flesh.

With certain cultivars and low initial salt contents (2–3%), the growth of LAB begins early, in spite of the pH control. Initially, their population is a mixture of species of the genera *Pediococcus* and *Leuconostoc*, which are less acid-tolerant. They persist up to around the 20th day. By this time, a population of *Lactobacillus* is already present and this remains up to about the 70th day. Its inhibition coincides with the salt levels reaching over 8%. In general, the production of acidity and a low pH favour safe preservation of the product during fermentation and storage. Their negative effect on olive surface colour is less than in the anaerobic process, since the presence of air and its colour enhancement effectively counteracts the discolouring effect of the low pH. There is also no negative influence on texture. Thus, in general, induction of lactic fermentation is advisable. Adequate initial pH control and periodic salt additions are absolutely necessary to prevent undesirable organisms and for a progressive salt increase to less risky levels.

Bulk aerobic fermented olive storage

Air injection is recommended mainly during the period of active fermentation. There is no drawback in maintaining aerobic conditions during the entire time that olives are in brine, although this can be costly. Other alternative storage conditions have also been tried. They consisted of decreasing aeration to only one weekly aeration (8 h at a rate of 1 l/h × l.c.). There was a slow increase in dissolved CO_2 with air interruption and with weekly aeration (Figure 9.11). The only difference between both methods was a slight decrease in the case of weekly aeration. However, in fermenters to which sodium sorbate was added and the microbial population inhibited, the gas concentration remained almost constant.

In the case of the other two storage procedures, the microbial population was similar to that found during fermentation. The CO_2 content was relatively low and indicated a reduced microbial activity. This does not mean, however, that fermentative yeasts cannot grow. The environmental conditions are not favourable and fermentable substrates scarce, but any accidental contamination or growth in the brines of fermentative species from the flesh (*Saccharomyces oleaginosus*) can give rise to a typical anaerobic fermentative yeasts population, the production of high levels of CO_2, and the formation of gas-pockets. Such undesirable changes may be detected by an unusual increase in the brine volume (overflow) as well as the evident release of CO_2 bubbles.

The effect of the different storage treatments on the main quality attributes and gas-pocket spoilage is shown in Table 9.19. Air interruption in all cases caused an increase in the reflectance values and a certain discoloration. The final values depended on the aerobic system applied. When

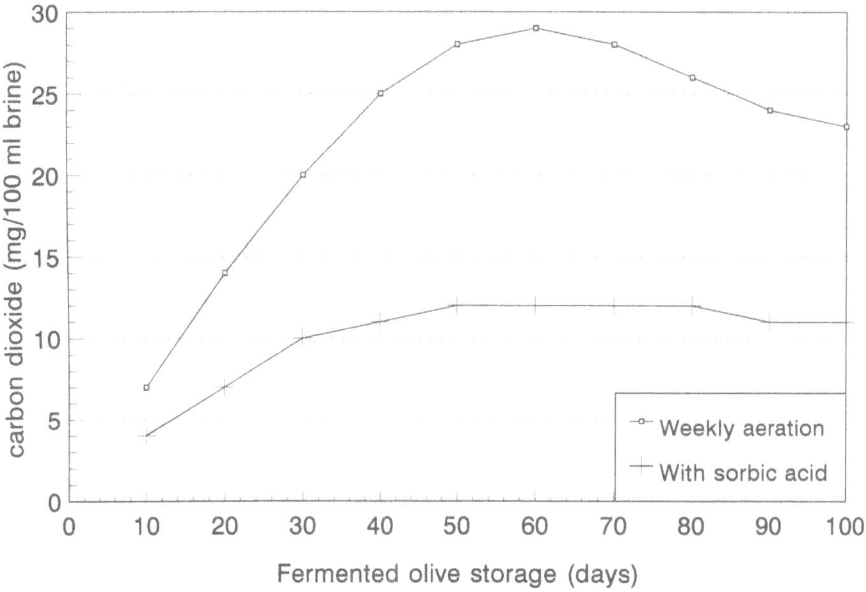

Fig. 9.11 Changes in carbon dioxide content in brine during storage of aerobically fermented naturally black olives. (From Garrido Fernández *et al.*, 1985b.)

this was carried out with a strong aeration, the final product still had good colour. When aeration was continued during storage at a rate of 1 l/l × l.c. for 8 hours each week, there was no discoloration and, normally, final colour improved during this period. In the case of sodium sorbate addition, changes were diverse, but no excessive quality loss was generally observed.

Changes in texture (Table 9.19) were of low significance except in some cases, where specific causes could not be detected. In general, interruption of aeration led to a decrease in texture during storage. This was observed to a lesser extent in the case of stabilization with sodium sorbate. The results suggest a slight tendency to better texture in those fruits preserved with sorbate.

The most important and significant changes were observed with gas-pocket spoilage during the storage period (Table 9.19). In some of the fermenters in which the air injection was stopped, there were significant increases in gas-pocket spoilage (between 10 and 30%). This can be a serious drawback for the interruption of air. The use of sodium sorbate completely prevented this spoilage.

In conclusion, it can be said that for adequate preservation of fermented olives, the options are rather limited. It can be achieved by a

Table 9.19 Changes in texture (shear-compression, N/100 g pitted olives), colour (percentage reflectance at 700 nm), and percentage of gas-pocket spoilage during the storage period of aerobically fermented naturally black olives (From Garrido Fernández *et al.*, 1985a.)

Fermentation system	Texture		Colour		Gas-pocket spoilage (%)	
	Initial	Final	Initial	Final	Initial	Final
Storage with total interruption of aeration						
Continuous aeration						
1.00 l/h × lc	872	786	20	25	2	15
0.50 l/h × lc	898	840	22	32	2	8
0.10 l/h × lc	900	600	33	36	2	25
Discontinuous aeration, 1 l/h × lc (1 : 2, air:stop)	866	750	19	26	1	30
Storage with weekly aeration						
Continuous aeration						
1.00 l/h × lc	873	620	20	13	2	8
0.50 l/h × lc	890	680	28	23	1	16
0.10 l/h × lc	886	490	22	21	3	13
Discontinuous aeration, 1 l/h × lc (8 : 16, air:stop)	960	790	24	23	8	34
Storage with addition of sodium sorbate (0.05%)						
Continuous aeration						
1.00 l/h × lc	914	1020	24	24	3	3
0.50 l/h × lc	864	830	25	33	2	9
0.10 l/h × lc	873	490	30	26	1	2
Discontinuous aeration, 1 l/h × lc (8 : 16, air:stop)	893	730	22	26	5	5

continuation of the aeration during the whole time that the olives are in brine, or by the addition of sodium sorbate. The choice depends on their respective costs.

9.2.5 SPOILAGE OF NATURALLY BLACK OLIVES

According to Balatsouras (1990), naturally black olives are delicate and subject to deterioration for three reasons:

1. Their advanced stage of maturity (and subsequent soft flesh) make them easily bruised or damaged.
2. They are picked under adverse conditions late in the fall or early in the winter, usually in rainy weather, so that they are easily contaminated with wet soil and other impurities.

3. A significant proportion are still produced in the rudimentary installations of growers who base their processing only upon experience and ignore even basic facts about raw material quality, the events that take place during processing and storage, and risks of spoilage or even complete loss of the crop.

The modernization of the Greek naturally black olives industry has improved markedly the quality of the production because of the improved control of the fermentation characteristics (salt, pH, acidity, etc.), the maintenance of conditions to prevent the formation of films with moulds and yeasts, the rapid handling of olives when they arrive at the factory, etc. The same is true also in Spain, where the industry uses almost exclusively fibre-glass vessels of 10 tonnes capacity. However, other countries producing naturally black olives are still dependent largely on the use of traditional containers or cement tanks, although most are rapidly transforming their technologies.

Naturally black olives are heavily contaminated with microbes, dust or even soil. Soil is the main source of clostridia, bacilli and coliforms. Even if the olives are washed, some of these, together with moulds and yeasts, are transferred into the brine and may start to grow. Good fermentation control must induce the growth of the desirable flora and prevent the growth of the undesirable.

Malodorous fermentations

Butyric acid fermentations are not common in naturally black olives because of the traditional use of high salt concentrations and the presence of other unfavourable conditions such as low temperature and high polyphenol content. Balatsouras (1990) described the inhibition of growth of saccharolytic clostridia by salt greater than 6.5%, temperatures less than 20°C, low pH and the presence of limited concentrations of sugars. Bearing in mind these conditions, the low incidence of this spoilage is understandable. The conditions that allow growth of these organisms hardly occurs in any of the diverse processing systems of naturally black olives. In fact, salt content only occasionally falls to 6%. This may happen during winter and by that time temperatures are usually below the minimum required for growth. In the case of initial low salt content, this practice is accompanied by pH control to about 4.0–4.2, which is also inhibitory. Another prerequisite for the growth of clostridia is the presence of high levels of sugars but this never happens in this type of fermentation because of the slow diffusion of substances from flesh into the brine and their rapid utilization by Gram-negative bacteria or yeasts. In the case of aerobic conditions, they too are inhibitory for *Clostridium*, so butyric fermentation cannot be expected when this system is used for processing naturally black olives.

Butyric fermentation commences (as noted in Chapter 7 on green olives) at the beginning of fermentation and manifests itself with a rancid and malodorous stench. Affected olives are unacceptable and any effort to recover them is impracticable, unwise and insalubrious. Exceptionally, only when the spoilage is noticed upon its onset, is it possible to get an acceptable product of inferior quality by replacing the brine and inoculating with pure starters of LAB.

Another malodorous spoilage is 'zapatería' which, similarly to green olives, is characterized by a mild but very penetrating odour that becomes progressively very objectionable. The smell of olives with 'zapatería' spoilage is not as rancid and malodorous as in the case of butyric acid fermentation. According to Balatsouras (1990), this spoilage is very common in the case of naturally black olives. Strangely enough some consumers consider the 'zapatera' smell as a quality attribute of these type of olives.

As already mentioned, the causative agents in the case of Spanish-style green olives are a group of microbes that includes mainly species of clostridia and propionibacteria. In olives produced in California the spoilage has been associated with sulphate reducers, which can reduce the sulphate salts present as impurities in the sodium chloride used for preparing the brines. The formation of H_2S imparts a highly objectionable smell, which is not typical of the first step of 'zapatería' spoilage, when the olives emanate a mild smell hardly perceptible to most people, although highly penetrating.

Appearance of 'zapatería' also occurs when low salt contents and high pH levels coincide. Such situations are frequently present in naturally black olives. In the traditional process, pH customary stabilizes around the value of 4.5–4.8 and salt contents below 8–10% are also common. Such conditions favour the onset of 'zapatería' spoilage. As in the case of green olives, organic acids such as propionic, butyric, succinic, formic, valeric and caproic are found in advanced forms of 'zapatería'. They impart to olives a malodorous stench and render then uneatable.

The use of the aerobic system with high flux of air, may lead to an excessively oxidative microflora, with the consumption of the acidity and a subsequent pH increase. If this coincides with a period of no air injection and low salt, the spoilage will probably start.

In general, the only effective way of avoiding malodorous spoilage is by good control of pH and the salt content. In cases of low initial salt content, to induce the lactic fermentation it is convenient to lower the pH with food-grade acetic acid (or other authorized organic or inorganic acids) to about 4.0. This, together with the progressive salt increase, can assure the product preservation. In this case, aeration must also be moderate to diminish acid consumption to a minimum. It is customary to increase the NaCl content in the final product of naturally black olives to

8% or higher. This is perfectly accepted by the consumer and is a safe tradition.

Galazoma or cyanosis

This spoilage ranks among the first causes of losses of Greek processors, though it is practically unknown in Spain. It manifests itself by a change in colour from faint-black to faint-blue or blue-grey and from deep black to deep-blue in most severe cases (Aligisakis, 1982). The final colour is a function of the maturation degree of the fruits. The name of galazoma, as it is commonly known, means blue coloration. The spoilage appears on black, red or turning colour olives during fermentation and storage. It appears at the end of winter and, mainly, during summer and autumn. It affects mainly olives whose processing has been defective, without adequate control or with low salt content. Colour changes are usually accompanied by a worsening of both taste and texture, so that the product becomes unsaleable. The blue colour persists even after drastic treatments with very strong chemicals and in advanced forms is accompanied by an undesirable smell.

The abnormality has been known to Greek processors for many years. According to Balatsouras, cyanosis was first induced deliberately in 1961 on a semi-industrial scale in trials by the technicians of the Cooperative Union 'Eleourgiki'. It was found to occur easily in black olives packed in dilute brine containing 6% NaCl or lower. Olives stored in a concentrated brine containing up to 11% NaCl also developed galazoma if the brine was added with 0.3–0.6% iron chloride. A possible microbial origin can be deduced from the fact that its incidence is prevented, even at low salt concentrations, by pasteurization.

Aligisakis and Patroclos (1971) mention also the following characteristics of this spoilage: (i) olives preserved in NaCl concentrations over 8% do not develop cyanosis when maintained in anaerobic conditions; (ii) olives from non-irrigated orchards are more resistant than those from irrigated trees; and (iii) small olives are less prone to the spoilage than bigger ones, while mature olives are more resistant than less mature fruits.

To recognize when naturally black olives are affected by cyanosis, it is enough to remove the brine and expose the fruits to the air for a period. There is at first a slight change of the normal colour followed by a later change to blue-black, blue-grey, etc. These changes are accelerated as temperature increases in the summer and are relatively slow at winter temperatures.

According to Balatsouras (1990) it is a common belief among Greek processors that galazoma is associated with unclean and unhygienic conditions in the processing plants, as well as low salt contents in brines and packs harvested in rainy weather.

Bearing in mind all the above factors it is probable that this spoilage might be produced by an association of microbial and physicochemical causes. The latter could play an even more important role than considered until now. This is suggested by the fact that naturally black olives, when processed as black-ripe olives, develop a similar colour to that of 'galazoma' in the presence of iron salts. Anthocyanins, the colouring compounds of naturally black olives, change their colour to blue tones as pH is increased. The presence of iron salts could enhance the changes when fruits had high pH due to the effect of oxidative microorganisms. Probably, the presence of iron in old wooden vats, released from the different devices used to handle the olives, could be a good source of this cation and the cause of a further spoilage outbreak.

Good sanitary conditions, proper handling of olives, appropriate salt concentrations and a good control of the fermentation have diminished the incidence of 'galazoma' in modern factories.

Softening

As previously mentioned, there is a sharp decrease in olive texture during ripening. Since naturally black olives are harvested when the fruits are in an advanced stage of ripeness, it is natural that they have a reduced texture compared with other table products and that they must be more liable to softening. An important role in olive texture is played by the pectic substances. These are polyuronides formed mainly of anhydrogalacturonic acid residues, which may be partially esterified with methanol, or salts of divalent cations (Ca^{++} and Mg^{++}). They form part of the middle lamella between cells and are also in the cell walls, together with cellulose, hemicellulose and lignin. Degradation of pectic substances (strongly related to softening) is accomplished by two enzymes, pectinesterase (which removes the methyl groups) and polygalacturonase (which splits the terminal glycosidic bonds, exo, or acts randomly, endo). The enzymes may come from the fruits or from other sources. Some authors have attributed the softening of pickles to pectolytic enzymes produced by moulds of the genera *Fusarium, Penicillium and Aspergillus* (Balatsouras and Vaughn, 1958), as well as from different species of *Bacillus* such as *B. mesentericus, B. polymyxa, B. megatherium* and *B. macerans* (Vaughn, 1954), or fermentative yeasts like *Saccharomyces oleaginosus, Sacch. kluyveryi* and *Hansenula anomala* (Vaughn *et al.*, 1972; Vaughn, 1982).

According to Mínguez Mosquera (1982), the anhydrogalacturonic acid content decreases progressively during ripening, as does the degree of esterification which is zero in over-ripe fruits. This last change parallels the pectinesterase activity in Hojiblanca fruits. Enzymatic activity in other cultivars (Gordal, Manzanilla and Lechín) was relatively low, but did show a similar tendency.

Polygalacturonase activity, which needs the previous action of pectinesterase to de-esterify carboxyl groups, was found in ripe and overripe olives (being higher when pectinesterase has already practically disappeared). The same author found that some of the Gram-negative bacteria and *Bacillus* isolated from fruits could produce polygalacturonase.

In spite of the presence of such enzymes in the ripe fruits (at least in some cultivars) and, possibly, in their fermentation brines, produced by microorganisms, the occurrence of softening in normal aerobic or anaerobic processes is limited. This is because the activity of polygalacturonase is very limited under fermentation conditions. Normally, polygalacturonase activity is much reduced at pHs of 4.0–5.0 and the salt also has a strong inhibitory effect. Similar effects are observed on pectinesterase. Thus, softening is produced only when lack of hygienic conditions is evident, or in the absence of any control during fermentation. In these cases, enzymes probably come from the growth of undesirable microorganisms such as those mentioned above.

Often, softening is also associated with gas-pocket or fish-eye spoilage, when the microorganisms simultaneously produce a great volume of gas.

Gas-pocket or fish-eye

This spoilage is characterized by the formation of a series of fissures in the interior of the flesh. They may begin close to the skin and continue down to the pit or vice versa. In most cases, there is a parallel accumulation of gas in the interior of the flesh, which is retained by the skin. This gives the fruits an aspect of a inflated balloon when pressed between the fingers. Many times, the skin is detached from the rest of the flesh, forming a transparent film. The space between this film and the flesh may also be filled with pressurized gas. In Spanish the last aspect is called 'afarolado' (Figure 9.12). Seriously infected and injured olives float on the surface of the brine and are known under the name of 'floater'. This spoilage in naturally black olives is rather different from that of Spanish-style green olives, probably due to differences in the gas permeability of the skin and the flesh structure and/or composition.

Losses due to this spoilage are substantial in Greece, although the deterioration is little known or understood by growers and small processors. Most of these olives, defective in appearance and texture but not in taste, are offered at lower price and consumed locally.

In Spain, fruits from Hojiblanca cultivar are easily affected and this spoilage has caused serious losses on occasion. Other cultivars like Manzanilla and Cacereña are also affected to varying extents. In Turkey, the high level of salt produces shrivelling in the olives, which can mask the defect and may also inhibit the microorganisms significantly, with the

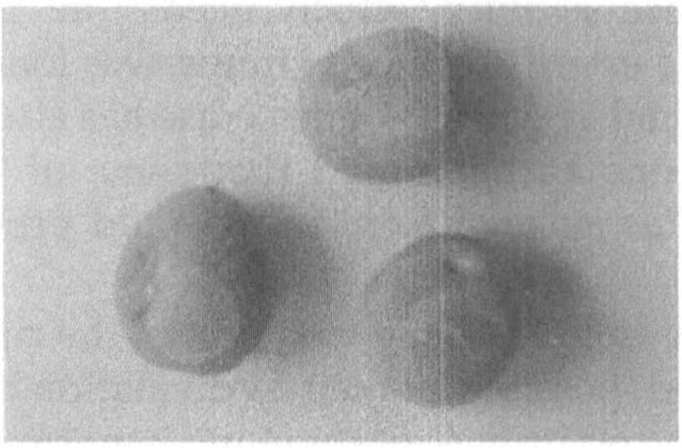

Fig. 9.12 Olives affected by 'Afarolado' spoilage. (Photograph courtesy of Dr Garrido Fernández.)

production of low levels of gas. The incidence of this spoilage is thus lower and data on its significance are not available.

Balatsouras (1990) associates the spoilage of naturally black olives with the early development of coliform bacteria and the tremendous amount of gas that is produced as soon as such olives are put into brine. For him, these organisms seem to play a more important role in processing black olives than in processing green ones. He explains the production of the spoilage by the fact that tissues of ripe olives are softer and more easily invaded by coliform bacteria, especially in the first stage of the fermentation. The great volume of gases liberated are trapped in the tissues of the flesh in the form of a gas-pocket, blisters and fissures (Figure 9.13). The severity of this gassy deterioration thus depends mainly on the number of coliforms gaining access to the olives, the environmental temperature, and the salt content of the brine.

This is in agreement with the fact that more or less similar deterioration in Spanish-style green olives has been shown to be caused by species of Gram-negative bacteria of the genera *Enterobacter*, *Klebsiella* and *Citrobacter* (Vaughn *et al.*, 1943; Borbolla y Alcalá *et al.*, 1959, 1960). In the washing stages of California ripe olive processing, species of the genera *Aerobacter*, *Escherichia*, *Paracolobactron*, *Aeromonas* and *Achromobacter* cause similar deteriorations (Vaughn *et al.*, 1969). However, it is obvious that these are not the only microorganisms able to provoke the gassy-deterioration. Gililland and Vaughn (1943) associated the formation of fissures in California ripe olives with *Clostridium butyricum* and Vaughn *et al.* (1972) found yeasts responsible for a similar spoilage during the storage of turning colour olives. Bloater formation in cucumbers, with some similarities to gas-pocket spoilage of olives, has been associated with yeasts or heterofermentative LAB (*Lactobacillus brevis*) (Etchells *et al.*, 1968) and even with the homofermentative *Lact. plantarum*, that produces only a very small amount of gas, according to Fleming *et al.* (1973). In

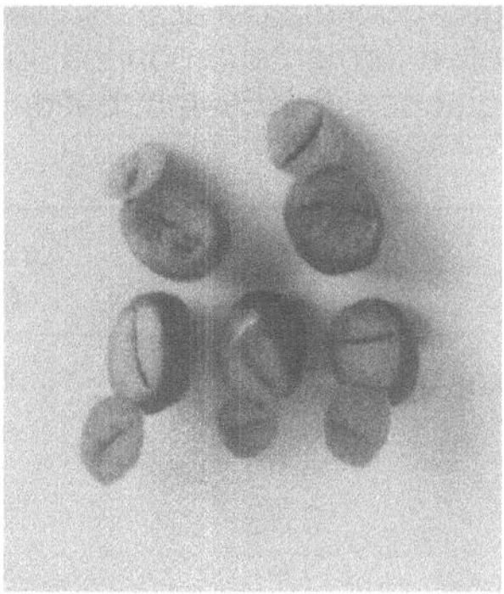

Fig. 9.13 Olives affected by gas-pocket spoilage. Some fissures extend to the pit. (Photograph courtesy of Dr Garrido Fernández.)

some cases, the gas produced by the cucumbers themselves caused the bloater (Fleming *et al.*, 1978).

In the case of naturally black olives of Spanish cultivars it has been shown that at least part of the spoilage is not due to the growth of Gram-negative bacteria, since some spoilage was also present in fermentations in which such bacteria were completely absent.

Trials to produce spoilage by different species of yeasts isolated from naturally black olives (Table 9.20) showed that most were able to produce the deterioration (Durán Quintana *et al.*, 1979). The two more relevant species in such fermentations, *Saccharomyces oleaginosus* and *Hansenula anomala*, produced the higher spoilage percentages. There was a clear relationship between the fermentative ability (higher consumption of sugar) and spoilage production. Those yeasts which had low fermentative capacity and left higher sugar residues yielded less defective fruits. From these results, it was evident that it was necessary to change the habitual yeast flora of the traditional fermentation if the appearance of the spoilage was to be reduced to a minimum.

The demonstration of the abundant growth of certain yeasts and LAB in the flesh of olives was also of interest. Normally, fermentation is followed by analysis and microbial control of brines. However, in the case of naturally black olives, due to the special maturation characteristics and structure of their tissues, some microorganisms can find a more friendly environment in the flesh rather than in the brine, especially when salt content and pH corrections have not yet reached equilibrium. They can also find abundant fermentable substrates which are scarce in the brine because of the slow diffusion through skin. The higher moisture content of mature fruits also facilitates this colonization.

Most of the yeasts found in fermentations of Spanish-type green olives and pimento were capable of producing this spoilage. This means that

Table 9.20 Production of gas-pocket spoilage by yeast species isolated from anaerobic naturally black olive fermentation brines. Initial conditions of essays: salt content, 8%; pH, 4.5; glucose, 2.00%. Incubation at 25°C) (From Durán Quintana *et al.*, 1979)

Species	Remaining sugar (%)	Growth in flesh	Gas-pocket spoilage (%)†
Saccharomyces oleaginosus	0.45	+++	98
Hansenula anomala	0.44	++	90
Hansenula subpelliculosa	0.80	+	76
Kluyveromyces veronae	0.35	+	86
Torulopsis candida	0.47	+	72
Torulopsis norvegica	0.75	++	70
Debaryomyces hansenii	0.42	++	67
Candida diddensii	0.45	+*	72
Candida boidinii	1.00	++	48
Candida valida	1.60	–	11
Candida krusei	1.90	–	7
Pichia fermentans	1.00	+++	62
Pichia farinosa	1.10	+	30
Pichia membranaefaciens	1.90	–	12
Control	2.00	–	13

* Contaminated after the 25th day.
† Analyses carried out after 30–40 days of inoculation.

this capacity was not exclusive to the more adapted species but related to the physiological characteristics of a wide range of fermentative yeasts. This may also be related to the ability of such microorganisms to grow in the flesh and to produce pectolytic and cellulolytic enzymes that facilitate utilization of the different olive components as carbon sources. The presence of abundant CO_2 is an indirect effect of the microbial growth which accompanies the spoilage; the latter effect does not occur unless the tissues have been weakened previously.

The spoilage was related to the presence of fermentable substrate and the growth of certain microorganisms, being appreciably reduced when the causative species were inoculated into already fermented fruits. However, the microorganisms were able to survive, grow slowly, and produce a progressive gas-pocket type of spoilage. Table 9.21 shows the effect of pH correction with two different acids (HCl and acetic) on the production of the deterioration by some of the most active yeasts in both fresh and pasteurized fruits. It is obvious that the use of acetic acid leads to a slightly lower percentage of gas-pocket spoilage. However, the most interesting feature from Table 9.21 is the formation of an important proportion of spoilage in sterile brined fresh fruits while those pasteurized had negligible spoilage. This was not the only time this was observed by the Spanish researchers and led to the conclusion that, in addition to the

Table 9.21 Effect of acid and pasteurization of fruits on gas-pocket production. Initial salt content, 10% NaCl; pH corrected to 4.0. Cultivar, Hojiblanca. Incubation temperature, 25°C (Reproduced with permission from Durán Quintana *et al.* 1979)

Species inoculated	Type of acid	Growth in flesh	Gas-pocket (%)
Saccharomyces oleaginosus	HCl	+++	88
Saccharomyces oleaginosus	Acetic	+++	86
Hansenula anomala	HCl	+	86
Hansenula anomala	Acetic	+ to ++	74
Debaryomyces hansenii	HCl	–	69
Debaryomyces hansenii	Acetic	–	71
Sterile	HCl	–	75
Sterile	Acetic	–	68
Pasteurized fruits			
Saccharomyces oleaginosus	HCl	+++	64
Saccharomyces oleaginosus	Acetic	+++	53
Hansenula anomala	HCl	+ to ++	62
Hansenula anomala	Acetic	+ to ++	29
Sterile	HCl	–	6
Sterile	Acetic	–	6

spoilage produced by microorganisms, the fruits themselves can give rise to a significant proportion of gas-pocket spoilage when held under anaerobic conditions and at a suitable temperature (25°C in this case). This may have a limited effect at the industrial level, since environmental temperatures during the naturally ripe olives harvesting are appreciably lower, about 15°C.

Gas-pocket formation is always associated with specific cultivars. In Spain, Hojiblanca, Cacereña and, to a lesser extent, Manzanilla are very susceptible. However, gas-pocket formation has never been observed in Lechín. Moderately high temperatures of up to 35–40°C markedly increase the spoilage, which can then affect all the olives. In this case, both spoilage from the own-fruit metabolism and that from microorganisms are maximized.

Traditionally, and guided only by experience, these fermentations were carried out in cool places, out of sunlight, and without application of heat to accelerate the fermentation. Thanks to these processing conditions the incidence of gas-pockets could be partially controlled. However, when fibre-glass or cement tanks, exposed to sunlight and to high temperatures were introduced for the traditional fermentation, results were discouraging, with significant losses occurring.

Fermented naturally black (and turning colour) olives must always be protected from sunlight and stored in cool places. Transportation must preferably be carried out during the night from spring to autumn, and even in winter if climatic conditions are warm.

Refrigerated storage in brine at 8°C totally prevented deterioration. However, if stored fruits are subjected to moderately high environmental temperatures, gas-pocket formation is rapid and abundant. This may be related to the vigorous fermentation process that is initiated as a consequence of the fermentable substrates which have accumulated during cold storage in the absence of microbial activity.

Initial salt content and pH correction can also have a certain influence on this deterioration. In addition to the results shown in the text on aerobic fermentation, Table 9.22 shows the average percentage of gas-pocket spoilage found in numerous trials carried out at 10% and 6% NaCl and with initial pH correction with acetic acid and HCl. It is obvious that low levels of salt, as well as the use of acetic for the initial pH correction, tends to diminish the incidence of the spoilage. Such effect must be related to the high inhibitory capacity of this acid, which is less dissociated in solution than is a strong acid such as HCl. It is of interest, however, to note that storage in 5% acetic acid did not prevent a certain proportion of gas-pocket spoilage. The acid could prevent the presence of microorganisms but was unable to prevent the spoilage due to fruit activity. The same was also observed when the pH was lowered to 2.8 with HCl or lactic acid, though with higher percentages of gas-pocket formation. The use of low salt must always be accompanied by pH correction to prevent malodorous and other deteriorations.

Lastly, any preceding treatment (especially pasteurization) which causes deterioration in the fruit structure and a more abundant availability of fermentable substrates may accelerate spoilage. This effect can be counteracted only if inhibitory conditions are used at the same time. Lye-treated or perforated olives did not show any increased sensitivity when brined in acetic acid/pH-controlled and moderately high NaCl solutions. Apparently, fruit metabolism is inhibited by the NaOH in the first case and by the rapid NaCl penetration in the latter.

Table 9.22 Average effect of salt concentration and type of acid used for the pH correction on gas-pocket formation (Reproduced with permission from Garrido Fernández *et al.*, 1979)

pH correction	Initial salt content 10%	Initial salt content 6%	Average for acid
With HCl	89	79	84
With acetic acid	56	57	57
Average for salt concentration	73	68	

Bearing such factors in mind, fermentation or storage conditions can be provided to diminish or prevent the spoilage. Table 9.23 shows some such conditions and their effect on the incidence on the percentage of gas-pocket formation. It is interesting that the addition of sodium sorbate, although diminishing the percentage of spoilage by inhibition of the yeast flora in the brines, still produces a moderate proportion. This could be due to the growth of some microorganisms in the olive flesh before equilibrium was reached and to the growth of small and sporadic counts of yeasts throughout storage. This is in agreement with the fact that the growth of microorganisms in the olive brine is very difficult to inhibit when high levels of nutrients are available. Additionally, the fruits remain rich in sugars, which may induce further secondary fermentation when packed. Storage in the presence of sodium sorbate has therefore a certain risk.

A simpler approach, and one which yields similar or even lower gas-pocket spoilage rates, is the application of the traditional process with pH correction with acetic acid. Yeast growth reduction due to the presence of the acid results in about 30% spoilage. The correction improves the traditional process but still give rise to an unacceptable level of gas-pocket spoilage.

Similarly effective are sterile storage (obtained by the addition to the brine of salicylic acid) and the induced lactic fermentation. The fact that sterile olives were able to develop a noticeable percentage of gas-pocket spoilage is further confirmation of production by fruit activity. Apparently, the induced lactic fermentation gives the same results. This supports the view that the LAB do not participate in the development of spoilage, as was demonstrated when they were tested in a similar way to

Table 9.23 Average percentage of gas-pocket formation under different fermentation (or storage) conditions (Reproduced with permission from García García *et al.*, 1982)

Fermentation (or storage) characteristics	Gas-pocket formation (%)
Spontaneous traditional (10% NaCl; no pH correction)	68
Spontaneous traditional (10% NaCl; pH 4.5; 0.05% sorbic acid)	34
Spontaneous (10% NaCl, pH corrected 4.5)	29
Sterile storage (10% NaCl, pH = 4.5; 2.5% salicylic acid)	22
Induced lactic (6% NaCl, pH = 4.5; inoculation with LAB)	22
Aerobic (6% NaCl; pH = 4.5; 1 l/h × lc)	6
Aerobic (10% NaCl; pH = 4.5; 1 l/h × lc)	5

LAB, lactic acid bacteria.
lc, fermenter capacity in litres.

yeasts. In the induced lactic process, the observed deterioration must also come from the fruits.

Sterile storage also has the inconvenience of excessive residual sugar, difficulties in final product stabilization, and the use of non-authorized additives. Thus, it is not a recommended procedure. Lactic fermentation, however, is advisable, the only objection being the relatively high proportion of deterioration that may still be produced.

Finally, the use of an aerobic system is a very efficient modification to avoid gas-pocket formation. Using appropriate fermentation conditions, its use may avoid the spoilage with both 10% and 6% initial NaCl. This process should be applied according to the information given in section 9.2.4.

9.2.6 COLOUR IMPROVEMENT AFTER FERMENTATION

By the time the period of active fermentation has subsided, the fermentable substrates have reached low levels in both the flesh and brine and sufficient oleuropein has diffused out of the flesh to render fruits edible, the olives are ready for marketing. Other soluble substrates pass into brine, including the anthocyanins. This dilution, together with the pH changes, decreases the olive colour intensity during storage.

According to Balatsouras (1975), to improve surface colour, olives are taken out of brine using cane baskets and are spread immediately over sorting tables. Unloading of tanks is a labour-intensive step in processing olives in Greece. Using fibre-glass vessels, olives can be transferred by vacuum pumps or adapted to olive-handling pumps. Pipe diameters used to be about 10 cm, the brine being re-circulated to the tanks or vessels until all fruits have been removed.

The colour of skin is markedly improved when olives are exposed to air for a period of 2–3 days. The skin of naturally ripe olives is usually resistant to shrivel due to fact that the lye treatment has been omitted. However, olives affected by gas-pockets or inflated lose the interior gas and become easily shrivelled. Oxidation regenerates the colour, at least partially. The olives become blacker as the pH increases. Immersion of olives in a 0.1% solution of ferrous gluconate favourably affects hue and depth of colour, although its utilization is only permitted in the case of black-ripe olives. Processors also believe that immersion of the olive with olive oil after oxidation enhances and stabilizes colour.

This step of aeration is not used currently in some industries, which rely on the air exposure during sorting and grading to produce the oxidation. This would be possible if the darkening speed of olives is enough to reach the desirable effect during the time of such operations.

Figure 9.14 shows the oxidation rate of the olive surface of Hojiblanca and Lechín cultivars when oxidized by air exposure or when fruits are

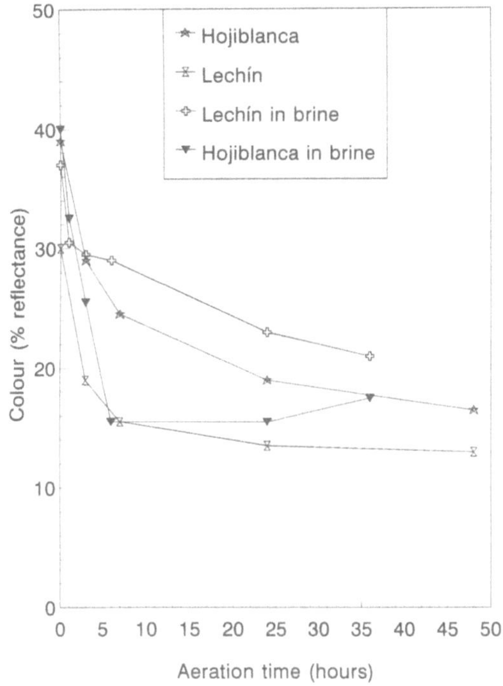

Fig. 9.14 Colour (percentage of reflectance at 700 nm) changes during the darkening of Hojiblanca and Lechín cultivars by exposure to air or by bubbling air through the suspension of olives and brine. (Adapted from Garrido Fernández and Fernández Díez, 1976.)

kept submerged in brine. In this figure higher reflectance values mean lighter colour. The colour after fermentation depends on pH and on the amount of colouring substances remaining in the olives. A fermentation by yeasts stabilizes the pH at around 4.5 and usually leads to a good olive colour in the case of Lechín cultivar. However, Hojiblanca, with the same type of fermentation, has markedly higher discoloration during processing due to its lower anthocyanin content. Darkening the olives is a rapid process. Seven hours of air exposure can cause a significant colour improvement, but colour may continue its improvement for appreciably longer periods. Accordingly, aeration during the sorting and grading operations, which normally take only a few hours, can improve the colour, but for deeper darkening it is less efficient than the traditional system.

Oxidation of fruits by injection of air through the suspension of olives in brine has the advantage of avoiding any moisture loss from the olives.

The oxidation rate is slower and colour recuperation takes longer. After 3 days of aeration, recovered colour is still worse than that of olives oxidized by 7 hours of air exposure (Figure 9.14). This method has the drawback of possible damage to olives. Some of these fruits were weakened during processing and may disintegrate during aeration. The losses can be important if softening or gas-pocket spoilage had seriously affected the olives. Overall, assessing the advantages and disadvantages of this system, it cannot be recommended.

With the aerobic system, olives have a better surface colour than that obtained with the traditional anaerobic fermentation. In this case, aeration is not necessary, especially when air injection has been maintained during the storage period or the time elapsed from the active fermentation is short.

In fresh, naturally black fruits, the total anthocyanin content increases as ripening progresses. This is mainly due to the increase in polymeric anthocyanins, which may then be associated with the progressive blackening of the olives (Vlahor, 1990). It is possible that colour recuperation and blackening during the fermentation of naturally black olives may also be due to the progressive polymerization of anthocyanins near the skin.

Marsilio *et al.* (1990) found significant colour deterioration with the traditional anaerobic process, but markedly better colour retention when olives were given a limited lye treatment to penetrate only the skin, followed by aeration.

During the aeration–oxidation process there is also a marked improvement of the homogeneity of the olive colour, which becomes more uniform, with a subsequent improvement in lot quality. Colour enhancement is a function of ripening degree and anthocyanin content. Oxidation cannot then mask any deficiency in the harvesting of the appropriate maturity.

Oxidation has no value if the colour improvement were not maintained after packing. Darkened olives lose part of the coloration gained when submerged in brine again, but the discoloration never reverts to the pre-oxidation colour levels. Table 9.24 shows the colour obtained when olives are packed without previous oxidation (Lechín) or after oxidation (Hojiblanca and Lechín). Storage colour, at 42% (expressed as percentage of reflectance), is slightly decreased after 4 months of packing, at 43–44%. Previously aerated fruits lost a certain proportion of colour (reflectance increased from 19 to 22–24%), but retained a fairly good colour when compared with that of olives packed without oxidation (43–44%). Good colour retention was also observed in the case of Hojiblanca cultivar (as shown in Table 9.24).

Table 9.24 Colour (percentage reflectance at 700 nm) maintenance in oxidized olives after 4 months packaging (Reproduced with permission from Garrido Fernández and Fernández Díez, 1976b)

	Hojiblanca cultivar		Lechín cultivar	
Type of packing brine	Before packaging	After packaging	Before packaging	After packaging
Re-used fermentation brine			42*	43
Fresh brine	23	27	42*	44
Re-used fermentation brine			19	22
Fresh brine	21	26	19	24

* Not previously oxidized.

9.2.7 SORTING AND GRADING

Olives after exposure to air are hand-sorted for bruised, mis-shapen, off-coloured, seriously spotted or otherwise blemished, insect-injured or defective fruits. They are then run through grading machines to be size-graded according to international standards or in compliance with local regulations. The grading machines that currently predominate are based on steel or rubber endless cables.

9.2.8 PACKING

The sorted and size-graded product is finally packed with an appropriately conditioned new brine. The most frequent containers (Fernández Díez *et al.*, 1985) are either wooden or plastic barrels with a fruit capacity of 130–150 kg, and tin-plate cans holding 10–15 kg fruit. Recently, small cans and plastic bags have begun to be used for this product, which is consumed mainly as a foodstuff and to a lesser extent as an appetizer. In these small containers the product is either pasteurized or preserved with the use of some permitted additive such as sorbic acid, added as potassium sorbate.

The most commonly recommended general limits for physical and chemical characteristics of the equilibrated packed product are as follows (Fernández Díez *et al.*, 1985; Unidad Estructural de Biotecnología de Alimentos, 1990):

pH value	3.6–4.5
Free acidity	0.3–1.0 (g/100 ml, as lactic acid)
Sodium chloride	8.0–10.0 (g/100 ml)
Residual reducing sugars	0.2–0.3 (g/100 ml, as glucose)

García García *et al.* (1986), working with naturally black olives in brine, fermented under aerobic conditions, have studied the treatment conditions for small consumer packs. Two methods are equally effective: (i) pasteurization at 80°C for 4 minutes; and (ii) the addition of potassium sorbate at 0.05%, expressed as sorbic acid at equilibrium. Typical values for those products, after 11 months of storage, were pH values of 4.40–4.45 and sodium chloride concentrations between 7.2 and 7.4%.

In Greece, the packed product is shipped to the local or international market under the designation of 'Greek naturally black olives in brine'. It never acquires a uniform colour and this is what makes the distinction from artificially black olives. Naturally black olives, if processed in the proper manner, are highly prized on local and international markets. They are a natural product that contains nothing but table salt, and have a slightly bitter but specific taste and flavour. They are used mainly as food, contain more oil than green olives and have a high nutritive value (approximately 205 cal/100 g flesh).

In packed olives, the colour is also dependent on the pH, which may reach lower values than in Spanish-style green olives due to the low combined acidity of naturally black olives. In this case, the colour becomes rather pink, although over certain pH ranges changes are minimal. Fernández Díez *et al.* (1975a) showed that the surface colour did not change for pH values ranging from 4.1 to 3.1. In this case, the use of lactic acid permits a lower final pH than with acetic, for the same final titratable acidity.

Further increase of texture during packing is difficult. Fernández Díez *et al.* (1975b) found the use of calcium chloride had no clear effect, although a tendency to improve consistency was observed.

Proper packing is essential, since naturally black olives in brine are prone to secondary fermentations. Only well-fermented, healthy, and well-washed olives should be used for the final product, coupled with the preservation methods already described.

9.2.9 COMPOSITION AND NUTRITIONAL ASPECTS OF THE FINAL PRODUCT

The main factors which can affect the composition of the processed Greek-style naturally black olives in brine are (Fernández Díez, 1983):

- cultivar;
- stage of maturity of the raw material;
- fermentation process and further storage in brine before packing;
- the final conservation of the packed product up to the time of consumption.

Table 9.25 shows typical values for processed olives of Hojiblanca

Table 9.25 Composition of Greek-style naturally black olives in brine (From Fernández Díez *et al.*, 1985)

Composition	Typical values in the wet pulp of Hojiblanca cultivar fruit
Major component (weight %)	
Moisture	60.34
Lipids	23.43
Protein	1.15
Fibre	2.18
Ash	6.88
Vitamins	
Thiamin (µg/100 g)	14.50
Riboflavin (µg/100 g)	91.00
Essential amino acids (mg/100 g)	
Valine	88.00
Isoleucine	37.00
Leucine	44.00
Threonine	42.00
Methionine	10.00
Phenylalanine	47.00
Lysine	5.00
Tryptophan	11.00
Minerals (mg/100 g)	
Phosphorus	3.00
Potassium	29.00
Calcium	28.00
Magnesium	8.00
Sodium	3740.00
Sulphur	6.00
Iron	0.30
Manganese	0.02
Zinc	0.25
Copper	0.06
Calorific value (cal/100 g)	240.00

cultivar, obtained by Nosti *et al.* (1984) and Fernández Díez *et al.* (1985). Quite similar results have been obtained by Balatsouras (1980) and Vamvoukas *et al.* (1980) for processed fruits of the Conservolea cultivar. In general, the water content is lower and that of oil is higher than in Spanish-style green olives. This is a consequence of the state of maturity of the fruit at harvest and does not vary greatly with the cultivar.

The changes previously produced during the fermentation process have been studied in detail (Durán Quintana *et al.*, 1971, 1973; Fernández Díez *et al.*, 1972; Garrido Fernández *et al.*, 1972; González Cancho *et al.*, 1975). In the typical alcoholic fermentation, produced in the traditional process, the following aromatic components have been mainly determined: ethanol (96–99% of the total volatile fraction) followed by small proportions of acetaldehyde, ethyl acetate and others.

Specific methods to determine polyphenols content (Garrido Fernández and Fernández Díez, 1976a) and compare texture (Fernández Díez, 1979; Fernández Díez *et al.*, 1985) of processed fruits have been developed. The subjective estimation of the superficial colour of the final product shows a high significance correlation ($P<0.01$) with reflectance measurements at 700 nm, which is valid at least for Hojiblanca, Lechín, Verdial and Gallega cultivars (Fernández Díez and Garrido Fernández, 1971).

Table 9.26 summarizes the main changes in composition during processing and the general characteristics found in the bulk commercial preparations.

9.2.10 INDUSTRIAL EQUIPMENT

Details on the different containers used in Greece and Turkey for the traditional process were given previously, to aid understanding of the fermentation process. From two decades ago, concrete tanks (Fernández

Table 9.26 Greek-style naturally black olives in brine. Changes in composition during processing and general characteristics of bulk commercial preparations (From Fernández Díez, 1983)

Operation	Change in composition
Prior phase	
None	None
Fundamental phase	
Spontaneous fermentation in brine. Yeasts predominate; lactic acid bacteria sometimes present.	Slow loss of sugars, tannins and oleuropein. Formation of organic acids, ethanol, acetaldehyde and ethyl acetate.
Final phase	
Storage in brine	None, under normal conditions.
Packaging	
General characteristics of bulk commercial preparations	
pH value	4.3–4.5
Free acidity	0.3–0.5 % (as lactic acid)
Sodium chloride	6–10 % according to commercial preparations

Díez, 1983) have been used habitually as vessels for the spontaneous anaerobic fermentation in Greece and Spain. They are sometimes partially and sometimes completely underground, hold 10–20 tonnes of fruit, and are lined with paraffin- or epoxy-type resin. However, during recent years fermenters of polyester and fibre-glass, similar to those used for the processing of green olives, have come into increasing use. They are generally buried underground to avoid the very high temperatures in factories located in the warmer zones. This is important because fruit that is very ripe cannot be processed at high temperatures. With both types of container, processors have always tried to maintain strict anaerobiosis in order to avoid the growth of film-forming yeasts and moulds which affect the texture as well as the flavour of the product.

In Spain at the present time the use of the aerobic fermentation procedure is being generally adopted, for which the fermenters are equipped with specially designed aeration devices. This (Garrido Fernández *et al.*, 1985a; Unidad Estructural de Biotecnología de Alimentos, 1990) maintains the oxygen dissolved in the brine at correct levels, to eliminate the CO_2 produced, and to provide homogeneous conditions throughout the fermenters (Spanish Patent 487077, December 1979).

The modification to apply the aerobic system consists of the introduction of a pipe with a perforated base through which air is bubbled. The bubbles of air are released from the bottom of the tube and, in their path upwards, carry away some brine which is replaced by fresh brine introduced into the column through the holes drilled in its base. At the same time, there is an exchange of gases (elimination of CO_2 and solubilization of O_2). The recirculated solution returns into the vessel through the screen situated at the vessel neck and which keeps the olives down (Figure 9.15). The diameter of the column can be optimized in terms of the volume of brine to be circulated and the volume of air introduced.

Complementary machinery, such as pumps, sorters and size-graders are similar to those used in other types and commercial preparations.

9.3 OTHER COMMERCIAL PREPARATIONS OF NATURALLY BLACK OLIVES TYPE

Among table olives obtained from fruits harvested when fully ripe, or just before, there are some other commercial preparations which are of some interest to food technologists and share an important fraction of the national and international olive market. Once again, it is necessary to differentiate between untreated and lye-treated olives.

Among the first, one of the most important is the Kalamata cultivar naturally black olives in brine, more simply known as Kalamata olives (Fernández Díez, 1969; Fernández Díez *et al.*, 1985). Black olives in dry salt and naturally shrivelled black olives in dry salt must also be mentioned in this group.

Fig. 9.15 Aerobic fermentation applied to underground fermenters. (Photograph from Dr Garrido Fernández; courtesy of Cooperative at Lora de Estepa, Sevilla.)

Among the treated naturally black olives, some commercial preparations are marketed in brine and others in dry salt. They are less important than the untreated group.

9.3.1 KALAMATA OLIVES

This is a very well-known and esteemed commercial preparation, mainly due to the excellent characteristics of the raw material which is one of the best cultivars used for table olives. Furthermore, this preparation is made with wine vinegar and olive oil which gives it a special flavour appreciated by a wide range of consumers (Fernández Díez, 1969).

These olives may be processed in two ways, known as short and long methods respectively. Figure 9.16 shows the general scheme for both procedures.

In the short method, fresh olives are sorted, size-graded, and cut lengthways twice on opposite sides of the fruit by automatic cutting machines. They are then debittered by immersion either in water or diluted brine (2%), changing the liquid two to three times each day, for a period of 5–8 days. Afterwards, the olives are either immersed in wine vinegar for 24–48 hours, or directly canned with a brine containing sufficient vinegar to obtain the desired acidity at equilibrium. All receptacles are filled with fresh brine containing 6–8% table salt and, very often, also

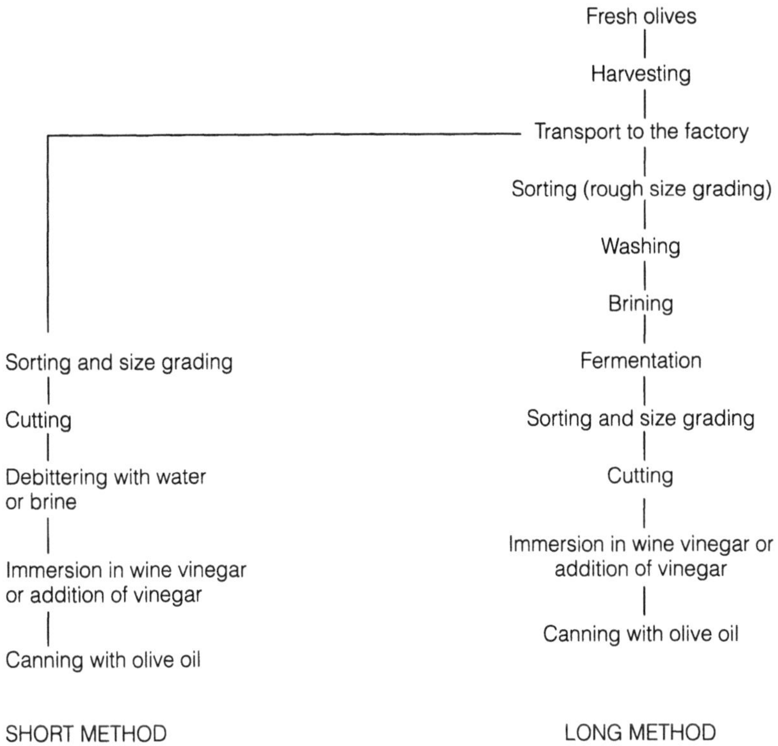

Fig. 9.16 General scheme for the processing of Kalamata olives. (Adapted from Fernández Díez *et al.*, 1985.)

pieces of lemon. To avoid corrosion, the tin-plate is of good quality and always lacquered at least on one face. Very often a quantity of good olive oil or refined vegetable oil is added, enough to form a thin layer on the surface of the brine.

As in the case of California-style black-ripe olives in brine (Chapter 8), not all the fresh fruit is directly prepared after harvesting. To regulate the market according to the consumer demand, processors keep the fresh olives in brine for some time before final processing. This previous storage in brine is analogous with the process of naturally black olives in brine and the same type of fermentation is, at least partially, produced. Stock olives during storage seem to undergo a weak lactic acid fermentation. Such olives are taken out of the mother brine and given the same treatment as the fresh-processed product. The finished product, however, lacks in colour and aroma, although debittering for the most part has been accomplished in the vats.

Acceptable levels of pH, titratable acidity and sodium chloride content for the equilibrated final packed product are:

pH value	3.80–4.00
Free acidity	0.74–1.20 (g/100 ml, as lactic acid)
Sodium chloride	6.0–8.0% (weight/volume)

Titratable acidity is expressed as lactic acid, even though it is mainly acetic acid.

Each lot of vinegar serves for two soakings and this makes processing expensive. Hand-sorting for defective fruits and size-grading must be done before incision.

The product packed as above is shipped to the market as 'Olives of Kalamata type'.

Occasionally, stock olives of the Conservolea cultivar are used instead of Kalamata olives. Such imitations always fail to acquire the good organoleptic properties and the intense aroma, since the genuine product depends mainly on the raw material for its good characteristics and eating qualities. In Spain, excellent results were obtained with Lechín and Hojiblanca cultivars, although they never reach the same consumer score as do Kalamata olives. All of them are known as Kalamata-style.

In some cases immersion in vinegar is omitted and the amount necessary to bring the acidity to the level of 0.50–0.75% calculated per recipient is added before packing.

Natural black olives in tins are also widely accepted in Greece thanks to their similarity to the Kalamata trade preparation. They are prepared with the same procedure and from the Conservolea cultivar, but without the longitudinal incision. Levels of salt and vinegar are similar, although in this case the last is directly added (about 25% of the holding solution). As with Kalamata, 20–60 g of oil are added to form a layer on the brine surface. This also adheres to the olive surface and gives it an attractive, brilliant aspect. Preservation is only accomplished due to the physico-chemical conditions (about 8% salt and pH around 3.8–3.9). These conditions make the product microbiologically safe, but do not prevent the growth of yeasts. Their activity produces gas, which causes bulging of the tin, though this defect can also result from lack of exhaustion and overfilling of the cans. To avoid this problem, a mild heat treatment is sometimes given to the sealed cans, although this may not be enough to completely prevent the bulging.

9.3.2 UNTREATED NATURALLY BLACK OLIVES IN DRY SALT

In general, these are obtained from fresh naturally black olives, or those partially dried at 65–70°C to a residual moisture of 20%, disposed in alternating layers of olives and dry salt, or by sprinkling dry salt over the olives. They retain a degree of bitterness and a characteristic fruity flavour and are generally packed in plastic bags, sometimes with potassium sorbate. Sometimes, a small amount of olive oil is added to cover the surface of the fruits. Analogous preparations are made from fruits which have become shrivelled on the tree. Two specific Greek preparations require special mention: Greek-style olives and Thruba-style olives.

Greek-style olives

In spite of its name, Greek-style olives are not a significant commercial product since they are mainly prepared by and consumed by the growers themselves. Rarely does this product reach the market. According to Balatsouras (1975), the olives used are from the Megaritiki cultivar, which grows in the dry region of Attica. Usually, olives are left in the tree to become fully ripe or over-ripe. Harvesting starts when part of the crop fall and the rest begin to shrivel.

Processing consists of a deep washing with pressurized water to remove soil, dust and other impurities adhered to the olive surface. The olives are then put in baskets and left for 2–3 days, apparently with the production of some type of fermentation, or treated immediately. In either case, olives are alternately layered with salt such that the final product reaches a concentration of about 15% or slightly lower. The high salt content produces the leaching of fruit juice containing oleuropein, sugar polyphenols, etc. This leads to a low-moisture product, partially shrivelled and salty. It is, in consequence, of higher nutritive value than olives in brine.

Thruba-style olives

These have also a very limited importance even at local level, but have some interesting peculiarities. They are prepared exclusively from the Thrubolea cultivar which grows in the Aegean islands, Crete and in some localities of Attica. These olives lose their bitterness completely when still on the tree.

According to Balatsouras (1990) it is claimed in the Greek literature that the loss of bitterness is due to the activity of an enzyme produced by the fungus, *Phoma oleae*. This has not been substantiated experimentally other than by the observation that loss of bitterness occurs only under specific meteorological conditions that might affect mould growth. The transformation of sound olives into Thruba-style olives proceeds progressively and is evidenced by a change of colour from black to 'copper-green'.

After harvesting, olives are deep washed and sun-dried to remove most of their moisture. The final product is added with a small quantity of salt to enhance the organoleptic characteristics and increase their keeping qualities until the time of consumption. However, the salt content is not high since it has no dehydrating function. Usually, the product is packed in plastic bags.

Part of the crop is also packed into tin cans in olive oil without salt and, as such, are highly prized by people suffering from heart or kidney disorders. The quantity of olives of this style produced each year is insufficient to meet even the local demand.

In Turkey, the final product is also sold without brine. Olives are put dry into plastic bags, which are then packed into cartons. These are sent to retailers who pass the olives to wooden boxes in which they are offered to consumers. A small part of the production is dry-packed in trays covered with plastic films. However, most of the exports of this country still rely on wooden barrels or plastic drums. Olives are brined in a 10% NaCl solution. Such olives have generally a high pH (4.5–5.0), good colour (which is also improved by air oxidation) and characteristic good taste. The product has a salty taste and always presents a certain surface shrivelling. They are mainly consumed in countries with a high Turkish immigrant population.

In Algeria there is another preparation system of naturally black olives partially brined. The general preparation system is similar to that of 'façon Grècque', although without alkaline treatment. It consists of putting completely ripe olives in wooden barrels of about 160–170 kg capacity with about 10 kg of salt. Salt extracts part of the olive juice and forms a concentrated brine. This solution is put into contact with all fruits by rotating the barrels (four turns per day). The olive sweetening is accomplished in 50–60 days or longer. The fruits maintain a relatively strong flavour and a characteristic taste with a salt content below 6%. Their exterior appearance is rather shrivelled due to the lost juice.

9.3.3 TREATED NATURALLY BLACK OLIVES IN BRINE

In this method, widely used in Algeria, fresh naturally black olives are treated with lye, washed with water and fermented in brine in the same way as pickled green olives in brine (Fernández Díez, 1972b). Due to the advanced stage of ripeness of the raw material, it presents some difficulties obtaining a final product of regular and high quality.

9.3.4 TREATED NATURALLY BLACK OLIVES IN DRY SALT

After a previous lye treatment and washing, olives are prepared with dry salt in the same way as indicated above for the untreated fruits (Fernández Díez, 1972b). The main difference between both final products is the flavour, which is sweeter because of the alkaline treatment. They are known as 'façon Grècque'. However, in spite of its name, this trade preparation is not prepared in Greece, where, up to now, the use of lye to treat black olives is forbidden.

A scheme of the process is presented in Figure 9.17. According to

Harvesting
↓
Transportation
↓
Sorting and grading
↓
(Storage)
↓
Lye treatment
↓
(Washing)
↓
Darkening
↓
Introduction in wooden barrels
(Salting)
↓
Curing (about 30–50 days)
↓
Darkening, grading, sorting
↓
Brining in the same wooden barrels

Fig. 9.17 General scheme for the preparation of treated naturally black olives in 'façon Grecque' style. (Adapted from Fernández Díez, 1972b.)

Fernández Díez (1972b), these olives are harvested when completely ripe and the olives have a deep black colour. Olives are picked by hand and transported in plastic boxes of 20–25 kg capacity. The cultivar Sigoise gives better quality.

After sorting and grading, the olives are directly treated or stored in brine. In this case, olives must be aerated and dried by air exposure before the lye treatment.

Fresh or dry olives are then introduced in a lye solution that penetrates one-half to one-third of the flesh. The sodium hydroxide concentration is fixed to achieve the treatment in about the same time as with green olives (8–10 hours). Olives are then exposed to air, sometimes after a washing to partially eliminate the excess NaOH, in perforated plastic boxes of 20–25 kg for 2–5 days. Olives are moved daily from one box to another to expose contact surfaces and to prevent the formation of brown spots where the oxidation is less intense. This operation provokes a deep darkening and a slight shrivelling.

When the oxidation process has finished, the olives are passed to wooden barrels of 145–200 kg capacity and packed in alternate layers with salt (about 6% by weight). Full wooden barrels are closed and placed horizontally in the factory yard. The solid salt extracts some liquid from the olives, which forms a concentrated solution that covers only part of the olives. The preserving effect of this solution is based on its high salt content, the abundant proportion of phenols, and usually the high pH provoked by residual NaOH when no washings are applied. To keep the liquid homogeneous and in contact with all the olives, the wooden barrels are rotated daily during the curing period. Each rotation must comprise about three to four complete turns of the barrel. This operation is slow and labour-intensive, taking 30–40 days in total. The olives are then darkened again by exposure to air in 20- to 25-kg perforated plastic boxes, sorted and graded. Finally, the olives are brined in the same wooden barrels. The original solution is then added with a 12°Bé brine to increase the salt content. The proportion in the final package is about 160–170 kg of olives and about 13–15 kg salt.

9.4 MOULD SPOILAGE OF NATURALLY BLACK OLIVES

There are numerous papers relating to mould growth and mycotoxin production in treated or untreated naturally black olives. This spoilage may occur when olives are not dry enough or sufficiently salted, or are not completely immersed in brine. Most of these cases occur when natural olives are processed according to the Greek-style or 'façon Grècque' (with lye treatment).

Mahjoub and Bullerman (1990) developed a method for determination of aflatoxin B_1, with a detection limit of 5 ng/g. Later, they showed the possibility of growth of *Aspergillus flavus* and *Aspergillus parasiticus*. The latter was able to produce aflatoxin B_1 in paste of fresh, black canned, green canned, and green pickled olives, while the former was able to produce it only in canned green and black olives. Lye treatment of olives led to the production of aflatoxin, while fresh olives inhibited toxin production. Whole olives supported growth but aflatoxin B_1 was not detected in these conditions (Mahjoub and Bullerman, 1987). Gourama and Bullerman (1988) identified *Aspergillus flavus*, *Aspergillus petrakii* and *Aspergillus ochraceous*. The first did not produce aflatoxin in olive paste. Only two out of five strains of *A. ochraceous* produced penicillic acid (11.4–30.2 mg/g) on olive paste but no ochratoxin, patulin or citrinin. However, biological toxicity of mould cultures was confirmed using chicken embryos and microbial tests. Other species isolated from Turkish-style naturally black olives (prepared by alternating layers of olives and salt) were *Aspergillus niger*, *Penicillium clavigerum* and *Penicillium expansum*.

Inhibition of mould growth by different agents has also been reported. Sorbic acid is one of the most efficient inhibitors, but when growth conditions are favourable (25°C), the presence of 6000 p.p.m. of sorbic acid could only delay the growth of *Aspergillus flavus* for 15 days (Mahjoub and Bullerman, 1986). Other possible agents of inhibitory effect are natamycin (Mahjoub and Bullerman, 1986) and methyl eugenol (Kivanç and Akgül, 1990). In general, the inhibition is always relative. The lower the temperature, the more time the mould took to initiate mycelium and spore development. At 25°C inhibition is always difficult. Tantoui-Elaraki *et al.* (1983) also found the occurrence of *Aspergillus flavus* and *Aspergillus ochraceous* spores in spoilt samples of 'façon Grècque' Moroccan olives. All the strains isolated were toxigenic when tested in laboratory media. Twelve spoilt samples contained aflatoxin B_1 and another five out of 103 were contaminated with ochratoxin A.

Most of the time toxin production in olives, especially lye-untreated olives, was not detected, but there is a potential health risk when moulds are present in the processing. Measures to avoid their growth must be adopted.

9.5 REFERENCES

Aligisakis, E.M. (1971) Fermentation et conservation des olives de table dans des cisternes en beton armé enduites d'un vernis plastique spécial. Tec. 18. II Congreso Internacional Técnicas Oleicolas, Torremolinos, Málaga, Spain.

Aligisakis, E.M. (1982) *Processing of Table Olives*, Athens (in Greek).

Aligisakis, E.M. and Patroclos, G.A. (1971) Observations sur la cyanosis des olives noires de table. Tec. 18, II Congreso Internacional Técnicas Oleicolas, Torremolinos, Málaga, Spain.

Balatsouras, G.D. (1964) Composition chimique des olives noires de Grèce. Variation de quelques constituants en fonction de la région de production. *Informations Oleicoles Internationales*, **28**, 131–56.

Balatsouras, G.D. (1966a) The chemical compositon of the brine of stored Greek black olives. *Grasas y Aceites*, **17**, 83–8.

Balatsouras, G.D. (1966b) *Contribution to the study of the chemical composition and the microflora for the stored in brine Greek black olives*. National Printing Office, Athens.

Balatsouras, G.D. (1967) Processing the naturally ripe (black) olives, in *Proceedings of the 'International Olive Oil Seminar'*, Perugia-Spoleto, Italy, pp. 491–510.

Balatsouras, G.D. (1969) Traitement des olives noires. *Informations Oleicoles Internationales*, **46**, 65–76.

Balatsouras, G.D. (1975) Preparación de aceitunas en negro. II Seminario Oleicola Internacional, 6–17 October, Cordoba, Spain.

Balatsouras, G.D. (1980) Nutritive and biological value of the Greek Table Olives, in *Proceedings of the III International Congress on the Biological Value of Olive Oil*, International Olive Oil Council, Canea, Crete, pp. 485–520.

Balatsouras, G.D. (1985) Taxonomic and physiological characteristic of the facultative rod type lactic acid bacteria isolated from fermenting green and black olives. *Grasas y Aceites*, **36**, 239–49.

Balatsouras, G.D. (1990) Edible olive cultivars, chemical composition of fruit, harvesting, transportation, processing, sorting and packaging, styles of black olives, deteriorations, quality standards, chemical analysis, nutritional and biological value of the end product, in *Proceedings of the 'Seminario*

Internazionale olio d'Oliva e Olive da Tavola: Technologia e Qualitá', Instituto Sperimentale per la Elaiotecnica de Pescara, Italy, pp. 291–330.

Balatsouras, G.D. and Polymenakos, N.G. (1964) Resultats preliminaires sur la fermentation des olives noires par acide lactique. *Informations Oleicoles Internationales*, **27**, 153–67.

Balatsouras, G.D. and Vaughn, R.H. (1958) Some fungi that might cause softening of storage olives. *Food Research*, **23**, 235–43.

Barret, A. and Bidan, P. (1972) Observation sur la technologie des olives noires de Nyons. III Congreso Oleícola Internacional, Torremolinos, Málaga.

Borbolla y Alcalá, J.M.R. de la, Fernández Díez, M.J. and González Cancho, F. (1959) Estudios sobre el aderezo de aceitunas verdes. XVI. Experiencias sobre el 'alambrado'. *Grasas y Aceites*, **10**, 221–34.

Borbolla y Alcalá, J.M.R. de la, Fernández Díez, M.J. and González Cancho, F. (1960) Estudios sobre el aderezo de aceitunas. XIX. Nuevas experiencias sobre el alambrado. *Grasas y Aceites*, **11**, 256–60.

Borcakli, M. Ozay, G., Alperden, I., Ozsan, E. and Erdek, Y. (1993) Changes in the chemical and microbiological composition of two varieties of olive during fermentation. *Grasas y Aceites*, **44**, 253–60.

Ciafardini, G., Marsilio, V., Lanza, B. and Pozzi, N. (1994) Hydrolysis of oleuropein by *Lactobacillus plantarum* strains associated with olive fermentation. *Applied and Environmental Microbiology*, **60**, 4142–7.

Durán Quintana, M.C. and Garrido Fernández, A. (1978) Aceitunas negras al natural en salmuera VI. Características de la fermentación en salmueras regeneradas. *Grasas y Aceites*, **29**, 197–201.

Durán Quintana, M.C. and González Cancho, F. (1973) Estudio microbiológico de la fermentación de aceitunas negras maduras en salmuera. *Microbiología Española*, **26**, 149–64.

Durán Quintana, M.C., Garrido Fernández, A., González Cancho, F. and Fernández Díez, M.J. (1971) Aceitunas negras maduras en salmuera. I. Estudio físico-químico y microbiológico de la fermentación. *Grasas y Aceites*, **22**, 167–77.

Durán Quintana, M.C., Garrido Fernández, A., González Cancho, F. and Fernández Díez, M.J. (1973) Aceitunas negras maduras en salmuera. III. Estudio físico-químico y microbiológico de la fermentación (continuación). *Grasas y Aceites*, **24**, 149–59.

Durán Quintana, M.C., González Cancho, F. and Garrido Fernández (1979) Aceitunas negras al natural en salmuera. IX Ensayos de producción de 'alambrado' por inoculación de diversos microorganismos aislados de salmueras de fermentación. *Grasas y Aceites*, **30**, 361–7.

Durán Quintana, M.C., García García, P. and Garrido Fernández, A. (1986) Fermentación en medio aeróbico de aceitunas negras maduras en salmuera con inyección alternante de aire. Estudio de la influencia de la adición de cloruro cálcico sobre la texture. *Grasas y Aceites*, **37**, 242–9.

Etchells, J.L., Borg, A.F. and Bell, T.A. (1968) Bloater formation by gas-forming lactic acid bacteria in cucumber fermentations. *Applied Microbiology*, **16**, 1029–34.

Fernández Díez, M.J. (1969) Elaboración de aceitunas de mesa en Grecia. *Grasas y Aceites*, **20**, 12–24.

Fernández Díez, M.J. (1972a) Elaboración de aceitunas de mesa en Turquía. *Grasas y Aceites*, **23**, 138–45.

Fernández Díez, M.J. (1972b) Elaboración de aceitunas de mesa en Argelia. *Grasas y Aceites*, **23**, 248–53.

Fernández Díez, M.J. (1979) Texture of table olives and pimientos. *Journal of Texture Studies*, **10**, 103–16.

Fernández Díez, M.J. (1983) Olives, in *Biotechnology*, Vol. **5** (eds H.J. Rehm and G. Reed), Verlag Chemie, Weinheim, Germany, pp. 379–97.

Fernández Díez, M.J. (1984) Changes in the chemical components during the processing of table olives and their relation to the quality. Proceedings of MOCA, Vol. I, Valencia, Spain, pp. 301–18.

Fernández Díez, M.J. (1991) Processing table olives, in *Encyclopedia of Food Science and Technology* (ed. Y.H. Hui), Wiley & Sons, New York, pp. 1910–25.

Fernández Díez, M.J. and Garrido Fernández, A. (1969) Ensayos de elaboración de aceitunas negras maduras estilo griego. *Grasas y Aceites*, **5**, 235–41.

Fernández Díez, M.J., and Garrido Fernández, A. (1971) Métodos de análisis y toma de muestras de las aceitunas de mesa. Determinación de color en aceitunas negras. *Hojas de Información del Consejo Oleícola Internacional*, **216**, 5–9.

Fernández Díez, M.J., Garrido Fernández, A., González Cancho, F., Durán Quintana, M.C. and Cordón Casanueva, J.L. (1972) *Elaboración de aceitunas negras de mesa*, Instituto de la Grasa, Sevilla, Spain.

Fernández Díez, M.J., Garrido Fernández, A. and Casares Merelo, R. (1975a) El pH y la acidez como factores determinantes del color en las aceitunas negras maduras sin oxidación alcalina. *Anales de Bromatología*, **XXVII**, 223–40.

Fernández Díez, M.J., Mínguez Mosquera, M.I. and Casares, Merelo, R. (1975b) Efecto del cloruro cálcico sobre la textura y el color de aceitunas negras. *Anales de Bromatología*, **XXVII**, 209–22.

Fernández Díez, M.J., Castro Ramos, R., Garrido Fernández, A., González Cancho, F., González Pellissó, F., Nosti Vega, M., Heredia Moreno, A., Mínguez Mosquera, M.I., Rejano Navarro, L., Durán Quintana, M.C., Sánchez Roldán, F., García García, P. and Castro Gómez Millán, A. (1985) *Biotecnología de la aceituna de mesa*, Consejo Superior de Investigaciones Científicas, Instituto de la Grasa, Madrid, Sevilla, Spain.

Fleming, H.P., Thompson, R.L., Etchells, J.L., Kelling, R.E. and Bell, T.A. (1973) Bloater formation in brined cucumbers fermented by *Lactobacillus plantarum*. *Journal of Food Science*, **38**, 499–503.

Fleming, H.P., Thompson, R.L. and Moroe, R. (1978) Susceptibility of pickling cucumber to bloater damage by carbonation. *Journal of Food Science*, **43**, 892–6.

García García, P., Durán, M.C. and Garrido, A. (1982) Modificaciones del proceso de fermentación de aceitunas negras para evitar alteraciones. *Grasas y Aceites*, **33**, 9–17.

García García, P., Durán Quintana, M.C. and Garrido Fernández, A. (1985) Fermentación aeróbica de aceitunas maduras en salmuera. *Grasas y Aceites*, **36**, 14–20.

García García, P., Durán Quintana, M.C. and Garrido Fernández, A. (1986) Envasado de aceitunas negras al natural fermentadas en medio aeróbico. *Grasas y Aceites*, **37**, 92–6.

Garrido Fernández, A. and Fernández Díez, M.J. (1976a) Estudio crítico de la determinación colorimétrica de polifenoles y su aplicación a las salmueras de aceitunas negras de mesa. *Grasas y Aceites*, **27**, 107–12.

Garrido Fernández, A. and Fernández Díez, M.J. (1976b) Aceitunas negras al natural en salmuera. IV. Oscurecimiento del color superficial de los frutos y su conservación en el envasado. *Grasas y Aceites*, **27**, 329–34.

Garrido Fernández, A., Durán Quintana, M.C., González Cancho, F. and Fernández Díez, M.J. (1972) Aceitunas negras maduras en salmuera. II. Estudio físico-químico y microbiológico de la fermentación (continuación). *Grasas y Aceites*, **23**, 22–31.

Garrido Fernández, A., Durán Quintana, M.C. and González Cancho, F. (1979) Aceitunas negras al natural en salmuera.VIII. Estudio de la influencia de diferentes tratamientos y formas de fermentación sobre la alteración denominada 'alambrado'. *Grasas y Aceites*, **30**, 301–7.

Garrido Fernández, A., García García, P. and Durán Quintana, M.C. (1985a) Conservación de aceitunas negras naturales procedentes de fermentación aeróbica. *Grasas y Aceites*, **36**, 313–316.

Garrido Fernández,A., García García, P. and Sánchez Roldán, F. (1985b) Nuevo proceso aeróbico de fermentación de aceitunas negras al natural. Optimización de la columna de aireación. *Alimentación, equipos y tecnología*, **IV**(3), 73–81.

Garrido Fernández, A., Durán Quintana, M.C. and García García, P. (1987) Fermentación en medio aeróbico de aceitunas negras al natural en salmuera. Efecto de algunas variables sobre la formación de alambrado. *Grasas y Aceites*, **38**, 27–32.

Garrido Fernández, A., Durán, M.C., García, P., Sánchez, F. and Brenes, M. (1992) Fermentation of olives in aerobic conditions. Processing and quality of foods, Vol. **2**, in *Food Biotechnology: Avenues to Healthy and Nutritious Products* (eds P. Zeuten, J.C. Cheftel, C. Eriksson, T.R. Gorniley, P. Linko and K. Panlus), Elsevier Applied Science, London.

Gilliland, J.R. and Vaughn, R.H. (1943) Characteristics of butyric acid bacteria from olive. *J. Bacteriol.*, **46**, 315–21.

González Cancho, F., Nosti Vega, M., Durán Quintana, M.C., Garrido Fernández, A. and Fernández Díez, M.J. (1975) El proceso de fermentación en las aceitunas negras maduras en salmuera. *Grasas y Aceites*, **26**, 297–309.

Gourama, H. and Bullerman, H.B. (1988) Mycotoxin production by molds isolated from Greek-style black olives. *International Journal of Food Microbiology*, **6**, 81–90.

International Olive Oil Council, COI (1980) Unified Qualitative Standard Applying to Table Olives in International Trade, Madrid.

Kivanç, M. and Akgül, A. (1990) Mold growth on black table olives and prevention by sorbic acid methy-eugenol and spice essential oil. *Die Nahrung*, **34**, 369–73.

Mahjoub, A. and Bullerman, L.B. (1986) Effects of nantamycin and potassium sorbate on growth and aflatoxin production in olives. *Archives Institut Pasteur Tunis*, **63**, 513–25.

Mahjoub, A. and Bullerman, L.B. (1987) Mold growth and aflatoxin production on whole olives and olive pastes. *Sciences des Aliments*, **7**, 629–38.

Mahjoub, A. and Bullerman, L.B. (1990) A method for aflatoxin B_1 determination in olives. *Revue Française des Corps Gras*, **37**, 245–6.

Marsilio, V., Brighigna, A. and Chichelli, A. (1990) Le caratteristiche del colore delle olive nere da tavola. Influenza della varietá e della tecnologia di transformazione. *Industrie Alimentari*, **XXIX**, 5–10.

Materassi, R., Midans, N. and Pelagatti, O. (1975) Sulla idrolisi della oleuropeina nei lieviti. *Annali Istitute Sperimentale per la Elaiotecnica*, **V**, 53–68.

Mínguez Mosquera, M.I. (1982) Evolución de los constituyentes pécticos y de las enzimas pectinolíticas durante el proceso de maduración y almacenamiento de la aceituna Hojiblanca. *Grasas y Aceites*, **33**, 327–33.

Nosti Vega, M., Castro Ramos, R. and Vázquez Ladrón, R. (1984) Composición y valor nutritivo de algunas variedades españolas de aceitunas de mesa. VI. Cambios debidos a los procesos de elaboración. *Grasas y Aceites*, **35**, 11–14.

Pelagatti, O. (1981–1983) L'impiego di lieviti oleuropeinolitici nella preparazione delle olive da tavola nere. *Annali Istitute Sperimentale per la Elaiotecnica.*, **IX**, 57–63.

Spanish Government Presidency (1983) Reglamentación técnico sanitaria para la elaboración, circulación, y venta de las aceitunas de mesa. *Spanish State Official Bulletin*, **108**, 12619–25.

Tantaoui-Elaraki, A., Le Tutour, B., Bouzid, M. and Keddani, M.J. (1983) Contamination des olives noires 'façon Grècque' par les spores d'*Aspergillus* toxigénes et leurs toxines. *Industries Alimentaires et Agricoles*, 997–1000.

Unidad Estructural de Biotecnología de Alimentos (1990) Instituto de la Grasa, Sevilla, Spain, 1990, *Table Olive Processing*, International Olive Oil Council, Madrid.

Vamvoukas, D., Stefanoudakis-Katsoudakis, E., Loupasakis-Androulakis, M. and Kiritsakis, A. (1980) Results from chemical analysis and determinations of the main cultivars and styles of Greek Table Olives, in *Proceedings of the III International Congress on the Biological Value of Olive Oil*, International Olive Oil Council, Canea, Crete, pp. 521–41.

Vaughn, R.H. (1954) Lactic acid fermentation of cucumbers, sauerkraut and olives, in *Industrial Fermentations*, Vol. II (eds L.A. Underkofles and R.J. Hickey), Chemical Publishing Co.Inc., New York, pp. 417–78.

Vaughn, R.H. (1982) The fermentation of olives, in *Industrial Microbiology*, 4th edn (ed. G. Reed), AVI Publishing Co., Westport, Connecticut, USA, pp. 207–36.

Vaughn, R.H., Douglas, H.C. and Gilliland, J.R. (1943) Production of Spanish-type green olives. *Univ. of Calif. Agri. Expt. Sta. Bull.*, no. **678**.

Vaughn, R.H., Yackulezy, K.T., McMillan, J.D., Thomas, E.H., Dave, B.A. and Crampton, V.M. (1969) Some pink yeasts associated with softening of olives. *Applied Microbiology*, **18**, 771–5.

Vaughn, R.H., Stevensen, K.E., Dave, B.A. and Park, H.C. (1972) Fermenting yeasts associated with softening and gas-pocket formation in olives. *Applied Microbiology*, **23**, 316–20.

Vlahor, G. (1990) Gli antociani polimeri delle olive nere. Loro determinazione in funzione del processo di maturazione. Nota I. *Industrie Alimentari*, **XXIX**, 11–12/18.

Industrial wastes 10

10.1 INTRODUCTION

The food industry – which has as its main purpose the transformation of animal and vegetable products – faces major problems through contributing to environmental pollution with the wastes it produces.

In order to design an appropriate wastewater management scheme for any food industry, it is convenient to follow a series of successive steps (Garrido Fernández, 1975). It is necessary first, to identify the different processing points where wastes are produced and estimate the volume and characteristics of each of them. With these data, the expert can usually introduce some measures to rationalize water use – which contributes to a substantial reduction in total waste volumes – and to plan any further activity. A careful review of the different stages of the processing is necessary to adjust not only the water utilization but also the amounts of chemicals used (NaCl, acetic acid, etc.). Special attention must be paid to possible alternatives to the current process. Most of them were designed when water was abundant and no restrictions on disposal were envisaged. Sometimes, results can be spectacular. A change in olive oil extraction technology to a process which separates only two phases from the olive flesh (oil and a semi-solid cake) has completely eliminated the production of 'alpechin', which for centuries has been one of the main water pollutants in all the Mediterranean olive oil-producing countries.

Another point that requires detailed study is the best way of handling the different wastewaters. Sometimes, it may be convenient to segregate them into two or more separate wastewater streams, according to their characteristics, pollution level or possible treatment. On other occasions, it can be preferable to combine them, to achieve more appropriate characteristics. Examples of both approaches can be found in the table olive industry, depending on the kind of industry, its situation, water availability, local municipal wastewater treatment facilities, etc.

Adoption of the measures described above will not prevent a final wastewater, but it will have the most appropriate physicochemical characteristics and will be of reduced volume. Methods for its elimination or treatment should then be implemented.

Scientists and technologists working in the field of table olives processing have devoted much effort over the past three decades to find a solution to some of the wastewater problems, and some promising results were obtained through two complementary approaches (Garrido Fernández, 1975; Fernández Díez *et al.*, 1985; Fernández Díez, 1991):

1. Reduction of the volume of industrial wastes and their re-use in processing operations.
2. Regeneration of the fermentation brines to be further used in the packaging of the final product.

In the first case, a series of internal control regulations have been adopted, such as the re-use of lyes for the treatment of olives; the reduction of the lye concentration to a minimum effective value; and the reduction, or even complete elimination of washing operations, by neutralization of the residual free lye which remains in the fruit after the alkaline treatment.

In terms of the regeneration of fermentation brines to be used for bottling and canning the final product, several procedures have also been studied and successfully applied.

Although the methods which have been applied to the different ways of processing table olives have important similarities, they also show specific differences (Brenes Balbuena *et al.*, 1989b; García García *et al.*, 1990a; Unidad Estructural de Biotecnología de Alimentos, 1990), which makes it more appropriate to consider them separately as they may be applied to the three most important commercial preparations described in this book.

10.2 VOLUME OF WASTEWATERS PRODUCED IN TABLE OLIVE PROCESSING

The volume of wastewaters produced by the various production processes differs widely according to treatments. An estimate of the liquid residues during the main processing steps of the most popular commercial presentations is shown in Table 10.1. Accurate estimation of the total volume of liquid wastes produced by each industry is impossible due to the specific water use practices applied by each of them, but in most cases the volumes are appreciably higher than those reported in Table 10.1. Obviously, the less-polluting systems are those based on dry olive processing, but they have a limited significance at both national and international level.

Directly brined olives (green, turning colour or naturally ripe) also have reduced pollution impact since they only give rise to wastewaters from their holding solutions. In contrast, lye-treated olives are always

Table 10.1 Approximate volumes (l) of wastewaters per kg olives produced during the main stages of various table olive commercial presentations

Commercial presentation	Lye	Washing waters	Holding solutions	Total
Spanish-style (green olives)	0.5	0.5–2.0	0.5	1.5–3.0
Untreated green and turning colour olives			0.5	0.5
California green-ripe olives	0.5	0.5–2.0	0.0–0.5	1.0–3.0
California black-ripe olives	0.5–2.5	0.5–3.0	0.5	1.5–6.0
Naturally black olives			0.5	0.5

associated with high pollutant potential due to the exhausted lye and the several washings needed to eliminate the excess of alkali. Within this group the California black-ripe olives are worst because of the several lye treatments frequently applied during the darkening step.

Some additional wastewaters are usually produced during canning the different products, such as container washing waters, water or steam used in the pasteurization processes, or cooling waters used in retorts. The characteristics of each of these are rather different. In general, those related to the debittering or darkening processes are heavily polluted, while those used for washing of glasses, cans cooking, etc. have a moderate or low charge. The pH is also variable: those containing NaOH have a high pH, but fermentation brines are usually acidic.

10.3 WASTEWATERS FROM SPANISH-STYLE PROCESSING

10.3.1 CHARACTERISTICS OF THE MAIN WASTEWATERS

The main post-harvest operations – lye treatment and washing – produce the largest part of the wastes. Considering that the liquid-to-fruit ratio in the tanks and fermenters is about 0.5 l/kg of olives, the traditional Spanish process can be estimated to produce a total volume of between 1.5 and 3.0 litres of wastewaters per kg of fruit, for both operations (one lye treatment plus two or three washing waters). Fermentation brine, which currently may be totally discarded before packaging operations, accounts for another 0.5 l/kg of fruit. All these wastewaters are highly polluting and are not easily treated by conventional methods. Their characteristics (Table 10.1) were studied by Garrido Fernández (1982a,b) who established the classification into two different groups: alkaline and acid wastewaters. This difference is central to their management and treatment.

Alkaline wastewaters (lye and washings) have a high pH, which is modified with difficulty due to their high combined acidity. In spite of their similar pH, the NaOH content is rather different. Exhausted lyes have an important residual alkali concentration after the treatment. As shown in Table 10.2 it may be about 11 g/l, but can vary widely, according to cultivar and, mainly, environmental temperature. In cases of low temperature (end of season or cold zones), it is necessary to increase the NaOH content to obtain adequate treatment. As the time in contact with the fruits is approximately the same (or slightly higher) and the flesh of olives is very similar, such higher NaOH lyes have also markedly higher NaOH in the exhausted solutions. Washing waters, however, contain only a limited amount of the chemical due to the fact that most of the hydroxide has reacted with the organic acids in the olive flesh and the remaining excess is really very low. When only one washing is used, the NaOH content may be 1.5–2.5 g/l. Most of this alkali comes from the lye adhered to the container and olive fruits. In the case of several washings, the first carries away almost the whole excess NaOH while the following washings had very low concentrations (about 0.5 g/l).

Other components in solution are the polyphenols. The sweeter the olives, the more of the original polyphenol content is eliminated. Their concentrations are very similar in lyes and washing waters and depend mainly on the contact time. The amount removed is about 1.5 g/kg for each immersion. In a debittering treatment that consists of one lye immersion and one washing, about 3 g/kg of polyphenols is eliminated

Table 10.2 Main general characteristics of wastewaters from processing Spanish-style pickled green olives in brine (From Fernandez Díez *et al.*, 1985 and García García *et al.*, 1990b)

Characteristic	Type of waste: Lye	Washing water	Fermentation brine
pH value	9.5–12.0	9.0–11.5	3.8–4.2
Free NaOH (g/l)	11	1.5	–
Free acidity (g lactic acid/l)	–	–	6–15
Polyphenols (g tannic acid/l)	2.5–4.0	2.5–4.0	4.0–6.0
Reducing sugars (g tannic acid/l)	6–9	6–9	–
Dissolved organic solids (g/l)	20–30	20–30	15–25
Dissolved inorganic solids (g/l)	20–35	7–25	90–110
COD (g O_2/l)	15–35	12–35	10–35
BOD_5 (g O_2/l)	9–20	9–20	8–20

from the whole olives (or 5 g/kg olive flesh). Such quantitative information indicates the importance of these operations to olive debittering and the final taste. From this point of view, as many as possible immersions are recommended, although the effect is counteracted by parallel solubilization of other beneficial components, such as reducing sugars. Due to their high water solubility, reducing sugars are present in lye and washing wastewaters up to about 10 g/l, depending on the contact time between fruits and solutions. In the entire debittering process, a substantial amount of reducing sugars may be lost, which is disadvantageous for the fermentation and also for the pollutant charge of the lyes and washing liquid wastes. Reducing sugars are easily assimilable compounds and their incorporation into public streams or municipal sewers causes a rapid drop in dissolved oxygen. In addition to phenols and reducing sugars, other organic substances are also present in these wastewaters, since their total organic matter content ranges from 20 to 30 g/l. This obviously leads to a high Chemical Oxygen Demand (COD) of 15.0–30.0 g O_2/l, and a five-day Biological Oxygen Demand (BOD_5) of the order of 9–20 g O_2/l.

Finally, inorganic contaminants are also very high, mainly in lyes due to the excess of NaOH in the used solution, but also from other inorganic components such as potassium, which are abundant in olives. Inorganic residues in washing waters occur in lower amounts because they only contain the residual NaOH adhering to surfaces and that removed from the interior of olives.

A detailed study of lyes and washing waters used during the debittering process of Spanish-style green olives was carried out by Sannia *et al.* (1978) who also analysed for a series of contaminant elements and extended the number of washing waters studied up to five. Naturally, the pollution charge (BOD_5 and COD) diminished markedly in the successive washings and was extremely low in the fifth, indicating that the remaining fermentable substrates may be insufficient for a normal process. As shown in Table 10.3, the concentration of some components (NH_3, B and Hg) did not depend on the type of wastewater, suggesting that their presence must be related to the water used in the processing. However, others apparently come from the olives since they (P, Fe, Cu, Zn and Ni) occurred in progressively lower concentrations.

Brines had a low pH of around 4.0 and a moderate free acidity (mainly a mixture of L- and D-lactic acid) formed during the fermentation. Although slightly reduced with respect to the sugars, they still contribute an important proportion to the pollutant charge. Brines also have high concentrations of polyphenols and, consequently, their total organic matter in solution is quite similar to that found in lyes or washing waters and they have practically the same COD and BOD_5 values. However, one of the brines' main characteristics is the high proportion of inorganic

Table 10.3 Pollution charge (including NH_3 and other elements) in the lyes and washing waters of Spanish-style green olives (From Sannia *et al.*, 1978)

Type of waste-water	Pollution charge													
	pH	BOD_5 (gO_2/l)	COD (gO_2/l)	NH_3 (mg/l)	B (mg/l)	P (mg/l)	Fe (mg/l)	Cu (mg/l)	Zn (mg/l)	Ni (mg/l)	Hg (mg/l)	Na (g/l)	K (g/l)	Cl^- (g/l)
Lye	12.7	4.8	24.0	1.3	0.14	18.3	5.75	0.23	0.48	0.12	0.08	96.5	18.5	2.9
1st washing	9.1	3.1	10.0	16.8	0.63	28.4	3.62	0.16	0.36	0.09	0.05	15.7	4.2	1.9
2nd washing	7.6	1.6	7.6	1.3	0.40	16.3	1.54	0.10	1.64	0.03	0.05	10.5	2.6	1.1
3rd washing	6.2	0.9	3.0	2.0	0.34	11.4	0.84	0.06	0.08	0.02	0.09	5.0	1.7	0.7
4th washing	6.5	0.6	2.0	2.3	0.28	10.6	0.82	0.04	0.07	0.01	0.06	3.7	1.1	0.5
5th washing	7.0	<0.1	0.2	3.5	0.16	4.8	0.40	<0.01	0.12	0.01	0.05	1.1	0.3	0.2

BOD_5, 5-day Biological Oxygen Demand.
COD, Chemical Oxygen Demand.

components, mainly Na and K salts from the added NaCl and from the high potassium content of the olive flesh.

The chemical composition of all of these wastewaters introduces two contradictory conditions. On the one hand, the presence of reducing sugars and lactic acid – which are easily used as microbial carbon sources – is responsible for a great part of the high BOD_5 and pollution problems; on the other hand, the high concentration of polyphenols in all wastewaters and the high level of NaCl in brines seriously interfere with the direct application of any biological purification procedure. They can be treated by traditional municipal wastewaters treatment plants only after a marked dilution. This could be possible if they were produced throughout the year and the contamination sources were widely distributed. However, the lyes and washing waters are generated over only a short period of time (about 2 months) and at least in Spain, processing factories are concentrated in a limited number of places. In this case, dilution is not a good option and their direction to municipal wastewater treatment facilities leads to an overload and to an inhibitory effect on the normal organisms responsible for purification. This has prevented municipal authorities from accepting lyes and washing waters into their sewers, other than in very large cities. Other authorities responsible for pollution control have also refused any possible discharge of them into public streams or rivers.

Brines are produced over a longer period of time, but packing factories are usually larger than those industries devoted to only the first stages of processing. So, the volumes of wastewater produced by the packing factories are a problem and, until now, they are not accepted into any public stream or sewer. However, in this case, a pre-treatment has sometimes been considered sufficient for their acceptance.

Thus, in general, Spanish-style green olives wastewaters are heavily contaminated, difficult to treat and present many problems for their handling and disposal. The two approaches outlined above have been followed to reduce their environmental impact: internal control measures (including recycling processes) and design of appropriate disposal or treatment systems.

10.3.2 INTERNAL CONTROL MEASURES

Measures to reduce both the volume and contamination during the debittering process are clearly possible. Fermentation brines, however, pose a different problem since they are essential to achieve the proper final product in the current process.

Lye and washing waters re-use

This potential modification of the traditional debittering process is suggested by the high residual NaOH content that remains in the used solutions currently. The only aspect to be studied previously is the effect of this re-use on the quality attributes of the final product as well as on the fermentation process. Re-use could also partially prevent sugar leakage and thus increase the sugars available for the fermentation.

Garrido Fernández *et al.* (1977) tried a series of modifications such as lye re-use, the preparation of lyes from washing waters, and the conversion of lye into brine by the addition of NaCl and neutralization of the excess alkali with HCl. Tables 10.4–10.9 show the influence of these modifications on the pollutant charge of the respective wastewaters. With respect to the residual NaOH in the exhausted lye solution (Table 10.4), its concentration does not follow any regular changes through the different modifications. Apparently, they are more closely related to other circumstances of the treatment such as immersion time (as this is longer, the residual NaOH decreases) and temperature (as the temperature increases, the residual NaOH is lower, in part because the initial content is also lower). A similar pattern can be seen in relation to the rinses and washing waters. In these, there is a distribution of the NaOH between the two solutions when two washing waters are applied (usually the first is of a shorter duration, 9–12 hours, than the second, 18–24 hours). However, when only one is used (of 18–24-hour duration), it contains a higher level of residual NaOH, although it rarely attains the total alkali removal achieved by two washings. This means that the volume of wastewaters can be reduced by using only one washing, but even by increasing the duration of this to 18–24 hours, this will always be slightly less effective than the traditional application of two washes. Nevertheless, a good result is obtained. Inorganic soluble solids in the wastewaters solutions from the different modifications (Table 10.4) show an increase with re-use of lye, 1st washing or 2nd washing water. Their concentrations are not, however, double that of the preceding cycle, which means that they have lost some of their capacity to leach these components, due mainly to those compounds already in solution. When the re-use lye is already heavily concentrated, it has a lower capacity for dissolving inorganic matter; the levels are therefore higher in the olive flesh and will be higher in the following rinse or washing. This is evident in the rinsing waters of the first re-use, but, in general, there is no clear trend. This must be related to the fact that washing waters are also strongly influenced by many other factors such as immersion time and temperature.

A similar comment can be made with respect to the organic matter and BOD_5 (Table 10.5). In general, the re-use of any solution leads to an

Table 10.4 Effect of lye and washing waters re-use on residual NaOH and total inorganic solids carried by the different waste-waters. Manzanilla cultivar (Reproduced from Garrido Fernández *et al.*, 1977)

	Residual NaOH content (g/l)				Inorganic soluble solids (g/l)			
Type of re-use	Lye	Rinse	1st washing	2nd washing	Lye	Rinse	1st washing	2nd washing
Traditional process								
Control	6.94	0.22	2.02	0.9	15.39	0.82	6.84	7.02
First re-use								
L (half)	5.01	0.20	0.87	0.90	18.95	0.78	5.00	4.38
L	4.93	0.25	0.91	0.81	20.00	1.30	4.78	11.01
1W	4.43	0.11	1.37	*	23.49	2.12	9.92	*
2W	3.90	0.20	0.84	*	25.71	2.90	8.20	*
Second re-use								
LL	4.65	0.25	1.37	*	22.44	1.00	5.12	*
1WL	5.27	0.22	0.98	1.07	19.26	0.87	3.68	7.14
Conversion of lye into brine								
LS	6.03	*	*	*	15.82	*	*	*

* Process without rinses or washing waters; L, re-use of lye from preceding cycle; 1W, re-use of 1st washing waters from preceding cycle; 2W, re-use of 2nd washing waters from preceding cycle; LS, conversion of lye into brine.

Table 10.5 Effect of lye and washing water re-use on dissolved organic matter and BOD_5 of different wastewaters from the Spanish-style green olive debittering process. Manzanilla cultivar (Reproduced from Garrido *et al.*,1977)

	Dissolved organic water (g/l)				BOD_5 (g O_2/l)			
Type of re-use	Lye	Rinse	1st washing	2nd washing	Lye	Rinse	1st washing	2nd washing
Traditional process								
Control	24.41	1.82	19.07	25.51	13.63	0.61	16.80	21.00
First re-use								
L(half)	26.35	2.00	12.49	11.29	22.75	0.88	6.58	11.59
L	42.86	2.31	13.20	17.90	26.70	0.92	9.15	16.89
1W	32.16	1.41	22.24	*	31.61	1.27	26.60	*
2W	47.10	1.83	20.98	*	30.69	1.50	11.36	*
Second re-use								
LL	47.51	1.79	17.67	*	33.91	0.91	25.30	*
1WL	41.90	1.76	8.34	17.51	30.41	0.90	6.29	23.44
Conversion of lye into brine								
LS	27.37	*	*	*	24.29	*	*	*

* Process without rinses or washing waters; L, 1W, 2W, re-use of the lye, 1st or 2nd washing water from preceding cycle, respectively; LS, conversion of lye into brine; BOD_5, 5-day Biological Oxygen Demand.

Table 10.6 Effect of lye and washing waters re-use on polyphenols and reducing sugar content of different wastewaters from the Spanish-style green olive debittering process. Manzanilla cultivar (Reproduced from Garrido *et al.*, 1977)

	Polyphenols (g tannic acid/l)				Reducing sugars (g glucose/l)			
Type of re-use	Lye	Rinse	1st washing	2nd washing	Lye	Rinse	1st washing	2nd washing
Traditional process								
Control	1.85	0.06	1.94	2.70	4.94	T	4.69	5.83
First re-use								
L(Half)	2.61	0.13	1.05	3.49	5.14	T	3.17	6.1
L	3.44	0.09	1.83	3.72	6.18	T	3.77	7.3
1W	3.35	0.15	1.82	*	4.09	T	7.26	*
2W	3.94	0.17	3.36	*	6.23	T	5.55	*
Second re-use								
LL	3.76	0.68	3.55	*	5.50	T	7.35	*
1WL	4.23	0.54	1.50	3.77	6.33	T	2.06	5.84
Conversion of lye into brine								
LS	2.85	*	*	*	5.00	*	*	*

* Process without rinses or washing waters; L, 1W, 2W, re-use of the lye, 1st or 2nd washing water from preceding cycle, respectively; LS, conversion of lye into brine. T, trace.

Table 10.7 Effect of lye and washing waters re-use on evolution of pH during fermentation. Manzanilla cultivar (Reproduced from Garrido *et al.*, 1977)

	Type of re-use							
Days in brine	Control	L(half)	L	1W	2W	LL	1WL	LS
0	5.80	5.90	5.70	6.90	6.80	6.80	7.15	7.60
2	5.03	5.50	5.35	5.75	5.60	5.95	5.80	6.60
4	4.90	5.00	5.25	5.25	5.40	5.55	5.50	4.75
8	4.78	4.00	5.00	4.70	4.65	4.80	4.60	4.60
15	4.53	4.50	4.70	4.45	4.40	4.60	4.45	4.40
25	4.33	4.35	4.50	4.35	4.25	4.35	4.25	4.35
28	4.23	4.20	4.35	4.30	4.15	4.25	4.05	3.95
36	4.03	4.05	4.15	4.10	4.05	4.10	4.05	3.95
46	3.96	3.95	4.00	4.00	4.00	4.00	4.00	4.00
85	3.80	3.75	3.80	3.80	3.85	3.85	3.75	3.95

L, re-use of lye from preceding cycle; 1W, re-use of 1st washing waters from preceding cycle; 2W, re-use of 2nd washing waters from preceding cycle; LS, conversion of lye into brine.

Table 10.8 Effect of lye and washing waters re-use on final titratable, volatile and combined acidity. Analysis after 85 days of fermentation. Manzanilla cultivar (From Garrido *et al.*, 1977)

Type of re-use	Titratable acidity (g lactic/l)	Volatile acidity (g lactic/l)	Combined acidity (m Eq/l)
Traditional process			
Control	7.7	2.8	79
First re-use			
L (half)	8.4	3.8	77
L	8.4	3.8	83
1W	9.3	3.8	100
2W	8.8	3.4	97
Second re-use			
LL	9.8	3.2	105
1WL	8.4	3.5	81
Conversion of lye into brine			
LS	12.1	3.8	172

L, 1W, 2W, re-use of lye, 1st or 2nd washing waters from preceding cycle.

Table 10.9 Effect of lye and washing waters re-use on pollutant charge of fermentation brine. Manzanilla cultivar (Reproduced from Garrido *et al.*, 1977)

Type of re-use	BOD_5 (g O_2/l)	Polyphenols (g tannic acid/l)	NaCl (g/l)	Inorganic(g/l) (except NaCl)	Organic (g/l)
Traditional process					
Control	15.11	1.86	52	26.5	25.2
First re-use					
L (Half)	17.00	12.08	53	16.0	19.9
L	17.32	2.30	53	56.1	22.4
1W	17.38	2.59	53	40.4	21.2
2W	17.23	2.80	50	54.4	27.3
Second re-use					
LL	18.51	2.80	53	44.4	28.9
1WL	15.20	2.21	56	23.0	22.1
Conversion of lye into brine					
LS	32.36	4.57	61	35.1	40.6

L, 1W, 2W, re-use of lye, 1st or 2nd washing waters; LS, conversion of lye into brine; BOD, Biological Oxygen Demand.

increase in its pollutant charge, although it also represents a lower waste-water volume. Thus, application of re-uses is convenient when methods to eliminate the remaining washing waters are not affected by this progressive concentration.

An important proportion of the organic matter comprises sugars and polyphenols. Both are leached into re-used lyes in lower concentrations than in fresh lyes. However, these differences can be modified in the following washing waters, which, in case of two washings, are able to correct any possible deficient elimination of polyphenols. The same applies to sugars, although these compounds should be retained as far as possible for the following lactic fermentation. A balance between sugars and polyphenols elimination must be sought, possibly through the use of a single long washing.

The different re-uses of lye and washing waters described above had little influence on the pH change during fermentation. Even those olives whose lyes were directly converted into brines reached low enough values for preservation of the product during storage in bulk. This pH behaviour was possible by virtue of the relationships between combined acidity, titratable acidity and pH. In effect, some of the treatments led to higher combined acidities (1W, LL and LS in Table 10.8). However, this higher buffering capacity was efficiently counteracted by higher acid production due to the higher level of reducing sugars remaining after re-use or elimination of washings. From this point of view, re-use or elimination of washing waters does not necessarily lead to unacceptable characteristics for good preservation. Furthermore, it was of interest to observe that the ratio of lactic to acetic acid – which can be considered as an index of the lactic acid process yield – was systematically higher in all the fermenters prepared with re-uses. This means that such modifications favoured more effective lactic acid production, probably through an indirect effect on the population of the different microorganisms. In general, the appearance and growth of Gram-negative bacteria was normal, although their presence was slightly less than in most of the spontaneous processes since they were present only for about 11 days, compared with up to 20 days or more in the traditional process. Growth of lactobacilli began after 4 days and was general at the 8th day. Their counts increased rapidly and caused the inhibition of the Gram-negative bacteria after a short period of time. On the other hand, this rapid growth of lactobacilli delayed the yeast appearance, since they were detected only after 28–30 days of fermentation. The high final lactic acidity reached also caused a decrease of yeast counts during the fermented olive storage phase. Thus, such modifications have no negative effects on the fermentation but exert, at least in these modifications, a favourable influence.

No negative effects were noted on texture, flavour, gas-pocket spoilage, blister formation or sloughing, and all treatments had similar behaviour in the pitting operations.

Re-use of lyes and washing waters may have some effects on the pollutant characteristics of the fermentation brines, in addition to the higher acid content already noted. As shown in Table 10.9, in those cases in which there was a conversion of the lye into brine, the general charge in organic matter and BOD_5 was markedly higher than in all other treatments. Compared with the traditional process, the BOD_5 was a little higher, mainly in the first cycle of re-uses, due to the higher acidity but a clear trend was not observed in the organic matter content. Inorganic dissolved solids tended to increase. These results confirm the fact that any possible effect of re-use on the solubilization of substances from the olive flesh can be adequately corrected in the following steps of washing or even fermentation, as will be seen later.

At the same time that pollution load can be reduced by lye re-use and reducing washing steps, it will also be markedly reduced if the concentration of pollutants does not increase proportionately with re-use. Changes in the organic and inorganic matter through re-uses are shown in Figure 10.1. There is a progressive increase in the content of lye-dissolved solids, but this is relatively slow and changes after four–five re-uses are of limited significance. This happens when the levels of dissolved solids are high enough to prevent osmotic exchange between lye and fruits. Table 10.10 shows the main physicochemical characteristics of a typical re-use cycle with up to 14 re-uses. At the end of these processes, the change was less than four times that of a single use lye. In this case, the volume of lye poured away was reduced to 1/14 and the pollution impact to about 4/14.

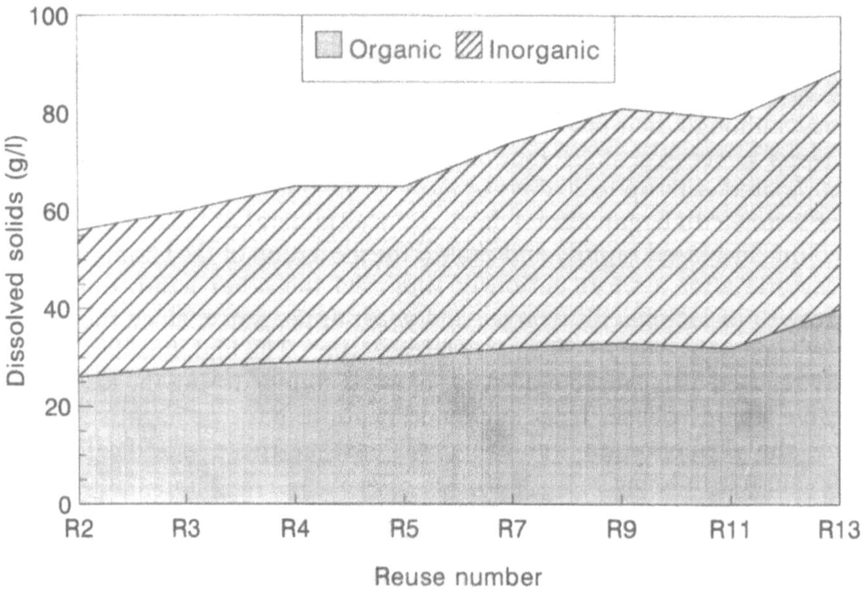

Fig. 10.1 Dissolved solid content changes in lyes with re-use. (From Garrido Fernández *et al*, 1979.)

Table 10.10 Pollutant characteristics of lye after the first and 14th re-use. Manzanilla cultivar (Reproduced from Garrido *et al.*, 1979)

Characteristic	1st lye	14th lye
pH	12.1	12.0
Residual NaOH (g/l)	8.4	9.1
Dissolved organic matter(g/l)	14.4	48.1
Dissolved inorganic matter(g/l)	16.2	35.4
Polyphenols (g tannic acid/l)	1.32	2.9
BOD_5 (g O_2/l)	18.13	39.8

BOD, Biological Oxygen Demand.

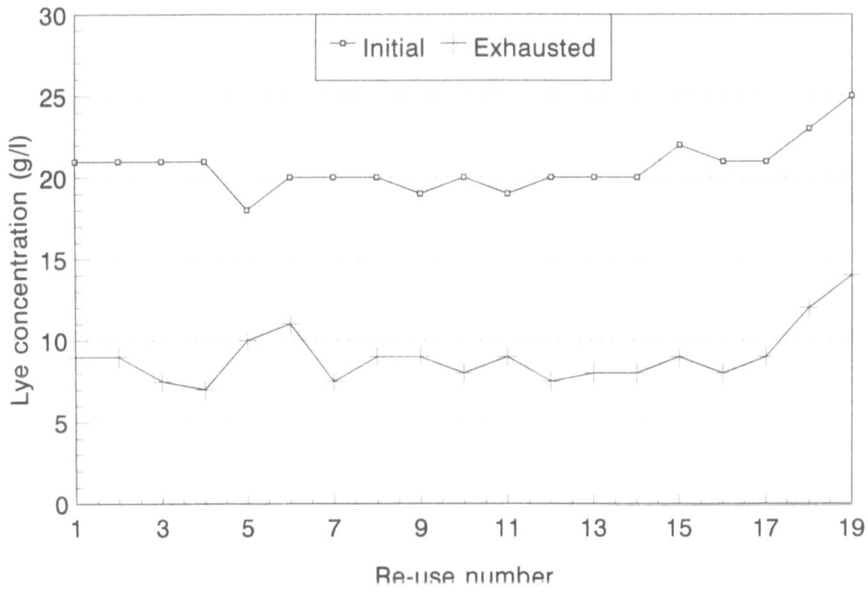

Fig. 10.2 Sodium hydroxide content changes in lyes with re-use. (From Garrido Fernández *et al*, 1979.)

These internal control measures also reduce costs due to the residual NaOH recovered. This may vary from factory to factory, according to the lye treatment characteristics, immersion time, temperature, etc. It may range from 25% to 60% or even higher. As an example, Figure 10.2 shows the initial and final NaOH concentrations of successive lye re-uses. Savings in a few seasons can possibly pay for any modification in plant needed to put the procedure into practice.

The influence of successive lye re-use on the fermentation is shown in

Table 10.11; pH values and combined acidity levels are similar to those obtained in the traditional process, while levels of acidity are markedly higher. Successive lye re-use also affects the pollutant charge of the fermentation brines. Table 10.12 shows an example in which this is observable mainly in the organic dissolved compounds and the corresponding BOD_5 (except in the last treatment with two washing waters). Polyphenol levels are also higher and this may lead to a higher sensitivity of this brine and olives to browning when exposed to air. This is not seen with normal pH values, but if the pH rises the oxidation rate will increase and the effect could be appreciable.

In general, lye re-use does not present any problems with respect to microbial growth due to the high NaOH content and pH (>12). Most spores are destroyed under these conditions. However, the risk must not be completely disregarded since there are literature examples of *Bacillus* species whose spores are resistant up to 1% NaOH. Thus, it is advisable to recondition such used lyes as soon as possible after their removal from the fruits. Once concentrations are again about 2.0–3.5%, the risk of any possible microbial spoilage is effectively eliminated.

Special attention must be paid to the procedure used to analyse and correct the NaOH concentration in the re-used solution. It has been customary to prepare either NaCl or NaOH solution by the use of a hydrometer and express the concentration in °Bé. In the case of lye re-use, it is not possible to use this system because the dissolved matter also contributes to the °Bé readings. It is thus necessary to calculate the residual lye content by titration with an acid.

Table 10.11 Effect of successive re-use in fermentation (after 130 days). Manzanilla cultivar (Reproduced from Garrido *et al.*, 1979)

Re-use number	pH	Titratable acidity (g lactic/l)	Combined acidity (m Eq/l)
R_2	3.90	10.1	110
R_3	3.85	9.9	110
R_4	3.85	10.7	107
R_5	3.80	10.5	110
R_6	3.80	11.5	110
R_7	3.80	12.0	114
R_8	3.70	12.4	106
R_9	3.75	11.8	106
R_{10}	3.70	14.1	111
R_{11}	3.80	12.4	115
R_{12}	3.75	11.5	106
R_{13}	3.80	12.2	114
R_{14}	3.85	13.2	122

Table 10.12 Effect of re-use of lyes and different washing water systems on the pollutant charge of brine. Manzanilla cultivar (Reproduced with permission from González Cancho *et al.* 1983)

Type of treatment	pH	COD ($g\ O_2/l$)	BOD_5 ($g\ O_2/l$)	Polyphenols (g tannic acid/l)	Titratable acidity (g lactic acid/l)	Dissolved solids(g/l)	
						Organic	Inorganic
Control*	4.1	9.12	8.55	6.0	6.1	14.95	94.39
L_1(longW)	4.0	10.98	14.77	7.2	7.9	23.01	91.20
L_2(shortW)(HCl)	4.1	32.92	13.79	8.8	4.8	27.83	94.63
L_3 (HCl)	4.0	31.08	13.80	8.1	7.9	28.61	93.03
L_4(WW)†	4.1	13.39	15.32	9.7	6.3	22.83	94.07

* Control processed with fresh lye and two washing waters; † Second washing was done with the neutralized first washing water; L_1, L_2, etc. indicate successive lye re-uses; W, washing; HCl, addition of this acid in substitution of washing.

Re-used lyes are heavily contaminated and their handling without proper care leads to formation of an abundant and persistent foam. This may be avoided by preventing the introduction of air in the pipelines. Thus, appropriate pumps must be used and the solution always introduced from the bottom of containers, below the solution surface. Some food-grade chemical compounds to prevent foam formation can also be used.

As re-use progresses, there is an accumulation of suspended solids which should be decanted or removed by filtration. Until now no special attention has been paid to this, but it may be helpful in diminishing the pollutant charge of these residues. Other gross suspended solids must of necessity be removed by using screens.

Polyphenols are easily oxidized in the alkaline medium of lyes and a progressive browning of solutions with re-use is evident. This property has given rise to the description of such solutions as 'chocolate' in the slang of Spanish table olive factories.

In summary, lye re-use is the most obvious, easily implemented, cost-effective pollution abatement measure that can be introduced in the Spanish-style green olive industry. The other re-use possibility studied – washing waters – has the same advantages except that they have a very low proportion of NaOH to be recovered and, after the first re-use, the washing water is converted into lye. Each olive debittering container produces one volume of lye and another of washing water, but only one is needed for the following lye treatment; thus, lye re-use is preferable.

Washing waters can be re-used as new washing waters after neutralization of the excess alkali. However, they have the disadvantage of a relatively high concentration of organic and inorganic matter in solution. Second and following re-uses could not extract sufficient of these substances to be effective. Furthermore, these waste solutions have high concentrations of easily degradable substrates and a low level of NaOH, insufficient to inhibit the growth of microorganisms. Its re-utilization, even after acidification, may result in inoculation with a series of microorganisms which could be undesirable spoilage organisms. In short, washing waters are not attractive solutions to be re-used either for preparing lyes or for other washing cycles. To diminish the volume of washing waters, the most recommendable option is to eliminate washing altogether or to reduce the number of washings to only one.

Reduction or elimination of washing waters

In some of the examples shown in the previous section, it was seen that some modifications consisted only of the elimination of some of the washing waters. In California, it was traditional to give many washings to lye-treated green olives to completely remove the alkali. In Spain, it was common to apply a quick rinse to remove the remaining lye adhering to olives and the container, followed by three washing waters. Modern technology and scarcity of fresh water have reduced the washing process to only two immersions. The first has a short duration, since it rapidly extracts a major proportion of alkali. The second is usually longer because in this case alkali removal is slower. This standard procedure was established to permit final combined acidities of around 100–120 mEq/l, since values over 140 mEq/l were considered excessive and incompatible with appropriate final pH values. As shown in Tables 10.4–10.6, it is possible to reduce the washing waters to just one, although the efficiency is not completely equivalent to that with two washes. Nevertheless, final characteristics of the fermented product may reach acceptable levels (Tables 10.7 and 10.8) because the higher acidity produced compensates for the possible excess of combined acidity left. However, when the fruit-to-brine ratio is high and the washings are completely eliminated or their efficiency low, it is necessary to modify the combined acidity in the fermentation brine to reach appropriate final pH values. One possible method may be to eliminate part of the combined acidity by incorporation of a strong acid, which partially liberates the organic acid and, consequently, reduces the remaining organic salts responsible for the buffering capacity. This acid substitutes the effect of the washing water with respect to alkali neutralization, but has the advantage of retaining the reducing sugars which are leached with the normal washing. Acid could thus be incorporated into the brine added after the debittering process. Some experience of its use was already available in the industry because of its sporadic use in cases of deficient debittering processes. The use of HCl or any other food-grade inorganic acid has the advantage that more sugar substrate is left in the medium. At the same time, it increases the acidity (Table 10.13). When this strong acid is added to provoke only slight combined acidity corrections, the effects are very limited and rarely lead to fermentation or flavour changes.

However, there was no previous knowledge about its effects when added in higher proportions. In some papers related to washing elimination (Garrido Fernández *et al.*, 1979; Castro Gómez-Millán *et al.*, 1983) special care was devoted to study its influence. A first obvious influence was the disturbance of the normal pH evolution of brines. The presence of HCl, the most widely used strong acid, produces a pH of about 2.0 in the initial brine. This is progressively increased as the alkali is leached from the olive fruits, but the sequence of microorganisms has been irretrievably altered by then. Gram-negative bacteria are practically eliminated, the presence of lactic cocci is sporadic, and the response of lactobacilli is variable. In those cultivars rich in fermentable substrates, which normally have good fermentability, lactobacilli usually grow without

Table 10.13 Effect of HCl addition on the fermentation evolution. (From de Castro *et al.* 1983 and González-Cancho *et al.* 1983)

	Gordal cultivar		Manzanilla cultivar	
Type of processing	pH	Titratable acidity (g lactic/l)	pH	Titratable acidity (g lactic acid/l)
Control	3.80	7.5	4.05	6.1
L_1(longW)			4.10	7.9
L_2(short)W(HCl)	3.70	10.3	4.05	6.0
L_3 (HCl)	3.80	12.6	4.80	4.3
L_4 (HCl)*	3.75	11.3	4.10	7.9
L_5(WW_2)(HCl)†			4.25	7.0

* Addition of HCl achieved in two phases: half in the initial brine and the other half after 48 h of brining; † second washing done with neutralized first washing water.

Control processed with fresh lye and two washing waters; L_1, L_2, etc, successive lye re-uses; W, washing; HCl, addition of this acid in substitution of washing.

any difficulty, but in those with some problems in the traditional process, brine colonization by lactobacilli is often doubtful. Table 10.13 shows the fermentation evolution in two examples of substitution of washing waters by HCl addition. In the case of Gordal cultivar, the process was completely normal while in Manzanilla there were some cases of stuck fermentations (L_3 HCl) or deficient processes (L_4 HCl, L_5 WW_2(HCl), and L_2 (short)W(HCl). Other effects of HCl on the fermentation brine of fruits are not yet well understood. The neutralizing effect is still not well known since a part of the initially added HCl is, in fact, not used in the NaOH neutralization. This lost HCl must have reacted with other organic matter present in brine or fruits. The negative effect on lactobacilli growth due to the HCl has sometimes been assigned to possible protein precipitation during the period of low pH (3–4 days). However, when the acid is added later (during or after fermentation) the liberation of combined acidity is that estimated by calculation.

The above problems were solved by González Cancho *et al.* (1984) who studied different procedures for HCl addition and established the most appropriate methodology to assure a normal further fermentation. Table 10.14 shows some aspects related to the fermentation evolution. In case of no initial addition (case A) there was a high pH for a rather long period of up to 24 days. This gave rise to abundant growth of Gram-negative bacteria. Since these bacteria are related to gas-pocket formation in Spanish-style green olives, the presence of an excessive number of such bacteria is not desirable. Thus, no addition at the beginning of the

Table 10.14 Changes in pH and titratable acidity (g lactic acid/l) during the fermentation of Spanish-style green olives in which washing waters had been sustituted by HCl addition. Manzanilla cultivar (From González-Cancho *et al.*, 1984)

Days after brining	A*		B*		C*	
	pH	Acidity	pH	Acidity	pH	Acidity
2	9.5		7.6		7.9	
3	9.2		4.8		7.8	
7	6.8		5.6		6.7	
14	5.4	2.8	5.3	2.3	5.3	2.3
24	4.6	6.9	4.3	7.5	4.5	7.2
58	3.9	12.5	4.2	10.5	3.9	12.5
135	4.0	12.8	4.0	12.4	3.9	13.1

* HCl, in substitution of washing waters, was added in the following sequence; A, 43 days after brining; B, half at 24 h after brining and the other half at 48 h; C, half in the initial brine and the other half 43 days after brining.

fermentation is inadvisable. Addition in two stages (24 and 48 hours after brining) led to successively low and moderately high pH that had an important impact on the Gram-negative population (which had markedly lower counts in this case). A similar result was observed when half of the acid needed was added to the initial brine. The growth of lactobacilli (*Lactobacillus plantarum*) was normal or even slightly improved in cases B and C. However with no initial acid, there was a slight delay due to the excessive presence of Gram-negatives. Use of a pure starter culture enhanced the lactic process in all cases, although the effect was limited, especially in cases B and C due to the good spontaneous fermentation processes. When all the acid was incorporated initially (data not shown) there was a major inhibition of the Gram-negative bacteria, but also a delay in lactobacilli growth and a increased presence of yeasts.

There were some differences in pH due to the different systems of HCl addition, which disappeared after all the acid had been added (Table 10.14). In general, the levels of pH and acidity may be considered as excellent in all cases, especially the high proportion of titratable acidity obtained by virtue of the HCl added and the elimination of washing waters and the consequent increased availability of fermentable carbohydrate. Combined acidity was markedly affected by the method used for the HCl addition but, finally, had values between 129 and 146 mEq/l, which are close to the maximum acceptable limit. Depending on the packing system used for the final product, combined acidity may need further correction.

In spite of the good fermentation of Spanish-style green olives processed without washing waters, the higher polyphenol content or perhaps the extended fermentation process give the fermented product a strong flavour, judged by most consumers as too intense. In this case, final packing needs a prior reduction of this flavour. Addition of HCl must thus be used carefully and only in the proportion needed to obtain the required combined acidity. Acid addition also needs careful handling by the personnel. Gas masks and other protective clothing must be provided.

Elimination of one of the two washings is sometimes used as a way of speeding up the debittering process during the peak season or in cases of water shortage. Partial neutralization with HCl or other corrective measures must then be applied. One of the most common ways of improving combined acidity levels and reducing strong flavours is to eliminate part of the fermentation brine and its substitution by a fresh brine with the appropriate concentration of NaCl and lactic acid. Usually, the brine is removed from the bottom of the fermenters, with the aim also of removing suspended solids that concentrate at the bottom. This operation has the additional advantage of preventing possible malodorous fermentations, since these solids comprise proteinaceous material from microbial lysis and other organic matter insoluble in acidic conditions. If its presence coincides with high pH values, then an abnormal fermentation is very probable.

This operation produces a highly contaminated waste liquid which, in contrast with the washing waters, can be poured away during a more favourable period (winter, rainy days, etc.), thus lowering its pollution impact.

When pasteurization is used during final packing, the usual pH values of unwashed olives can be appropriate for the holding solutions and no correction of the combined acidity is needed, except when flavour modifications are also necessary. In this case, the only effective known method to enhance flavour is brine dilution. This is because of the lack of knowledge about the components responsible for the fermented olive flavour. As stated earlier, information is available on volatiles, the influence of acid and salt, etc., but a clear assignment to each particular flavour sensation is currently absent. Nothing can be said about the compounds responsible for the extraneous flavour of green olives produced without washings.

In short, elimination of the second washing can be recommended while increasing the duration of the remaining washing to approach as far as possible the results obtained by the traditional process. It should not be extended for more than 18–24 hours. If necessary, it is possible to eliminate the washing completely, neutralizing (at least partially) the excess NaOH by addition of HCl to control the combined acidity. It is recommended that the amount of HCl does not exceed 0.75% of the brine volume and that the addition is undertaken in two stages: half of the total amount of acid has to be added to the initial brine, and the second half once the active fermentation is over. Any further correction of the combined acidity can be achieved by partial dilution, which can also improve the flavour. A summary of the effect of the different options on the fermentation process as well as in the contaminant charge of the fermentation brines can be seen in Table 10.15 for olives of the Gordal cultivar. Elimination of washing waters increased the titratable acidity as well as COD and organic and inorganic matter with respect to the traditional process. Polyphenol content, however, was only slightly affected in this case because it is low in this cultivar.

Debittering with low-concentration lyes

As mentioned earlier, the main inconvenience of washing water elimination is the subsequent high combined acidity. Lower-concentration NaOH lyes will lead to lower final combined acidities and lower pH values for the same titratable acidity. If NaOH contents were low enough, even the conversion of lye into brine might be possible. When concentrations of NaOH in lyes are about 8–12 g/l, they could be transformed into brine after the debittering process, yielding, theoretically, combined acidities of 100–150 mEq/l, respectively. This last value is only slightly over the maximum acceptable level for obtaining good final pH values. Concentrations greater than 12 g/l do not yield suitable combined acidities if converted into brines. In these cases, debittering must be achieved by removing the lye and its replacement with a fresh brine (perhaps with HCl, depending on the residual NaOH).

Rejano Navarro *et al.* (1983) tested these possibilities using NaOH contents from 8 to 20 g/l, where the higher limit can be considered as a normal concentration. Duration of the debittering process and the combined acidity obtained after converting lyes into brines or replacing with fresh brines are shown in Table 10.16. A concentration of 8 g/l needs a long

Table 10.15 Main physicochemical characteristics of Spanish green olives fermentation brines, processed as traditionally, with and without single washing. Gordal cultivar (From Castro Gómez Millán *et al.* 1983)

Characteristic	Normal process	Process with one washing	Process without washing
pH	3.85	3.85	3.90
NaCl content (g/l)	56	50	52
Titratable acidity (g lactic/l)	7.3	10.6	12.4
Polyphenols (g tannic acid/l)	3.4	4.1	4.7
Inorganic dissolved solids (g/l)	20.9	26.3	35.8
DQO (g O_2/l)	9.4	15.9	19.5

Table 10.16 Characteristics of debittering process with different (low) NaOH concentrations in lyes. Manzanilla cultiver (From Rejano Navarro *et al.* 1983)

Type of process	Lye concentation (g NaOH/l)	Duration (h : min)	Temperature (°C)	Combined acidity (m Eq/l)
Lye converted into brine L_1	8.0	48:00	21	142
Lye converted into brine (L_2)	8.0	48:00	38*	145
Lye converted into brine (L_3)	12.0	24:00	21	178
Lye converted into brine (L_4)	12.0	24:00	38*	205
Lye followed by brining (LB_1)	13.6	8:50	20	136
Lye followed by brining (LB_2)	16.4	7:00	20	125
Lye followed by brining (LB_3)	18.0	6:45	20	173
Lye followed by brining (LB_4)	19.6	5:10	20	171
Lye followed by two washing and brining (control)	19.6	5:10	20	88

* Temperature left to change freely after reaching this level.

alkaline treatment of 2 days. This was not accelerated when the initial temperature was increased to 38°C and then left to cool to room temperature, but fruits were softened by this treatment. Treatment with lyes of 12 g/l NaOH content had a duration of 24 hours, independent of the initial temperature. Concentrations higher than 13 g/l gave reasonable treatment times working at 20°C. Conversion of lyes into brine (treatments with 8 and 12 g/l NaOH) gave rise to markedly higher combined acidity than those expected from the NaOH content. This supports the view that the combined acidity not only comes from the remaining alkali but also from the organic matter leached to brine when the lye-treated olives were immersed in the brines. With about 14 g NaOH /l it was possible to obtain moderate and acceptable combined acidities when olives were brined directly after lye treatment. Above this level and up to 20 g/l there was no consistent trend. The 20 g/l NaOH lye and two washing waters treatment gave rise to normal values of combined acidities. Normal growth of Gram-negative bacteria and spontaneous appearance of lactobacilli were observed only in heated fermenters. Nevertheless, the rest of the treatments supported lactobacilli growth when a starter culture of *Lactobacillus plantarum* was used. Growth, however, was slow and the lactic bacteria counts were always lower than that in the normal process. Changes in physicochemical characteristics (Table 10.17) paralleled the microbial changes. Heated olives (L_2 and L_4) showed good acidity development, although the pH remained moderately high (about 4.30). With the lower NaOH levels (8 and 12 g/l NaOH) limited acidity was formed and the pH was excessively high. In those fermenters where lye was replaced by a fresh brine there was a clear lactic process,

Table 10.17 Fermentation changes in Spanish-style green olives during debittering with different low NaOH concentrations in lyes. Manzanilla cultivar (From Rejano Navarro *et al.* 1983)

Days after brining	L_1		L_2		L_3		L_4		LB_1		LB_2		LB_3		LB_4		Control	
	pH	Acid	pH	Acid	pH	Acid	pH	Acid	pH	Acid	pH	Acid	pH	Acid	pH	Acid	pH	Acid
4									8.25		9.45		9.60		9.45		7.70	
5	7.05		6.10		8.75		7.35											
12	6.90	1.9	5.85	2.5	7.60	1.7	5.35	4.9	5.80	1.6	6.05	1.6	6.20	1.6	6.00	2.2	5.65	1.3
27	6.70	1.6	5.25	4.2	7.15	1.5	5.00	6.8										
46							4.30	7.3	4.50	6.5	4.65	5.5	6.1	4.3	4.3			
54	6.40	2.1	4.85	4.7	6.60	1.5	4.55	10.5										
81									4.10	8.9	4.30	8.6	4.40	7.0	4.40	6.6	4.15	4.3
108	5.35	2.7	4.50	7.4	5.9	2.6	4.40	11.5										
118									4.20	9.5	4.40	8.8	4.40	7.6	4.4	7.8	4.15	5.5
140	4.85	3.4	4.30	7.9	5.75	3.3	4.35	10.9										

L_1, L_2, L_3, and L_4 indicate treatments in which lyes of 0.8, 0.8 (38°C), 1.2, 1.2 (38°C)% NaOH, respectively, were converted into brines. LB_1, LB_2, LB_3, and LB_4 indicate treatments in which olives were treated with lyes of 1.2, 1.4, 1.6, and 2% NaOH, respectively and then added to fresh brine. Control was treated with 2.0% lye and given two washing waters.

Acidity is expressed as g lactic acid/l.

developing higher acidities than in the control. The pH was similar to that of the control in the case of lower combined acidity (about 14 g/l NaOH in lye).

The organoleptic characteristics of the fermented olives showed a clear difference between the fruits in which the lye was converted into brine and those olives brined after the alkaline treatment. The first had an appearance and flavour that resembled untreated directly brined olives more than Spanish-style green olives. Heated olives were, in addition, softer. However, olives immersed in fresh brine after the debittering process had normal appearance and flavour. Treatments with lower NaOH concentrations showed a strong flavour that, in some cases, could be identified as being more bitter than those which followed the normal process (with two washing waters).

A relationship between flavour and brine composition was difficult to establish (Table 10.18). CODs of all treatments were fairly similar and gross organic matter ranged between roughly the same limits. Differences in polyphenols were also not important and flavour could not be related to acidity. This again emphasizes the lack of knowledge on the components responsible for the olive flavour and their origin.

Good results were obtained with Manzanilla cultivar using low NaOH concentrations (15–18 g/l) at controlled temperature, 30°C (Sánchez Gómez *et al.*, 1990). There were significant combined acidity reductions with respect to the control. The process produced high acidity levels and a pH sufficiently low to ensure preservation (Table 10.19). In Hojiblanca, slightly higher concentrations (18–20 g/l NaOH) were necessary and, consequently, the remaining combined acidity was higher. In spite of this, good acidity development was obtained and final pHs were very close to those suitable for an appropriate preservation. No negative effect on colour and texture was observed. The method, however, must be used carefully because inadequate temperature control can lead to blisters and softening, but the system did produce a drastic reduction in wastewaters and pollutant charge since the procedure involves lye re-use and no washings.

It can thus be deduced that the use of low concentrations of lye for the debittering process is possible. A reasonable fermentation is obtained when olives are put into fresh brine after lye treatment and inoculated with a starter culture of *Lactobacillus plantarum*. Minimum NaOH concentrations are about 14 g/l. This represents a good wastewater reduction, which can be increased further with lye re-use, as mentioned earlier. An example of the expected results with different concentration lyes during the debittering process followed by immersion in fresh brines is shown in Table 10.20. The final pH values are slightly higher than the usual ones and a further correction of this characteristic could be convenient. Additions of up to 70 mEq/l of HCl have not shown any negative effect

Table 10.18 Characteristics of storage brines after different debittering processes with low NaOH lyes and fermentation. Manzanilla cultivar (From Rejano Navarro *et al.*, 1983)

	L_1	L_2	L_3	L_4	LB_1	LB_2*	LB_3	LB_4	Control
pH	4.6	4.5	4.8	4.4	4.3	4.5	4.4	4.4	4.6
NaCl (g/l)	51.5	75.0	53.0	73.5	52.0	62.0	52.0	53.0	53.0
Titratable acidity (g lactic acid/l)	4.3	5.9	4.4	8.1	9.0	3.9	6.3	7.0	3.5
Polyphenols (g tannic/l)	5.4	5.1	7.0	5.7	5.6	3.6	4.8	5.6	5.8
Organic dissolved solids (g/l)	19.7	21.9	26.7	27.0	27.0	14.3	21.2	25.7	13.6
Inorganic dissolved solids (g/l)	56.4	81.7	58.9	77.8	59.7	66.4	60.2	59.6	54.3
COD (g O_2/l)	14.6	13.9	16.8	15.3	15.0	9.8	11.8	9.4	14.0

* Brine partially changed during fermentation.

L_1, L_2, L_3, and L_4 indicate treatments in which lyes of 0.8, 0.8 (38°C), 1.2, 1.2 (38°C)% NaOH, respectively, were converted into brines. LB_1, LB_2, LB_3, and LB_4 indicate treatments in which lyes of 1 2, 1.4, 1.6 and 2% NaOH, respectively were added to fresh brine. Control treated with 2.0% lye and given two washing waters.

Table 10.19 Characteristics of debittering storage and final product of Manzanilla olives processed with low NaOH lyes without washing waters (including control.) (From Sánchez-Gómez *et al.*, 1990)

Lye concentration (g NaOH/l)	Temperature (°C)	Combined acidity (m Eq/l)	pH	Titratable acidity (g lactic acid/l)
Hojiblanca				
24 (Control)	24	223	4.30	11.5
17	30	181	4.15	13.2
20	30	201	4.30	11.9
24 (Control)	25	177	3.89	12.4
17	30	149	3.98	11.6
19	30	158	4.04	11.2
Manzanilla				
21	27	175	3.96	13.8
15	31	137	3.84	15.3
18	30	155	3.91	14.2

Table 10.20 Characteristics of Spanish-style green olive processing with low NaOH lyes. Manzanilla cultivar (From González-Cancho *et al.*, 1984)

		Final characteristics	
Lye concentration (g NaOH/l)	Duration h : min	pH	Titratable acidity (g lactic/l)
15.6	08 : 55	4.4	9.5
14.0	14 : 44	4.6	7.0
12.0	17 : 15	4.8	5.0

on the quality or flavour of green olives (Castro Gómez Millán *et al.*, 1989). However, there is no experience on the effect of high acidity and polyphenol contents on the stability of the fermented product at these pH levels. It is possible that it might be better than that of the traditional fermented product and that pH correction is unnecessary. Nonetheless, until this is established, pH corrections are recommended.

Pollution abatement by lye re-use and elimination of washing waters

Based on the pollution charge of the different wastewaters and the volume produced during the debittering process, it is possible to estimate the reduction in both volume and contaminants that can be obtained through successive application of the modifications discussed above. Such calculations are only approximations since specific situations and procedures can vary. Figure 10.3 gives the percentages of volume reduction and the pollutant charge saved with successive re-uses of lyes and elimination of one or two of the washing waters. A simple calculation can show that re-use of lye and a single washing water tends to produce a maximum volume reduction of 66.66%. Without washings, this figure approaches 100%. These considerations are important when the water discharge facilities are reduced. Bearing in mind that lye re-use is accompanied by only a small increase in the pollutant load, it too decreases in parallel with the reduction in volumes. Since both reductions are similar, this can mean in practice that application of internal control measures can also reduce the absolute pollution caused by the factory. This will also reduce the final costs charged for effluent disposal. The total number of re-uses is variable and must be determined by each factory in accordance with its working schedule and facilities. However, it is evident from Figure 10.3 that an excessive number of cycles produces only slight benefit and may, on the contrary, increase the handling difficulties. Re-use seven to ten times is considered suitable by many processors, while others prefer to use the same lye throughout the whole season. In this case, measures must be taken to prevent excessive suspended solids in the re-used lyes (decantation, filtration, etc.). When such lyes are low in NaOH, additional treatments (such as pasteurization) must also be used to prevent undesirable microbial growth.

Regeneration and re-use of Spanish-style green olive fermentation brines

Use of fermentation brines is unavoidable for Spanish-style green olives. Currently, these solutions are discarded when the olives are packed and the brine substituted by a fresh one. Practically, all the final commercial presentations of this type are also packed in holding solutions, otherwise when the green olives are removed from the brine they become progressively brownish, making the product unmarketable. The main characteristics of the fermentation brines have been shown in Table 10.2 and the influence of the different modified debittering processes on pollutant charge in Tables 10.9, 10.12, 10.15 and 10.18. Their pollutant charges are similar to those of other fermented products like sauerkraut (Hang, 1977) or cucumbers (Glaros and Geisman, 1980). The main difference between fermentation brines and the alkaline wastewaters is the high level of salt and lactic acid and the acidic pH (3.8–4.4). The presence of NaCl is a serious drawback for their incorporation into municipal sewers and treatment by conventional purification facilities. Disposal in evaporation ponds, as is usual for other wastewaters from the olive industry, has the problem that, if not well constructed, the NaCl can leak out and pollute the adjacent land and ground water. In some places this practice has been forbidden.

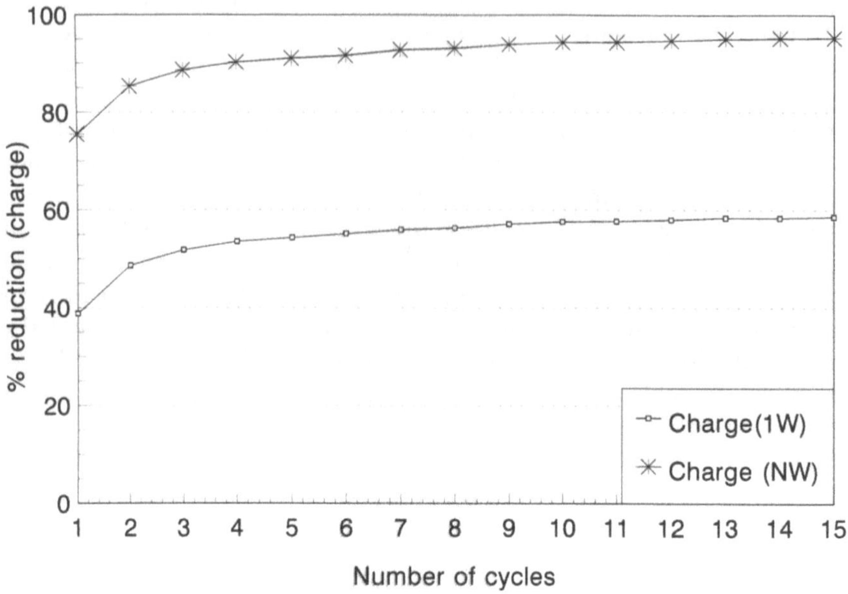

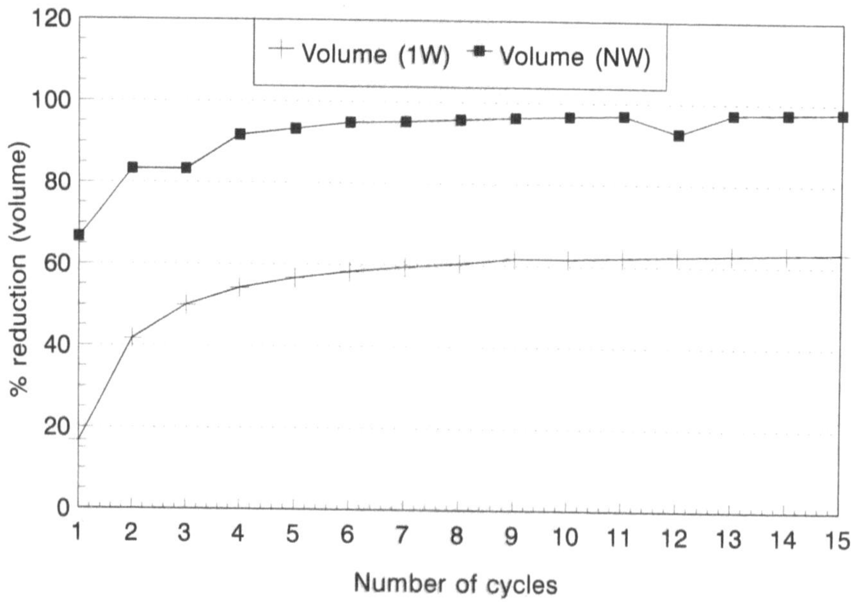

Fig. 10.3 (a) Pollution charge diminution with lye re-use and elimination of washing waters. (b) Wastewater reduction (volume) with lye re-use and elimination of washing waters. 1W, processes with one washing water; NW, process without washing waters. (Redrawn from Garrido Fernández *et al*, 1979.)

Brine regeneration can recover the NaCl and lactic acid dissolved in the solution. These compounds must, in any case, be incorporated into the fresh brine in which the final product is packed. Another possibility is to use these brines for fermentation of new fruits during the following season. If the fermentation brine is re-used either for a new fermentation or as a holding solution for the final product, fresh chemicals need not be used and the saving may, at least in part, pay for the treatment. For these reasons, many researchers have looked for ways to recondition the different brines.

Durkee *et al.* (1974) used recycled salt, obtained by submerged combustion brine evaporation followed by incineration, in the fermentation of salt stock cucumbers. The results indicated substantially no differences in the fermentation or product quality with either fresh kiln-dried salt or salt reclaimed by this procedure. The test data showed no advantage in clarifying or neutralizing the starting brine. Elimination of these two operations, naturally, reduces the costs of salt recovery. The method was, however, expensive due to the price of the gas needed.

Little *et al.* (1977) proposed a reduction of wastes from cucumber pickle processing by the use of a controlled culture fermentation process. Apparently, the process proceeded more rapidly and gave higher levels of acidity than natural fermentation. It produced superior quality product and a dramatic reduction in the amount of salt to be discharged. McFeeters *et al.* (1977) developed methods using pasteurization and chemical treatments to recover spent brines. Their main objective was to assure the processors that the recycled brine was free of pectinase activity, which can cause product softening. The first method consisted in heating the brine up to 88°C for 30 seconds. The chemical procedure consisted in raising the pH above 10 and holding for 24–48 hours. At the end of this period, the sludge layer was removed and the pH of the brine was re-adjusted to around 7.0 with HCl or vinegar (acetic acid). The use of fermentation brines for three cycles of fermentation showed no significant differences from control brines. The differences found in pH, titratable acidity and time of fermentation did not alter the way in which the fermentations proceeded and the salt stock produced by recycled brines was equal to that obtained in control fermentations. There were no differences between the cucumbers obtained from pasteurized and chemically treated brines that may indicate a preference for either treatment method. The chemical treatment did have the advantage of degrading pesticide residues in those materials tested in the experiments. Glaros and Geisman (1980) proposed a refinement of the technique for recycling spent curing pickling brine. The process involves the addition of 6 p.p.m. of anionic polyelectrolyte, which reduces the settling time to about 2 hours. The large floc produced excellent removal of suspended solids and a high retention of salt in a re-usable brine and was low cost. The reconditioned brine had no adverse effect on quality of the cured product. To ensure the absence of softening problems, the authors recommended that each lot of brine to be recycled is tested for the presence of pectinase enzymes.

The recycling of fermentation brines in lemon curing has been proposed by Cancel and Hernàndez (1975). The process consisted of centrifugation of the brine at 12 000 r.p.m. to remove suspended solids and the addition of salt to the desired level. Eleven consecutive fermentations using the recycled brine continued to produce good cured lemons. Brine re-use in lemon curing provides three important contributions: (i) it shortens fermentation time to about 2 days; (ii) it can save about 78% of the salt used; and (iii) it eliminates the waste disposal problem.

Fermentation brines of Spanish-style green olives differ from those of the other fruits and vegetables in their high buffering capacity due to the combined acidity, the presence of high levels of lactic acid, and a marked polyphenol content. Usually, no enzymatic contamination is present. Brine regeneration for a new fermentation increases its combined acidity and, consequently, low pH values cannot be obtained, even in the case of vigorous fermentation. The formation of new acidity and the higher polyphenol content interferes seriously with the growth of new lactic acid bacteria and a vigorous lactic process cannot then be expected. These problems have, in fact, been found in tests studying the possibility of recycling brines for fermentations the following season. Values of pH remained too high and fermentations were frequently malodorous.

Packing has traditionally used a fresh brine for which it is necessary to buy salt and lactic acid. Both chemicals are already present in fermentation brines so their reconditioning as holding solutions for the final product will reduce costs, provided that regeneration costs are not too high. In view of this, most of the regeneration methods studied have been directed to re-using the fermentation brine for packing the final product.

Decoloration procedures

The regenerated brine must be free of suspended solids and, ideally, colourless. To achieve these two requirements, several procedures have been tried. These have been mainly focused on colour removal since elimination of suspended solids is relatively straightforward.

Brenes Balbuena and Garrido Fernández (1988a) studied the use of ion exchange to remove the colour. Anionic resins preferentially retained acidity versus colour, but non-ionic resins offered interesting possibilities. Treatments needed about 2–5 g/100 ml of brine. The procedure was successfully applied in a series of three columns with an optimum flux of 0.36 column volume per hour. Resin was successfully regenerated with acetone and maintained the same efficiency over 20 cycles. An example of

Table 10.21 Regeneration of Spanish-style green olive fermentation brines with ion-exchange resins. Effect of different concentrations and types of resins. (From: Brenes Balbuena and Garrido Fernández, 1988a)

Characteristics	Fermentation brine	Regenerated brine			
		3 g resin/100 ml		7 g resin/100 ml	
		XAD-2	XAD-4	XAD-2	XAD-4
Colour ($A_{440}-A_{700}$)	0.450	0.260	0.223	0.140	0.100
Polyphenols (g tannic acid/l)	3.05	2.50	1.67	1.65	0.9
Titratable acidity (g lactic/l)	7.7	7.4	6.7	6.5	6.0
pH	4.00	4.00	4.02	4.05	4.10

the results obtained is shown in Table 10.21. As shown, resins selectively remove colour and the polyphenols while acidity is only slightly reduced. Efficiency was relatively unaffected by pH. The process can be applied directly to any fermentation brine.

Decolouring earths The use of decolouring earths (Garrido Fernández, 1979; Brenes Balbuena and Garrido Fernández, 1988b), which are widely applied in olive oil decoloration during the refining process, gave acceptable colour removal in only some cases (Table 10.22). However, their use introduced a greenish colour in the brine that could seriously interfere with their use as holding solutions. In addition, the levels needed to reach satisfactory decoloration were about 4% in the case of the better adsorbents (X-63 and Tonsil Optimum). In view of their drawbacks, the use of decolouring earths is not recommended.

Activated carbon Another classical adsorbent is activated carbon, the performance of different types of which has been compared by Garrido Fernández (1979). The behaviour of progressive concentrations of activated carbon on polyphenols, responsible for the colour of Spanish-style green olive brines, and total COD is shown in Table 10.23. As can be observed, polyphenol elimination is more rapid than COD removal. Polyphenols had been practically eliminated with a treatment of 50 g/l while COD had reduced to only one-half of its initial content. The behaviours of different activated carbons against brine colour is diverse and it is recommended that trials are conducted to select the best for particular local conditions. Figure 10.4 shows the adsorption isotherms expressed as a function of the colour. As seen in the figure, the activated carbon type GA is the best tested and therefore the most appropriate to regenerate these brines. Table 10.24 shows their effect on the main characteristics of

Table 10.22 Effect of different decolouring earths* on regeneration of Spanish-style green fermentation brines (From Brenes Balbuena and Garrido Fernández, 1988b)

Type of earth	Colour ($A_{440}-A_{700}$)	Titratable acidity (g lactic acid/l)
No treatment	0.59	6.7
Gador OD	0.56	6.7
Gador	0.90	6.7
Gador SV	0.55	3.5
Bentonita	0.50	5.0
Gador C	0.35	6.7
Tonsil Optimum	0.34	6.6
Tolsa X-63	0.29	6.6

* Concentration of earths was 2% in all cases.

Table 10.23 Effect of treatment with progressive concentrations of activated carbon on the polyphenol and Chemical Oxygen Demand (COD) of regenerated brines (From Garrido Fernández, 1979)

Carbon concentration (g/l)	Residual polyphenol (mg/l)	COD (g O_2/l)
0 (Initial brine)	990	9.51
10	917	7.83
20	187	6.54
30	64	5.70
40	28	5.10
50	5	4.50

the brine when this is treated with 5 and 10 g/l of brine. GA type activated carbon reduces the colour by 76% with only 5 g/l and up to 92% when raised to 10 g/l (residual colour values of 0.15 and 0.05, respectively) while F-400 and CPG only could reach 60% and 30% removal even at the higher levels.

The adsorbent also has the potential to remove other organic components present in the fermentation brines such as lactic acid. Consequently, pH levels can diminish after the activated carbon treatment. If the adsorption were excessive, these characteristics should be re-adjusted. Affinity of different activated carbon types towards lactic acid is shown in Table 10.24. In general, all the carbon types remove progressively more lactic acid as the proportion of carbon used increases. However, GA performs well in this respect since the percentages of lactic removal by GA,

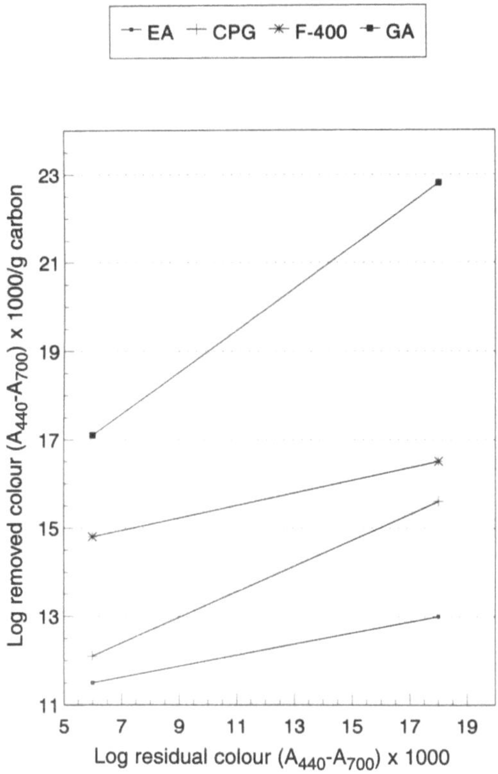

Fig. 10.4 Adsorption isotherms for different types of activated carbon (EA, CPG, F-400, GA) with green table olive brines.(From Brenes Balbuena *et al.*, 1988b.)

F-400 and EA are 10, 20, 15 and 15% respectively. The effect on the salt is practically negligible.

It is possible to conclude that the GA type is an acceptable adsorbent for the regeneration of Spanish-style green olives fermentation brines since with about 10 g/l around 90% of the initial colour (sufficient almost to clarify the brine) can be removed while leaving about 90% of the initial acidity. Other types of activated carbon (not tested) could prove even better and experiments with all commercially available types are thus necessary before any industrial application is initiated.

The influence of pH down to 2.6 (the level at which the combined acidity disappears) and temperature on activated carbon performance are negligible.

Nowadays, there is great diversity in the characteristics of fermentation brine due to differences in the re-use of lyes and washing waters,

Table 10.24 Effect of using different types and percentages of activated carbon for regeneration of Spanish-style green olive fermentation brines.(From Brenes Balbuena and Garrido Fernández, 1986b)

Activated carbon type	Colour (A_{440}–A_{700})	Polyphenols (g tannic/l)	Titratable acidity (g/l)	pH
O (original brine)	0.640	0.354	7.50	4.20
Regeneration with 5 g/l				
GA	0.150	0.206	7.00	4.20
EA	0.570	0.251	7.00	4.25
F-440	0.440	0.213	6.70	4.22
CPG	0.480	0.206	6.90	4.23
Regeneration with 10 g/l				
GA	0.050	0.162	6.70	4.24
EA	0.450	0.165	6.40	4.32
F-400	0.260	0.099	6.00	4.25
CPG	0.310	0.101	6.40	4.30

and in storage methods. Consequently, regeneration must be applied in very different initial conditions. In these cases, variable amounts of activated carbon must be used. Amounts may vary between 4 and 7 g/l to obtain residual values of colour (A_{440}–A_{700}) between 0.03 and 0.09 in the regenerated brines. In these conditions, acidity losses can oscillate between 7 and 13%. Changes in pH are thus around 0.1 units or slightly higher.

Contact times are relatively short since the adsorption equilibrates completely in about 10 minutes. The efficiency of the adsorption process increases when several counter-current contacts are applied. In this case, the amount of activated carbon used can be diminished drastically, and costs markedly reduced. Table 10.25 shows some examples. Its application in columns needs granulated types which have lower adsorption capacity than powdered charcoal. Use of granulated activated carbon is only possible if adequate regeneration procedures and installations for the spent material are available. Since, as far as we are aware, there is no appropriate method currently available to recover those from brine regeneration, powdered activated carbon should be used. When this is used as regenerating adsorbent, a slurry is produced containing the brine-suspended solids. These can be eliminated by the traditional filtration procedures and by cross-filtration.

Brine filtration

Suspended solids in the fermentation brine of Spanish-style green olives are composed of olive pieces, dead microorganisms, their residues after partial or total lysis, oil drops, etc. Their contents range from 0.2 to 1.8 g/l. They decant slowly, but there is usually an accumulation in the

Table 10.25 Relative efficiencies and effect of counter-current treatment on proportion of activated carbon required for the generation of Spanish-style green olive fermentation brines. Data based on polyphenols adsorption (From Garrido Fernández, 1979)

		Activated carbon needed for 95% purification	
Activated carbon type	Relative efficiency	One contact	Two contacts
N2G	1.00	4.7	2.4
NS	0.96	5.1	2.1
N2S	1.36	3.6	1.8
NG	1.11	4.4	1.9
AC35(1.8 mm)	0.30	16.1	5.3
AC35 (powder)	1.81	2.7	1.6

bottom of fermenters. Some industrial practices include a total or partial elimination of the decanted solids during fermentation, because an accumulation of organic matter, when combined with a high pH, can give rise to malodorous fermentations. Often these solids form a difficult-to-remove colloid. Most colloids in the agro-food industry are electronegative. However, attempts to characterize the charge of those found in the brines have failed since Zeta potential meters were unusable because of the high brine conductivity or because colour interfered with the determinations. However, it is reasonable to assume that the colloid has a strong negative charge.

Until now, only bentonite has been shown to reduce the colloids in suspension. Effects of different concentrations are shown in Table 10.26. As is obvious, the amount of moisture added to the bentonite (imbibition degree) has a marked influence on the further effect when added to brines. If imbibition is 10% the turbidity decrease is only 17%. However, when imbibition was 2%, cloudiness diminished by up to 75%.

Bentonite is also able to eliminate the turbidity formed after the addition of powdered activated carbon. In fact, final turbidity of brines treated with 1 or 2 g/l of bentonite had the same turbidity, irrespective of whether or not they were treated with activated carbon. Settling of solids formed by bentonite addition is relatively slow, although after 1 hour more than 80% of the depth was already free of precipitate. Further settling was very slow; after 24–48 hours, flocs occupied about 15–20% of the depth and retained a large volume of brine.

Use of membrane filters is impossible because filter flux collapsed rapidly due to the suspended flocs. A better option is conventional or cross-filtration. Brenes Balbuena *et al.* (1988b) found that an appropriate turbidity reduction was obtained only after filtration through a 0.45 μ pore-size membrane. To use conventional filtration requires previous selection of the filtering earths. Among the wide variety commercially available, Diatom 185 gave good results; this forms a filtering cake with a

Table 10.26 Effect of bentonite imbibition and concentration on removal of turbidity in Spanish-style olive fermentation brine* (From Brenes Balbuena *at al.*, 1988c)

	Bentonite concentration (g/l)			
Imbibition(%)	0.2	0.5	1	2
2	300	225	150	95
6	320	275	250	225
10	380	360	345	325

* Initial turbidity 390 units.

pore size which only permits the passage of solids under 0.1–0.2 μ. However, this is not suitable for the pre-cake since the amount of filtered liquid is relatively low. It is thus necessary to incorporate a filtering earth that forms a cake with a higher pore size. Results showed that Speedplus (with a pore size of 0.6–0.8 μ) was the best. A mixture of equal amounts of both filtering aids was considered a good compromise. The volume of brine that can be filtered depends on the suspended solids content and diminishes as these increase. A mixture of Diatom 185 and Diaflow 2, which retains particles of size between 0.4 and 0.6 μ when mixed in a proportion of 50 : 50, also gave good results when added to brines not excessively charged with suspended solids (around 0.26 g/l). The amount needed in this case was about 0.8–1.0 g/l of the mixture.

Cross-flow filtration

This technique consists of passing the suspension through a ceramic filter with a pore size between 0.20 and 0.45 μ. To avoid rapid membrane bulking, brine is circulated rapidly along the membrane channels. From time to time there is an inversion of the pressure direction by the injection of pressurized air through the circuit. These pulses loosen the solids adhering to the membrane, which are re-incorporated in the circulating brine that is being regenerated. The flux decline after 9 hours of use was from 130 $l/m^2 \cdot h$ to 80 $l/m^2 \cdot h$ when working with a brine of 6% NaCl. The method works with about the same efficiency with normal brine as with brine treated previously with activated carbon. When temperature increases, flux through the membrane also increases; by 40% if the temperature rises from 30°C to 50°C. This effect is due to solution viscosity decreasing with temperature. At 25°C viscosity is 1.12 cp, while at 40°C it is only 0.83 cp.

Experiments carried out at industrial level (Garrido Fernández *et al.*, 1992a) showed a similar behaviour. Good recovery rates after each working cycle were observed, indicating a remarkable influence of membrane cleaning on permeability recovery. Maximum recovery was about 90–95%. The concentrated slurry had a high organic content, in addition to the activated carbon. Although no treatment was tried to eliminate this, the technical staff of the company thought it might be burned after sun drying, as was usually done with other solid olive residues. Cleaning the membranes after each working day took about 2.5 hours. During this time one worker was assigned to the operation, which must be performed carefully to give satisfactory flux recovery during further use.

Regeneration by ultrafiltration

Ultrafiltration (UF) and reverse osmosis (RO) are increasingly being used in the food industry. After passing Spanish-style green olive fermentation brines through different pore-size membranes, it was recognized that a 1000-Da pore size would eliminate most of the brine colour (Brenes Balbuena *et al.*, 1988d). This means that the polymerized polyphenols forming the colour have different degrees of polymerization with a wide range of molecular weights. The 1000-Da membrane was chosen as appropriate to green olive brine regeneration since it provides good salt and lactic acid recoveries while giving an acceptable colour in the permeate. The membrane should be of polysulphone, since this is the only substance that showed good resistance to polyphenol adherence and bulking. However, in time, fouling is always observed.

Pilot experiments showed that suspended solids were eliminated without any problem. Permeate flux increased as working pressure was increased, but above 20 bar the effect declined. A pressure of 18 bar was then chosen for operation because at this pressure the flux curve almost levelled out. The flux showed little decline during the first 4 hours, but after that a moderate decline was observable. This effect was less pronounced at the higher pressure (18 bar).

Not surprisingly, flux reduction was greater when the equipment was running in a partial recycling mode (only retentate was returned to the feed tank). In any case, the flux can be estimated to be around 32 $l/m^2 \cdot h$. Flux was strongly influenced by polyphenol concentration in the initial brine as shown in Table 10.27. When temperature increases flux also increases, but at about 50°C brine browning occurs. It is therefore necessary to fit the equipment with a heat exchanger to cool the brine when its temperature exceeds 30–40°C, the level at which flux is high enough but no significant browning occurs.

The effect of UF on the main physicochemical characteristics of interest in green olive brines is shown in Table 10.28. Decreases in the levels of the different parameters were as follows: titratable acidity, 7.5%; combined

Table 10.27 Effect of polyphenol concentration on permeate flux in Spanish-style green olive fermentations brines regenerated by ultrafiltration (Adapted from Brenes Balbuena et al., 1988d)

	Flux ($l/m^2 \cdot h$)	
Operating time (h)	Low polyphenol*	High polyphenol†
0.5	58	36
1.0	55	32
2.0	53	30
3.0	52	29
4.0	51	29
5.0	50	29

* 3.05 g tannic acid/l; † 6.10 g tannic acid/l

Table 10.28 Effect on the physicochemical characteristics of Spanish-style green olive fermentation brine regenerated by ultrafiltration through a 1000 Da membrane (Adapted from Brenes Balbuena *et al*, 1988c)

Characteristic	Original brine	Permeate	Concentrate
Concentration grade		84%	16%
pH	3.7	3.7	3.7
Titratable acidity (g lactic/l)	8.1	7.5	9.8
Combined acidity (m Eq/l)	63	54	68
Salt (g NaCl/l)	80	78	88
Polyphenols (g tannic acid/l)	2.10	1.45	4.5
Colour ($A_{440}-A_{700}$)	0.44	0.06	> 2

acidity, 14.5%; polyphenols, 31%; colour, 86%; and salt, 2.5% (Brenes Balbuena *et al.*, 1988d). Consequently, the procedure can yield, after the necessary dilution, a practically colourless brine which retains the original lactic acid and salt and slightly less combined acidity, which favours its use in canning.

When regenerated brines must be stored for several days after their treatment, darkening can be prevented by addition of antioxidants. According to Brenes Balbuena *et al.* (1990), the greatest antioxidant action was exerted by 150 p.p.m. of SO_2, the maximum level permitted for olives in Spain, followed by the addition of 300 p.p.m. of ascorbic acid. Keeping the solutions in the dark was also favourable. There was a synergistic effect of ascorbic acid and SO_2.

The effect of UF on the polyphenol content in brine can be summarized as follows: there was a decrease in their concentration in the permeate ranging from 10 to 30% and an increase of 20 to 50% in the concentrate. However, important exceptions were vanillin and caffeic and 4-hydroxybenzoic acids, which were reduced by 55–80% during the UF process while their concentrations in the permeate were increased by about 450%. The brown colour of Spanish-style green olive brines, as deduced from the oxidation of the responsible compounds by CuO might be formed by polymers of catechin, vanillin and vanillic, caffeic and *p*-coumaric acids.

Industrial experiments (Figure 10.5) carried out by Garrido Fernández *et al.* (1992a) showed similar results. Additionally, previous flocculation of brines with 2 g/l of bentonite (from a 2%, w/v solution) was shown to increase flux by about 30% and maintain it better than with untreated brines. The industrial use of this will depend on the balance between the treatments costs and reduced investment (less filtering surface) that may be achieved.

Fig. 10.5 Ultrafiltration apparatus used in industrial experiments. (The apparatus was purchased from Paterson Candy Membrane Systems, Whitchurch, UK; illustration courtesy of Internacional Olivarera, S.A.)

In all cases, the cleaning operation took about 1.5 hours. An operator was required only at the beginning and end of the process. After this, flux recoveries showed considerable fluctuation, indicating that the correct application of the cleaning had a major influence on this and, consequently, must be carried out carefully.

A complete comparative study of the application of cross-filtration (after adsorption with activated carbon) and UF to the regeneration of Spanish-style olive fermentation brine is presented in Garrido Fernández *et al.* (1992a,b).

Influence of brine recycling on packing characteristics

Some physicochemical characteristics of activated carbon-treated, recycled brines used to pack olives are shown in Tables 10.29 and 10.30. A complete discussion of the effects of recycling on the characteristics can also be found in Garrido Fernández *et al.* (1992a,b). In general, combined acidity and pH are higher when the proportion of recycled brines increases and the proportion of activated carbon decreases. Values for packing solutions containing 70% recycled brines are excessively high, indicating that to use these levels would require a correction of combined acidity to reduce both combined acidity and pH to the levels usual with fresh brines. However, using 35% of regenerated brines, values were

Table 10.29 Effect of packing with activated carbon-treated regenerated brines on the physicochemical characteristics of the final product. Hojiblanca cultivar. Pasteurized, 2700 g cans. Plain olives (From Garrido *et al.*, 1992b)

	Regenerated brine				
	0.6%*		0.0%*		
Characteristics	70†	35†	70†	35†	Fresh brine
pH	4.55	4.40	4.57	4.42	4.20
Salt (g NaCl/l)	59	56	59	57	57
Titratable acidity (g lactic acid/l)	5.8	5.8	6.0	5.8	5.7
Combined acidity (mequiv/l)	82	81	99	93	76
Colour (A_{440}-A_{700})	0.36	0.37	0.59	0.48	0.34

* Percentage of activated carbon used in the regeneration; † proportion (v/v) of regeneration brine used in the final holding solution.

lower and could sometimes be acceptable, although a slight correction is advisable. In any case, the product must be preserved by pasteurization.

Brine colour was also affected, especially in those cases in which activated carbon was not used. However, when 6 g/l of carbon was used for regeneration, no practical differences were found, even at the higher proportion of recycled brine usage (70%). Sensory analysis applying preference ranking tests with both plain and pitted olives showed that there were no significant differences between those olives packed with recycled brines and those with fresh brines. Thus, it may be concluded that brine regeneration by activated carbon adsorption followed by a cross-filtration (or conventional filtration procedures) could be applied successfully for packing purposes. For packing with 70% of regenerated brine in the final holding solution about 6 g/l carbon is recommended. Depending on debittering and washing practices, the amount should be modified accordingly. A correction of excessively high initial combined acidity is always advisable. Lower levels of regenerated liquids in the packing solutions would require a lower proportion of adsorbent and combined acidity reduction.

When regeneration is accomplished by UF, effects on the physicochemical characteristics of brines and olives are similar to those for the brines regenerated with activated carbon. Table 10.30 shows some examples. In this case, olive colour was also studied and shown to be unaffected. However, there is a clear tendency to improve texture in cases where recycled brines were used. Addition of 150 p.p.m. of SO_2 had no significant effect on colour improvement in both fruits and brines. Ascorbic and citric acids also had no favourable effect. A complete description of the

Table 10.30 Effect of packing with ultrafiltration-treated regenerated or fresh brines on physicochemical characteristics of holding brines and fruits. Manzanilla cultivar (From Brenes Balbuena *et al.*, 1988c)

	5, 0.5%*		6, 0.6%*		7, 0.7%*	
Characteristics	Regenerated	Fresh	Regenerated	Fresh	Regenerated	Fresh
Polyphenols in brine (g tannic acid/l)	1.51	1.13	1.67	1.10	1.83	1.09
Brine colour (A_{440}-A_{700}	0.20	0.18	0.21	0.18	0.21	0.16
Olive colour index †	24.93	25.15	25.08	24.48	24.92	26.69
Texture (N/100 g) ‡	12.3	11.6	12.7	11.5	13.4	11.3

* First figure (5, 6, 7) refers to salt content; second figure (0.5, 0.6, 0.7) refers to titratable acidity.
† As defined in Chapter 12, in function of reflectance percentages at 560, 590 and 635 nm.
‡ Pitted olives.

effect of brine recycling on brine colour, polyphenol content, texture, and use of additives to improve brine and fruit colour is found in Brenes Balbuena *et al.* (1989a).

10.4 WASTEWATERS FROM TURNING COLOUR OLIVES

This type includes a wide range of olive processing systems. Those that are not based on an alkaline treatment produce less wastewater comprising only fermentation or storage brine and, sometimes, water from washing before packing. Thus, the brine regeneration procedures and conditions already described for Spanish-style green olives can also be applicable to the turning colour variety. Washing waters from the packing industry may be treated in the same way as those from the green olive packing industry.

However, the alkali treatment of turning olives is among the most polluting of processes, due mainly to the numerous washings applied to eliminate the excess alkali. California-style green-ripe olives have only very limited production and, consequently, their impact is relatively small. Their wastewaters are similar to those produced from California black-ripe olives processing, so that information relating to the liquid residues from these can be applied equally to the California green-style olives.

10.4.1 WASTEWATERS FROM BLACK-RIPE OLIVE PROCESSING (CALIFORNIA-STYLE)

The essential difference between the wastes from the Spanish-style and California-style processing is that with the former the greatest accumulation of wastes is produced during the harvest season, whereas black-ripe olive processing produces wastes more regularly throughout the year. The California-style does, however, give rise to a higher volume of wastewaters as a result of the greater number of lye treatments and washings used in the traditional process. So in a typical process, using a previous holding brine, three washings and immersion in ferrous gluconate or lactate solution, a total waste volume of 3.5 to 4.0 l/kg of fruit is likely. When different internal measures of control are applied, this volume can be reduced, but some processors prefer to use a higher number of washings because this is thought to enhance the quality of the final product. However, this increases the range of wastewater volume produced to between 1.5 to 6 l/kg of olives or even higher (Table 10.1).

Characteristics of the main wastewaters

Table 10.31 shows the general characteristics of the wastewaters from this preparation (García García *et al.*, 1990a), expressed in terms of

Table 10.31 Contaminant charge (COD) for liquid wastes from different stages of typical processing of ripe olives in brine. (From García García *et al.*, 1990a)

Type of waste	COD (g O_2/l)
Holding brine	35.0
1st lye	2.0
2nd lye	2.5
3rd lye	2.6
1st washing water	3.5
2nd washing water	3.8
3rd washing water	3.7
Ferrous gluconate solution	1.5

COD. The highest pollutant charge comes from the holding brine, while lyes and washing waters are relatively moderate. Other characteristics have been summarized in Table 10.32. In general, washing waters always carry a higher proportion of organic matter than lyes. This is directly related to the fruit contact time, which is longer for the washing waters (18–24 hours) compared with the lye (3–6 hours). Successive colour lyes must be in contact with the fruits for longer periods because they must penetrate deeper into the olive flesh. Consequently, third lyes are usually more heavily charged with pollutants than second lyes, which are more charged than first lyes. The behaviour of washing waters is similar.

Usually, those processes carried out with fresh fruits produce more polluted wastewaters. Previously stored fruits have already lost part of their water-soluble constituents when they arrive at the darkening process and thus the dissolution by lyes and washing waters is appreciably reduced. Increased temperature also increases the pollutant charge. Also, higher ratios of fruit to lye or washing water produce higher concentrations of pollutants.

A common characteristic of all the wastewaters produced during the darkening stage of black-ripe olives is an alkaline pH. Residual NaOH is found in the exhausted lye but is negligible in washing waters. This is due to the fact that the initial lye concentrations are lower than in green olive processing and most is neutralized by acids from the flesh or from the storage solutions.

The fermentation–storage brine has characteristics very similar to those from Spanish-style green olives discussed earlier.

Table 10.32 Characteristics of the main wastewaters from a typical darkening stage during processing of black-ripe olives. Raw material was fresh fruits. (From Garrido Fernandez, 1983)

Characteristics	1st treatment		2nd treatment		3rd treatment		Gluconate solution
	Lye	Wash	Lye	Wash	Lye	Wash	
pH	12.11	6.90	11.85	8.00	11.70	8.55	3.70
Free NaOH (g/l)	8.90	–	4.60	–	4.30	–	–
Dissolved inorganic solids (g/l)	2.38	3.18	4.34	7.21	6.88	8.14	7.328
Dissolved organic solids (g/l)	19.27	3.15	12.25	32.21	12.08	36.64	43.380
Polyphenols (g/l)	0.18	0.22	0.59	0.31	0.48	0.64	0.800

Internal control measures

Holding brines

Although the essential part of this process is the darkening stage, it is impossible to handle all the fruits during the harvesting season and it is necessary to store fruits for varying periods of time. Storage or holding brines present the highest level of contaminant charge of the process, although their volume is relatively low (0.5 l/kg of fruit) in comparison with the total amount of wastewater.

Dry storage was originally studied by Maxie (1964) and Woskow and Maxie (1965) but has recently been reassessed (Kader *et al.*, 1990). The obvious advantage here is that no storage brine is needed, reducing the overall pollution charge. It was found that Manzanilla olives cannot be stored below 5°C because of their high susceptibility to chilling injury. Green Manzanilla olives can be stored in air at 5, 7 or 10°C for 8, 6 or 4 weeks, respectively, before chilling injury or excessive ripening occur. The olives can be stored in 2% oxygen at 5 and 7°C for about 12 and 9 weeks, respectively, without significant quality losses. The relative humidity during storage must be between 90 and 95% to minimize water loss. Exposure to 5% or more CO_2 may cause physiological injury to olives. This system, unfortunately, needs proper cold storage facilities that are expensive and does not guarantee storage for the total period of time that the black-ripe olive industry requires. For these reasons, the method has not been adopted and most of the fruits for this type of processing are still stored in brine.

The main problem for brine disposal is its high level of NaCl. Possible alternatives to reduce this could be brine regeneration or the substitution of the salt by other compounds. Ralls (1967, 1968) and Ralls *et al.* (1971) looked at utilization of ammonium salts in an attempt to provide an alternative storage system which would reduce the pollution potential of the olive canning industry. Chemical salts used as fertilizers were partially successful for holding olives for several months without spoilage. Ammonium nitrate brines could be used for up to 15 weeks but it was difficult to remove completely the ammonium salt during preparation for canning. Brines containing ammonium sulphate (with the pH controlled by the addition of sulphuric acid) held olives for six months. The difficulties in completely removing ammonium sulphate from the olives limited the commercial potential of this type of storage. Food-grade glycols containing bacteriostats were used to store olives for periods of over 6 months. Olives submerged in propylene glycol–diethyl pyrocarbonate mixtures showed severe shrivelling on subsequent processing. Satisfactory results were observed with only the smallest-sized olives. Similar effects were seen with olives stored in a mixture of glycol–benzoic acid–sorbic acid. The most promising results were obtained with a mixture of water–propylene glycol–benzoic acid–potassium sorbate–HCl. By using KOH in place of NaOH for the chemical curing preparation steps, it was possible to prepare canned olives with essentially no generation of liquid wastes containing sodium ions.

Much work has been devoted to regeneration of brines for the storage of turning colour olives used to prepare this style. In the USA, experiments were carried out with activated carbon (Ginnette, 1967, 1968; Ralls, 1969) to regenerate the brine for further re-use in processing. In this case the objective was different from that already described for Spanish-style green olives brines. Popper *et al.* (1967) designed an activated carbon column to regenerate the black-ripe olive holding brines, taking the appearance of hydrogen sulphide as an indicator of depletion of the activated carbon. When this happened, the exhausted carbon was removed and replaced with fresh. Olive brine purified in this way was returned to processors to be re-used in brining fresh olives for the following season. There was no significant difference between olives stored in the recycled brine and controls.

Exhaustive work was carried out by Ginnette (1967, 1968) who studied the corresponding adsorption isotherms as well as the behaviour of the different organics during the adsorption process. Changes can be summarized as follows: titratable acidity decreased; pH rose; COD fell dramatically at the beginning of the regeneration and, later, after each new addition of fresh activated carbon; colour was also removed very efficiently. However, suspended solids were about the same before and after the treatment due, at least in part, to the contribution from the activated carbon itself. The level of all characteristics increased progressively with the working time. This problem can be solved by working in countercurrent with a series of columns.

The most important drawback with this regenerated brine was its instability. In spite of the activated carbon treatment and the removal of a great proportion of the organic matter, a considerable amount of this still remains in the reconditioned brine. When it was stored, microbial activity developed in the brine, sometimes generating off-flavours. To increase its stability, several additional treatments have been tried. One of these consisted of vigorous aeration of brine before regeneration. The process eliminated all the sugars and acids and diminished the COD by about 20%. However, treatment also produced an increase in the content of suspended solids due to the microbial growth, which was mainly by moulds, and markedly darkened the solution. Application of UF before or after decoloration gave an important reduction in the organic charge, reducing it by two-thirds. Despite the noticeable charge decrease due to these treatments, the regenerated liquid still contained about 2% of dissolved organic charge. This can be used slowly by environmental microbial contamination as a carbon source and, in this sense, the brine stability was not very different from that of untreated storage brines. To achieve good preservation, it was necessary to add preservatives. Sodium benzoate at a level of 0.1%, SO_2 at 500 p.p.m., or increasing the salt content to 20% were effective at ensuring brine stability. Alternatively, problems related to brine storage can be solved by regenerating the brine just before use in the new process.

Many industrial trials were conducted by Ralls (1969). Black-ripe olives (California-style) prepared after storage in regenerated brines were of similar quality to those prepared from fresh brines. Use of the method, which is technically viable, has not been implemented in the California industry because of the risk of off-flavour production and, more importantly, because of the development of the salt-free storage referred to in Chapter 8.

Salt-free storage

A solution to the problems of holding brine disposal is its substitution by salt-free, acidified solutions (Vaughn, 1967, 1969) which apparently do not dissolve too much organic matter; in addition, the dissolved matter is easily biodegradable (Fernández Díez *et al.*, 1985; Unidad Estructural de Investigación de Biotecnología de Alimentos, 1990). The method consists of using solutions of lactic and/or acetic acids with preservatives such as sodium benzoate or sorbate (Vaughn *et al.*, 1969). The pH value is adjusted to around 3.5. Although these solutions do not have the problem of the presence of NaCl, they have a pH of about 3.5–4.0 and the acids themselves represent a high BOD_5. Although this can be degraded easily under aerobic conditions by moulds and yeasts, the presence of preservatives could interfere. Re-use of these acidic solutions is possible when the acids are adjusted to their initial levels. Nevertheless, some problems related to their storage have been observed. The high concentrations of acids are not enough to guarantee their stability during the period from the end of one season to the beginning of the next. Furthermore, the presence of preservatives does not prevent all microbial growth. Vaughn (1969) reported the possibility that some micro-organisms could use sodium benzoate up to 0.3% as a carbon source for their growth.

Storage must be under strictly anaerobic conditions to avoid growth of oxidative yeasts and moulds. Their presence could increase the pH through utilization of added organic acids and thus open the way to other malodorous spoilage. When their growth is rapid, these micro-organisms can also excrete pectinolytic and cellulolytic enzymes, the presence of which can cause softening during storage.

Lye and washing waters re-use in the darkening stage

As was shown in Chapter 8 and Table 10.32, the exhausted lye solutions retain a proportion of their initial NaOH content, especially after the first treatment, which only penetrates the skin. The residual level in the rest of darkening lyes is lower because the initial NaOH level is also lower and because it must neutralize the acids in the flesh and those added to the storage brine or formed during storage. In general, NaOH decrease may range from 10 to 50%. Re-use of such solutions may thus save chemicals and costs and has become a general practice in the United States and Spain (Vaughn, 1974; Garrido Fernández, 1983, 1984).

Re-use affects the pollutant charge of some of the remaining wastewaters. As an example, Table 10.33 shows the characteristics of some washing waters from black-ripe fresh fruit processing in which lye and lye and washing waters were re-used. The behaviour of the dissolved organic and inorganic solids, polyphenols, etc. followed a similar trend to that found with green olives. After three uses, the lye organic matter had increased substantially by up to three-fold. As with green olives, the increase declined with the number of uses and further re-uses produced only a slight increase with respect to that of the preceding cycle. After the eighth or ninth re-use, the organic content remained almost constant.

Pilot plant experiments and industrial trials have shown that the exhausted first lye retains a NaOH concentration similar to that needed for the second one. Thus, it could be used directly in the following cycle. Further cycles require addition of about another 50% of the initial NaOH

Table 10.33 Effect of lye and and washing waters re-use on the pollutant charge of wastewaters from the darkening stage of black-ripe olives. Fresh fruits (From Garrido Fernández, 1983)

	Lye re-use*			Lye and washing waters re-use†	
Characteristics	Lye	2nd washing	Gluconate solution‡	Lye	Gluconate solution
pH	11.5	7.8	7.4	10.8	5.0
Free NaOH (g/l)	6.7	–	–	5.4	–
Organic dissolved solids (g/l)	9.68	5.32	8.32	7.52	7.48
Inorganic dissolved solids (g/l)	26.93	31.42	38.74	27.32	34.56
Polyphenols (g/l)	0.34	0.26	0.91	0.45	0.75

* Process in which only lye was re-used; † Process in which lyes and washing waters were re-used; ‡ Gluconate solutions prepared from last washing which in † had been re-used from the initiation of the process.

content. Specific re-use practices must, however, be implemented in accordance with the industrial facilities and washing schedule.

Lye re-use should be applied taking into consideration those basic measures already mentioned with respect to green olives. Suspended solids (leaves, pits, pieces of olives, etc.) should be separated before the exhausted lye is re-incorporated into the lye stock tank. Filtration is strongly recommended. The holding tank should be provided in the bottom with a device to separate the settled solids. Special attention must be paid to microbial contamination. These solutions are less concentrated than lyes from green olives and have, in general, slightly lower pH. NaOH concentrations should be raised as soon as possible and solutions returned to stock solution containers directly, without any intermediate storage. Despite the potential risk, there are no reports of spoilage caused by lye re-use. Re-used lyes must, in any case, be handled carefully to prevent any environmental contamination and the NaOH concentration maintained as high as compatible with their use.

Re-use of washing waters is more difficult because the pH and NaOH contents are markedly lower than those of lyes and microbial growth is therefore easy. In fact, as commented in section 9.4 on black-ripe olive spoilage, most of the spoilage in this style is produced during the washing periods and is favoured by a long immersion in the washing waters. If this coincides with high temperatures, then spoilage is almost certain. Nevertheless, re-use is possible, but strict measures related to pH control and microbial contamination must be implemented. A drastic pH decrease with a strong acid is not enough to prevent completely the growth of yeasts and moulds (on the surface). A low pH could also be a serious drawback for washing waters usage in following cycles because it can cause discoloration of the olives and interfere in the darkening process. A moderate pH (4.0–4.5) would not be enough to assure the stability of these solutions. In this case, the only possibility would be to apply a pasteurization before addition to lye-treated olives. The use of continuous pasteurization could improve yield and reduce the cost and handling problems. Although the procedure is viable, it must be emphasised that washing water re-use is an operation with a high spoilage risk, if not properly conducted.

Table 10.34 shows the colour and texture of black-ripe olives darkened by the normal process and with re-use of lyes only and with re-use of lyes and washing waters. As is evident no effect on quality is apparent; only fresh fruits showed a worse colour, but this was not consistent throughout all treatments.

Neutralization of washing waters

Washing waters could be reduced if the excess of alkali were progressively neutralized as it is leached from the fruits. Traditionally, a strong

Table 10.34 Effect of lye and washing solution re-uses on colour and texture of black-ripe final product. Hojiblanca cultivar (From Garrido Fernández, 1984)

	Normal process		Lye re-use		Lye and washing re-use	
Characteristics of raw material	F	G	F	G	F	G
Colour*						
Fresh fruits	25.6	23	26	20	32	23
Storage 4 months	15	13	18	14	18	13
Texture†						
Fresh fruits	76	75	75	77	73	76
Storage 4 months	78	82	78	80	81	73

* Expressed as percentage of reflectance at 700 nm; † Expressed as N/100 g pitted olives; F, final packing with fresh brine; G, final packing with ferrous gluconate solution.

acid such as HCl or H_2SO_4 has been used. Their use has the disadvantage of producing large and rapid pH changes. Any excess may drop the pH too low and this may discolour the fruits and interfere with the darkening process. By using such acids, however, only one washing can be given after each lye treatment while, without acid, the number of washings, especially after the third lye, are numerous if complete elimination of the alkali is sought.

Other neutralizing agents have been introduced in the past 20 years. Of special interest is carbon dioxide (Moffat, 1975; Mondok, 1979), which has the following advantages:

- it is a weak acid, which does not produce a rapid pH decrease and stabilizes the pH at around 6.8–6.9, releasing the excess to the atmosphere;
- it is very efficient in neutralizing the alkali, even in the olive flesh, because – as a gas – it diffuses easily through the skin and flesh while any other acid diffuses only by osmotic pressure forces; and
- it has an inhibitory effect on some undesirable microorganisms that usually grow during the washing period.

The use of CO_2 can diminish the total volume of wastewater appreciably, as well as the time needed. A large time reduction, however, is not possible because, in this case, the neutralization may occur more rapidly than darkening. Fruits could be well neutralized but have developed a poor colour. A balance between both reactions is necessary. According to Mondok and Moffat, application of CO_2 neutralization can reduce the darkening process to only 3 days, increasing production by about 30% and reducing wastewaters by about 80%, at least in California. In other countries, where the water is still available and where there are few restrictions on wastewater generation, the use of carbon dioxide has not been introduced.

Wastewaters have been reduced drastically in the past two decades. Nowadays it is difficult to imagine a process that might produce about 34 l of liquid residues per kg of processed fruits, as was usual in California before environmental control. In Spain, many industries can generate about 2–3 l/kg because of the reduced number of lyes and washings and the partial neutralization of these with strong acids. Finally, it must be emphasized that water reduction in black-ripe olive darkening must be accomplished with care because these waters also dissolve both organic and inorganic compounds with important influences on the olive flavour. A higher than normal concentration in fruits or brines could lead to changes in flavour and eventual quality losses. Because of this, some processors are reluctant to employ re-use and neutralization with acids.

Equilibrium brines and ferrous gluconate or lactate solutions

The characteristics of the ferrous gluconate or lactate solutions are shown in Tables 10.31–10.33. As with other wastewaters from black-ripe olive processing, they have an important organic and inorganic charge that is difficult to remove, and, in addition, they carry a proportion of ferrous or ferric salts. The presence of the cation represents a new element which is difficult to separate because some is complexed with the polyphenols, giving a very dark colour. Usually, authorities in charge of waste control have serious prejudices against the disposal of such wastes. One internal control measure to save a volume of wastewater per kg of olives is the addition of the iron salts into the brine in which the fruits are usually immersed before or after the colour fixation. Sometimes, the last washing water is added with salt and the ferrous salt to prepare simultaneously the equilibrium brine and the fixing colour solution.

A greater reduction of waste liquids is achieved by the addition of ferrous gluconate or lactate directly to the packing solution. In this case, the disposal of the fixing colour solution is completely avoided. This method, however, gives the holding solution a more intense brown to black colour. Its use will depend on consumers' reaction to this brine coloration when cans are opened. Some industries use this procedure habitually. The same precautions mentioned with other internal wastewater measures must also be considered here. Any microbial contamination may be dramatic in its effect in this last processing stage. Thus, any re-use must be done only after a careful consideration of the advantages and risks.

Wastewaters from the retorts

Condensation waters from retorts, can sterilization water and cooling waters are not at any time in contact with the olives and so do not carry any organic pollution. A first method of saving water in these processes is the correct selection of the equipment. Some new retorts are computer-controlled and can use a reduced volume of water during sterilization. Also, such waters may be separated from other factory wastewaters, cooled and re-used in subsequent operations. The heat absorbed by the water can be used in operations such as brine preheating and heating the factory during winter.

In any event, the excess or those with excessive coloration, normally due to the presence of iron salts or hydroxides, can be accepted in public sewers without problems in most countries, at least currently. The direct incorporation of such wastewaters into the receiving waters may increase their temperature and damage the microflora.

Process modifications to reduce wastewaters

Some modifications of the darkening process to reduce the volume of wastewater are possible. In general, the reduction in the number of lyes implies a parallel reduction in the number of washings. Garrido Fernández (1979) studied the use of only one lye treatment using different lye concentrations, aeration and washing conditions. A slow colour development was observed with a less intense final black tone. The process needed about the same aeration period that is necessary when a higher number of lyes are used. Best results were obtained when lye was substituted, after pit penetration, by fresh water or weak brine. The best pH range in the surrounding liquid was 4.5–5.0 and by maintaining it at this level an adequate black colour was obtained. Optimum lye concentration was about 1%, since lower levels had problems with penetration into the flesh. Different times of addition of the gluconate solution, process duration, etc. did not improve the results. It was found that good results were obtained only with contents between 0.06 and 0.10% (w/v) ferrous gluconate. Re-using the aeration waste solutions as holding solutions in the final packing gave a marked enhancement in the colour of the packed olives.

The process with only one lye needs less washing waters because the quantity of excess alkali in the flesh is lower. Maintaining the pH between 4.5 and 5.0, it is possible to control growth of Gram-negative bacteria and have little spoilage risk. Re-use of the washing water as packing solutions also gives a drastic wastewater reduction. Thus, the overall process is markedly less polluting.

It is applied industrially, especially to fruits stored for 3 months or over. Fresh or recently harvested fruits develop colour more slowly and need several lye treatments.

In the light of current knowledge on colour precursors and oxidation kinetics, an optimization of this process may be obtained. Some possible modifications are envisaged after the work carried out by García García *et al.* (1991) who reported on the oxidation rate of olive darkening with only one lye with different lye and washing conditions. They have found that, with appropriate control of pH and temperature, the black surface colour is rapidly obtained and that the whole process can take about only 1 day. From the point of view of the washing waters such modifications represent a marked reduction in the volume generated.

10.5 WASTEWATERS FROM NATURALLY BLACK OLIVE PROCESSING

This is a case of low wastewater production during processing. The initial washing waters, used to clean the fruits before brining, are disposed of without difficulty due to their reduced volume. Fermentation brines are the only waste liquids produced during this type of processing that pose a problem.

10.5.1 CHARACTERISTICS OF THE FERMENTATION BRINES

Parameters describing the pollution potential of these wastewaters are shown in Table 10.35. In general, they are liquid residues of moderately acidic pH, with a high content in dissolved polyphenols (about 5 g/l, mostly anthocyanins), and higher levels of organic matter than in any other type of table olives processing. This is due to the maturation stage of the raw material. Usually, as fruits mature, the proportion of water-soluble materials increases. The organic concentration of these brines is

Table 10.35 Pollutant characteristics of fermentation brines from naturally black olives (From Garrido Fernandez, 1978)

	Brines from different origins			
Characteristics	6N-3	6N-12	H-8	H2-8
pH	4.40	4.40	4.25	3.60
Salt (g NaCl/l)	56.0	59.0	77.0	76.0
Organic dissolved solids (g/l)	95.4	118.8	95.7	95.3
BOD_5 (g O_2/l)	37.9	34.8	38.3	–
Polyphenols (g tannic acid/l)	4.4	4.2	5.1	3.2

close to those reached in olive oil mill wastewaters, although these may be higher in cases where very traditional systems are applied and where the concentration can reach up to 150–175 g/l. Bearing in mind that all the residual organic matter of these brines has undergone fermentation over a long period (3–9 months), it might be thought that the remaining substances are difficult to degrade by any conventional purification process. This is not the case however, as shown by their high BOD_5 and the fast decrease of the dissolved oxygen when they are disposed of into a watercourse. They are thus heavy potential contaminants.

The salt content of these brines can vary from factory to factory, but, in general, it is higher than in any other process, ranging from about 6 to 10% (w/v) or even higher. Such concentrations without dilution can cause serious problems for conventional municipal wastewater treatment facilities. Authorities are then reluctant to accept such liquids into their sewers.

10.5.2 RE-USE OF NATURALLY BLACK OLIVE BRINES

There are numerous cases of washing waters re-use in the processing of many vegetable products. The same techniques can be applied to the production of naturally black olives. The regeneration procedure must remove mud, leaves, broken olives, etc. Screens are useful for this purpose. Sedimentation can effectively eliminate most of the other contaminants. Recuperated waters must be chlorinated to diminish their microbial population, as well as that on the olive surface.

Bearing in mind the characteristics of the naturally black olive fermentation brine and its reduced volume, an immediate internal control measure could be its re-use either for new fermentation or for packing the final product. In any case, some treatment to reduce the dissolved organic matter is advisable, especially when brine must continue to dissolve the bitter principle. Garrido Fernández (1978) tried its partial purification with chlorine, a chemical oxidant, and adsorption with adsorbent earths and activated carbon.

Dissolved polyphenols were easily oxidized by the addition of chlorine. In the case of naturally black olives, it produced an abundant precipitate as the addition progressed. Apparently, the partially oxidized compounds are insoluble. Their progressive elimination as they are produced could diminish the chlorine needed for their complete oxidation. Efficiencies of polyphenol destruction diminished as the residual content was lower and was always higher at acidic pH values. Working under these conditions, the final concentration of polyphenol can be reduced to about 0.4 g/l, 100-fold lower than the initial content. By using this method with the appropriate amount of chlorine, complete removal of the red colour produced by the anthocyanins is possible. Regenerated brines have a slight yellow colour, a moderate odour of chlorine, and a characteristic sweet flavour.

Use of this procedure at the industrial level is still conditional on a deeper study of possible chlorinated residual compounds remaining in the brine and their potential toxic effects. Fermentation trials of naturally black olives using brines regenerated by this method (Durán Quintana and Garrido Fernández, 1978) showed a marked inhibition of the normal yeast microflora, even with respect to other processes with recycled brines. Apparently, some of the compounds formed must be inhibitory to the yeasts.

Treatment with decolorizing earths was slightly more efficient than in the case of green olive fermentation brines, but the amount required to obtain enough decoloration was so high that the process was not applicable.

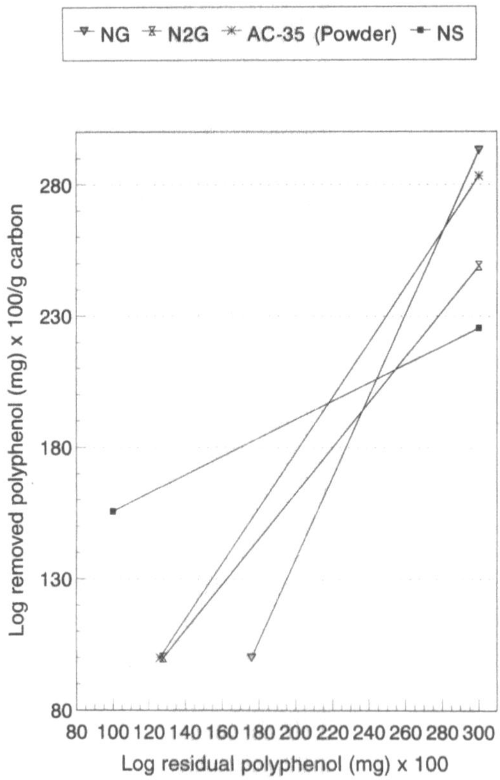

Fig. 10.6 Adsorption isotherms for different types of activated carbon with naturally black olive brines. (From Garrido Fernández, 1978.)

Treatment with activated carbon gave good results. Figure 10.6 shows the adsorption isotherms for some commercially available activated carbons at room temperature and pH 4.5, which is a normal level for brines of this olive type. They are expressed with respect to the polyphenol elimination. There are appreciable differences between the different activated carbons, so preliminary trials are essential before applying one in practice. Slopes are high in all cases, which means that the counter-current treatment in series, both batch or column, could cut costs. To study this point in detail, Table 10.36 shows their relative efficiencies and the proportion of carbon needed to obtain a 75% reduction in the polyphenol content. The highest efficiency was found in the case of NS, which, in consequence, needs the lowest proportion in the case of a single contact in batch. However, when two counter-current contacts are given, calculations showed that AC-35 (powdered) was better due to its higher adsorption isotherm slope. Bearing in mind the higher organic content of these brines, the proportions of carbon needed are not too high. Nevertheless, it must be emphasized that the residual polyphenol content of 1 g/l is enough to give an intense red to purple colour to brines.

Apparently, there is a selective adsorption of polyphenols compared with COD, as happened in green olives, since the first are removed at a faster rate than the second. Differences in colour are not easily appreciable in the naturally black olive brines because of the high content in colouring substances and the fact that this characteristic is only affected when the residual polyphenol content is very low. Consequently, the carbon adsorption phenomenon is expressed better by means of its effect on polyphenols.

No studies of possible selective adsorption of polyphenols are available, but analysis of oleuropein showed that its elimination paralleled that of polyphenols in general and that it was reduced in the same proportion. Thus, brines treated in this way may also be suitable to continue the debittering of olives immersed in them. Due to the adsorption of

Table 10.36 Relative efficiencies of different activated carbons and proportions (g carbon/100 ml solution) needed for a 75% reduction* in polyphenol content. Naturally black-ripe olives (From Garrido Fernandez, 1978).

		Carbon proportion needed	
Activated carbon	Relative efficiency	One contact	Two contacts†
NG	1.00	12.56	4.40
N2G	1.92	6.52	3.06
NS	3.42	3.67	2.27
AC-35 (powder)	2.98	4.22	1.81

* Reduction from 4 g/l to 1 g/l; † Counter-current process used.

acidic components, pH increases as the brine is treated with increasing proportions of adsorbent. Re-use may thus need a pH correction with lactic acid or HCl. Acetic acid can only be used in moderate concentrations, except when it has to contribute to the final flavour. Sodium chloride levels are not affected by regeneration.

Modifications of the adsorption conditions produced only slight changes in the corresponding isotherms compared with those in Figure 10.6. This is of interest because it means that no prior conditioning of the brine is necessary, which can save costs and handling operations.

Fermentative processes carried out with regenerated or diluted brines were normal, although the microbial counts (yeasts) were lower than in fresh brine fermentations. A higher proportion of gas-pocket formation was also observed. Final brines were more highly contaminated than those from the first cycle and from regenerated brines used as initial brines. The following year, problems could be far greater than those found with brines of only one cycle. Regeneration of brines, to be used for a third fermentation, would yield an excessively concentrated reconditioned brine which could not be used. Consequently, the best use for the activated carbon regenerated brines is for the final packing. If regenerated brines are used with only this aim, the regeneration process would require only those treatments necessary to achieve a final product with similar characteristics to that obtained by the traditional process. The proportion of activated carbon used may thus vary accordingly. With enough sweet fruits, the reconditioning process could consist simply of a filtration through a 0.45 μ filter to remove the microbial charge and suspended solids. This is the case with fruits prepared by the aerobic system, which are usually very sweet after fermentation.

10.6 WASTEWATER PURIFICATION METHODS IN THE TABLE OLIVE INDUSTRY

In the preceding paragraphs, we have described the characteristics of the different table olive wastewaters and internal control measures and process modifications which can be used to reduce the volume or charge of such liquid wastes. As was noted, the characteristics of the olive wastewaters limit the effectiveness of classical purification systems. The two most important of these were the high polyphenol content (oleuropein and derivatives with green or colour turning olives, and anthocyanins in the case of naturally black olives) and sodium (from the NaOH in the lyes and from the NaCl in the brines). This has made it impossible, at least until now, to discharge such wastewaters into city sewers – the easiest disposal method from the processors' point of view. The only report of this possibility was by Batina (1969), who treated mixtures of municipal wastewaters containing up to 7% of table wastes by conventional methods, obtaining a 90% reduction in BOD_5. However, in view of the concentration of these industries in only a few cities and the high BOD_5 of most of the effluents, it is clear that, under normal conditions, it would be impossible to achieve the necessary dilution.

10.6.1 PURIFICATION OF ALKALINE WASTEWATERS

Castro Gómez Millán *et al.* (1983) tried partial purification of re-used lyes from debittering the Gordal cultivar by lowering the pH to 7.0, 4.5 and 2.7. In all cases, an abundant precipitate was formed. Once the solids were separated, the remaining solution (acidified to pH 2.7) contained about 46% of the initial solids in solution. Inorganic solids were separated in a higher proportion (about 90%) than the organic (36%). Reducing sugars were eliminated by about 40%, but polyphenols declined by only 19%. The overall COD removal was about 80%. Effectiveness increased as the pH was lowered.

García García and Garrido Fernández (1984) studied the same procedure in lyes and washing waters from the normal processing of the Manzanilla cultivar. Results (Tables 10.37 and 10.38) were essentially the same as with Gordal, although decrease of the pollutant charge was lower. At room temperature the organic matter removed varied between 2.4 and 3.8 g/l, with an average of about 20%. The COD followed about the same pattern. When the treatment was done at 0°C (in order to reduce solubility) the organic matter removal ranged from 4.5 to 8.9 g/l, which is a fairly good removal efficiency. The average percentage of pollutant charge elimination was between 36 and 39%.

It was interesting to note that inorganic solids decreased slightly when the treatment was at 25°C, but markedly diminished at 0°C, probably due to the temperature effect on solubility. In the case of washing waters, a higher elimination of inorganic matter was observed both at 25°C and 0°C, but was more pronounced at the lower temperature. In contrast, organic charge was almost not removed at all at the lowest temperature.

García García *et al.* (1989) applied wet oxidation to the aqueous effluents from the olive industry. Working in a pressurized reactor at 200°C and 35.15 kg/cm^2 pressure, 99% of the dissolved polyphenols disappeared in 30 minutes in the presence of excess NaOH. The reaction rate was increased when agitation and mass transfer were improved. Other organic matter, however, was not oxidized in the same proportion since organic carbon was only reduced by 75%. In order to increase the reaction of other organics in solution, García García *et al.* (1990c) also tested the jet-reactor technology. In these cases, when pure oxygen was used, the oxidation was more intense. Polyphenols were degraded at the same rate as in the reactor, but the rest of the organics were oxidized more quickly because the final treated lye or washing water had only about

Table 10.37 Effect of pH correction (reduction by up to 4.5 units) and temperature on the pollutant charge of lyes from Spanish-style green olive debittering. Manzanilla olives. Normal process (From García García and Garrido Fernández, 1984)

		Treated lyes							
		26°C				0°C			
Characteristics	Original lye	A	B	C	D	A	B	C	D
pH	12.3	5.0	5.8	5.7	4.6	4.6	4.6	4.5	4.4
COD (g O_2/l)	14.6	11.7	11.1	11.1	13.8	11.9	10.2	12.2	10.0
Dissolved inorganic solids (g/l)	17.3	13.5	14.1	14.7	14.9	12.8	11.2	11.4	8.4
Dissolved organic solids (g/l)	13.1	14.4	14.6	13.9	13.8	11.9	10.9	12.3	11.4
Total dissolved solids (g/l)	30.4	28.3	28.6	28.6	28.7	24.7	22.1	23.7	19.8

A, treated only with HCl; B, treated also with a coagulant; C, treated also with a flocculant; D, treated also with a mixture of coagulant and flocculant
COD, Chemical Oxygen Demand.

Table 10.38 Effect of pH correction (reduction by up to 2.7 units) and temperature on the washing waters from Spanish-style green olives. Manzanilla olives. Normal process (From García García and Garrido Fernández, 1984)

		Treated washing waters							
		26°C				0°C			
Characteristics	Original washing water	A	B	C	D	A	B	C	D
pH	8.5	3.0	2.7	2.7	3.8	2.4	3.0	2.6	3.3
COD (g O_2/l)	22.2	20.4	20.0	20.3	19.5	20.5	21.0	20.4	21.0
Dissolved inorganic solids (g/l)	54.3	53.5	53.7	54.1	50.7	53.5	54.4	53.6	55.0
Dissolved organic solids (g/l)	26.5	14.6	14.7	16.7	13.6	13.9	13.8	13.9	13.9
Total dissolved solids (g/l)	80.8	68.1	68.4	70.8	64.3	67.4	68.3	67.5	68.9

A, treated only with HCl; B, treated also with a coagulant; C, treated also with a flocculant; D, treated also with a mixture of coagulant and flocculant
COD, Chemical Oxygen Demand.

10% of the initial organic carbon content. This was in agreement with previous results that have shown that the reaction was limited by the rate of oxygen transfer. An example of the effectiveness of this method is shown in Table 10.39. Despite its high efficiency, jet-reactor technology still produces a treated liquid with a COD and BOD_5 about 10% and 17–26% respectively of the initial values. Higher temperatures may have led to a more complete removal of contaminants by completing the oxidation.

Chemical oxidation of wastewaters is still not available at the industrial level in the table olive industry, but is a promising technology. Partial precipitation of the dissolved solids in re-used lyes and washing waters from green olives can be used in those cases where municipal authorities are tolerant, but this entails pre-treatment by the processor. In those cases where wastewater treatment is the responsibility of the processor (evaporation ponds, forced evaporation, etc.) precipitation only adds cost. Normally, in such situations, lyes and washing waters are disposed of without any previous correction.

The same methods can be applied to lyes and washing waters from the darkening process of black-ripe olives, but efficiencies are significantly lower since the total organic charge and inorganic solids in solution are lower.

10.6.2 TREATMENT OF WASTEWATERS FROM SPANISH-STYLE GREEN OLIVE PACKING

There are two major stages in the production of wastewaters during the processing of Spanish-style green olives: alkaline waters (already discussed) and acidic brines. When olives are packed, the acidic fermentation brines are usually discarded into the general sewer of the factory where they are mixed with the other wastewaters coming from the washing of olives, jars, etc. and from the sanitary installations. The characteristics of such brines were shown in Tables 10.2, 10.15 and 10.18. However,

Table 10.39 Reduction of contaminant charge of treatment lyes by wet oxidation in a jet-reactor, at 250°C and 35.1 kgf/cm^2 (From García García *et al.*,1990c)

Characteristic	Initial lye	Final lye
Polyphenols (mg tannic acid/l)	2700	29
Organic carbon (g/l)	18.1	2.9
COD (g O_2/l)	47.3	5.6
BOD_5 (g O_2/l)	9.6	2.5

BOD_5, 5-day Biological Oxygen Demand.
COD, Chemical Oxygen Demand.

after dilution in the sewer, the general wastewaters of the packing factory have a lower pollutant load which is shown in Table 10.40. Titratable acidity has been drastically reduced. Chloride and sodium remain high (about 7 g/l), far above the upper limits fixed for these parameters in the legislation of the USA and EU. COD and BOD_5 have been reduced, but are also very high with respect to normal municipal wastewaters.

A first operation with all wastewaters from the table olive industry where pitting, pitting and stuffing, and slicing operations are involved is the elimination of suspended solids, including oil. This is achieved by screens or sieves (Figure 10.7) and equipment especially designed to remove oil by flotation. There are many options available on the market and no special emphasis will be given to this operation.

Brine treatment in other fermented products such as cucumbers does not use the lactic acid produced in the final packing brine since this is usually prepared with vinegar. Thus, regeneration processes for these products were designed without considering the lactic acid in solution and some purification processes in these industries are also thought to remove lactic acid. Stevenson *et al.* (1979) investigated the aerobic fermentation of pasteurized pickle process brine by *Candida utilis* NRRL Y-900. They found that after a lag period, the organism grew well in the brine. The lag period was reduced substantially by the addition of 0.05% Na_2PO_4. Under optimum conditions, the culture reduced the BOD_5 by 91% within 24 hours with a concomitant production of 1–2 g yeast dry matter/l of processed brine. In experiments using unpasteurized brines, *C. utilis* was unable to grow as rapidly as some of the naturally occurring yeasts. Two yeasts identified as *Pichia* spp. grew in the brine with a shorter lag period and a faster growth rate than *C. utilis*. A similar process could be applied to Spanish-style green table olive brines to remove the

Table 10.40 Average pollutant charge of wastewaters generated in Spanish-style green table olive packing (Adapted from Brenes Balbuena *et al.*, 1989b)

Characteristic	Value
pH	3.8–4.5
Titratable acidity (g lactic/l)	0–1.7
Suspended solids (g/l)	0.7
Oil (g/l)	0.6
COD (g O_2/l)	3.4
BDO_5 (g O_2/l)	1.6
Chloride	7.0

Fig. 10.7 Screen used to remove suspended solids in the table olive industry. (Picture courtesy of Cooperativa Sor Angela de la Cruz de Estepa, Sevilla.)

lactic acid, which accounts for 75–85% of the BOD_5 of such brines. The treated brine will probably need further treatment, together with the rest of the wastewaters, before appropriate pollutant limits could be obtained, particularly those applying to inorganic contaminants.

García García *et al.* (1990b) adjusted the pH to 11 with NaOH or $Ca(OH)_2$. In both cases the formation of heavy, slow-settling flocs was observed. The effect of this treatment on the pollutant charge is described in Table 10.41. Reduction of the organic charge increases with the pH, although differences between pH 7.0 and 8.5 were negligible (averages of 19% and 22%, respectively). However, when the final pH was 11.0, the increase was more evident (average charge reduction of 26.4%). Addition of coagulants (aluminium or ferric sulphate) at the same time as the pH correction produced only a slight improvement of the removal efficiency, between 2 and 10% with respect to that obtained with just pH correction.

Table 10.41 Effect of pH increase on pollutant charge elimination from original wastewaters produced by a Spanish-style green table olive packing industry (From García García *et al.*, 1990b)

	Percentage COD reduction after each treatment		
COD (g O_2/l)	7.00*	8.5*	11.0*
2.22	10.9	23.3	
2.26	5.9	16.0	20.2
4.33	28.0	29.7	36.3
4.44	24.9	26.7	27.9
4.82	21.9	24.3	26.7
Average	19 ± 5	21.8 ± 4.6	26.4±4.5

* pH.

Optimum levels of addition were about 200 p.p.m. When the chemical added as flocculant was Prodefloc-AC-190, the optimum level of addition was higher (up to 500 p.p.m.). The response of wastewaters was diverse and removal efficiencies were variable, and did not correlate with any of the characteristics studied. Addition of cationic or strong polyelectrolytes as flocculants did not improve the precipitation rate when used up to 8 p.p.m. However, a moderately anionic flocculant (Praestrol 2.540 TR) at a concentration of 0.3 p.p.m. appreciably reduced the settling time, independently of the final pH. This effect is significant because it could mean a reduction in the decanter capacity and in retention times. A final treatment of the resultant solutions with ozone or chlorine may be of interest from the sanitary point of view, but causes only a limited further reduction in the soluble organic matter, measured at about 3%.

From the above data, it is evident that this physicochemical treatment alone does not reduce the pollutant charge from the packing industry to acceptable levels, according to current EU and US environmental laws. Additional or complementary measures must be implemented for a more complete treatment. One of these is based on the fact that fermentation brines represent only about 22% of the total final wastewaters, but account for approximately 70% of the pollutant charge. Thus, if these brines are segregated from the rest of the wastewaters, a substantial decrease in most of the indices (COD, BOD_5, dissolved organic solids, etc.) will be expected. Also, sodium and chloride contents will be lower, something that is of great interest because these chemicals are not removed by biological or physicochemical treatments. In this case, the physicochemical treatment described above could be sufficient to achieve a pre-treated wastewater acceptable to local authorities. Brines could be regenerated by any of the previously described procedures and the remaining concentrated solutions evaporated in shallow ponds, conventional evaporators, etc.

Another option is a biological treatment to remove the excess organic charge, though it does not remove inorganic components. There is no experience of this up to now, but the main problems will be related to polyphenol degradation, which will need the appropriate dilution or organism, and the salt level, which will have to be reduced or require halotolerant micro-organisms. On this last aspect, Leoni *et al.* (1976) tested the purification of the wastewaters from fruit and vegetable packing with salt levels of up to 10%, obtaining a BOD_5 reduction of up to 90%. When salt levels were below 3–4%, organisms were of the genera *Epistylis*, *Opercularia* and *Verticella* at the beginning, followed by *Paramecium*, *Stylonychia* and *Aspidisca*. As the salt levels were increased, the microorganisms were reduced and above 5% salt only *Aspidisca* was present. At 10% salt no protozoa were present and only lower organisms grew. This, however, only reduced the system performance from about 95% to 84%. Consequently, it is possible that the table olive industry wastewaters could also be treated by a well-adapted biological method. Its optimization, however, needs further work.

10.6.3 TREATMENT OF WASTEWATERS FROM THE TURNING COLOUR OLIVE INDUSTRY

Directly brined (lye-untreated) olives produce a smaller volume of liquid wastes. The comments relating to brines in the Spanish-style green olives section can also be applied here and to the holding solutions of the black-ripe (California-style) olives. If the latter are acidic solutions, their components are easily degraded in an aerobic process.

The main problem within this olive style arises from the wastewaters produced during the darkening and packing operations. Their characteristics were given in Tables 10.31–10.33. Purification treatments of these wastewaters have been minimally studied, in part because of their relatively low organic contents and the difficulty of their degradation. The only treatment applied consisted of the separation of the suspended solids and oil, usually by flotation.

The main problem with such wastewaters relates to their intermediate organic charge (about 2.5 g/l COD), which is too high for aerobic treatment and too low for anaerobic processing. Also the high content of inorganic solids, mainly sodium, interferes with any treatment system.

Recently, an anaerobic approach has been reported by Borja *et al.* (1992). The bioreactors used held various suspended micronized clay supports of different chemical composition (sepiolite and bentonite). Assuming that the overall anaerobic digestion process conforms to first-order kinetics, the experimental data showed that the support had a marked influence on the kinetic constants of the process, sepiolite giving the highest values. The average k_0 values obtained in the COD range 0–1 g/l were 1.83 $days^{-1}$ (sepiolite) and 1.23 $days^{-1}$ (bentonite). For values of COD greater than 1 g/l, the specific rate constant decreased when the volume of wastewater added or substrate concentration was increased. The yield coefficients of methane were 333 and 316 ml CH_4/g COD for the sepiolite and bentonite, respectively. COD reduction in both cases was more than 90%.

The study showed that anaerobic treatment was possible, but should be limited to COD concentrations under 1 g/l. The decrease of the rate constants was attributed to high COD, but could also have been due to an excess of inorganic solids, which increased in parallel with COD.

Similar studies (Borja *et al.*, 1993) carried out in a continuous-flow, completely mixed bioreactor operated at 35°C and containing sepiolite-immobilized biomass at a concentration of 10.5 g volatile suspended solids (VSS) per litre worked satisfactorily at hydraulic retention times of 2.5 to 10 days and eliminated in all cases more than 93% of the initial COD. Yield coefficient of biomass was 0.035 g VSS/g COD, with a specific rate of substrate uptake for cell maintenance of 0.078 g COD/g VSS per day. The rate of substrate uptake (g COD/g VSS) was correlated with the concentration of the biodegradable substrate (g COD/l) through a Michaelis–Menten-type equation.

In general, the macroenergetic parameters (Borja *et al.*, 1994b) of the anaerobic digestion decreased six-fold (g VSS/g COD) and increased five-fold (g COD /g VSS per day) when the influent substrate increased from 1.1 to 4.4 g COD/l. Prediction of growth yield can be done by using the Guiot kinetic model, which relates substrate utilization to product formation.

In light of these results, anaerobic treatment for the black-ripe-style olive industry wastewaters can be envisaged, but COD concentration in the influent must be reduced a couple of times for adequate process kinetics. Dissolved inorganic solids should also be diminished. However, until now there is no industrial plant to purify these wastewaters using this system.

Black-ripe olive wastewaters are easily degraded by a completely mixed activated-sludge treatment system (Borja *et al.*, 1994a). The biokinetics indicated that the effluent COD was proportional to the product of the influent COD concentration and the specific growth rate.

10.7 DISPOSAL METHODS FOR TABLE OLIVE INDUSTRY WASTEWATERS

In general, the table olive industry has traditionally paid little attention to wastewater treatment. For most, this aspect has been limited to the elimination of suspended solids and oil and liquid residue disposal has been done directly to public sewers. However, during the past 20–25 years the environmental authorities of the more developed countries have been urging these factories to implement effective measures of pollution abatement.

Different methods have been proposed. Batina (1969) suggested that the California olive industry inject their wastewaters into deep pools in a similar way to some oil refineries. It is necessary to drill a well 600–1000 m in depth, according to the geological structure of the ground, far away from any aquifer. The method seems clean and simple but, in practice, is difficult and expensive. In particular, finding a suitable location is not easy.

Since the most contaminating residues are the brines, many of the systems have been directed at the specific elimination of these streams. Popper (1970) suggested the use of ion-exchange systems to remove excess sodium and render the brines treatable together with the rest of the table olive wastewaters. The method is however expensive and has some unsolved technical problems. Stepakoff *et al.* (1974) developed a eutectic freezing process for brine disposal. At the eutectic point, both ice and salt crystallize from salt water and the brine volume is minimized. The work demonstrated the technical feasibility and gave a conceptual design of a two-stage eutectic freezing process.

Durkee *et al.* (1973) developed in Albany (California) new equipment for brine concentration. This consisted of a submerged combustion evaporator in which the salt from the used brine solution was crystallized. Heat was transferred rapidly from rising gas bubbles to the liquid through the bubble interface. Since the vapour pressure in the rising bubbles is less than one atmosphere, boiling temperature is considerably lower than 100°C. For a saturated NaCl solution, the observed temperature was about 93°C, corresponding to a depressed vapour pressure of about 443 mm. Conventional evaporators can be used to crystallize the salt, but capital costs are several times greater. Because there are no heat transfer surfaces in a submerged combustion heater, equipment costs are comparatively low and there is no fouling risk. The disadvantage of a submerged combustion compared with a multiple-effect evaporator is a lower thermal efficiency. The slurry obtained from the crystallizer contained about 57% solids of which 6% was combustible organic matter that must be eliminated before the salt is suitable for re-use. This is achieved by burning the slurry at a temperature slightly lower than that of the NaCl fusion, leaving only a small carbon residue. The reclaimed salt can thus be stored as a solid from one season to the next. Brines prepared from such salt have a high pH (around 10.0) that must be corrected by HCl before their re-use. Industrial trials showed the method to be viable and that its use depended on the economic balance of its operation.

However, in general, the value of the recovered salt is slight and the procedure is expensive.

In California, heavily contaminated and difficult-to-treat wastewaters have been disposed of traditionally into shallow ponds, which were unlined until 1967 when this was made a requirement. When building them, the processor must be aware of the risk of aquifer contamination and adopt the measures necessary to prevent such a possibility. Impermeabilization by clay was used for many years, but it has been noted that eventually such ponds leaked significantly. This is also thought to be a significant problem with lined ponds. Leaks from the ponds (Martin, 1992) have resulted in an area of very high groundwater salinity in their vicinity. In some cases, both shallow groundwater (concentration exceeds 3 g NaCl/l) and the deeper aquifers (about 0.6 g NaCl/l) have been affected. This has created an important problem that authorities are currently trying to remedy by washing the ground around and below the ponds and desalting all the contaminated aquifer. This will take many years and considerable cost, and illustrates the dangers of not considering the ultimate fate of waste streams produced by industrial processes.

Currently, there is no alternative to eliminate lyes and washing waters and even brines, if they are not re-used; but careful consideration must be paid to the construction of evaporating ponds. Leakage must be completely prevented and measures installed to ensure that this condition is always maintained. At least double-lined ponds are suggested. In addition, evaporation ponds must be situated at a distance from inhabited places and in the opposite direction to prevailing winds in order to avoid any trouble to residents. Ponds must also be protected by fences to prevent entrance of unauthorized personnel and eliminate drowning risks. The capacity of ponds must be calculated bearing in mind the volume of wastewater produced by the factory, annual rain and the evaporation rate, which is specific for each region. An approximate value could range between 5 to 10 mm per day.

Prevention of malodorous fermentations is effectively controlled by maintaining an alkaline pH, which is the norm if ponds receive only lyes and washing waters. If the equilibrium pH is acidic, it should be corrected with alkali, but aeration of the solution by a surface or submerged aerator is usually a better option. Any commercially available equipment may be used for this operation.

Evaporation rate can be improved by using sprays (Batina, 1969), addition of surfactants agents, or enhancing solar energy absorption. Some devices recently introduced in Spain called 'Nucleos de interface' have been shown to be very effective in increasing the evaporation rate. The equipment can take advantage of the wind but some models have electric fans that can be used when the wind is absent or insufficient. Industrial experience is favourable to its utilization.

Fig. 10.8 Evaporation pond in the table olive industry. (Picture courtesy of Dr Garrido Fernández.)

Researchers are not satisfied with disposal in evaporating ponds and consider it an intermediate stage while other cleaner methods are found. However, when used with the appropriate precautions, it is a satisfactory and economical system, especially for those regions with low rainfall and plenty of sunshine. Figure 10.8 shows a typical evaporation pond.

10.8 DISPOSAL OF SOLID WASTES

Solid wastes were not important in the traditional processing industry but the progressive introduction of pitting and pitting and stuffing, as well as the new commercial presentations such as slices and wedges generate large amounts of pits. Their characteristics have been studied by Fernández-Bolaños Guzmán *et al.* (1990) and Guillén Bejarano *et al.* (1990). Most of these are bought for extraction of oil from the seed. When this demand is slack, they can be used as boiler fuel. This is easy in countries where the cellulolytic residues of olive oil extraction are also used as fuel in the same extraction factories or in the table olive processing facilities. In all the olive-producing countries, there are numerous models of burners that can be used successfully.

A higher yield from an energy point of view can be obtained when burning is carried out in a fluidized bed. Sowards (1977) made a detailed analysis of the advantages of such systems and also pointed out the potential problems that may arise during their use. The main problem can be from the potassium and sodium oxides formed during burning. Some modifications in furnaces and the use of a certain proportion of clay together with the residues can enhance the process and facilitate its operation.

Pits are very widely used as a source of energy in the California olive industry. The major part of the steam used for sterilization and other heating needs are obtained from their combustion. However, it must be emphasized that this be considered as a complementary energy source.

Those industries that separate oil and other suspended solids also produce important quantities of solid wastes. Such waste has a very high moisture content, so some method for its concentration is advisable. The wastes are rich in olive oil and are sometimes used for feeding animals (pigs). Another option is their incorporation into exhausted olive cakes. The mixture can be extracted by organic solvents to obtain an oil that should be refined. Lastly, if no other use is made, they can be sun-dried and burned.

Those industries (mainly packing) that apply physicochemical treatment to their wastewaters produce large amounts of sludge. This must be concentrated and sent to specially selected disposal sites where it cannot cause environmental damage. Authorities in charge of the environmental control usually advise on this.

10.9 REFERENCES

Batina, R.T. (1969) Practical approaches to canning wastes treatment problems, in *48th Technical Report*, California Olive Association, Sacramento, California, USA, pp. 67–72.

Borja, R., Garrido Fernández, A. and Durán Barrantes, M.M. (1992) Estudio cinético del proceso de digestión anaerobia de las aguas residuales de la elaboración de aceitunas negras de mesa. *Grasas y Aceites*, **43**, 317–21.

Borja, R., Martin, A. and Garrido, A. (1993) Anaerobic digestion of black olive wastewaters. *Bioresource Technology*, **45**, 27–32.

Borja, R., Banks, C.J. and Garrido, A. (1994a) Kinetics of black olive wastewater treatment by the activated sludge system. *Process Biotechnology*, **29**, 587–93.

Borja, R., Garrido, A., Martín, A. and Alonso, V. (1994b) Influence of substrate concentration on the macroenergetic parameters of anaerobic digestion of black olive wastewaters. *Biotechnology Letters*, **16**, 321–6.

Brenes Balbuena, M. and Garrido Fernández, A. (1988a) Regeneración de salmueras de aceitunas verdes estilo español con resinas intercambiadoras de iones. *Grasas y Aceites*, **39**, 22–7.

Brenes Balbuena, M. and Garrido Fernández, A. (1988b) Regeneración de salmueras de aceitunas verdes estilo español con carbón activo y tierras decolorantes. *Grasas y Aceites*, **39**, 96–101.

Brenes Balbuena, M., García García, P. and Garrido Fernández, A. (1988c) Regeneration of Spanish-style green table olives brines by ultrafiltration. *Journal of Food Science*, **53**, 1733–6.

Brenes Balbuena, M., Sánchez Roldán, F. and Garrido Fernández, A. (1988d) Coagulación-filtración de salmueras de aceitunas verdes estilo español. *Grasas y Aceites*, **39**, 264–71.
Brenes Balbuena, M., García García, P. and Garrido Fernández, A. (1989a) Influencia en el envasado de aceitunas verdes estilo español del reuso de salmueras regeneradas. *Grasas y Aceites*, **40**, 182–9.
Brenes Balbuena, M., Vicente González, J., García García, P. and Garrido Fernández, A. (1989b) Características de las aguas residuales en la elaboración de aceitunas de mesa. *Grasas y Aceites*, **40**, 287–90.
Brenes Balbuena, M., Montaño Asquerino, A. and Garrido Fernández, A. (1990) Ultrafiltration of green table olive brines: influence of some operating parameters and effect on polyphenols composition. *Journal of Food Science*, **55**, 214–17.
Cancel, L.E. and Hernández, E.L. (1975) Recycling of fermented brine in citron curing. *The Journal of Agriculture of the University of Puerto Rico*, **LIX**, 199–207.
Castro Gómez Millán, A., Durán Quintana, M.C., García García, P., Garrido Fernández, A., González Cancho, F., Rejano Navarro, L., Sánchez Roldán, F. and Sánchez Tebar, J.C. (1983) Elaboración de aceitunas verdes estilo español de la variedad Gordal con reutilización de lejías y supresión de lavados. Estudio de sus aguas residuales y algunos ensayos de purificación. *Grasas y Aceites*, **34**, 162–7.
Castro Gómez Millán, A. de, Montaño Asquerino, A., Sánchez Gómez, A.H., Rejano Navarro, L. and Garrido Fernández, A. (1989) Influencia de la adicion de HCl y de la inoculación con levaduras en la fermentación y características organolépticas de las aceitunas verdes estilo sevillano. *Grasas y Aceites*, **40**, 376–84.
Durán Quintana, M.C. and Garrido Fernández, A. (1978) Aceitunas negras al natural en salmuera. VI. Características de la fermentación en salmueras regeneradas. *Grasas y Aceites*, **29**, 197–201.
Durkee, E.L., Lowe, E., Baker, K.A. and Burgess, J.W. (1973) Fields tests of salt recovery system for spent pickle brine. *Journal of Food Science*, **38**, 507–11.
Durkee, E.L., Lowe, E. and Toocheck, E.A. (1974) Use of recycled salt in fermentation of cucumber salt stock. *Journal of Food Science*, **39**, 1032–3.
Fernández-Bolaños Guzmán, J., Guillén Bejarano, R. and Heredia Moreno, A. (1990) Características del material lignocelulósico del hueso de aceitunas, in *Proceedings of the Seminar on Valoración de los residuos lignocelulósicos producidos en la extracción del aceite y en la elaboración de aceitunas de mesa*, Instituto de la Grasa, Sevilla, Spain, pp. 70–85.
Fernández Díez, M.J. (1991) Processing table olives, in *Encyclopedia of Food Science and Technology*, Vol. III (ed.Y.H. Hui), Wiley & Sons, New York, pp. 1910–25.
Fernández Díez, M.J., Castro Ramos, R., Garrido Fernández, A., González Cancho, F., González Pellissó, F., Nosti Vega, M., Heredia Moreno, A., Mínguez Mosquera, M.I., Rejano Navarro, L., Durán Quintana, M.C., Sánchez Roldán, F., García García, P. and Castro Gómez Millán, A. (1985) *Biotecnología de la aceituna de mesa*, Consejo Superior de Investigaciones Científicas, Instituto de la Grasa, Madrid, Sevilla, Spain.
García García, P. and Garrido Fernández, A. (1984) Depuración parcial por precipitación de las lejías y aguas de lavado de la elaboración de aceitunas verdes estilo sevillano. *Grasas y Aceites*, **35**, 295–9.

García García, P., Garrido Fernández, A., Chakman, A., Lemonier, J.P., Overend, R.P. and Chornet, E. (1989) Purificación de aguas residuales ricas en polifenoles: aplicación de la oxidación húmeda a los efluentes acuosos derivados de las industrias olivareras. *Grasas y Aceites*, **40**, 291–5.

García García, P., Brenes Balbuena, M. and Garrido Fernández, A. (1990a) Características de las aguas residuales producidas en la elaboración de aceitunas de mesa. *Proceedings of the Seminar on Aplicación de la tecnología Jet-reactor a la depuración de los efluentes acuosos derivados de las industrias olivareras*, Instituto de la Grasa, Sevilla, Spain, pp. 1–6.

García García, P., Brenes Balbuena, M., Vicente González, J. and Garrido Fernández, A. (1990b) Depuración de las aguas residuales de las plantas envasadoras de aceitunas verdes mediante tratamientos físico-químicos. *Grasas y Aceites*, **41**, 263–9.

García García, P., Garrido Fernández, A., Chakman, A., Lemonier, J.P., Overend, R.P. and Chornet, E. (1990c) Aplicación de la tecnología jetreactor a la oxidación húmeda de las aguas residuales de las industrias derivadas del olivo. *Grasas y Aceites*, **41**, 158–62.

García García, P., Brenes Balbuena, M. and Garrido Fernández, A. (1991) Effect of oxygen and temperature on the oxidation rate during darkening of ripe olives. *Journal of Food Engineering*, **13**, 259–71.

Garrido Fernández, A. (1975) Tratamiento de las aguas residuales de la industria del aderezo. Métodos para su eliminación o reacondicionamiento para su posterior empleo. *Grasas y Aceites*, **26**, 237–44.

Garrido Fernández, A. (1978) Aceitunas negras al natural en salmuera. V. Ensayos de regeneración de salmueras de fermentación. *Grasas y Aceites*, **29**, 111–18.

Garrido Fernández, A. (1979) Regeneración de salmueras de aceitunas verdes estilo español mediante la adsorción con carbón activo. *Grasas y Aceites*, **30**, 5–10.

Garrido Fernández, A. (1982a) New wastewater free process to produce Spanish-green olives, in *Proceedings of International Symposium on Food Industries and Environment*, Budapest, pp. 107–14.

Garrido Fernández, A. (1982b) Aguas residuales del aderezo de aceitunas. Posibilidades de tratamiento y reutilización. Seminario sobre tratamiento y reciclado de aguas residuales de bajos costos economicos y energeticos, MOPU/OCDE, Madrid, Spain.

Garrido Fernández, A. (1983) Estudio de las aguas residuales del proceso de elaboración de aceitunas negras por oxidación y de sus posibilidades de reutilización. *Grasas y Aceites*, **34**, 317–22.

Garrido Fernández, A. (1984) Estudio de las aguas residuales del proceso de elaboración de aceitunas negras por oxidación y sus reutilizaciones (y II). Influencia sobre la calidad del producto final y sus salmueras de envasado. *Grasas y Aceites*, **35**, 165–71.

Garrido Fernández, A., González Pellissó, F., González Cancho, F., Sánchez Roldán, F., Rejano Navarro, L., Cordón Casanueva, J.L. and Fernández Díez, M.J. (1977) Modificaciones de los procesos de elaboración y envasado de aceitunas verdes de mesa en relación con la eliminación y reuso de vertidos. *Grasas y Aceites*, **28**, 267–85.

Garrido Fernández, A., Cordón Casanueva, J.L., Rejano Navarro, L., González Cancho, F. and Sánchez Roldán, F. (1979) Elaboración de aceitunas verdes estilo español con reutilizaciones de lejías y supresión de lavados. *Grasas y Aceites*, **30**, 227–34.

Garrido, A., García, A. and Brenes Balbuena, M. (1992a) The recycling of table olive brine using ultrafiltration and activated carbon adsorption. *Journal of Food Engineering*, **17**, 291–305.

Garrido Fernández, A, Brenes Balbuena, M. and García García, P. (1992b) Tratamiento de salmueras de fermentación de aceitunas verdes. *Grasas y Aceites*, **43**, 291–8.

Ginnette, L.F. (1967) Rehabilitation of olive brines by activated carbon, in *46th Annual Technical Report*, California Olive Association, Sacramento, California, USA, pp. 51–7.

Ginnette, L.F. (1968) Reconditioning and reusing brines by carbon column adsorption, in *47th Annual Technical Report*, California Olive Association, Sacramento, California, USA, pp. 12–17.

Glaros, T. and Geisman, J.R. (1980) Organic flocculants speed spent pickle curing brine recycling. *Journal of Food Science*, **45**, 402–3.

González Cancho, F., Rejano Navarro, C., Durán Quintana, M.C., Sánchez Roldán, F., García García, P., Castro Gómez Millán, A. de and Garrido Fernández, A. (1983) Influencia de la adición de HCl en la fermentación de aceitunas estilo Sevillano. *Grasas y Aceites*, **34**, 375–9.

González Cancho, F., Rejano Navarro, L., Durán Quintana, M.C., Sánchez Roldán, F., Castro Gómez Millán, A., García García, P. and Garrido Fernández, A. (1984) Elaboración de aceitunas verdes estilo sevillano sin lavados. Solución a los problemas de adición del HCl y efecto del tratamiento con lejías bajas. *Grasas y Aceites*, **35**, 155–9.

Guillén Bejarano, R., Heredia Moreno, A., Jiménez Araujo, A. and Sánchez, C. (1990) El hueso de la aceituna como fuente de material lignocelulósico. *Alimentación, Equipos y Tecnología*, **IX**, 109–15.

Hang, Y.D. (1977) Waste control in sauerkraut manufacturing. *Processes in Biochemistry*, **April**, 96–101.

Kader, A.A., Nanos, G.D. and Kerbel, E.L. (1990) Storage potential of fresh Manzanilla olives. *Calif. Agric.*, **44**, 23–4.

Leoni, C., Grischot, F. and Bellucci, G. (1976) Sulla depurazione degli scarichi idrici delle industrie conserviere. Influenza della concentrazione del cloruro sodico sul rendimiento di depurazione in un impianto biologico funzionante ad aeratione prolungata. *Industria Conserve*, **LI**, 175–82.

Little, L.W., Harrison, R., Davis, J., Harris, J. and Dunn, S.J. (1977) Reduction of wastes from cucumber pickle processing by use of the controlled culture fermentation process, in *Proceedings of 8th National Symposium on the Food Processing Wastes*, Seattle, Washington, USA.

Martin, C.J. (1992) Remediation of groundwater contaminated by olive processing waste. *Desalination*, **88**, 253–63.

Maxie, E.C. (1964) Cold storage systems, in *43rd Annual Technical Report*, California Olive Association, Sacramento, California, USA, pp. 12–15.

McFeeters, R.F., Palnitkar, M.P., Velting, M., Fehringer, N. and Coon, W. (1977) Reuse of brines in commercial cucumber fermentations, in *Proceedings of 8th National Symposium on the Food Processing Wastes*, Seattle, Washington, USA.

Moffat, W.P. (1975) Neutralization of caustic effluents, in *53rd Annual Technical Report*, California Olive Association, Sacramento, California, USA, pp. 45–50.

Mondok, F.J. (1979) Carbon dioxide gas application in the olive industry, in *57th*

Annual Technical Report, California Olive Association, Sacramento, California, USA, pp. 42–51.

Popper, K. (1970) A new desalting process utilizes recoverable solid regenerant ion exchange. *Water and Wastes*, **7**, 61–3.

Popper, K., Camirand, W.M., Waters, G.G. Bouthilet, R. J. and Boyle, F.P. (1967) Recycled process brines prevent pollution. *Food Engineering*, **April**, 78–80.

Ralls, J.W. (1967) Alternative storage systems, in *46th Annual Technical Report*, California Olive Association, Sacramento, California, USA, pp. 58–60.

Ralls, J.W. (1968) Alternative storage systems, in *47th Annual Technical Report*, California Olive Association, Sacramento, California, USA, pp. 31–3.

Ralls, J.W. (1969) The use of activated carbon to renovate brines, in *48th Annual Technical Report*, California Olive Association, Sacramento, California, USA, pp. 24–34.

Ralls, J.W., Mangdenberg, H.J., Lemoine, G. and Mercer, W.A. (1971) Alternate storage systems for the production of canned ripe olives. *Journal of Food Science*, **36**, 408–12.

Rejano Navarro, L., González Cancho, F., Sánchez Roldán, F., Durán Quintana, M.C., Castro Gómez Millán, A., García García, P., Sánchez Tebar, J.C. and Garrido Fernández, A. (1983) Utilización de lejías de baja graduación en la preparación de aceitunas verdes. *Grasas y Aceites*, **34**, 311–16.

Sánchez Gómez, A.H., Rejano Navarro, L., Durán Quintana, M.C., Castro Gómez Millán, A. de, Montaño Asquerino, A. García García, P. and Garrido Fernández, A. (1990) Elaboración de aceitunas verdes con tratamiento alcalino a temperatura controlada. *Grasas y Aceites*, **41**, 218–23.

Sannia, M., Pelosi, N. and Pietrini, R. (1978) Acque di rifiuto degli stablimenti di preparazione delle olive da tavola. *Industri Alimentari*, **Gennaio**, 17–24.

Sowards, N.K. (1977) Olive pits as an energy source in the olive packing, in *55th Annual Technical Report*, California Olive Association, Sacramento, California, USA, pp. 66–71.

Stepakoff, G.L., Siegelman, D., Johnson, R. and Gibson, W. (1974) Development of an eutectic freezing process for brine disposal. *Desalination*, **14**, 25–38.

Stevenson, K.E., Black D.E. and Costilow, R.N. (1979) Anaerobic fermentations of pickle process brine by *Candida utilis*. *Journal of Food Science*, **44**, 181–5.

Unidad Estructural de Biotecnología de Alimentos (Instituto de la Grasa) (1990) *Table Olive Processing*, International Olive Oil Council, Madrid, Spain.

Vaughn, R.H. (1967) Saltless storage, in *46th Annual Technical Report*, California Olive Association, Sacramento, California, USA, pp. 43–50.

Vaughn, R.H. (1969) Progress report on saltless storage, in *48th Annual Technical Report*, California Olive Association, Sacramento, California, USA, pp. 40–6.

Vaughn, R.H. (1974) Progress in year's research, in *52nd Annual Technical Report*, California Olive Association, Sacramento, California, USA, pp. 12–18.

Vaughn, R.H. Martin, M.H., Stevenson, K.E., Johnson, M.G. and Crampton, V.M. (1969) Salt free storage of olives and other produce for future processing. *Food Technology*, **23**, 832–4.

Woskow, M. and Maxie, E.C. (1965) Cold storage studies with olives, in *44th Annual Technical Report*, California Olive Association, Sacramento, California, USA, pp. 6–8.

Trade regulations 11

11.1 INTRODUCTION

The table olive is identified in many countries with a high standard of living (Fernández Díez, 1983, 1991). It is used as an appetizer with all kinds of alcoholic and non-alcoholic beverages, or as a decorative or nutritional element in various dishes, such as salads and pizzas. Its use is worldwide. It also constitutes a basic food for the majority of people in countries, mainly located in the Mediterranean area.

Table olives prepared by different processes, and particularly by the three main processes (Spanish, Californian and Greek methods) described in this book, are traded throughout the whole world, including both developed and less-developed countries.

For transactions in this market, the International Olive Oil Council (IOOC), through a Committee of Experts in which all the member countries are represented, has evolved a 'Unified Qualitative Standard Applying to Table Olives in International Trade' (International Olive Oil Council, 1980), which is revised periodically during one or two annual meetings of the Committee.

In the same manner, the Joint FAO/WHO Codex Alimentarius Programme, with the cooperation of the International Olive Oil Council through a Mixed Commission, has also evolved a nutritional standard for table olives (Codex Alimentarius Commission, 1974).

Both standards agree quite well in their content; the first constitutes an important aid in international trade, and the second emphasizes consumer protection aspects. The European Union (EU) is taking actions to adopt these standards as official ones for its member countries, although agreement has not as yet been reached.

Apart from these standards, each producer or importer nation has its own set of domestic rules which often concur exactly, or at least partially, with international standards (Fernández Díez, 1985). Unfortunately though, differences in food legislation between countries do exist, giving rise to some important marketing problems. Although the IOOC standard is recommended to be applied to all transactions, it is also necessary for exporter countries to consider the specific regulations of importer countries. This is an area in which international cooperation between scientists and technologists can help to solve the problems created through suitable research work.

Spain, which is the biggest producer and exporter of table olives, has developed its own standards (Ministry of Economy and Finances of Spain, 1983; Ministry of Health, 1983, 1984; Spanish Government Presidency, 1983), which are more detailed and more restrictive in certain aspects than those described above, though they are in agreement with them. In the same way, other exporting countries have developed their own standards. The United States, as the fourth producer in the world and the biggest importer, has also developed very detailed regulations (Office of the Federal Register, 1981a,b), which are gradually revised and published in the Code of Federal Regulations and Federal Register. Spain, as the main supplier for the United States, regulates its exports of table olives, in common agreement with both standards.

11.2 INTERNATIONAL STANDARDS

11.2.1 INTERNATIONAL OLIVE OIL COUNCIL STANDARD

General principles

As indicated above, this is the recommended standard to be followed in international trade. It starts with the definition of the product, the table olive, and some general recommendations to all the member countries to authorize the use of olive varieties which fulfil the following requirements:

- good size and shape of the fruit
- good flesh-to-pit ratio
- delicacy, flavour and firmness of the flesh
- ease of separation of the flesh from the pit
- delicacy of the skin
- small size and smoothness of the pit.

Types of olives, commercial preparations, and presentation forms are fully and precisely described, in those definitions included in Chapter 2 of this book. They try to clarify the differences between the multiple products which are found in international trade.

Sizing

The olives are size-graded according to the number of fruits per kilogram. The size scale is as follows:

60/70	121/140	261/290
71/80	141/160	291/320
81/90	161/180	321/350
91/100	181/200	351/380
101/110	201/230	381/410*
111/120	231/260	

*Above 410 the interval is 50 fruits

This scale is used for whole olives; when stuffed olives are concerned, the interval is 20 fruits from 210/220 up to 401/420. Size-grading is compulsory for whole, pitted, stuffed and halved olives. In the case of pitted olives, the size shown shall be the one corresponding to the original whole olive, which is calculated according to coefficients set by each producing country. The method to establish those coefficients has recently been revised (Rejano Navarro and Sánchez Gómez, 1990).

A maximum tolerance, expressed as percentage of fruits of the size immediately above or below that mentioned on the pack, is allowed:

10% for sizes with a 10-fruits interval
5% for sizes with a 20-fruits interval
2% for sizes with a 30-fruits interval or more.

In addition, it is also stipulated that the difference between the horizontal diameters among olives must not exceed 4 mm, once the olives having the largest and smaller horizontal diameter have been removed from a sample of 100 fruits. The same tolerances allowed in the number of fruits are also applicable to this characteristic.

Essential composition and quality factors

Once selected and packed, table olives shall be sound, clean, exempt of abnormal odour and taste, sufficiently ripe according to the type, without defects and foreign matter, except those admitted by tolerances, without any sign of deterioration or abnormal fermentation, correctly size-graded, uniform in colour (except for turning colour olives), and belonging to only one variety.

As optional ingredients, water, salt, vinegar, olive oil, sugars, spices and aromatic herbs, or their natural extracts, and any edible material used as stuffing, such as, pimento, onion, almond, celery, anchovy, caper, or their prepared natural pastes are permitted.

Besides sodium chloride and water, covering brines may contain sugar, vinegar, lactic acid, oil and other authorized substances, with or without aromatic substances such as spices or plants. In the case of green olives having undergone natural lactic fermentation, the brine shall have a minimum of 0.4% acidity expressed as lactic acid. According to types and

Table 11.1 Limits of sodium chloride content and pH value for different types and preparations of table olive (From International Olive Oil Council, 1980)

Type and preparation	Minimum sodium chloride content (%)	Maximum pH value
Green olives in brine, pickled or untreated		
In hermetically sealed containers	5	4.0
In non-hermetically sealed containers	6	4.5
Seasoned green olives:		
In hermetically sealed containers	4	4.0
In non-hermetically sealed containers	6	4.5
Turning colour olives		
All preparations	6	–
Naturally black olives		
In brine	7	–
In dry salt	10	–

commercial preparations, some limits for salt content and pH value are established (Table 11.1).

For pasteurized olives of all types and preparations, the sodium chloride content of the brine, may be reduced to 2%, and the maximum pH is 4.3. If the covering liquid does not contain any salt, the maximum pH limit is reduced to 4.0. For sterilized olives, sodium chloride and pH limits are not specified in the standard.

In olives packed without brine the flesh juice shall comply with the sodium chloride and pH stipulations relating to the same preparations of olives in brine preserved in hermetically-sealed containers.

Qualitative classification

Table olives are classified in three trade categories:

Extra or Fancy

High-quality olives endowed to the maximum extent with the cultivar's specific characteristics are considered as belonging to this category. Notwithstanding, and providing this does not affect the overall favourable aspect or each fruit's organoleptic characteristics, they may have slight colour, shape, flesh-firmness or skin defects. In this category, only the whole, bruised, split, pitted and stuffed olives of the best cultivars may be exported, providing their size exceeds 351/380 fruits per kilogram.

First, Choice or Select

In this category are included good-quality olives with a suitable degree of ripeness and endowed with the cultivar's specific characteristics. Providing that this does not affect the overall favourable aspect or each individual fruit's organoleptic characteristics, they may have slight colour, shape, skin or flesh-firmness defects. In this trade category all the types, preparations and styles of table olives may be exported, except for 'chopped' or 'broken' olives and 'olive paste' (see Chapter 2 for definitions).

Second, Choice or Standard

This category includes good-quality olives which, although they cannot be classified in the two higher categories, comply with the general conditions defined for table olives in this standard.

Defects to which the above definitions refer are fully described in the standard. They are related to harmless extraneous material, skin blemishes not affecting the flesh, shrivelled, soft or fibrous flesh, abnormal colour, cryptogamic damage, damage caused by insects (other than *Dacus oleae* with exit hole), *Dacus* damage (with exit hole), damage caused by abnormal cultivation practices, stems, defective stuffing (in the case of stuffed olives) and stone or stone fragments (except for the whole olives). The maximum tolerances, expressed as a percentage of fruits are indicated for each specific defect, each type of olives (green, turning colour and naturally black olives), and each one of the categories mentioned above, Extra, First and Standard. Table 11.2 summarizes the total tolerance figures.

Table 11.2 Total defect tolerances* for different trade categories of table olives (green, turning colour, black-ripe, and naturally black) (From International Olive Oil Council, 1980)

Category	Total defect tolerance for all types (% of defective fruits)
Extra or Fancy	7
First, Choice or Select	12
Second or Standard	17

* Tolerances laid down for stones (pits) and stone fragments are not included in the above tolerance figures.

Food additives

There is a list with the preservatives and processing aids which may be used, as well as the maximum permitted levels. Additives not mentioned in that list but allowed by the FAO/WHO Codex Alimentarius Commission may also be used. Likewise, coagulants, hardeners and other additives used in prepared natural pastes may be used if they are authorized by the Codex Commission.

Olive packing

Olive containers shall be made of any material capable of ensuring correct olive preservation and not transmitting harmful substances. Among others, are permitted: wood, metal, tin, glass and polymeric materials.

Metal drums must be internally covered with epoxyphenol or similar varnishes. Tin-plate containers for black olives shall be varnished on the inside at least of both covers. Transparent containers shall not produce optical effects liable to change the appearance of the product and contents. Lastly, plastic containers shall comply with prevailing technical health stipulations and be sufficiently resistant for transport purposes. Returnable containers (wooden or plastic) may be used providing they are in good condition.

Filling tolerances

Net content (olives and covering brine) shall in no container be less than 90% of its total filling capacity. The tolerance concerning the net drained weight mentioned on the container shall not exceed the following percentage scale, providing the sample mean net drained weight is equal to, or in excess of, said declared weight:

5% for containers with drained weight less than 200 g.
4% for containers with drained weight between 200 and 500 g.
3% for containers with drained weight between 500 and 1500 g.
2% for containers with drained weight in excess of 1500 g.

General specifications on containers and labelling are similar to those indicated for other food products. The specific trade preparation (nature of the product, cultivar, trade preparation and style) must be included. In those containers prepared for the USA, labels must also include information on nutritional aspects, according to current legislation.

A copy of the IOOC Standards can be obtained from: International Olive Oil Council, c/o Principe de Vergara, 154. 28002. Madrid, Spain.

11.2.2 CODEX ALIMENTARIUS COMMISSION STANDARD

Since its first publication in 1974 (Codex Alimentarius Commission, 1974), the evolution of this standard, through the work of the IOOC/Codex Alimentarius Mixed Commission, has led to both international standards reaching almost complete concurrence in 1981 (Mixed Commission IOOC/Codex Alimentarius, 1981). The Mixed Commission revised this standard further, in 1984 and 1986, mainly in relation to the use of new additives due to important changes in the technology of table olives and the introduction of new olive products in the international market (Rejano Navarro and Fernández Díez, 1986).

The current Codex Standards for Table Olives (CODEX STAN 66–1981. Rev., 1987) was published in Supplement number 1 of the Codex Alimentarius, Vol II 'Codex Standards for Fruits and Processed Vegetables and Edible Mushrooms'. In March 1990, the Codex Alimentarius Commission related to Food Additives and Contaminants accepted the use of ferrous lactate as a colour stabilizer in black-ripe olives. The different additives permitted at present are shown in Table 11.3.

The main differences between the Codex Alimentarius Standards and that from IOOC are related to additives. The current Codex Standards include a more exhaustive list while the Standards from IOOC only mention some of them and refer to Codex Alimentarius for the rest. At the present time, all the included and proposed additives are generally recognized as safe (GRAS) in the United States (Office of the Federal Register, 1977). Codex Alimentarius establishes the minimum quality criteria required for Second or Standard olives in order to protect the consumers while the IOOC Standards are related to all the qualities defined for the International Trade. Codex Standards also include, as an appendix, the qualitative classifications established by the IOOC. A copy of the Codex Standards can also be obtained from IOOC.

11.3 NATIONAL STANDARDS

11.3.1 UNITED STATES STANDARDS

Part 932 of the Title 7, Chapter IX, of the Code of Federal Regulations, under the heading 'Olives Grown in California', constitutes an important order (Office of the Federal Register, 1981b) regulating handling of different olive cultivars which are divided into two groups according to fruit size:

- **Group 1**: this includes, among other cultivars of minor technological importance, the following important ones: Ascolano, Azapa, Barouni, Carydolia, Cucco, Gigante di Cerignola, Gordale, Grossane, Sevillano

Table 11.3 Food additives and processing aids for olives (From International Olive Oil Council, 1980)

Food additive	Maximum level (expressed as net weight over total weight of olives and brine)
Preservatives	
Benzoic acid and its sodium and potassium salts	1 g/kg (as benzoic acid)
Sorbic acid and its sodium and potassium salts	0.5 g/kg (as sorbic acid)
Acids	
Lactic acid	15 g/kg
Citric acid	15 g/kg
L-tartaric acid	15 g/kg
Acetic acid	GMP
Carbon dioxide	GMP
Antioxidants	
Ascorbic acid	0.2 g/kg
Stabilizers	
Ferrous gluconate or lactate (only to stabilize the colour of treated olives darkened by oxidation)	0.15 g/kg (as Fe in fruit)
Flavourings	
Natural flavours defined by the Codex Alimentarius	
Flavour enhancers	
Monosodium glutamate (only for olives stuffed with anchovies)	5 g/kg
Thickeners and agglutinates	
(Only for the pastes to prepare stuffing streams)	
Sodium alginate	5 g/kg
Carragenate	GMP
Garrofin gum	GMP
Guar gum	GMP
Xanthan gum	3 g/kg
Hardeners	
(only for olives stuffed with pimento paste)	
Calcium chloride	1.5 g/kg as Ca*
Calcium lactate	1.5 g/kg as Ca*
Calcium citrate	1.5 g/kg as Ca*
Potassium chloride	1.5 g/kg as K*
Other additives	
Sodium chloride	GMP
Hydrochloric acid	GMP
Processing aids	
Lactic starter cultures	GMP
Nitrogen	GMP
Carbon dioxide	GMP
Contaminants	
Lead	1 mg/kg
Tin	250 mg/kg

* In the stuffed final product.
GMP, Good Manufacturing Practice

and Sant Agostino. The main characteristics of these cultivars were indicated in Chapter 3. This group includes those cultivars which in California generally give higher average sizes.

- **Group 2**: this includes Manzanillo, Mission, Nevadillo, Oblitza and Redding Picholine, which in general present smaller average sizes than those belonging to Group 1, and which were also described in Chapter 3.

This regulation establishes the composition of the Olive Administrative Committee, consisting of 16 members, to administer the terms and provision of this part which regulates handling of olives within each 'crop year' or 'fiscal year'; that is, from 1st September of any year to 31st August of the following year, both dates inclusive.

The order establishes outgoing and incoming regulations for the inspection of 'natural condition olives' or olives in their fresh harvested state, and for 'packaged olives' or processed olives as ripe, green olives. These mainly attend to average and minimum size, size-grade and 'grade', which means classification of olives as to quality, according to the grading specifications established.

The Olive Administrative Committee may also, with the approval of the secretary, establish or provide for the establishment of production research, marketing research, and development projects designed to assist, improve, or promote the marketing, distribution or efficient production of California olives.

Subpart 2852.5441–5447, in Title 7, Chapter XXVIII (Food Safety and Quality Service) of the CFR (Office of the Federal Register, 1981a), refers to 'United States Standards for Grades of Green Olives'. It starts with definitions of the product, types of green olive packs, and styles of green olives, which are essentially in concurrence with those definitions found in international standards, although some semantic differences may be detected in the application of the terms. There follows a classification of olives in four different quality grades:

- **US Grade A (or Fancy)**: this is the quality of whole, pitted, stuffed, halved and sliced green olives that have similar varietal characteristics; normal flavour and odour; good colour; are practically uniform in size in whole, pitted and stuffed styles of single size; are practically free of defects; have good character; and score **not less than 90 points** when scored in accordance with the outlined scoring system.
- **US Grade B (or Choice)**: this is the quality of whole, pitted, stuffed, halved, sliced and chopped or minced styles of green olives that have similar varietal characteristics; have normal flavour and odour; have reasonably good colour; are reasonably uniform in size in whole, pitted and stuffed styles of single sizes, are reasonably free of defects;

have reasonably good character; and score **not less than 80 points** when scored in accordance with the outlined scoring system.

- **US Grade C (or Standard)**: this is the quality of whole, pitted, stuffed, halved, sliced, chopped or minced and broken pitted styles of green olives that have similar varietal characteristics; normal flavour and odour; fairly good colour; are fairly uniform in size in whole, pitted and stuffed styles of single sizes; are fairly free of defects; have fairly good character; and score **not less than 70 points** when scored in accordance with the outlined scoring system.
- **Substandard**: this is the quality of any style of green olives that fails to meet the applicable requirements of Grade C.

The relative importance of each factor which is scored is expressed numerically on the scale of 100, with 20 being the maximum number of points for colour and uniformity of size, and 30 the maximum value for absence of defects and character, which means general texture characteristics of the flesh and skin. Sizes are calculated as counts per pound or their equivalents as counts per kilogram, and different terms are for each particular size interval (Table 11.4). The applied scoring system is shown in Table 11.5.

Table 11.4 Sizes for green olives, according to USDA standards (From Office of the Federal Register, 1981b)

Size designation	Counts per pound	Counts per kg
Sub–Petite	181–220	400–420
Petite or Midget	141–180	380–400
		340–360
		300–320
Small or Select or Standard	128–140	280–300
Medium	106–127	240–260
Large	91–105	200–220
Extra Large	76–90	180–220
		160–180
Mammoth	65–75	150–160
		140–160
Giant	53–64	130–140
		120–130
Jumbo	42–52	110–120
		100–110
		90–100
Colossal	33–41	80–90
		70–80
Super Colossal	32 or less	60–70

Table 11.5 Scoring system for green olives, according to USDA Standards (From Office of the Federal Register, 1981b)

Factor	Grade	Score points
Color	A	18–20
	B	16–17
	C	14–15*
	SStd	0–13*
Uniformity of size	A	18–20
	B	16–17†
	C	14–15‡
	SStd	0–13§
Absence of defects	A	27–30
	B	24–26*
	C	21–23*
	SStd	0–20*
Character	A	27–30
	B	24–26*
	C	21–23*
	SStd	0–20*

* Limiting rule (regardless of the total score for the product); † limiting rule for Sub-petite size; ‡ limiting rule for sizes counting 221 to 275 per pound; § limiting rule for sizes counting more than 275 per pound.

Recommended minimum drained weights are also established, depending upon the olive sizes and container sizes. A detailed qualitative description of the main defects, such as blemishes, wrinkles, gas-pockets and mutilated olives, and the allowances for each kind of those defects, complete the standards.

Subpart 2852.3751–3763 (in the same Title and Chapter above mentioned) refers to 'United States Standards for Grades of Canned Ripe Olives' (Office of the Federal Register, 1981b). It has the same structure as the standards for green olives and starts with definition of the product itself, types and styles for olives of 'ripe type' and 'green-ripe type'. As indicated in Chapter 8, the main difference between both types is that green-ripe olives are not oxidized in processing and ripe olives are oxidized to produce a typical dark brown to black colour.

There are also four quality grades in this standard:

- **US Grade A**: this is the quality of canned ripe olives of whole, pitted, halved, segmented, sliced and chopped styles that has good flavour; that has a good colour; that is practically free from defects; that has a good character; and that for those factors which are rated in accordance with the scoring system outlined in the subpart, the total score is **not less than 90 points**. Provided, that for the styles of whole and pitted olives that are declared to be of single size, the variation in diameters does not exceed 8 mm, and of the 80%, by count, of the most uniform in size, the diameter of the largest does not exceed the diameter of the smallest by more than 4 mm.

- **US Grade C**: this is the quality of canned ripe olives of whole, pitted, halved, segmented, sliced, chopped and broken pitted styles that has a reasonably good flavour; that has a fairly good colour; that is fairly free from defects; that has a fairly good character; and that for those factors which are rated in accordance with the scoring system outlined in this subpart, the total score is **not less than 70 points**. Provided, that for the styles of whole and pitted olives that are declared to be of a single size, of 60%, by count, of the most uniform in size, the diameter of the largest does not exceed the diameter of the smallest by more than 4 mm.
- **Substandard**: this is the quality of canned ripe olives of any style that fails to meet the applicable requirements for US Grade C.

Uniformity of size is not included here in the scoring system, and size limitations are specifically incorporated to the definition of each grade. The rated factors are distributed, over a total of 100 points, in the following manner: colour, maximum 30 points; absence of defects, 40 points; and character, 30 points. Size designations and the corresponding count per pound are indicated in Table 11.6, and the applied scoring system is shown in Table 11.7.

As in the case of the green olives standards, recommended minimum drained weights for different olive sizes and container sizes, a qualitative description of the main defects, and their allowances are also included in this standard.

Both the green and canned ripe olives standards indicate the need for compliance with the provisions of the Federal Food, Drug and Cosmetic Act. Important and complete information about additives is found in Title 21, Chapter 1, parts 182 and 184 of the CFR (Office of the Federal Register, 1977).

Table 11.6 Sizes for canned whole and pitted ripe olives, according to USDA Standards (From Office of the Federal Register, 1981b)

Size designation	Counts per pound
Small	128–140
Medium	106–121
Large	91–105
Extra Large	65–88
Jumbo	51–60
Colossal	41–50
Super Colossal	40 or less

Table 11.7 Scoring system for canned ripe olives, according to USDA Standards (From United States Department of Agriculture, 1983)

Factor	*Grade*	*Score points*
Colour	A	27–30
	B	24–26
	C	21–23*
	SStd	0–20*
Absence of defects	A	36–40
	B	32–35*
	C	28–31*
	SStd	0–27*
Character	A	27–30
	B	24–26*
	C	21–23*
	SStd	0–20*

* Limiting rule (regardless of the total score for the product).

Copies of the Unites States Standards for Grades of Green and Canned Ripe Olives can be obtained from: Chief, Processed Products Branch, Fruit and Vegetable Division, AMS, US Department of Agriculture, Washington, DC 20250, USA.

11.3.2 SPANISH STANDARDS

Spanish standards for table olives were revised in 1983, and adapted to the International Olive Oil Council Standard mentioned above. Two regulations were provided by the Government for that purpose: (i) 'Normas de calidad para la exportación de aceituna de mesa', from the Ministry of Economy and Finances; and (ii) 'Reglamentación técnico sanitaria para la elaboración, circulación y venta de las aceitunas de mesa', from the Spanish Government Presidency. Both were prepared by a mixed commission composed of technical representatives from the Ministries of Agriculture, Health, Education and Science, and Economy and Finances, as well as by olive growers and processors.

The only important difference with respect to the IOOC Standards is found in the classification of the types of olives, which in this case consider four types (green, turning colour, black-ripe and naturally black) instead of the three types adopted by the international standard. Qualitative classification of olives is also in agreement with the one expressed in section 11.2.1, that is, three commercial categories are defined: Extra or Fancy, First or Select, or Choice and Second or Standard.

The main difference between the two Spanish standards is that the first also pays special attention to aspects related to export regulations, such as characteristics of the containers, labelling procedure and inspection mechanisms, and the second particularly specifies the requirements for the working conditions at the processing plants, mainly affecting facilities, equipment, cleaning operations, sanitary conditions and personnel responsibilities.

Subsequently, the Ministry of Health, after consulting with the mixed commission, published a list of additives which may be used in processing table olives, with the limitations expressed in each particular case (Ministry of Health, 1983, 1984). Although all of them are generally recognized as safe (GRAS) in United States legislation, and the majority are accepted by the European Union, there are some importer countries which present either qualitative or quantitative reservations on some of them (Rejano Navarro and Fernández Díez, 1986). Thus, special consideration of the national regulations of those countries is required in the case of exports. Table 11.8 shows the additive list and allowances. Copies can be obtained directly from the Spanish State Official Bulletins mentioned in the References.

Table 11.8 Food additives and processing aids for table olives, according to the Spanish Standards (From Ministry of Health, 1983, 1984)

	Maximum level (expressed as net weight of olives and brine)
Food additives	
Benzoic acid and its sodium and potassium salts	1 g/kg (as benzoic acid)
Sorbic acid and its sodium or potassium salts	0.5 g/kg (as sorbic acid)
Lactic acid	5 g/kg
Citric acid	15 g/kg
Ascorbic acid	0.2 g/kg
Acetic acid	GMP
Tartaric acid	GMP
Carbon dioxide	GMP
Sulphur dioxide or sodium or potassium sulphite	100 p.p.m. in the final product as SO_2
Monosodium glutamate	5 g/kg
Ferrous gluconate or ferrous lactate	0.15 g/kg (as total Fe in the fruit)
(only to stabilize the colour of treated olives darkened by oxidation)	

(*Continued*)

Table 11.8 *Continued*

	Maximum level (expressed as net weight of olives and brine)
Processing aids	
Sodium hydroxide, when used in the preparation of the alkaline lye	Not limited
Pure culture of lactobacilli, as starter of lactic fermentation	GMP
Food-grade hydrochloric acid, to neutralize the residual lye	GMP
Sodium alginate*	GMP
Carragenate*	GMP
Carob bean gum*	GMP
Guar gum*	GMP
Xanthan gum*	GMP
Calcium lactate†	350 p.p.m. as Ca in final product
Calcium citrate†	350 p.p.m. as Ca in final product
Calcium chloride†	350 p.p.m. as Ca in final product

* Used as thickener in the preparation of stuffing pastes; † used as hardener in the preparation of stuffing pastes
GMP, Good Manufacturing Practice.

In 1995, a new Directive relating to additives was published by the EU (Directiva 95/2/CE, 1995). In it, the use of sodium or potassium benzoate and sulphur dioxide are not included. Currently, it is too early to judge the impact of this decision on the table olive legislation of Spain and other EU countries, but it is likely that these additives will be removed from the corresponding Standards in the near future.

11.4 REFERENCES

Codex Alimentarius Commission (FAO/WHO) (1974) *Norma international recomendada para las aceitunas de mesa*, Rome, Italy.

Directiva 95/2/CE (1995) Directiva del Parlamento Europeo y del Consejo de 20 de Febrero de 1995. *Diario oficial de las Comunidades Europeas*, no. **L61** (18/3/95), 1–32

Fernández Díez, M.J. (1983) Olives, in *Biotechnology*, Vol 5 (eds H.J. Rehm and G. Reed), Verlag Chemie, Weinheim, Germany, pp. 379–97.

Fernández Díez, M.J. (1985) Table Olives. Quality through Research. *Olivae*, **II**(9), 11–15.

Fernández Díez, M.J. (1991) Processing table olives, in *Encyclopedia of Food Science and Technology*, Vol. III (ed Y.H. Hui), Wiley & Sons, New York, pp. 1910–25.

International Olive Oil Council (IOOC) (1980) *Unified Qualitative Standard Applying to Table Olives in International Trade*, Madrid, Spain.
Ministry of Economy and Finances of Spain (1983) Normas de calidad para la exportación de aceitunas de mesa. *Spanish State Official Bulletin*, **18**, 1568–72.
Ministry of Health (1983) Lista positiva de aditivos y coadyuvantes tecnológicos para uso en la elaboración de aceitunas de mesa *Spanish State Official Bulletin*, **241**, 27404–5.
Ministry of Health (1984) Modificación de la lista positiva de aditivos y coadyuvantes tecnológicos para la elaboración de aceitunas de mesa. *Spanish State Official Bulletin*, **299**, 27718.
Mixed Commission COI/Codex Alimentarius (1981) *Norma internacional recomendada para las aceitunas de mesa*. Document CODEX STAN 66.
Office of the Federal Register (1977) *Code of Federal Regulations, Title 21, Chapter 1, Parts 182 and 184.* Food and Drugs, US Government Printing Office, Washington, DC.
Office of the Federal Register (1981a) *Code of Federal Regulations, Title 7, Chapter IX, Part 932. Olives Grown in California.* US Government Printing Office, Washington, DC.
Office of the Federal Register (1981b) *Code of Federal Regulations, Title 7, Chapter XXVIII, Subpart 2852.5441–5457. United States Standards for Grades of Green Olives* and *Subpart 2852.3751–3763. United States Standards for Grades of Canned Ripe Olives*, US Government Printing Office, Washington, DC.
Rejano Navarro, L. and Férnandez Díez, M.J. (1986) Necesidad de nuevos aditivos en la elaboración de aceitunas de mesa. *Alimentación, Equipos y Tecnología*, **V**, 97–102.
Rejano Navarro, L. and Sánchez Gómez, A.H. (1990) Clasificación por tamaños y diferenciación de variedades de las aceitunas de mesa. *Alimentación, Equipos y Tecnología*, **IX**, 133–8.
Spanish Government Presidency (1983) Reglamentación técnico sanitaria para la elaboración, circulación, y venta de la aceituna de mesa. *Spanish State Official Bulletin*, **108**, 12619–25.
United States Department of Agriculture (1983) *United States Standards for Grades of Canned Ripe Olives*, Agricultural Marketing Service, Washington, DC.

Control methods 12

12.1 INTRODUCTION

The purpose of this chapter is not to present an exhaustive account of all the analytical procedures which are used in table olives work. It will only summarize the main methods which the table olives technologists must use routinely in their daily work of process control, especially during the fermentation and storage phases, and those recommended in the international trade.

It was indicated in Chapter 1 that the role that Nature has assigned to each microorganism may be partially modified or substituted by the technologist varying a very limited number of factors. Among these, pH, free acidity, combined acidity and sodium chloride content may be considered of prime importance. The presence or absence of reducing sugars in fermentation brines as the main components responsible for the fermentation and safe storage of the product was also emphasized in Chapters 7, 8 and 9. A high volatile acidity may be the red light to announce the development of an abnormal fermentation and possible spoilage of Spanish-style green olives. Iron content in ripe olives is a very specific and important determination for control purposes, and for that reason will also be included.

The sequential order of growth and persistence of the most important groups of microorganisms must be suitably controlled, either with the help of starters or simply by changing the physical and chemical characteristics of the medium.

All these methods will be described below, as they are generally used at most of the control laboratories and factories, and are verified as accurate and precise enough for routine controls. Some recent modifications of these procedures, which are now under experimental research, will also be described. The Unified Qualitative Standard Applying to Table Olives in International Trade does not establish specific control methods but recommends those from the Codex Alimentarius. Some of these are in the current Revision of the Table Olives Standards (Codex STANN 66, 1990) and will be commented on in this Chapter.

12.2 PHYSICAL CHEMICAL METHODS

12.2.1 MEASUREMENT OF pH AND COMBINED ACIDITY

The term combined acidity or residual lye (Fernández Díez *et al.*, 1985; Unidad Estructural de Biotecnología de Alimentos, 1990) refers to the organic acid salts present in the brine. These are mainly sodium acetate and lactate, as a result of the action of the sodium hydroxide which penetrates the olive pulp during the alkaline treatment. It is measure of the buffering capacity of the medium and is obtained by potentiometric titration of the brine with hydrochloric acid to a pH of 2.6, which is the average minimum pH value of the buffer system curve.

Equipment and reagents

- Potentiometer to measure pH value, provided with glass electrode and calomel electrode as reference
- Magnetic stirrer
- Hydrochloric acid, 2 M standardized solution

Procedure

In a 50–100 ml beaker, titrate exactly 25 ml of brine with the hydrochloric acid standardized solution. Add the acid slowly and shake until a pH of 2.6 is reached.

Calculation

Results are expressed in normality, that is, equivalents of sodium hydroxide per litre of brine. Milliequivalents /l (mEq/l) is also sometimes used.

12.2.2 TITRATABLE ACIDITY

This represents the sum of the different free organic acids which are present in the brine (Fernández Díez *et al.*, 1985; Unidad Estructural de Biotecnología de Alimentos, 1990). The determination is performed on the fermentation brines or covering brines used for packing, once the fruit–brine equilibrium is reached. Otherwise, it can also be performed on the juice obtained by squeezing the crushed pulp, once the oily phase has been separated by decantation.

Procedure

Titrate 10 ml of brine or juice with a 0.2 N standardized solution of sodium hydroxide, using a 1% alcoholic solution of phenolphthalein as indicator. Shake for a few seconds until a pink colour persists. Add 0.1 ml more of the alkaline solution. If the brine turns an intense reddish-pink colour, the titration is correct. If the change in colour does not occur, add more alkaline solution, always 0.1 ml at a time, until the pink colour is obtained, taking the last reading minus 0.1 ml as the final volume consumed. In the case of highly coloured brines, such as those from naturally black olives in brines, it is advisable to dilute the brine with distilled water before titration.

Calculation

Conventionally, free acidity is expressed as grams of lactic acid per 100 ml of brine or juice:

$$\text{FA} = 0.9 \times N \times V \text{ (as g of lactic acid/100 ml brine)}$$

where N is the normality of the standardized sodium hydroxide solution and V is the volume of that solution consumed during titration of 10 ml of brine or juice.

12.2.3 VOLATILE ACIDITY

Volatile acidity is present in all brines to a greater or lesser extent, depending upon the alkali treatment of the olives made in the process (Chapters 7–9) and the fermentation and storage phases. For a normal process, acetic acid is the only important component of the volatile acidity and the ratio of volatile acidity to free acidity is relatively low and less than 1. However, some abnormal fermentations are characterized by the presence of an elevated volatile acidity in the brine.

Gas chromatography and high-performance liquid chromatography have recently been applied (Montaño Asquerino *et al.*, 1992) to normal and 'zapateria' olive brines from green type, and several C_2 to C_6 fatty acids were identified; among them, acetic, propionic, butyric, isovaleric, valeric, caproic and cyclohexane carboxylic acids.

For these reasons, the determination of total volatile acidity is a useful and practical aid for control purposes and the rapid detection of spoilage.

Equipment

- Conventional micro-Kjeldahl apparatus of about 20-ml load capacity, provided with a 25-cm length condenser
- 500-ml conical flask, graduated at 250 ml.

Procedure

A sample of 10 ml of brine is placed into the micro-Kjeldahl apparatus with approximately 1 g of tartaric acid to release the combined volatile acids. Place the 500-ml conical flask, containing 50 ml of distilled water under the condenser in such a way that the bottom of the latter is submerged in the distilled water. Run steam through the inside of the apparatus, allowing it to bubble through the brine until the 250 ml mark is reached in the flask by condensed liquid. Titrate the condensate with a 0.2 N standardized solution of sodium hydroxide using a few drops of phenolphthalein as indicator.

Calculation

To allow comparison with free acidity, results of total volatile acidity are conventionally expressed as grams of lactic acid per 100 ml of brine, which are obtained by the same formula indicated above in section 12.2.2.

12.2.4 SODIUM CHLORIDE CONTENT

Among the different parameters of the brine to be controlled during fermentation, storage, and packing of table olives, sodium chloride concentration and pH value may be considered as the most important because of their influence on the growth/inhibition of the microorganisms involved in the process.

Sodium chloride determinations are made by the Volhard titration method, using a standardized solution of silver nitrate and potassium chromate as indicator. The brine or the juice of the flesh are diluted with distilled water before titration (Fernández Díez *et al.*, 1985).

Use of flame photometric and ion-selective electrode (sodium or chloride) methods to determine NaCl concentration in olive brines has recently been studied (García García *et al.*, 1991). According to the results, using standard solutions or brines from olives which have not been previously treated with alkali, these methods have the same accuracy as the Volhard titration. However, when they are applied to brines from Spanish-style olives, it is necessary to use a correction factor due to the effect of sodium organic salts present in the residual lye (García García *et al.*, 1992).

Reagents

- Approximately 0.1 N solution of silver nitrate, previously standardized against a standard 0.1 N solution of sodium chloride

- Potassium chromate, 5% solution (weight/volume) in distilled water, as indicator

Procedure

Dilute 5 ml of brine, or juice from pulp, up to 100 ml with distilled water. Titrate 10 ml of this diluted solution with the standardized silver nitrate solution in a 250-ml conical flask, filled in advance with about 100 ml of distilled water, using 1 ml of the potassium chromate solution as indicator.

Calculation

$$\% \text{ NaCl (weight/volume)} = (N \times V) \times 5.84 \times 0.5$$

where N is the normality of the silver nitrate solution; V is the volume of silver nitrate solution used in titration; 0.5 is the equivalent original volume of brine or juice used in titration; and 5.845 is one-tenth of the molecular weight of sodium chloride.

12.2.5 REDUCING SUGARS

Reducing sugars, as the major components of soluble sugars in the olive flesh (see composition and Chapters 7–9), are the main carbon sources for the development of lactobacilli and other microorganisms present in storage and fermentation brines. The semi-quantitative method (Fernández Díez *et al.*, 1985; Unidad Estructural de Biotecnología de Alimentos, 1990) described below is generally applied and constitutes an important routine tool for control purposes.

Reagents

- Fehling's A solution: dissolve 69.28 g of crystalline copper sulphate, $CuSO_4{\cdot}5H_20$, in 400–600 ml of hot distilled water. Cool, transfer the solution to a graduated flask of 1-l capacity, bringing it up to the mark. After 48 h, filter into a topaz-coloured flask.
- Fehling's B solution: dissolve 346 g of sodium-potassium tartrate, $C_4H_4KNaO_6{\cdot}4H_20$, in approximately 300 ml of hot distilled water. Dissolve 100 g of caustic soda pellets in approximately 300 ml of cold distilled water and then boil for 5 minutes. Add the tartrate solution to the soda and leave the mixture to cool. Transfer to a flask of 1-l capacity, bringing it up to the mark.
- Methylene blue, 1% solution in distilled water, as indicator.

Procedure

For the titration, mix and then homogenize equal volumes of Fehling's A and B solutions. Transfer 5 ml of the mixture to a 250-ml flask. Add the appropriate quantity of brine (determined according to the supposed sugar concentration and the equivalence table indicated below) and 100 ml of distilled water. Place the flask on a tripod, apply a naked burner flame and boil for 2 minutes. The solution turns a reddish colour that is typical of reduced copper. Add three drops of methylene blue indicator solution. If the blue colour disappears after a few seconds of boiling, the estimated quantity of sugar is correct. If the colour disappears immediately after adding the indicator, the actual quantity is higher, and if the bluish or green colour persists, the quantity is lower than that calculated. In the two last cases, the titration must be repeated with an appropriately smaller or larger volume of brine.

Calculation

The reducing sugars concentration is obtained from the equivalence table, expressed as grams of glucose per 100 ml of brine.

ml of brine	Reducing sugars content (g/100 ml)
3	0.95
4	0.75
5	0.60
6	0.50
7	0.40
8	0.35
10	0.30
12	0.25
14	0.20
30	0.10

As glucose is the main component of reducing sugars in olives, a very quick method for its determination is now being used for experimental research. An automatic YSI (Yellow Spring Inc.) Model 2000 Analyzer is used. It is based on diffusion of the sample through a thin polycarbonate membrane and the action of immobilized glucose oxidase on glucose. This method has also the advantage of using the same brine sample to determine simultaneously the amount of L-lactate present, by the action of L-lactate oxidase.

12.2.6 IRON CONTENT IN OLIVE FLESH

As indicated in Chapter 8, iron in the form of ferrous gluconate or lactate is used to stabilize the colour of black-ripe olives, elaborated as

California-style in brine. The iron level in the flesh must not exceed 0.15 g/kg (Chapter 11), calculated as total Fe in the fruit.

Iron determination can be made by two methods: colorimetric and atomic absorption spectrometry.

Colorimetric method

Quantitative determination of iron is made by a procedure which is based on the reaction of ferrous irons with *o*-phenanthroline, adapted to be used on the olive pulp (Garrido Fernández *et al.*, 1973).

Equipment and reagents

- Blender
- Flat-bottomed clear quartz dishes, 8-cm diameter with a 100-ml capacity
- Graduated flasks of 25 ml
- Glass funnels, 4-cm diameter, or suction hood filtering
- Visible spectrophotometer with 1-cm glass trays
- Litmus paper for pH (2.6–4.1) Lyphan Ref. L 656 or equivalent
- Magnesium nitrate: 25% $Mg(NO_3)_2{\cdot}6H_20$ (AR) solution in 96% ethanol
- Hydrochloric acid, 6 N
- Hydroquinone: 2% solution in a 0.1% ascorbic acid solution in double-distilled water
- *o*-Phenanthroline: monohydrochloride, 0.5% solution in double-distilled water
- Bromophenol blue: 0.1% solution in 20% ethanol
- Sodium citrate: 25% solution in double-distilled water

Preliminary cleaning

The glass equipment must be carefully washed to remove all traces of iron. This is done by filling with or immersing in 6 N hydrochloric acid then bringing the solution to boil and maintaining it for about 10 minutes. If boiling is not suitable, leave them in the solution for 24 hours. Then wash them three times with double-distilled water and, if possible, boil in this water for a further 10 minutes, rinsing three times afterwards.

Procedure

The determination is performed on pitted olives. To obtain the sample, mix 100 g of flesh with the same weight of double-distilled water, then homogenize using a blender. In a quartz dish place 200 µl of the magnesium nitrate solution and weigh out 4 g of the mixture (equivalent to

2 g of flesh). The magnesium nitrate has the dual role of facilitating combustion and carrying the small amounts of ash obtained. Place the dish on an electric block and gradually raise the temperature to about 250°C and then more slowly up to the indicated maximum level, which is then maintained for 8–10 hours.

Lightly moisten the greyish-white ash and dissolve in three portions of 2 ml each of 6 N hydrochloric acid. Pour each portion in turn through filter paper into a 25-ml graduated flask. To aid the dissolution, gently heat the dish every time the hydrochloric acid is added. An easier method is to filter with a suction hood. Add double-distilled water to make up to 25 ml.

The above mineralization procedure, and the operations that follow, must be carried out simultaneously on three samples of flesh, owing to its lack of homogeneity, and must be supplemented with a blank consisting of just 200 µl of magnesium nitrate solution.

Take 5 ml (or more or less, as governed by the iron content) of the resulting solution and place in a 25-ml graduated flask. Add 1 ml of hydroquinone, 1 ml of *o*-phenanthroline and the necessary amount of sodium citrate to raise the pH value to 3.5 (approximately 1 ml). Shake after each addition. The exact volume of sodium citrate is determined by extracting an aliquot of the solution (2 ml for example) and placing it in a test tube. Add one drop of bromophenol blue and then carefully add citrate until the colour changes from yellow to blue (this change can be confirmed by dipping a strip of litmus paper into the test tube).

Leave the mixture for 1 hour to allow the colour to develop. Fill the 25-ml graduated flask with double-distilled water and take a reading at 508 nm, using a 1-cm cell.

The calibration curve should have been obtained in advance by taking 5 ml of solutions containing 0.4, 0.8, 1.6 and 2.0 µg of iron per ml and performing the test with the *o*-phenanthroline as described above.

The **standard iron solution** (0.1 mg/ml) to determine the calibration curve is prepared following Hanson's recommendations (Hanson, 1973). Dissolve 3.512 g of $Fe(NH_4)_2(SO_4)_2.6H_2O$ in double-distilled water, add 5 ml of HCl, transfer quantitatively to a 500-ml graduated flask, bring up to the mark with double-distilled water and homogenize.

Calculation

The iron content of the analysed sample expressed in mg/kg of pulp (p.p.m.) is expressed as follows:

$$C \text{ (p.p.m.)} = (X/U) \times (Z/T)$$

where X = mg of Fe obtained in the standard curve; Z = ml to which the ashes have been brought in HCl; T = ml used to make the colorimetric reaction; and U = g of sample used.

Atomic absorption spectrophotometry method

Materials and apparatus

- Grinder. Ultra-Turrax T25 (Janke & Kunkel GmbH Co.) or similar
- Quartz capsules of 50-ml capacity
- Incineration oven
- Volumetric flasks of 25-ml capacity
- Glass funnels or suction hood for vacuum filtration
- Atomic absorption spectrophotometer

Preparation of the sample

The preparation is similar to that described for the colorimetric method. The determination is also carried out in pitted olives, that is in the pulp of the fruit. Once pitted and ground, 7 g of fruit (accurately weighed) are taken in a quartz capsule containing 1 ml of magnesium nitrate which has the double function of facilitating the combustion and is the carrier of the small amount of ash, avoiding possible losses by adsorption on the walls of the vessel.

Put the capsule over an electrical heating plate, increasing the temperature gradually to 350°C. Put it in a muffle furnace and incinerate at 550°C. To this end, the temperature is quickly raised up to 250°C, and is slowly increased until reaching the calcination temperature which is maintained for approximately 8–10 hours.

Calculation

The iron concentration in the analysed sample expressed in mg Fe/kg of pulp (p.p.m.) is given by:

$$C\ (\text{p.p.m.}) = (X/U) \times Z$$

where X = p.p.m. of the iron obtained in the standard curve or in the measurement of the apparatus; Z = ml at which the ashes have been levelled when dissolved in HCl; and U = g of sample taken.

12.2.7 OLIVE CONSISTENCY (TEXTURE)

Texture is a quality attribute of great importance for olives. Its subjective assessment is always difficult and can lead to discrepancies in commercial transactions. So, its objective determination has long been of interest. An additional advantage of objective methods is the possibility of studying the influence of the numerous processing factors on this attribute. It is convenient that there is a good correlation between objective texture determinations and subjective appreciations. Such methods can be instrumental, based on physicochemical measurements (force or pressure) related to fruit consistency or chemical, based on the analysis of certain components of the fruit related to cell structure and the mechanism of its degradation. In olives, the first approach has been more widely used, although the chemical studies of cell wall and pectic substances have been very useful from the research point of view.

The most widely used items of equipment for texture determination are the Texture Test System and the Instron Universal Testing Machine. The procedures for the Texture Test System are summarized in Table 12.1. The procedure for the Instron Testing Machine must be similar to the following: cross-head speed 200 mm/min; shear-compression is normally measured in two to four pitted olives and expressed as kgf or N/100 g pitted olives. Compression is applied by application of a similar procedure, although it is scarcely used in practice. In both cases the machine must be properly prepared since differences in texture between flesh and pit are very marked. The machine must then be prepared in case of compression to support the higher resistance offered by the pit. If appropriate, a texture index can be obtained by dividing shear-compression/compression measures.

There are also other manual texturometers which are useful to compare, within the factory, relative textures of olive lots or the effects of treatments. Comparisons with other laboratories are, however, difficult

Table 12.1 Selected Kramer shear press conditions for olives (From Fernández Díez and Cordón Casanueva, 1966 and Fernández Díez and Vidal Sigler, 1968)

Compression test	Shear compression test
Type of sample	
Size-graded	
Unpitted fruits	Pitted fruits
Size of sample	
1–4 fruits placed in a single layer	100 g placed randomly into the cell
Speed of the piston	
Constant, 24.3 s	Constant, 24.3 s
Number of replicates	
10	10
Reading	
First stop of dynamometer; or first relative maximum when using electronic recorder	Maximum point reached by dynamometer or recorder
Results	
kgf or N per 100 g unpitted fruits	kgf or N per 100 g pitted fruits

because of the strong influence of the operator. A complete discussion of the use of one of these instruments can be found in Etchells *et al.* (1975) and Márquez Gónzalez *et al.* (1975), where there is information on the influence of several variables, such as plunger size, olive cultivar, brine acidity, olive size, storage time, and heat-shock treatment on olive firmness readings.

12.2.8 OLIVE SURFACE COLOUR

This is one of the most important quality attributes. Methods for its determination in green and ripe or naturally black olives are different.

Spanish-style green olives

The most appropriate and easiest is that proposed by Sánchez Gómez *et al.* (1985). It consists of measuring the percentage reflectance at 560, 590 and 635 nm and then applying the following formula:

$$\text{Colour index (C.i.)} = (4 \times R_{636} + R_{590} - 2 \times R_{560})/3$$

A lot measure is obtained as the average of 10 determinations.

Alternatively, green olive colour can be expressed as: L* (value of lightness or darkness); a* (redness or greenness); and b* (yellowness or blueness) or Yxz (C.I.E.). C.I.E. values can be plotted on a chromaticity diagram and the colour of the object may thus be located in colour space, in terms of officially and internationally recognized colour specification. All these systems are widely applied in colour determinations of many foods and non-food items. Comparisons are, in this case, more difficult.

Apparatus

Any equipment may be used which will measure reflectance.

Ripe and naturally black olives

Only a few studies have been devoted to colour determination in naturally black olives. Good results in research have been obtained with the method proposed by Garrido Fernández and Fernández Díez (1970). With black-ripe olives, Fernández Díez and Garrido Fernández (1971) applied the same method based on the measurements of the percentage of reflectance of the olive surface at 700 nm. Recent research has also been carried out by Marsilio *et al.* (1990). They studied the colour changes according to the C.I.E. method or using other characteristics (L*, a* and b*). Results of the spectrophotometric measurements to determine the chromaticity characteristics of fresh black olives, naturally black olives in brine and treated (oxidized) black olives in brine led them to conclude that the olive colour was not very saturated while the brightness values were useful to determine the fruits' external colour. In any case differences were slight and thus difficult to correlate to processing conditions, treatments, etc. No further work has been done in this respect. In consequence, only those methods based on the percentage of reflectance at 700 nm will be described. This is an easy characteristic to analyse because it can be determined directly, uses low-cost equipment, and has been widely used to evaluate the influence of the different processing variables on colour.

Apparatus

Any spectrophotometer provided with reflectance accessory.

Procedure

Each olive is cut horizontally into two halves. The half which does not contain the pit is put in the bottom of a small transparent plastic cylinder. The olive half is carefully pressed towards the bottom until the olive skin is flat and extended on the cylinder base. This is done using a cork of the appropriate diameter, or any other suitable device, which is pushed down the cylinder. No air must be trapped between the cylinder base and the olive skin, which must also remain intact. The prepared sample is then put into the place where reflectances are measured. Interference by stray light must be minimized by covering samples with a box with a matt black interior. The colour of a homogeneous lot is expressed as the average of the analysis of ten olives. The coefficient of variation is also useful to indicate the lack of homogeneity. The two scales prepared for both naturally black and black-ripe olives are shown in Tables 12.2 and 12.3.

Table 12.2 Suggested scale of colour (% reflectance at 700 nm) for naturally black olives (From Garrido Fernández and Fernández Díez, 1970)

Subjective appreciation	% Reflectance interval
Black	< 10.0
Very dark purple	10.0–14.5
Dark purple	14.5–21.5
Purple	21.5–31.5
Clear purple	31.5–36.0
Very clear purple	> 36.0

Table 12.3 Suggested scale of colour (% reflectance at 700 nm) for black-ripe olives (From Fernández Díez and Garrido Fernández, 1971)

Subjective appreciation	% Reflectance interval
Very black	< 6.0
Black	6.0–9.5
Dark brown	10.0–12.0
Brown	12.5–17.0
Clear brown	>17.5

12.2.9 BRINE COLOUR OF GREEN OLIVES

The method most used up to now for measuring this parameter, which is of great importance in packed green olives, is that proposed by Montaño Asquerino *et al.* (1988). It consists of filtering the brine and measuring its absorbance at 400 and 700 nm. Colour of brines is obtained as:

$$\text{brine colour} = (A_{400} - A_{700})$$

In non-homogeneous brines, the measurement is repeated with several samples to obtain an average value. When appropriate, samples should be diluted.

Colour of naturally ripe or black-ripe olives is not usually determined as those products are packed in non-transparent containers. In the case of naturally black olives the brine colour has been expressed as absorbance at 700 nm.

Apparatus

Any spectrophotometer.

12.3 MICROBIOLOGICAL METHODS

12.3.1 SAMPLING

All the microorganisms related to the processing of table olives may be found, at different times in fermentation and storage brines (Fernández Díez *et al.*, 1985). For that reason, in the majority of the cases the brine is used in the microbiological study of the process. However, under certain circumstances, it may be better to examine the fruit itself. Such cases might be the presence of 'yeast spots', produced by the growth of lactic acid bacteria under the skin of the olive, or the production of gas-pockets by Gram-negative bacteria.

Sampling from the fermenters must always be performed aseptically, with perfectly clean special plastic tubes. In general, it is enough to take the sample from the middle part of the fermenter. However, if it is suspected that an abnormal microflora is present, then the bottom of the container must also be analysed. In both cases, the samples are placed in sterile flasks.

It is convenient to carry out the microbiological analysis as soon as possible, particularly when the process is in the active fermentation phase, and in any case within a period of 24–48 hours after sampling. If the analysis is not performed immediately, samples should be kept in a refrigerator or at the lowest temperature possible. When it is necessary to check the presence of microorganisms in the pulp itself, this is done in the following way.

Procedure

Take eight or ten fruits and place in distilled water containing 40 p.p.m. of chloride. Leave them for 60 minutes. With the aid of a sterile lance, remove the pits and then crush the pulp of all the olives in a sterile mortar. Transfer the resulting paste to a flask containing 100 ml of sterile water, which is weighed before and after the addition of the pulp, to determine the pulp weight. Homogenize the mixture, leave it to settle for 10 minutes, and them remove the required amount of the upper part of the suspension to make the corresponding culture.

12.3.2 MICROSCOPIC EXAMINATION

It is very useful to make a microscopic examination of brines and pulp suspensions before proceeding to the microbiological analysis of samples. This allows properties of the microorganisms, such as shape, cellular aggregates, motility, and spore formation to be determined rapidly – this may be an important help to the further analysis. This examination is generally made with a phase-contrast microscope.

12.3.3 MICROBIOLOGICAL ANALYSIS

The nutrient needs of the microorganisms present during the fermentation and storage of table olives are diverse and depend on the different genera present. It is therefore necessary to use several culture media. Those organisms which generally appear during the first days of the fermentation, such as *Bacillus*, *Micrococcus*, *Enterobacter*, *Citrobacter*, *Klebsiella* and *Escherichia*, develop quite well in a normal medium. However, lactobacilli, mainly responsible for the fermentation process, yeasts, and other microorganisms such as those belonging to the genera *Propionibacterium* and *Clostridium*, which may be present and produce different types of spoilage, are nutritionally more fastidious and require a richer medium.

Culture media

The most suitable culture media for an effective microbiological inspection are as follows (Fernández Díez *et al.*, 1985; Unidad Estructural de Biotecnología de Alimentos, 1990):

Gram-negative bacteria
- Violet red bile agar
- Eosin–methylene blue agar

Lactic acid bacteria.

(i) *Pediococcus*, *Streptococcus*, and *Leuconostoc*
- Agar–Lemco
- APT agar

(ii) *Lactobacillus*
- Rogosa agar
- MRS agar

For the growth of lactic acid bacteria in general, yeast extract–glucose agar also gives very good results.

Yeasts
- Malt extract agar
- Yeast extract–glucose agar

Clostridia
- S.A.T. medium
- DCRM broth
- Liver broth

Propionibacteria
- Lactate–yeast extract broth (as enrichment medium)
- Yeast extract–glucose agar (as selective medium)

Incubation conditions

The optimum incubation temperatures and incubation times for the different microorganisms involved in the process are as follows.

Microorganism	Temperature (°C)	Time (days)
Gram-negative bacteria	30	2
Lactic cocci	30	3
Lactobacilli	30	3
Yeasts	25	3
Propionibacteria	30–35	8–15
Clostridia	30–37	8–15

12.3.4 USE OF LACTIC STARTERS

When the spontaneous lactic acid bacteria population is not sufficient to obtain a quick and efficient lactic fermentation, a 'starter' is frequently used. Two different procedures are equally effective. In the first, lactobacilli are isolated on Rogosa agar from a well-fermenting brine, and grown in a liquid medium buffered at pH 5.57. Yeast extract–glucose, with acetate, citrate or sodium phosphate solution gives very good results. Frequently 4% of sodium chloride is added to the medium to get a culture which will be well adapted to the brine to be inoculated. After an incubation time of 48–72 hours at 30°C the culture may be used as an inoculum at about 2% of the liquid capacity of the fermenter. Very effective freeze-dried pure strains of lactobacilli and the corresponding liquid medium to grow them for further inoculation of the brines are now supplied by different commercial companies.

Very good results are also obtained by the inoculation of the fermenters with a portion of 'mother brine', that is, brines from olives which are in the active phase of the lactic fermentation. This method has the advantage that lactobacilli are already adapted to the brine. The inoculum level is this case is between 2 and 5% of the liquid capacity of the fermenter and is dependent on the number of organisms present.

In both methods, the starter is added to the fermenter by means of a funnel and plastic tube so that the inoculum stays between one-third and one-half the depth from the top of the fermenter. The inoculum must remain there for at least 48 hours. After 3 days, brine is recirculated to homogenize the whole contents of the fermenter. The optimum pH of the medium for this operation is about 5.5, and the temperature must be kept between 20 and 30°C, with the help of a heat exchanger, if necessary.

Before proceeding to the inoculation it is quite important to check the presence of sufficient fermentable matter in the brine, and to add it, if necessary.

12.4 METHODS OF ANALYSIS AND SAMPLING RECOMMENDED BY THE IOOC AND CODEX ALIMENTARIUS

Pending adoption of methods of analysis and sampling by the International Olive Oil Council (IOOC), those recommended by the FAO/WHO Codex Alimentarius Commission shall apply and constitute the international arbitration methods. The sampling and analysis methods included in the Codex Alimentarius (Codex STANN 66, 1990) follow.

12.4.1 SAMPLING METHODS

Sampling must be carried out according to the Sampling Methods for Prepacked Foods (1969), CA/RM 42–1969 amended (paragraph 8, part II, Appendix I of this publication).

12.4.2 METHODS OF ANALYSIS

Drained weight

- According to the Codex Alimentarius method (Methods of Analysis of Codex Alimentarius FAO/WHO for Prepared Fruits and Vegetables, CA C/RM, 36–1970, Drained weight determination, Method I). Results should be expressed as % (w/w) calculated with reference to the distilled water contained in the same container at 20°C when this is completely full.

Salt content in brine (Type II)

- According to the Codex general method of reference to determine chlorides (expressed as sodium chloride) in foods, ALINORM 99/23, Appendix IV. Results should be expressed as g NaCl per 100 g of sample.
- According to the alternative method (Type III). Norma ISO 3634–1979 Vegetable Products; Determination of chloride content. This method is still to be approved by the Codex Committee for Methods of Analysis.

Titratable acidity determination

Put 25 ml brine, using a pipette, into a conical flask of 250-ml capacity and add a few drops of phenolphthalein as indicator. Titrate the solution with another solution of sodium hydroxide (0.1 N) until a permanent pink colour is obtained. The sodium hydroxide solution could be standardized by titration with potassium acid phthalate (potassium hydrogen phthalate) of analytical reagent degree, applying the necessary factor:

1 ml of 0.1 N NaOH = 0.0090 g lactic acid

Determination of pH (Type I)

Note that the presence of acidic food additives may lead to false results.

Fit the pH meter with the glass and calomel electrodes, according to the supplier's instructions, to use the equipment at 20°C. Standardize it with a known buffer solution of pH = 4.0 at 20°C. Rinse the electrodes thoroughly with abundant distilled water to remove any residue of the buffer solution. Submerge the electrodes in the sample, and adjust temperature to 20°C. Read the pH to the nearest 0.05 unit.

Determination of the container water capacity (Type II)

- According to the Methods of Analysis of Codex Alimentarius FAO/WHO for Prepared or Elaborated Fruits and Vegetables (Ref. CAC/RM 46–1972).

Results are expressed as the volume of water that the container can contain.

Determination of tin content

- According to the Official Methods of Analysis of the AOAC, 13th Edition, 1990. 'Tin; Atomic Absorption method (28). Interim Official First Action. 25.136–25.138'. Results are expressed in mg Sn/kg product.
- According to the alternative method (Type III) ISO NORMA 2447–1974 Fruits and Vegetable Products. Determination of tin.

Determination of lead (Type II)

- According to the Official Methods of Analysis of AOAC, 13th Edition 1980. 'Lead, 25.061–25.067'. Results are expressed as mg Pb/kg product.

12.5 REFERENCES

Codex Alimentarius (1990) *Norma del Codex para las Aceituna de mesa* (Codex STANN 66–1981, Revis. 1 (1987), Addition, April, 1990.

Etchells, J.L., Kitted, I.D., Kelling, R.E., Bell, T.A., Monsalves, R.J. and Fleming, H.P. (1975) The olive pressure tester: an instrument for measuring the firmness of Spanish-type green olives. *Grasas y Aceites*, **26**, 139–46.

Fernandez Díez, M.J. and Cordón Casanueva, J.L. (1966) Determinación de textura en aceitunas. *Grasas y Aceites*, **17**, 88–94.

Fernández Díez, M.J. and Garrido Fernández, A. (1971) Aceitunas negras por oxidación en medio alcalino.I. El color como criterio de madurez y calidad en el grodiente elaborado. *Grasas y Aceites*, **22**, 193–9.

Fernández Díez, M.J. and Vidal Sigler, A. (1968) Determinación de textura en aceitunas II. Las pruebas de compresión-cizallamiento. *Grasas y Aceites*, **19**, 199–204.

Fernández Díez, M.J., Castro Ramos, R. de, Garrido Fernández, A., González Cancho, F., González Pellissó, F., Nosti Vega, M., Heredia Moreno, A., Minguez Mosquera, M.I., Rejano Navarro, L., Durán Quintana, M.C., Sánchez Roldán, F., García García, P. and Castro Gómez Millán, A. (1985) *Biotecnología de la aceituna de mesa*, Consejo Superior de Investigaciones Científicas, Instituto de la Grasa y sus Derivados, Madrid, Sevilla, Spain.

García García, P., Brenes Balbuena, M. and Garrido Fernández, A. (1991) Métodos instrumentales para la determinación de NaCl en las salmueras de aceitunas. *Grasas y Aceites*, **42**, 261–6.

Gárcia García, P., Brenes, M. and Garrido, A. (1992) Rapid indirect method for determining the sodium content of table olives. *Analyst*, **117**, 173–6.

Garrido Fernández, A. and Fernandez Díez, M.J. (1970) Determinación del color en las aceitunas negras maduras estilo griego. *Grasas y Aceites*, **21**, 141–7.

Garrido Fernández, A., Albi, M.A. and Fernández Díez, M.J. (1973) Aceitunas negras por oxidación en medio alcalino.II. Determinación del contenido en hierro y evolución de diversos factores durante el proceso de elaboración. *Grasas y Aceites*. **24**, 287–92.

Hanson, N.W. (1973) *Official Standardized and Recommended Methods of Analysis*. The Society for Analytical Chemistry, London.

Marquez Gónzalez, J., Monsalves Cortes, J. and Fernández Díez, M.J. (1975) Empleo del penetrometro FF2 para determinar la textura de aceituras verdes estilo español. *Grasas y Aceites*, **26**, 221–32.

Marsilio, V., Brighigna, A. and Chichelli, A. (1990) Le caratteristiche del colore della olive nere da tavola. Influenza della varietá e della tecnologia di trasformazione. *Industrie Alimentari*, **XXIX**, 5–10.

Montaño Asquerino, A. Sánchez Gómez, A.H. and Rejano Navarro, L. (1988) Método para la determinacíon del color en las salmueras de aceitunas verdes de mesa. *Alimentaria*, **193**, 78 –82.

Montaño Asquerino, A., de Castro, A., Rejano, L. and Sánchez, A.H. (1992) Analysis of zapatera olives by gas and high performance liquid chromatography. *Journal of Chromatography*, **594**, 259–67.

Sánchez Gómez, A.H, Rejano Navarro, L. and Montaño Asquerino, A. (1985) Determinación del color en aceitunas verdes aderezadas de la variedad manzanilla. *Grasas y Aceites*, **36**, 258–61.

Unidad Estructural de Biotecnología de Alimentos (Instituto de la Grasa y sus Derivados) (1990) *Table Olive Processing*, International Olive Oil Council, Madrid, Spain.

Index

MIX
Papier aus verantwortungsvollen Quellen
Paper from responsible sources
FSC® C105338

If you have any concerns about our products,
you can contact us on
ProductSafety@springernature.com

In case Publisher is established outside the EU,
the EU authorized representative is:
Springer Nature Customer Service Center GmbH
Europaplatz 3, 69115 Heidelberg, Germany

Printed by Libri Plureos GmbH
in Hamburg, Germany